AF676904

Yeasts in food

Beneficial and detrimental aspects

Edited by
T. Boekhout and V. Robert

ELSEVIER

AMSTERDAM • BOSTON • HEIDELBERG • LONDON
NEW DELHI • NEW YORK • OXFORD • PARIS • SAN DIEGO
SAN FRANCISCO • SINGAPORE • SYDNEY • TOKYO
Woodhead Publishing is an imprint of Elsevier

WP
WOODHEAD
PUBLISHING

Yeasts in food- Beneficial and detrimental aspects
Boekhout and Robert

Woodhead Publishing Ltd.
An imprint of Elsevier
Abington Hall, Abington,
Cambridge CB21 6AH, England

First published 2003, Woodhead Publishing Ltd and CRC Press LLC
© 2003, B. Behr's Verlag GmbH & Co. KG, Averhoffstrde 10, 22085 Hamburg.
The authors have asserted their moral rights. This edition is published by arrangement with B. Behr's Verlag GmbH & Co. KG, Hamburg, Germany.

ISBN: 978-1-85573-706-8

First Indian Print 2014

Indian ISBN: 978-93-5107-279-9

This edition has been authorized by Elsevier for sale in the following countries: India, Pakistan, Nepal, Sri Lanka and Bangladesh. Sale and purchase of this book outside these countries is not authorized and is illegal.

No part of this publication may be reproduced, stored in a retrieval system, or transmitted in any form or by any means electronic, mechanical, photocopying, recording, or otherwise without the prior written permission of the publisher.

Notice
No responsibility is assumed by the publisher for any injury and/or damage to persons or property as a matter of products liability, negligence or otherwise, or from any use or operation of any methods, products, instructions or ideas contained in the material herein.

Published by Elsevier, a division of Reed Elsevier India Private Limited.

Registered Office: 305, Rohit House, 3 Tolstoy Marg, New Delhi-11 00 01
Corporate Office: 14th Floor, Tower 10B, DLF Cyber City, Phase-II, Gurgaon - 122 002, Haryana

Digitally printed and bound at Avantika Printers Pvt. Ltd., New Delhi (India)

Dedicated to the memory of

Prof. Dr. Herman Phaff

Editors preface

The production and maintenance of good quality food products contribute to the quality of life. Yeasts and food are intimately related since the early days of human civilization. Early humans discovered that fermented foods and drinks had added nutritional value and, in various cases, could be better preserved. Consequently, fermented foods contributed to human survival during historical times. The workhorse among the yeasts, *Saccharomyces cerevisiae*, which is rare in natural environments, may be considered as a domesticated microbe. Since the discovery of yeasts by Antonie van Leeuwenhoek, the recognition of the biological nature of fermentation reactions by Pasteur, and the isolation of pure yeast cultures by Hansen, our knowledge of yeast biodiversity has increased enormously. About 800 species of yeast are presently known, and several play significant roles in the food, brewing, wine and beverage industries. This is clearly illustrated by the various chapters of this book. The contribution of yeast to the food industry can be either beneficial or detrimental. In many cases the relationship between these two aspects is a fragile balance, which depends on the interplay between various biotic and abiotic factors. In this sense, the study of yeast-food interactions can be really seen as applied ecology. Considerable progress has been made in the detection and identification of yeasts from food, due to the introduction of various molecular methods, and the development of extensive genome databases and advanced identification tools. Various protocols have been developed to selectively isolate yeasts from different sources of food and drinks, because of the increased knowledge on the ecology of food-related yeasts and the physico-chemical properties of the various foods. The genomics era already yielded significant progress in our understanding of the effects of the preservation of food on the yeast transcriptome. New insights will arise in the near future, and we are happy to present a comprehensive overview of the first genomic studies in this field. The physiological background of spoilage by yeasts, and the detection and management of spoilage incidents require utmost attention in the food industry. Yeasts cause a spoilage risk as many species are able to grow at low temperatures and low pH values. Only a few years ago a new yeast species was discovered, which was found to be resistant to commonly used preservatives in the food industry, and thus poses a serious spoilage threat.

The second part of this volume is dedicated to the various foods, fermented drinks and beverages. It is noteworthy that so many yeast species are involved in the manufacturing of the various foods and drinks. The diversity of foods and drinks involved is impressing as well. In many cases, yeasts interact with other microbes, such as filamentous fungi and bacteria, in temporarily and spatially differentiated, but balanced, physiological processes. This is the case in the production of e. g., soy sauce, coffee, cocoa, cheeses, kefyr, and the various traditional fermented products discussed. The production of wine, beer and bread are among the best-understood fermentation processes. Yeasts do not only contribute by the production of ethanol or CO_2, but are responsible for the production of a huge variety of olfactory and gustatory important compounds. These largely contribute to the value of the existing, and appreciated variety of wines, beers and breads occurring worldwide. Soft drinks present a niche for a specific yeast flora, which in most cases is detrimental to the product quality.

Due to the studies performed on this specific environment, the spoilage problem of beverages can be controlled in most cases.

Various authors emphasized the differences between yeast populations of processed and non-processed foods, in particular of dairy-, fruit- and meat-related products. The introduction of environmental yeasts into the food chain poses a potential spoilage risk, e. g., in products such as fruit yogurts. In contrast, environmental yeasts are indispensable in other fermentation processes.

We hope that many students of yeast biology, fermentation biology, food processing, brewing, viniculture and beverage industries will use this book, both educationally and professionally. Finally, we want to thank all authors for their pleasant and cooperative collaboration in the preparation of the book.

Teun Boekhout and Vincent Robert (Utrecht, November 26, 2002)

Editors and authors

Editors

Teun Boekhout, Centraalbureau voor Schimmelcultures (CBS), Utrecht, The Netherlands (Boekhout@cbs.knaw.nl)

Teun Boekhout has been working for 15 years at the CBS Yeast Division, where he is mostly involved in research on various basidiomyceteous yeasts. His research interests include systematics, evolution, phylogeny and biodiversity. Recently, he explored possibilities to develop electronic means for the identification of yeasts using various sources of data. He is adjunct editor in chief of FEMS Yeast Research, and an editor of the 5th edition of the standard work on yeast systematics 'The yeasts, a taxonomic study'.

Vincent Robert, Centraalbureau voor Schimmelcultures (CBS), Utrecht, The Netherlands (Robert@cbs.knaw.nl)

Vincent Robert has been the head of the laboratory of food microbiology at the University of Burundi for several years where he developed semi-automated methods for computer-based identification of yeasts. He then moved to Belgium where his researches were mainly focusing on yeast biodiversity and bioinformatics. He is presently appointed at the CBS as curator of the yeasts collection. As a bioinformatician, he has developed many programs, including the BioloMICS software package.

Authors

Bruno Blondin, Equipe de Microbiologie et Technologie des Fermentations, UMR Sciences pour l'Œnologie, INRA-AgroM-UMI, Montpellier, France (blondin@ensam.inra.fr)

Bernard Bonjean, Gelka International, Andenne, Belgium (BBonjean@gelka.com)

Stanley Brul, Swammerdam Institute for Life Sciences, University of Amsterdam, Amsterdam, The Netherlands (stanley.brul@unilever.com)

Peter Coote, Center for Biomolecular Sciences, University of St. Andrews, North Haugh, St. Andrews Fife, U.K. (pjc@st-and.ac.uk)

Tibor Deak, Department of Microbiology, Szent Istvan University, 14-16 Somloi ut, 1118 Budapest, Hungary (tdeak@omega.kee.hu)

Sylvie Dequin, Equipe de Microbiologie et Technologie des Fermentations, UMR Sciences pour l'Œnologie, INRA-AgroM-UMI, Montpellier, France (dequin@ensam.inra.fr)

Guy Derdelinckx, Centre for Malting and Brewing Science, Department of Food and Microbial Technology, Katholieke Universiteit Leuven, Leuven, Belgium (guy.derdelinckx@agr.kuleuven.ac.be)

Jean-Pierre Dufour, Department of Food Science, University of Otago, Dunedin, New Zealand (Jean-Pierre.Dufour@stonebow.otago.ac.nz)

Jack W. Fell, School of Marine and Atmospheric Sciences, University of Miami, Key Biscayne, Florida, U.S.A. (jfell@rsmas.miami.edu)

Graham H. Fleet, Food Science and Technology, School of Chemical Sciences, University of New South Wales, Sydney, New South Wales, Australia (g.fleet@unsw.edu.au)

Rosane Freitas-Schwan, Department of Biology, Federal University of Lavras, 37 200-00 Lavras, MG, Brazil (rschwan@ufla.br)

Marie-Therese Fröhlich-Wyder, Swiss Federal Dairy Research Station (FAM), Liebefeld, Bern, Switzerland (Marie-Therese.Froehlich@fam.admin.ch)

Luc-Dominique Guillaume, Puratos N.V., Groot-Bijgaarden, Belgium. (ldguillaume@puratos.com)

Yoshiki Hanya, Kikkoman Corporation, Imagami, Noda City, Japan. (hanya@tky.3web.ne.jp)

Bob J. Hartog, Department of Risk Management and Microbiology, TNO Nutrition and Food Research Institute, Zeist, The Netherlands (Hartog@voeding.tno.nl)

Klaas J. Hellingwerf, Swammerdam Institute for Life Sciences, University of Amsterdam, Amsterdam, The Netherlands (k.hellingwerf@science.uva.nl)

Stephen A. James, National Collection of Yeast Cultures, Institute of Food Research, Norwich Research Park, Colney, Norwich, U.K. (Steve.James@bbsrc.ac.uk)

Frans M. Klis, Swammerdam Institute for Life Sciences, University of Amsterdam, Amsterdam, The Netherlands (klis@bio.uva.nl)

Cletus P. Kurtzman, National Center for Agricultural Utilization Research, USDA, Peoria, Illinois, U.S.A. (kurtzman@mail.ncaur.usda.gov)

Tadanobu Nakadai, Research Department Division, Kikkoman Co., Noda City, Chiba Pref., Japan (7555@mail.kikkoman.co.jp)

Huu-Vang Nguyen, Collection de Levures d'Intérêt Biotechnologique (CLIB), Laboratoire de Génétique Moléculaire et Cellulaire, INRA, Thiverval-Grignon, France (nguyen-hv@grignon.inra.fr)

Monique W.C.M. de Nijs, Department of Risk Management and Microbiology, TNO Nutrition and Food Research Institute, Zeist, The Netherlands (denijs@voeding.tno.nl)

Hans de Nobel, Swammerdam Institute for Life Sciences, University of Amsterdam, Amsterdam, The Netherlands (presently Genencor International B.V., Leiden, The Netherlands) (hdenobel@genencor.com)

M. J. Robert Nout, Department of Agrotechnology and Food Sciences, Wageningen University, Wageningen, The Netherlands (Rob.Nout@wur.nl)

Suus J. C. M. Oomes, Swammerdam Institute for Life Sciences, University of Amsterdam, Amsterdam, The Netherlands (Suus.oomes@unilever.com)

Herman J. Phaff †, Department of Food Science, University of California, Davis, U.S.A.

Hakim Rahaoui, Department of Risk Management and Microbiology, TNO Nutrition and Food Research Institute, Zeist, The Netherlands (Rahaoui@voeding.tno.nl)

Jean-Michel Salmon, Equipe de Microbiologie et Technologie des Fermentations, UMR Sciences Pour l'Œnologie, INRA-AgroM-UMI, Montpellier, France (jmsalmon@ensam.inra.fr)

John Samelis, National Agricultural Research Foundation, Dairy Research Institute Katsikas, Ioannina, Greece (jsam@otenet.gr)

John N. Sofos, Department of Animal Sciences, Colorado State University, Fort Collins, Colorado, U.S.A. (John.Sofos@colostate.edu)

Malcolm Stratford, Unilever R & D, Colworth House, Sharnbrook, Bedford, U.K. Malcolm.Stratford@unilever.com)

Kevin Verstrepen, Centre for Malting and Brewing Science, Department of Food and Microbial Technology, Katholieke Universiteit Leuven, Kasteelpark Arenberg, Leuven, Belgium (Kevin.Verstrepen@agr.kuleuven.ac.be)

Jos M. B. M. van der Vossen, Department of Risk Management and Microbiology, TNO Nutrition and Food Research Institute, Zeist, The Netherlands (vandervossen@voeding.tno.nl)

Alan E. Wheals, Department of Biology and Biochemistry, University of Bath, Bath, BA2 7AY, U.K. (bssaew@bath.ac.uk)

Contents

1 Yeast biodiversity

TEUN BOEKHOUT and HERMAN J. PHAFF †

1.1 Introduction

Identifying, naming and placing yeasts in their proper evolutionary framework is of importance to many areas of science, including agriculture, medicine, the biological sciences, biotechnology, food industry, and for determining industrial-property rights. At present, approximately 750 yeast species are recognized, but only a few are frequently isolated. Relatively few natural habitats have been thoroughly investigated for yeast species. Consequently, we can assume that many additional species await discovery. Because yeasts are widely used in traditional and modern biotechnology, the exploration for new species should lead to additional novel technologies.

Several definitions have been used to describe the yeast domain. According to GUILLIERMOND [53] and LODDER [88], yeasts are fungi reproducing unicellularly by budding or fission. In this sense only true unicellular fungi are regarded as yeasts. However, many yeast species are dimorphic and produce pseudohyphae and hyphae in addition to unicellular growth. Similarly, many hyphal fungi are dimorphic and are usually referred to as yeast-like. Because of the overlap in morphological appearance, some authors regard yeasts merely as fungi that produce unicellular growth, but that otherwise are not different from filamentous fungi [42], or as unicellular fungal growth forms which have resulted as a response to a commonly encountered set of environmental pressures [67]. OBERWINKLER [102] placed the yeasts in a phylogenetic framework and defined them as unicellular, ontogenetic stadia of either asco- or basidiomycetes [140]. In summary, yeasts are ascomycetous or basidiomycetous fungi that reproduce vegetatively by budding or fission, with or without pseudohyphae and hyphae, and forming sexual states that are not enclosed in fruiting bodies.

Some yeasts may reproduce sexually, resulting in an alternation of generations with the formation of characteristic cells in which reduction division (meiosis) takes place. In ascomycetous yeasts this cell is the ascus, in which ascospores are formed. In basidiomycetous yeasts the site of meiosis is called a basidium, on which basidiospores are exogenously formed.

Asexually reproducing yeasts are referred to as imperfect, mitosporic or anamorphic yeasts (e. g., *Cryptococcus neoformans*, *Candida utilis*), and sexually reproducing yeasts are called perfect, meiosporic or teleomorphic yeasts (e. g., *Filobasidiella neoformans, Pichia jadinii*). The combination of both states is called the holomorph, and for this the name of the sexual stage (teleomorph) is being used (in these examples *F. neoformans* and *P. jadinii*).

Molecular comparisons show the ascomycetous yeasts to be phylogenetically distinct from the filamentous Ascomycetes [78, 80, 81], whereas the basidiomycetous yeasts belong to the three main classes of Basidiomycetes, namely the Urediniomycetes, Ustilaginomycetes and Hymenomycetes [39].

1.2 Developments in yeast systematics

Three main periods can be discerned in yeast taxonomy in which new concepts were developed, largely based on technological and scientific innovations. The first period (until approximately 1960) was characterized by a thorough study of morphology, comparative nutritional physiology, and conventional genetics. Important workers in this period were M. Reess (morphology), E. C. Hansen (application of pure cultures and physiology), A. J. Kluyver (physiology), L. J. Wickerham (physiology, genetics, ecology), and A. Guilliermond, Ö. Winge and C. C. Lindegren (genetics). Comparative taxonomic studies performed at the CBS Yeast Division [31, 87, 127], resulted in a series of monographs, which created the 'Delft School'.

Initially, responses on only a limited number of carbon and nitrogenous compounds were used for taxonomic purposes. Wickerham [150] extended this series, and today approximately 60 tests are being performed routinely, including fermentation and assimilation of carbon compounds, assimilation of nitrogenous compounds, vitamin requirements, resistance to cycloheximide, temperature requirements, etc. (see Chapter 3).

Genetic studies revealed the presence of different sexual strategies. Sexual cycles of ascomycetous yeasts may be haplontic, diplontic or diplohaplontic. Yeast species were found to be homothallic, heterothallic, or a combination of these. Incompatibility systems of basidiomycetous yeasts are bipolar, tetrapolar, or modified tetrapolar, and mating factors biallelic or multiallelic [7, 34–36, 84, 155].

The second period of yeast systematics (from 1960 until c.2000) was characterized by an extension of morphological characteristics because of the introduction of the electron microscope, the application of biochemical criteria, and the introduction of molecular studies. Transmission electron microscopy revealed differences between ascomycetous and basidiomycetous yeasts. Ascomycetous yeasts have electron-transparent cell walls and a thin electron-dense outer layer, whereas basidiomycetous yeasts have lamellate and electron-dense cell walls [70]. Bud formation is also different in these two groups of yeasts. Ascomycetous yeasts show holoblastic budding, i. e., the entire cell wall seems to be involved in the formation of the newly formed wall of the bud, while basidiomycetous yeasts have enteroblastic budding in which only the inner cell wall layer is involved in this process.

Septal ultrastructure shows important differences between the two groups of yeasts. Septa of many ascomycetous yeasts have one or several micropores. These are very thin electron-dense connections between two adjacent cells. Additionally, diaphragm-like pores occur as well, and Woronin bodies may be present. Pores of *Ambrosiozyma* species are swollen around the pore, thus resemble somewhat the dolipores of basidiomycetes. Basidiomycetous yeasts show a greater variation in septal ultrastructure. In the cytoplasm, a structure made up of modified endoplasmic reticulum, the parenthesome or septal pore cap (SPC) may be present. The parenthesome can have different morphologies. Hymenomycetous yeasts usually have dolipores in which the septum is swollen around a central pore. *Filobasidiella* and *Bulleromyces*, have a parenthesome made up of U-shaped vesicles (Tremellales-type). Other basidiomycetous yeasts, currently classified in the order Cystofilobasidiales [37] lack such a parenthesome. The urediniomycetous yeasts have diaphragm-like pores reminiscent of those found in the higher ascomycetes, but without Woronin bodies. The ustilaginomycetous yeasts may have micropore-like structures [17]. The fine structure of septa is in full accordance with phylogenies based on ribosomal DNA (rDNA) data [39].

Biochemical characteristics, such as carbohydrate composition of cell walls and capsules [112, 146, 147, 131], proton magnetic resonance spectra of cell walls [125, 126], number of isoprene units of the coenzyme Q [156, 157, 159], cytochromes [23, 41, 98], fatty acid composition [25,145, 148], and isozyme patterns [160, 161] have been used for taxonomic purposes as well.

The introduction of comparative DNA studies in the late sixties of the last century provided, in principle, an objective parameter for estimating evolutionary distances between taxa. Different methods offer resolution at different taxonomic levels. The taxonomic value of nucleic acid base composition (mol% G+C) is mainly exclusionary (see Chapter 3). Phenotypically similar strains differing by more than ca. 2–3 % in their base composition are usually regarded as different species [72, 76, 107], while strains with the same base composition do not necessarily represent one and the same species. The range of nucleic acid base compo-

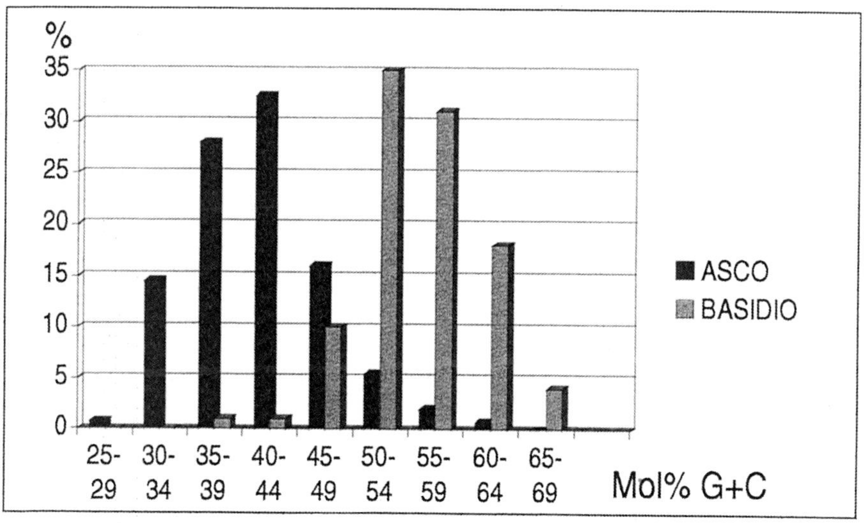

Fig.1.2-1 Distribution of percentage Guanine plus Cytosine (Mol% G+C) of the DNA among asco- and basidiomycetous yeasts

sitions differs for ascomycetous and basidiomycetous yeasts. Most ascomycetous yeasts have a mol% G+C lower than 50, whereas most basidiomycetous yeasts have a mol% G+C above 50 (Fig. 1.2-1).

DNA hybridization studies are used to determine DNA similarity between species [3, 20, 73]. Commonly used methods (for reviews see 60, 74, and see Chapter 3) include spectrophotometric analysis of heteroduplex formation, membrane-bound reassociation techniques using isotopes or fluorochromes, and methods in which the reassociated heteroduplex is bound on hydroxylapatite columns [20]. More recently, fluorometric and colorimetric methods using microtitre plates have been developed [33, 55], which are being applied in yeast taxonomy [66]. DNA-binding percentages above 65–70 % are generally interpreted to indicate conspecificity [73, 106]. A proportional relationship has been suggested between the occurrence of gene flow and high values of DNA similarity [86]. A positive correlation seems to exist between DNA complementarity and interfertility [73, 106]. However, low values of DNA similarity, even up to ca. 25 %, do not necessarily exclude gene exchange [72, 76, 82, 142]. Consequently, a rigid application of any lower limit up to ca. 25 % DNA sequence similarity as the sole criterion for species delimitation does not seem justified. Intermediate values of DNA similarity (40–70 %) sometimes are interpreted to indicate the presence of infraspecific taxa [82, 109].

Amplified fragment length polymorphism (AFLP) is a multilocus genotyping method combining universal applicability, high discriminative power and reproducibility for which only small amounts of DNA are required. The method is based on digestion of genomic DNA, e. g., with *Mse*I and *Eco*RI, ligation with two primers complementary to the restriction sites, and 2 rounds of PCR. The primers used in the final PCR have an increased specificity because they are elongated with 1–3 additional (selective) primers. The resulting fragments are being separated by acrylamide electrophoresis and analyzed. The method has been used for the genotyping of both ascomycetous and basidiomycetous yeasts [5, 19, 26, 136]. A large number of strain-specific genetic characters is generated, and this renders AFLP a sensitive tool allowing differentiation of genetically related strains.

Pulsed-field gel electrophoresis provides information on the composition (number and size) of individual chromosomes. Many species show a considerable variation in number and size of chromosomal DNAs [1, 6, 12, 15, 16, 18, 63, 143]. Chromosomal length polymorphisms occur in most species studied so far, but appear absent in several species of *Malassezia* [14]. Other molecular approaches applied at or below the species level are restriction analysis [63, 85, 89, 95, 96, 124, 137, 144], and random amplification of polymorphic DNA (RAPD) [91, 97, 99, 111]. Sequence variation of the internal transcribed spacer (ITS) of the rDNA is becoming another tool for recognizing and identifying species, e. g., in the genus *Trichosporon* [129] and *Saccharomyces* [100].

In our opinion the third era of yeast systematics started with the publication of the full genome sequence of *Saccharomyces cerevisiae* [48, http://genome-www.stanford.edu/Saccharomyces]. The genome analyses of the following yeasts are near completion: *Eremothecium (Ashbya) gossypii*, *Schizosaccharomyces pombe* [http://sanger.ac.uk/Projects/

S_pombe], *Candida albicans* [http://www-sequence.stanford.edu/group/candida] and *Cryptococcus neoformans* [56, http://sequence-www.stanford.edu/group/C.neoformans/index.html]. Comparative genomic studies yielded new insights in the evolutionary dynamics of the *S. cerevisiae* genome. It has been proposed that the entire or large parts of the genome of *S. cerevisiae* duplicated during evolution followed by subsequent loss of parts of it [110, 123, 154]. Recently, it has also been demonstrated that yeast genomes may be made up of composite genomes of up to three species [51]. In a comparative sequencing project of 13 ascomycetous yeast species it was concluded that functional clustering of genes of these species corresponded reasonably well with the phylogenies based on rDNA. The authors suggested that gene sequence drift has been the main driving force in the evolution of ascomycetous yeast biodiversity and they considered losses or duplications of genes less important [45].

1.3 Species concepts

Evolutionary uniqueness can be expressed at all levels between gene and the whole organism. Consequently, comparative investigations also need to be performed at various levels of biological development and diversification, and include morphology, physiology, genetics, biochemistry, ecology, molecular genetics and genomics. Taxonomic concepts change as the result of developments in science and philosophy. As a consequence, several different species concepts have been proposed in yeast systematics.

The phenetic species concept is based on discontinuities of phenotypic characteristics. In the past, delimitation of yeast taxa was based mainly on morphological and physiological differences between strains or groups of strains. KREGER-VAN RIJ [69], for instance, defined yeast species as an assemblage of clonal populations. The reliability of the phenetic approach depends largely on the quality and number of characters investigated. Interpretation of physiological data is complicated, because many of the carbon sources used in discriminatory growth tests can be metabolized by common pathways [9; 49]. Furthermore, the metabolism of many mono-, di- and trisaccharides is controlled by only one or a few genes [8, 152], and physiological characteristics are not always genetically stable and reproducible [121, 122]. Extension of the series of carbon and nitrogen compounds used in growth tests may increase the significance of this approach to yeast classification. Phaff [107] suggested that compounds with a complicated metabolic route will be most useful in this respect. The weakness of the phenetic approach has been stressed by several authors [106, 107, 76, 140]. However, from a practical point of view, fermentation and assimilation reactions are still widely used for identification.

The biological species concept assumes the existence of arrays of Mendelian populations, which are reproductively isolated from other population arrays [32]. The occurrence of a perfect (sexual) state following mating of complementary strains is commonly interpreted to indicate conspecificity of the mated strains. However, unless viability of the F_1 and F_2 generations is verified, the presumption of conspecificity may be incorrect because some closely related species may mate but the progeny is not viable [73]. Among basidiomycetous yeasts, gene exchange has only been documented in a limited number of heterothallic species such as *Rhodosporidium toruloides* [7], and *Filobasidiella neoformans* [83]. By definition, the biological species concept cannot be applied to asexual (anamorphic) yeast species such as the genus *Candida*, but measurements of DNA relatedness can lead to an approximation.

The evolutionary or phylogenetic species concept [151], regards a species as a single phylogenetically derived lineage. The increasing number of molecular evolutionary studies of yeasts, particularly those using sequence analysis of ribosomal RNA and other genes, may result in a better understanding of the applicability of the phylogenetic species concept to yeasts.

At present, many taxa are best interpreted as genetically uncertain entities whose definition needs to be tested by the analysis of independent phylogenetically derived character sets. Ideally, these approaches, together with a critical evaluation of phenetic and genetic data, should lead to a more stable species concept.

1.4 Phylogeny of yeasts

Most present-day yeast taxonomists have the opinion that a proper taxonomy must represent the phylogeny of the group of organisms concerned. One has to assume an orthologous relationship between the characters studied and the actual, but unknown, evolutionary relationship of the group of organisms concerned [113]. Phylogenetic reconstructions based on sequence analysis of ribosomal RNA (rRNA) and ribosomal DNA (rDNA), recently received much attention. These molecules are considered to be chronometers because of their universal occurrence, functional constraints, and the presence of both variable and less variable regions [153]. Application of PCR and universal primers [149] makes it easy to compare different species. The most frequently used numerical methods for sequence comparison are parsimony, distance methods, and maximum likelihood methods. Resulting phylogenetic trees need to be statistically tested to set confidence limits for the branching order, eg. by bootstrap or jackknife analysis [40].

Nucleotide sequences of 5S rDNA (ca. 120 nucleotides) are highly conserved, and were found to correlate well with septal ultrastructure within the basidiomycetous yeasts [135]. Most phylogenetic studies of yeasts have been based on the 18S or partial sequences of the 26S rDNA [39, 57, 58, 77–81, 130, 132–134, 139]. Some genera, such as *Pichia, Candida, Rhodotorula* and *Cryptococcus* are found to be phylogenetically heterogeneous. Partial 26S rDNA have been studied of almost all currently accepted yeast species [39, 77–81]. The resolution of this method was examined studying sibling species. The sibling pairs *Pichia mississippiensis/P. amylophila, P. americana/P. bimundalis, Issatchenkia scutulata* var. *scutulata*/var. *exigua, Saccharomyces cerevisiae/S. bayanus/S. pastorianus* showed identical sequences in four areas of the 18S rDNA, whereas differentiation was found to occur in the most variable region of the 26S rDNA [105]. Internal transcribed spacer (ITS) sequences are becoming an important tool to distinguish species [129, 100]. Sequence analysis of the highly variable intergenic spacer (IGS) demonstrated a huge variability within the pathogenic yeast *Cryptococcus neoformans* [30].

1.5 Classification of yeasts

The classification scheme presented (Table 1.5-1) is based mainly on the results of KURTZMAN and FELL [39, 75, 78, 80, 81]. This classification scheme certainly will change in the future. We have tried to combine the asexual (anamorphic, imperfect or mitosporic taxa) and the sexual taxa (teleomorphic, perfect or meiotic taxa) into a single classification. A large number of anamorphic basidiomycetous yeast genera, e. g., *Cryptococcus, Sporobolomyces, Bensingtonia* and *Rhodotorula* are polyphyletic. Some ascomycetous genera, particularly the anamorphic genus *Candida* and the teleomorphic genus *Pichia,* are polyphyletic as well.

Tab. 1.5-1 Overview of classification of yeasts [after 39, 75, 81]

Fungi (Kingdom)

1. Eumycota (Division)
    1. Ascomycotina (Subdivision)
        1. Archiascomycetes (Class)
            1. Neolectales (Order)
                1. Neolectaceae (Family)
                    1. *Neolecta*
            2. Pneumocystidales (Order)
                1. Pneumocystidaceae (Family)
                    1. *Pneumocystis*
            3. Protomycetales (Order)
                1. Mitosporic Protomycetales
                    1. *Saitoella*
                2. Protomycetaceae (Family)
                    1. *Protomyces*
            4. Schizosaccharomycetales (Order)
                1. Schizosaccharomycetaceae (Family)
                    1. *Schizosaccharomyces*
            5. Taphrinales (Order)
                1. Taphrinaceae (Family)
                    1. *Taphrina* (including mitosporic members of *Taphrina* classified in *Lalaria*)
        2. Euascomycetes (Class)
            1. Meiosporic Euascomycetes
                1. *Endomyces* p.p. (see also Endomycetaceae)
            2. Mitosporic Euascomycetes
                1. *Oosporidium*

Tab. 1.5-1 Continued

- 3. Hemiascomycetes (Class)
  - 1. Saccharomycetales (Order)
    - 1. Ascoideaceae (Family)
      - 1. *Ascoidea*
    - 2. Cephaloascaceae (Family)
      - 1. *Cephaloascus*
    - 3. Dipodascaceae (Family)
      - 1. *Dipodascus*
      - 2. *Galactomyces*
      - 3. ? *Hyphopichia*
      - 4. ? *Kodamaea*
      - 5. ? *Sporopachydermia*
      - 6. ? *Starmerella*
      - 7. ? *Stephanoascus*
      - 8. ? *Wickerhamiella*
      - 9. ? *Yarrowia*
      - 10. ? *Zygoascus*
    - 3. Endomycetaceae (Family)
      - 1. ? *Endomyces* p.p. (*E. decipiens*)
    - 4. Eremotheciaceae (Family)
      - 1. ? Coccidiascus
      - 2. *Eremothecium*
    - 5. Lipomycetaceae (Family)
      - 1. *Babjevia*
      - 2. *Dipodascopsis*
      - 3. *Kawasakia*
      - 4. *Lipomyces*
      - 5. *Smithiozyma*
      - 6. *Zygozyma*
    - 6. Metschnikowiaceae (Family)
      - 1. *Clavispora*
      - 2. *Metschnikowia*
    - 7. Phaffomycetaceae (Family)
      - 1. *Phaffomyces*
    - 8. Saccharomycetaceae (Family)
      - 1. *Arxiozyma*
      - 2. ? *Citeromyces*
      - 3. ? *Cyniclomyces*
      - 4. ? *Debaryomyces*
      - 5. ? *Dekkera*
      - 6. ? *Issatchenkia*
      - 7. *Kazachstania*
      - 8. *Kluyveromyces*
      - 9. *Lodderomyces*
      - 10. ? *Pachysolen*

Tab. 1.5-1 Continued

- 11. ? *Pichia* (polyphyletic)
- 12. *Saccharomyces*
- 13. ? *Saturnispora*
- 14. ? *Starmera*
- 15. *Tetrapisispora*
- 16. *Torulaspora*
- 17. ? *Williopsis*
- 18. *Zygosaccharomyces*

9. Saccharomycodaceae (Family)
    1. ? *Hanseniaspora*
    2. ? *Nadsonia*
    3. *Saccharomycodes*
    4. ? *Wickerhamia*
10. Saccharomycopsidaceae (Family)
    1. ? *Ambrosiozyma*
    2. *Saccharomycopsis*
11. Candidaceae (Family, mitosporic members of Saccharomycetales)
    1. *Aciculoconidium*
    2. *Arxula*
    3. *Blastobotrys*
    4. *Botryozyma*
    5. *Brettanomyces*
    6. *Candida* (polyphyletic)
    7. *Geotrichum*
    8. *Myxozyma*
    9. *Schizoblastosporion*
    10. *Sympodiomyces*
    11. *Trigonopsis*

2. Basidiomycotina (Subdivision)
    1. Hymenomycetes (Class)
        1. Cystofilobasidiales (Order)
            1. Cystofilobasidiaceae (Family)
                1. *Cystofilobasidium*
                2. *Mrakia*
                3. *Xanthophyllomyces*
            2. Mitosporic members of Cystofilobasidiaceae (Family)
                1. *Cryptococcus* (polyphyletic)
                2. *Udeniomyces*
                3. *Phaffia*
                4. *Trichosporon pullulans*
        2. Filobasidiales (Order)
            1. Filobasidiaceae (Family)
                1. *Filobasidium*
            2. Mitosporic members of Filobasidiaceae

Tab. 1.5-1 Continued

        1. *Cryptococcus* (polyphyletic)
    3. Tremellales (Order)
        1. Sirobasidiaceae (Family)
            1. *Fibulobasidium*
            2. *Sirobasidium*
        2. Tremellaceae (Family)
            1. *Asterotremella* (nom. nud.)
            2. *Bulleromyces*
            3. *Filobasidiella*
            4. *Holtermannia*
            5. *Sterigmatosporidium*
            6. *Tremella*
    4. Mitosporic members of Tremellales
            1. *Bullera*
            2. *Cryptococcus* (polyphyletic)
            3. *Fellomyces*
            4. *Kockovaella*
    5. Trichosporonales (Mitosporic order)
        1. Trichosporonaceae nom. prov. (Family)
            1. *Trichosporon*
            2. *Cryptococcus* (polyphyletic)

2. Urediniomycetes (Class)
    1. *Agaricostilbum* clade
        1. Meiosporic members of Agaricostilbum clade
            1. *Agaricostilbum*
            2. *Chionosphaera*
            3. *Kondoa*
        2. Mitosporic members of *Agaricostilbum* clade
            1. *Bensingtonia*
            2. *Kurtzmanomyces*
            3. *Sporobolomyces* (polyphyletic)
            4. *Sterigmatomyces*
    2. *Erythrobasidium* clade
        1. Meiosporic members of *Erythrobasidium* clade
            1. *Erythrobasidium*
            2. *Occultifur*
            3. *Sakaguchia*
        2. Mitosporic members of *Erythrobasidium* clade
            1. *Rhodotorula* (polyphyletic)
            2. *Sporobolomyces* (polyphyletic)
    3. *Sporidiobolus* clade
        1. Meiosporic members of *Sporidiobolus* clade
            1. *Sporidiobolus*
            2. *Rhodosporidium*

Tab. 1.5-1 Continued

2. Mitosporic member of *Sporidiobolus* clade
   1. *Rhodotorula* (polyphyletic)
   2. *Sporobolomyces* (polyphyletic)

4. *Microbotryum* clade
   1. Meiosporic members of *Microbotryum* clade
      1. *Colacogloea*
      2. *Heterogastridium*
      3. *Leucosporidium*
      4. *Mastigobasidium*
   2. Mitotic members of *Microbotryum* clade
      1. *Rhodotorula* (polyphyletic)
      2. *Reniforma*
      3. *Sporobolomyces* (polyphyletic)

3. Ustilaginomycetes (Class)
   1. Exobasidiomycetidae (Subclass)
      1. Malasseziales (Mitosporic order)
         1. *Malassezia*
      2. Microstromatales (Order)
         1. *Rhodotorula* (polyphyletic)
         2. *Sympodiomycopsis*
   2. Ustilaginomycetidae (Subclass)
      1. Ustilaginales (Order)
         1. *Ustilago*
      1. Mitotic members of Ustilaginaceae (Family)
         1. *Pseudozyma*
         2. *Rhodotorula* (polyphyletic)

4. Unclassified Basidiomycetes
   1. *Tausonia*

? = Species of uncertain affinity

1.6 Morphology of yeasts

Morphological characteristics are still of great importance in yeast systematics. Genera are usually delimited by morphological characteristics, e. g., type of budding (conidiogenesis), cellular morphology, characteristics of ascus formation, morphology of ascospores, teliospores, basidia (basidium) etc. Some morphological characteristics indicate whether imperfect (anamorphic) yeasts belong to the ascomycetes or the basidiomycetes, e. g., mode of budding (enteroblastic versus holoblastic budding), presence of ballistoconidia (ballistoconidium) or clamp connections, and fine structure of cell walls and septal pores. Because of the variable character of many yeasts and yeast-like organisms the use of standardized experimental conditions for the investigation of morphological features is strongly recommended.

1.6.1 Vegetative reproduction

Yeasts show different modes of vegetative reproduction (usually referred to as budding or fission, but a more universal term is conidiogenesis). Vegetative or asexual reproduction occurs in yeasts by budding, by fission, and by the production of conidia on short stalks called sterigmata (Figs 1.6-1, 1.6-2). Knowing how the buds (conidia) are formed helps in the identification of a strain.

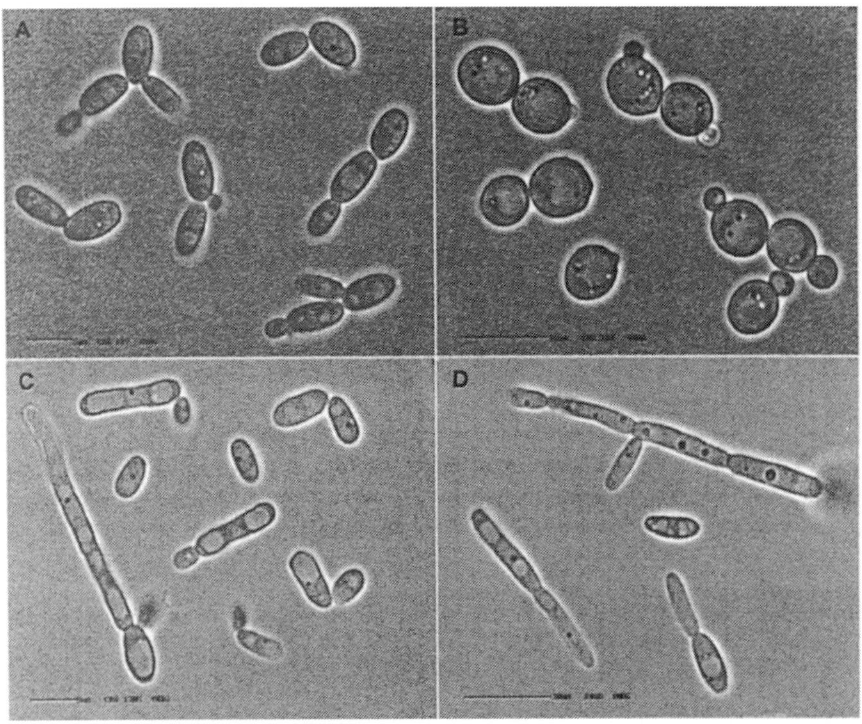

Fig. 1.6-1 Cell morphology of different yeast species (Bar = 10 μm). Fig. 1.6-1A, Ellipsoid cells of *Pichia membranifaciens* var. *membranifaciens* CBS 107; Fig. 1.6-1B, Subglobose cells of *Torulaspora delbrueckii* CBS 133; Fig. 1.6-1C, Ellipsoidal to cylindrical cells of *Pichia silvicola* CBS 1705; Fig. 1.6-1D, Ellipsoidal to cylindrical cells of *Candida boidinii* CBS 2428.

Buds may arise either on yeast cells or on hyphal cells. Budding is termed holoblastic or enteroblastic, depending on how the bud is formed in terms of the fine structure of the cell wall. All layers of the wall of the mother cell are involved in the formation of a holoblastic bud, and the bud separates, usually on a narrow base. Enteroblastic budding is characteristic of basidiomycetous yeasts and their anamorphic states. Here the inner layers of the cell wall rupture the cell wall at the site of bud formation. Subsequent budding at the same site leaves distinct scars. The site of budding is eventually surrounded by a collarette due to the recurrent formation and abscission of a succession of buds arising from the inner layer of the wall of the cell.

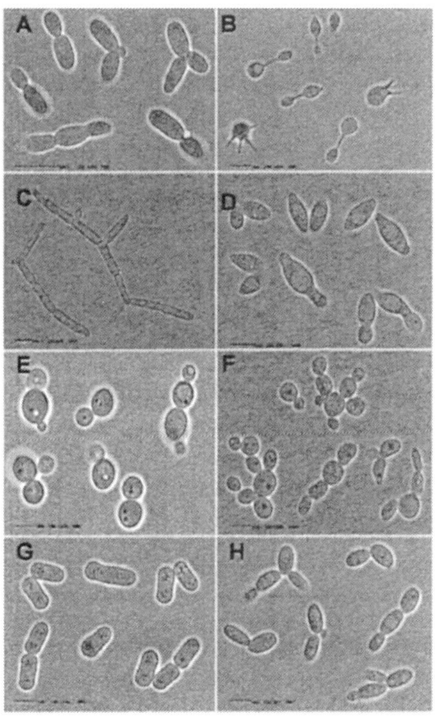

Fig. 1.6-2 Vegetative reproduction of yeast cells (Bar = 10 µm). Fig. 1.6-2A, Polar and sympodial budding in *Cryptococcus macerans* CBS 2206 ; Fig. 1.6-2B, Budding on stalks in *Fellomyces polyborus* CBS 6072; Fig. 1.6-2C, Filaments and pseudohyphae of *Metschnikowia gruessii* CBS 611; Fig. 1.6-2D, Bipolar budding of *Hanseniaspora osmophila* CBS 313; Fig. 1.6-2E, Multipolar budding of *Debaryomyces vanrijiae* var. *vanrijiae* CBS 3024; Fig. 1.6-2F, Multipolar budding in *Pichia nakasei* CBS 5141; Fig. 1.6-2G, Fission (arthroconidiogenesis) in *Schizosaccharomyces pombe* var. *pombe* CBS 356; Fig. 1.6-2H, budding cells of *Pichia membranifaciens* CBS 107.

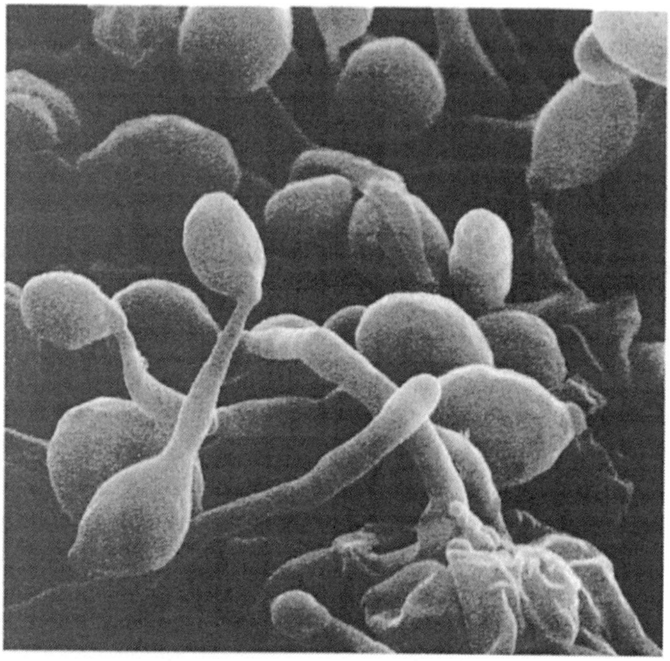

Fig. 1.6-3 Scanning electron image of ballistoconidia of *Bulleromyces albus* CBS 501.

Budding can also be subdivided in terms of the position of the site where it occurs (Figs 1.6-2, 1.6-3, 1.6-4). Budding restricted to one pole of the cell is termed monopolar (Fig. 1.6-4); budding at both poles of the cell is termed bipolar. The buds are often abstricted on a rather broad base by the formation of a cross wall, which is referred to as 'budding on a broad base' or 'bud fission'. Bipolar budding is characteristic of the apiculate yeasts. Budding from various sites on the cell is termed multilateral or multipolar, e. g., *Saccharomyces cerevisiae*. In many basidiomycetous yeasts the buds occur only near the poles of the cell, usually on a narrow base, which is referred to as polar budding. Sympodial budding is the process in which new buds appear just behind and adjacent to a previous bud site. Acropetal budding is the formation of successive buds in a chain with the youngest at the apex. Basipetal budding is the formation of successive buds with the oldest at the apex.

Reproduction by fission is the duplication of a vegetative cell by means of a septum growing inwards from the cell wall to bisect the long axis of the cell. The newly formed fission cells, which are arthroconidia (arthrospores), elongate and the process is repeated. Recurrent fission by a cell may give rise to transverse multiple scars or annellations [108, 128]. This manner of reproduction is characteristic of the genus *Schizosaccharomyces*.

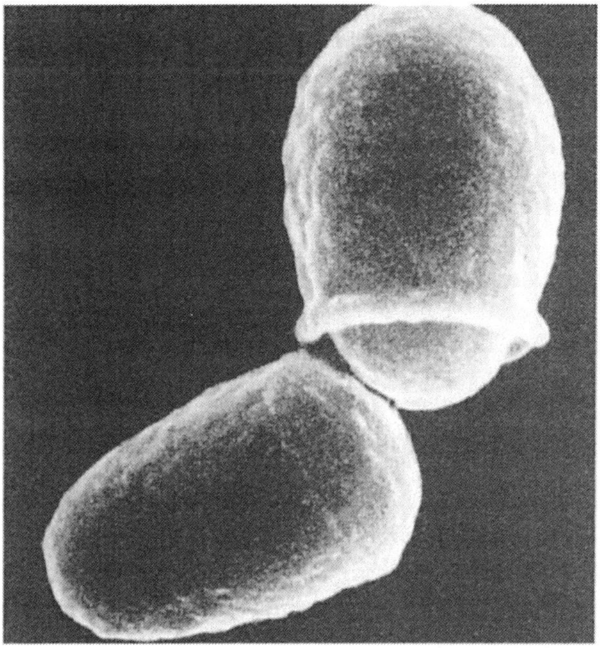

Fig. 1.6-4 Scanning electron image of monopolar budding in *Malassezia pachydermatis* CBS 1879.

Lateral conidia are formed on hyphae of some species. They may occur on specialized cells, the so-called conidiogenous cells (e. g., in *Ambrosiozyma cicatricosa*). Conidia formed on denticles are characteristic of *Stephanoascus* species and *Pichia burtonii.*

Conidia can also be formed on stalk-like structures, usually referred to as sterigmata. It entails the formation by a mother cell of one or more tubular protuberances, each of which gives rise to a terminal conidium. On maturation the conidium is disjointed at a septum either in the mid-region of the tube (*Sterigmatomyces*) or close to the bud (*Fellomyces*). The conidia are not forcibly discharged.

Ballistoconidia are formed on tapering stalk-like structures (sterigmata) and are forcefully discharged. They can be bilaterally symmetrical or more or less rotationally symmetrical (Fig. 1.6-3).

Hyphae are not constricted at their septa, whereas pseudohyphae show distinct constrictions. Pseudohyphae are formed when more or less elongate budding yeast cells adhere in branched or unbranched chains. Proliferation usually occurs acropetally, so that the youngest cell is formed at the apex of the chain of cells. Anastomoses between hyphae may occur in *Ambrosiozyma platypodis*. Dikaryotic hyphae of basidiomycetous yeasts may have clamp connections, whereas monokaryotic hyphae may have incomplete clamp connections (in which the clamp connection is not completely fused with the hypha) or lack these structures. Hyphal septa of both the ascomycetous and the basidiomycetous yeasts may have distinct pore structures.

Hyphae of some yeasts disarticulate into arthroconidia (arthroconidium), which when formed on solid media may remain arranged in a zig-zag position.

Dimorphism, the alternate occurrence of unicellular and hyphal or pseudohyphal phases occurs in a number of yeasts (e. g., *Candida albicans*). Many basidiomycetous yeasts have dimorphic life cycles. Vegetative monokaryotic yeast cells alternate with dikaryotic hyphae on which the sexual form of sporulation may be formed.

Chlamydospores are defined as thick-walled, non-deciduous, intercalary or terminal, asexual spores formed by the rounding off of a cell or cells [2]. The asexual nature of the chlamydospore distinguishes it from the teliospore of the Uredinales and Ustilaginales from which the basidium is produced. Chlamydospores are characteristic of *Candida albicans* and *Metschnikowia* species, but are occasionally noticed in old cultures of other taxa on agar, including some *Trichosporon* and *Cryptococcus* species.

Endospores occur in some yeasts such as *Candida, Cryptococcus, Trichosporon, Cystofilobasidium*, and *Leucosporidium*. They are vegetative cells formed endogenously inside other cells and may occur in long standing cultures.

1.6.2 Generative reproduction

1.6.2.1 Ascomycetous yeasts

In homothallic ascomycetous yeasts with a diploid vegetative phase, a single diploid vegetative cell may undergo meiosis and become an unconjugated ascus. The diploid condition can be restored by conjugation of ascospores inside the ascus, or by fusion of daughter nuclei [108].

Conjugation takes place in one of several ways:

1. Parent cell-bud conjugation: two haploid nuclei, one each from the parent cell and the bud, fuse and give rise to ascospores in an ascus (e. g., *Debaryomyces*). The asci typically have a small protuberance.

2. Gametangial conjugation: short protuberances (gametangia) frequently develop adjacent to a septum, fuse and form an ascus. After meiosis one to four, or sometimes more, ascospores are formed (e. g., *Saccharomycopsis, Galactomyces, Dipodascus)*.

3. Conjugation between two cells (heterothallism): cells of complementary mating type each form a conjugation tube that grow toward each other, fuse, and form an ascus with

ascospores formed in either one or both of the conjugating cells (e. g., *Pichia* and *Zygosaccharomyces*).

4. Conjugation between hyphae: conjugation tubes are formed between hyphae followed by ascus formation (e. g., *Zygoascus*).

The form of asci can be characteristic of certain genera (Fig. 1.6-6). In *Lipomyces* the asci are sac-like appendages; in *Metschnikowia* they are long and clavate; in *Zygosaccharomyces* they are dumbbell shaped. Asci can be persistent (e. g., *Saccharomyces* and *Zygosaccharomyces*) or they can be evanescent (e. g., *Kluyveromyces* and *Clavispora*).

Ascospore morphology has often been used for generic delimitation. However, recent molecular studies have shown that ascospore morphology is not a reliable character. For most yeasts, the number of ascospores varies from 1–4(–8). However, multispored asci occur in several genera (e. g., *Ascoidea, Lipomyces, Dipodascus,* and *Schizosaccharomyces*). Ascospores are usually hyaline, but occasionally pigmented (e. g., *Lipomyces, Nadsonia,* and *Pichia*), and can be globose, ellipsoidal, hat-shaped (Fig. 1.6-5), saturn-shaped, or needle-shaped, with or without whip-like appendages. The spore surface may be smooth, verrucose or ridged (Fig. 1.6-6).

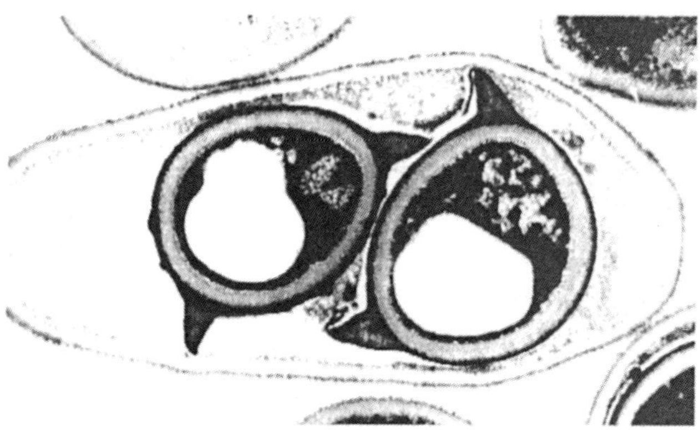

Fig. 1.6-5 Transmission electro micrograph of hat-shaped ascospores of *Pichia delftensis* CBS 2109.

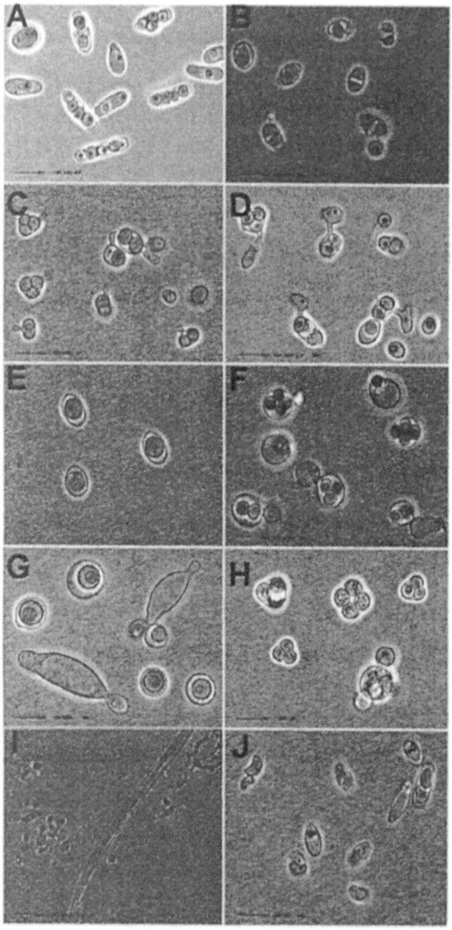

Fig. 1.6-6 Ascus and ascospore morphology of yeasts (Bar = 10 μm). Fig. 1.6-6A, Asci and globose ascospores of *Pichia scaptomyzae* CBS 1329; Fig. 1.6-6B, Asci and hat-shaped ascospores of *Pichia canadensis* CBS 1992; Fig. 1.6-6C, Parent cell-bud conjugation in *Debaryomyces pseudopolymorphus* CBS 2008; Fig. 1.6-6D, Conjugating asci of *Kodamae ohmeri* CBS 2037; Fig. 1.6-6E, 1-spored ascus of *Saccharomyces transvaalensis* CBS 2186; Fig. 1.6-6F, Multispored asci of *Lipomyces kononenkoae* CBS 2514; Fig. 1.6-6G, 1-spored asci on tip of cell of *Nadsonia fulvescens;* Fig. 1.6-6H, Ascus of *Saccharomyces paradoxus* CBS 432; Fig. 1.6-6I, Needle-shaped ascospores of *Metschnikowia hawaiiensis* CBS 7432; Fig. 1.6-6J, Curved ascospores of *Kluyveromyces marxianus* CBS 712.

1.6.2.2 Basidiomycetous yeasts

Many basidiomycetous yeast species have dimorphic life cycles in which monokaryotic yeast phases alternate with dikaryotic hyphal phases. Clamp connections are frequently present. The mating system (or incompatibility system) of basidiomycetous yeasts can be bipolar, tetrapolar or modified tetrapolar. The presence of complementary mating factors (e. g., A_1B_2 x A_2B_1) results in conjugation followed by formation of dikaryotic hyphae usually with clamp connections. Meiosis usually occurs in a specialized cell, the basidium (also

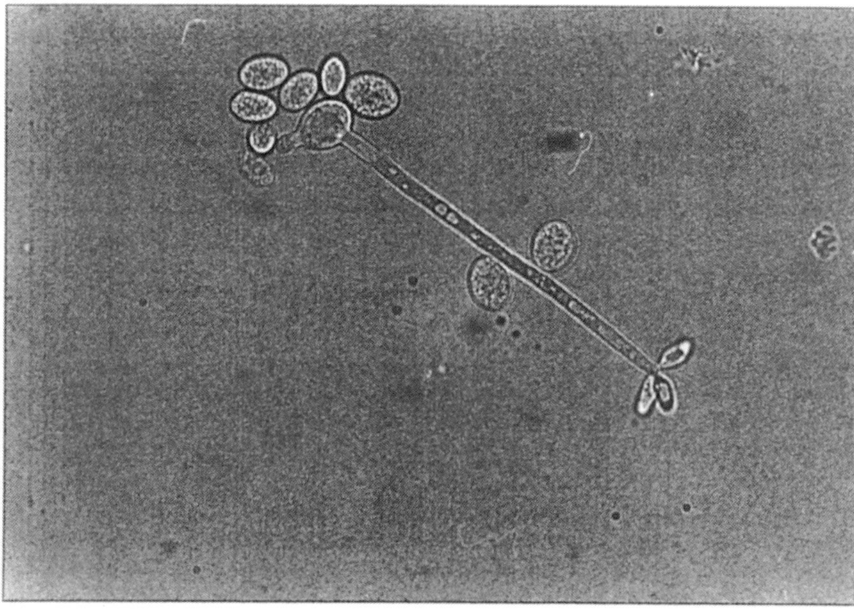

Fig. 1.6-7 Basidium and basidiospores of *Xanthophyllomyces dendrorhous* CBS 7919.

called metabasidium). Many species form thick-walled teliospores, in which nuclear fusion occurs (e. g., *Sporidiobolus, Rhodosporidium, Leucosporidium, Cystofilobasidium*). They can only be differentiated from vegetative chlamydospores by karyology (karyogamy and meiosis) and typically germinate with basidia. Teliospores may be intercalary or terminal, single or in small clusters, (sub)globose or angular, hyaline or pigmented, and are usually smooth. However, teliospores of the yeast-like fungus *Tilletiaria anomala* are covered with warts. Teliospore germination is often enhanced by soaking them in water for several weeks [38], and occurs by transversely septate or one-celled basidia on which basidiospores are formed.

Some species do form basidia directly at the dikaryotic hyphae (examples: *Filobasidiella, Filobasidium, Bulleromyces*) or yeast cells (*Xanthophyllomyces*, Fig. 1.6-7). The basidia of *Filobasidiella, Filobasidium* and *Xanthophyllomyces* are one-celled, clavate, capitate to cylindrical ('Filobasidiales'-type), whereas in *Bulleromyces* they are longitudinally or obliquely septate ('Tremellales'-type). Basidiospores in *Filobasidiella* are formed basipetally with the youngest spore at the base of the chain of spores.

1.7 Where do yeasts occur

Yeasts grow in many different habitats. For the purpose of this book we distinguished:

1. yeasts occurring in natural biota and habitats.
2. yeasts occurring on man and other animals.
3. yeasts occurring in man-made environments and with applied importance.

1.7.1 Yeasts from natural substrates

Many yeasts occur in soil. These include a variety of soils and related substrates such as decomposing litter, humous layers, sandy, clayey or loamy soils, podzolic soils, and permafrosts. Insects (e. g., *Drosophila* spp. and beetles) and insect related products (e. g., frass in tunnels of insects) are a rich source of yeasts as well. On plants, yeasts occupy their own niches such as leaves, flowers, pollen, and roots. Succulents have their own yeast species associated with them. Trees also provide diverse and rich substrates for yeasts, e. g., in the phyllosphere, non-degraded and degraded wood, and in tree exudates. Many yeasts originate from fruits, like grapes, apples, figs, dates, olives, and fruit products such as fruit pulp and fruit juices. Other natural substrates where yeasts have been found are fish, shrimps, birds, lichens, mushrooms, mosses, and algae. Seawater is a rather rich habitat for certain yeasts species, such as those belonging to the genera *Rhodosporidium, Leucosporidium* and *Mrakia.* Yeasts occur also in fresh water, mud and swamps, and the atmosphere.

1.7.2 Yeasts from clinical and animal sources

Clinical and veterinary material is an important habitat for several yeast species. Three main clinical habitats and associated clinical phenomena can be discerned:

a. superficial substrates such as skin, nails and hairs (e. g., dandruff, pityriasis versicolor, eczema, white piedra, onychomycosis etc.).

b. deep mycoses (organs and fluids inside the human body, e. g., blood, cerebrospinal fluid, lungs, brains).

c. mucous membranes (e. g., in mouth, vagina etc.).

Among the most important medical yeasts are *Candida albicans, C. glabrata, C. guilliermondii, C. krusei* (= *Issatchenkia orientalis*), *C. lusitaniae, C. parapsilosis, C. tropicalis, Cryptococcus neoformans* (sexual state *Filobasidiella neoformans*), *Trichosporon cutaneum, T. inkin, T. mucoides, Geotrichum candidum, Malassezia furfur, M. pachydermatis,* and *M. globosa.* However, many fungi that are usually considered to be no pathogenic appear as opportunists nowadays, such as e. g., *Saccharomyces cerevisiae* [21, 24, 90].

Information on the classification of pathogenic hazard groups can be found on: http://bio-safety.ihe.be/; http://europe.osha.eu.int/; http://www.who.int/emc/biosafety.html; http://www.orcbs.msu.edu/biological/bmbl/bmbl-1.htm.

Yeast strains used in the food industry usually are considered as GRAS-organisms (Generally Recognized As Safe).

1.7.3 Yeasts from man-made and related habitats and/or with practical importance

Wine, most, cider and beer are important sources for yeast species, as are fermented foods such as miso, tea-beer, tepache, tempeh, fermented dairy products, cocoa, coffee and tobacco. Fruit juices and fruit concentrates, dairy products, soft drinks, molasses, sugar and honey, meat, pickles and vinegar, and baking products and doughs are rich sources for yeast isolates. Fewer yeasts occur in sewage, active sludge, effluents, mud, asphalt soils and oil spills.

1.8 Appendix: Overview of yeast genera of importance to the food industry

1.8.1 Teleomorphic ascomycetous genera

Arxiozyma van der Walt & Yarrow
One species: *A. telluris* (van der Walt) van der Walt & Yarrow.
Anamorph: *Candida pintolopesii* (van Uden) S.A. Meyer & Yarrow.
The species ferments sugars and is isolated primarily from poultry and other birds.

Ashbya Guilliermond, see Eremothecium

Citeromyces Santa María
One species: *C. matritensis* (Santa María) Santa María.
The species is a strong fermenter of sugars and is usually isolated from sugar concentrates and slime fluxes.

Clavispora Rodrigues de Miranda
Two species: *C. lusitaniae* Rodrigues de Miranda and *C. opuntiae* Phaff, Miranda, Starmer & Barker
The species occur in a variety of habitats, such as cactus, prickle pear (*Opuntia* spp.), insects, effluent of a chocolate factory, and clinical samples. The species is also known from raw milk (Chapter 8).

Debaryomyces Lodder & Kreger-van Rij, Taxon 27, 306 (1978).
The presently accepted 15 species are described in KURTZMAN & FELL [75] and BARNETT et al. [10]. Comparisons of ribosomal RNA sequence similarities resulted in the transfer to *Debaryomyces* of several species from other genera, namely *Schwanniomyces occidentalis* [77], *Wingea robertsii* [79], and *Pichia carsonii* and *P. etchellsii* [158]. The species give a weak to strong fermentation of sugars. The species formerly classified as *Schwanniomyces occidentalis* shows high amylase activity [29]. The species are commonly found in soil, plant products, food and in clinical specimens. Important food-related habitats are dairy products (Chapter 8), meat and meat products (Chapter 9), grapes (Chapter 10), fermenting coffee beans (Chapter 16), and traditionally fermented products such as pulque, idli and turuk (Chapter 17).

Dekkera van der Walt
Two species: *D. anomala* Smith & van Grinsven, *D. bruxellensis* van der Walt.
Anamorph: *Brettanomyces.*
The species may show a variable fermentation of sugars, which is stimulated by the presence of oxygen (Custers effect). Species are noted for a vigorous production of acetic acid and this causes early cell death in cultures. The species are usually isolated from beer, wine and soft drinks. *Dekkera* yeasts are important spoilage organisms in non-alcoholic beverages (Chapter 12). They have, however, a beneficial effect in the production of lambic-type beers (Chapter 13).

Endomyces Reess
Four species: *E. cortinarii* Redhead & Malloch, *E. decipiens* Reess, *E. polyporicola* (Schumacher & Ryvarden) de Hoog, M.Th Smith & Guého, *E. scopularum* Helfer.
Species of *Endomyces* are parasitic on mushrooms.

Eremothecium Borzi
Five species: *E. ashbyi* (Guilliermond ex Routien) Batra, E. coryli (Peglion) Kurtzman, *E. cymbalariae* Borzi, E. gossypii (Ashby & Nowell) Kurtzman, E. sinecaudum (Holley) Kurtzman.
Cultures of *E. ashbyi* are often yellow-orange in color from formation of riboflavin, and the species has been used for the production of this vitamin. *E. coryli* causes diseases of hazelnuts, cotton bolls, and various beans. The species have been isolated from cotton bolls, citrus, insects and mustard seed.

Galactomyces Redhead & Malloch
Three species: *G. citri-aurantii* E.E Butler, *G. geotrichum* (E. E. Butler & L. J. Petersen) Redhead & Malloch, *G. reessii* (van der Walt) Redhead & Malloch.
Anamorph: *Geotrichum.*
The species are common in soil, plant material, food (dairy products), and clinical specimens. *Galactomyces geotrichum* (= *Geotrichum candidum*) may occur on cheese (Chapter 8).

Hanseniaspora Zikes
The six species are discussed by KURTZMAN & FELL [75] and BARNETT et al. [10].

Anamorph: *Kloeckera*.
The species are most frequently isolated from soil, fruits and plant exudates. The species occur on grapes and processed fruit (Chapter 10). *Hanseniaspora* species are important in the first phase of grape fermentation and they are supposed to play a role in the production of certain flavours beneficial for the quality of the wine (Chapters 10 and 14).

Issatchenkia Kudryavtsev emend. Kurtzman, Smiley & Phaff
Four species are recognized in KURTZMAN & FELL [75]. *Issatchenkia occidentalis* is known from a tea fungus, *I. orientalis* originates from fruit juice, tea beer, bread, dairy products, fermented foods (Chapter 17), and clinical samples, whereas *I. scutulate* is known from cherry juice, old wine and pressed grapes [75]. *Issatchenkia orientalis* causes spoilage of non-alcoholic beverages (Chapter 12).

Kluyveromyces van der Walt emend. van der Walt
Fifteen species are described by Kurtzman & Fell [75], and 17 in Barnett et al. [10].
Because of their ability to ferment lactose, *K. lactis* (anamorph *Candida sphaerica*) and *K. marxianus* (anamorph *C. kefyr*) have been used industrially to produce ethanol from waste dairy products such as whey. Lactose utilization from whey has also been reported for *K. marxianus* (as *K. fragilis*) [62]. *Kluyveromyces marxianus* shows a high inulinase activity [141], and this species is also known for the production of extracellular polygalacturonase. Species are isolated from soil, water, fruit and other plant materials, tree fluxes, dairy products, *Drosophila* and occasionally from clinical specimens. *Kluyveromyces lactis* and *K. marxianus* are important dairy yeasts (Chapter 8), and the former species occurs in cocoa fermentations (Chapter 16).

Lodderomyces van der Walt
One species: *L. elongisporus* (Recca & Mrak) van der Walt.
Isolates are isolated from soil and orange juice, occur in cocoa fermentations (Chapter 16). The species occasionally causes spoilage of soft drinks (Chapter 12).

Pachysolen Boidin & Adzet
One species: *P. tannophilus* Boidin & Adzet.
Isolates of *P. tannophilus* are isolated from tanning liquors and leather. The species ferments a variety of sugars. Of interest to biotechnology is the ethanolic fermentation of D-xylose, D-galactose and glycerol, and the bioconversion of wheat straw. *P. tannophilus* is also noted for production of an extracellular polysaccharide.

Pichia E.C. Hansen emend. Kurtzman
The presently accepted ca. 90 species are described in KURTZMAN & FELL [75] and BARNETT et al. [10].
Pichia appears to be extremely heterogeneous. The genus nearly doubled in size with the transfer of nitrate-positive *Hansenula* species to *Pichia* [71]. Besides *Yamadazyma*, the genus *Hyphopichia* von Arx & van der Walt is presently considered a synonym of *Pichia*. *Pichia stipitis* (whose anamorph is *Candida shehatae*) is of biotechnological importance, because of its ability to ferment xylose, a major component of plant biomass. *Pichia naka-*

zawae was found to have high amylase activity [29]. Members of the genus are commonly isolated from soil, water, tree exudates, fruits, necrotic cactus tissue, insects and clinical specimens. Several species occur on fruits, fruit concentrates (Chapter10), and traditionally fermented products (Chapter 17). *Pichia* (= *Hyphopichia*) *burtonii* causes spoilage of bread (Chapter 11), and *P. membranifaciens* is a spoilage organism of soft drinks, dairy and meat products (Chapters 8, 9, 12).

Saccharomyces Meyen ex Reess
The presently accepted species are described in KURTZMAN & FELL [75] and BARNETT et al. [10].
The species are strongly fermentative, and are commonly isolated from soil, fruits, foods, beverages, but also from clinical samples. Four of the species in this genus, which form the *S. cerevisiae* sibling species complex, are widely used for bread making (Chapter 11), and the production of beer (Chapter 13), wine (Chapter 14), distilled beverages and fuel alcohol. *S. cerevisiae* occurs on fruit, in processed fruits, dairy products (Chapter 8), and plays a role in the fermentation of kefir (Chapter 8), coffee, cocoa (Chapter 16), and the production of traditional fermenting products (Chapter 17). *S. cerevisiae* and *S. bayanus* cause spoilage of soft drinks (Chapter 12). *Saccharomyces* yeasts are also involved in sour rot of grapes (Chapter 10).

Saccharomycopsis Schiönning
Ten species are recognized in KURTZMAN and FELL [75]. *Saccharomycopsis fibuligera* has a strong amylolytic activity, and is known from flour, bakery products (Chapter 11), dry fermentations of coffee (Chapter 16), and traditionally fermented products (Chapter 17).

Schizosaccharomyces P. Lindner
Three species: *S. japonicus* Yukawa & Maki, *S. octosporus* Beyerinck, *S. pombe* Lindner.
The species are isolated from fruits and fruit juices, wines, tequila fermentation and high-sugar substrates. The species are strong fermenters of sugars and have been used for the production of ethanol. The genus is only distantly related to *Saccharomyces*. *Schizosaccharomyces pombe* occurs in fruit concentrates (Chapter 10), wet fermentations of coffee (Chapter 16) and causes spoilage of soft drinks (Chapter 12).

Torulaspora P. Lindner
Three species: *T. delbrueckii* (Lindner) Lindner, *T. globosa* (Klöcker) van der Walt & E. Johannsen, *T. pretoriensis* (van der Walt & Tscheuschner) van der Walt & E. Johannsen.
The species strongly ferment sugars. Separation of *Torulaspora*, *Saccharomyces* and *Zygosaccharomyces* has been problematic. Strains are frequently isolated from soil, fruits, fruit juices and other plant products, and occasionally from human and animal sources. *Torulaspora delbrueckii* occurs in a variety of fermented products (Chapter 17), including kefir and cheese (Chapter 8),

Yarrowia van der Walt & von Arx
One species: *Y. lipolytica* (Wickerham et al.) van der Walt & von Arx.
Anamorph: *Candida lipolytica* (F.C. Harrison) Diddens & Lodder.

Isolates are from soil, agricultural and industrial processing wastes, fatty and proteinaceous materials, and animal and human clinical specimens. *Y. lipolytica* is an important non-fermentative industrial yeast. It is markedly proteolytic and lipolytic, and because of its ability to grow on hydrocarbons, has been used to produce single-cell protein from petroleum. More importantly is its capability to produce high yields of citric acid [44]. For a review on the molecular genetics and biotechnological aspects of the species the reader is referred to Heslot [59]. *Yarrowia lipolytica* is an important dairy yeast of cheese (Chapter 8), and causes spoilage of meat products (Chapter 9).

Zygosaccharomyces Barker
Nine species are listed by KURTZMAN & FELL [75] and BARNETT et al. [10].
Isolates are commonly obtained from wine, various foods, fruit, trees, and *Drosophila* species. Species of *Zygosaccharomyces* are among the most important spoilage organisms (Chapters 6 and 7). The species cause spoilage of bread (Chapter 11), non-alcoholic beverages (Chapter 12), acidified foods and condiments, and are involved in sour rot of grapes (Chapter 10). *Zygosaccharomyces rouxii* is involved in the fermentation of soy sauce (Chapter 15), and traditionally fermented products (Chapter 17).

1.8.2 Anamorphic ascomycetous genera

Arxula van der Walt, Smith & Y. Yamada
Two species, *A. adeninovorans* and *A. terrestris*, are recognized [75]. The first species is known from ensiled maize, soil, intestines of a lizard, and *A. terrestris* is known from soil. *Arxula adeninovorans* seems to be an important species during dry fermentations of coffee (Chapter 16).

Brettanomyces Kufferath & van Laer
The five species are listed in KURTZMAN & FELL [75] and BARNETT et al. [10].
Teleomorph: *Dekkera.*
The species may show variable fermentation of sugars, which is stimulated by the presence of oxygen (Custers effect). Species are noted for vigorous production of acetic acid and this causes early cell death in cultures. The species have been isolated from beer, wine and soft drinks (see *Dekkera*).

Candida Berkhout
The most recent published summary of the ca. 160 Candida species is given by KURTZMAN & FELL [75] and BARNETT et al. [10]. However, ca. 200 species are currently being recognized (V. ROBERT, unpubl. observ.)
The genus *Torulopsis* Berlese is considered to be a synonym of *Candida*, and consequently, *Candida* includes those species that form pseudohyphae and hyphae as well as those that do not. Several *Candida* species are of medical importance, e. g., *C. albicans, C. glabrata, C. parapsilosis* and *C. tropicalis*. Many *Candida* species are involved in food products, either beneficial or detrimental. *Candida papapsilosis* is also an opportunistic spoilage

yeast occurring in soft drinks (Chapter 12), and *C. tropicalis, C. parapsilosis* and *C. pelliculosa* occur in wet fermentations of coffee (Chapter 16). *Candida davenportii* causes spoilage of soft drinks, and *C. valida* (= *Pichia membranifaciens*) and *C. holmii* (= *Saccharomyces exiguus*) are important spoilage yeasts of soft drinks (Chapter 12). *Candida guilliermondii* and *C. parapsilosis* are spoilage species of dairy products (Chapter 8). *Candida versatilis* contributes to the flavour of soy sauce (Chapter 15). *Candida rugosa* is involved in cocoa fermentations, and *C. boidinii* degrades pectin thus having a beneficial effect on coffee fermentations (Chapter 16). *Candida utilis* (= *Pichia jadinii*) and *C. maltosa* are used for biomass production from carbohydrate and hydrocarbon substrates, respectively.

Kloeckera Janke
This genus represents the anamorphic state of *Hanseniaspora* and only one species, *K. lindneri* (Klöcker) Janke, remains for which the ascosporic state has not yet been found. The species ferments glucose and has been isolated from soil in Java and from plant material in Taiwan.

Trigonopsis Schachner
One species: *T. variabilis* Schachner.
The species is non-fermentative. Isolates are from beer and grape must.

1.8.3 Teleomorphic heterobasidiomycetous genera

Erythrobasidium Hamamoto et al.
One species: *E. hasegawianum* Hamamoto et al.
Anamorph: *Rhodotorula.*
Isolated from spent brewers's yeast.

Filobasidium Olive
Five species [10, 75].
Anamorphs: *Cryptococcus.*
Filobasidium floriforme, F. elegans and *F. globisorum* are known from plant material, *F. uniguttulatus* is known from clinical specimens, and *F. capsuligenum* is isolated from soil and wine-making equipment. *Filobasidium capsuligenum* shows extracellular amylolytic activity because of production of α-amylase and glucoamylases [27, 29].

Leucosporidium Fell et al., Antonie van Leeuwenhoek 35, 438, (1969).
Four species: *L. antarcticum* Fell et al., *L. fasciculare* Bab'eva & Lischkina, *L. fellii* Giménez-Jurado, *L. scottii* Fell et al.
Anamorph: *Candida.*
Isolated from soil, fresh water, seawater, seaweeds, food, and trees. The degradation of L(+)-tartaric acid by *L. fellii* may be of interest to the wine industry [47]. Aromatic compounds are assimilated by *L. scottii* [92].

Mrakia Y. Yamada & Komagata
One species, *Mrakia frigida*, which is known from soil, mosses, lichens, algae, frozen food [75]. The psychrophilic species may grow on foods stored at low temperatures (Chapter 4).

Rhodosporidium Banno
Nine species are currently known [10, 75].
Anamorphs: *Rhodotorula.*
Isolated from various substrates such as soil, fresh water, seawater, mangrove swamps, air and wood pulp. *Rhodosporidium lusitaniae* has been reported to degrade phenolic compounds [43], and *R. toruloides* accumulates large amounts of lipids [114–117].

Sporidiobolus Nyland
Five species: *S. johnsonii* Nyland, *S. microsporus* Higham ex Fell et al., *S. pararoseus* Fell & Tallman, *S. ruineniae* Holzschu et al., *S. salmonicolor* Fell & Tallman.
Anamorph: *Sporobolomyces.*
The species occur mainly on leaves, but have also been isolated from air, fruit, skin, fodder, seawater, wood chips, and oil. Torularhodin was found to be the main carotenoid pigment, but β-carotene, torulene and β-carotene occur as well [138].

Xanthophyllomyces Golubev
Only 1 species: *X. dendrorhous*
Anamorph: *Phaffia rhodozyma.*
The species ferments D-glucose [93] and occurs in slime fluxes of deciduous trees [50]. Its main pigment is astaxanthin, which is an important dietiery source for aquaculture and poultry industries [64, 65].

1.8.4 Anamorphic heterobasidiomycetous genera

Bullera Derx
Eightteen species are known [75].
Teleomorph: *Bulleromyces.*
The species have been isolated mainly from leaves, but also from fruit, plants, larvae of beetles, wood, frozen salmon and air in a dairy.

Cryptococcus Kützing
Thirty-four species are listed in KURTZMAN & FELL [75].
The species have been isolated from diverse substrates such as soil, fruit, water, man, animals, wine, leaves, fungi. *Cryptococcus flavus* is amylolytic [29]. *Cryptococcus laurentii* and *Cryptococcus curvatus* have been reported to accumulate large amounts of lipids [104, 114–117]. Various species of the genus (e. g., *C. laurentii, C. albidus*) occur on plant surfaces, including fruits (Chapter 12), and fresh meat (Chapter 9).

Fellomyces Yamada & Banno
Eleven species are listed in BARNETT et al. [10].
Isolated from diverse substrates, such as food, flowers, tree, lichens and fungi.

Kurtzmanomyces Yamada et al.
Three species: *K. insolitus* Sampaio & Fell, *K. nectairii* (Rodrigues de Miranda) Yamada et al., *K. tardus* Giménez-Jurado & van Uden.
The species have been isolated from cheese and water.

Phaffia Miller et al.
See *Xanthophyllomyces*.

Pseudozyma Bandoni emend. Boekhout
Several basidiomycetous yeast-like organisms, presently classified in the genus *Pseudozyma* represent anamorphs of Ustilaginales [11, 13, 39]. *Candida* 107 (NCYC 911), which apparently is closely related to or identical with *Ps. antarctica*, accumulates large amounts of fatty acids (ca. 41 % of the dry weight, and ca. 40 % of the total fatty acid content as saturated triglycerides) [46]. *Pseudozyma antarctica* is reported to produce extracellular mannosylerythritol lipids when grown on soybean oil as a carbon source [68]. *Pseudozyma flocculosa*, initially described in the genus *Sporothrix and Stephanoascus,* is a biocontrol organism against powdery mildews [4, 54, 61]. This antagonistic activity appears to be caused by toxic extracellularly produced fatty acids [22].

Rhodotorula F.C. Harrison
Thirty-seven species are listed by BARNETT et al. [10], but currently 41 are known (V. ROBERT, unpubl. observ.).
Many *Rhodotorula* species form torulene or torularhodin as the main pigment, but β-carotene and β-carotene are usually present as well and some species also form neurosporene and/or lycopene [138]. Degradation of phenol is reported for *R. glutinis* var. *glutinis* and *R. rubra. Rhodotorula glutinis, R. minuta, R. rubra,* and *R. aurantiaca* are reported to utilize a variety of aromatic compounds [92, 94]. *Rhodotorula graminis, R. glutinis* var. *glutinis, R. gracilis* and *R. mucilaginosa* are oleaginous [101, 114–118, 120]. The species have been isolated from a wide variety of substrates, including clinical specimen. Many species occur in the phyllosphere and on fruits, such as apple (Chapter 12), *Rhodotorula mucilaginosa* is known from raw milk and cheese (Chapter 8), and *Rhodotorula* species occur on fresh meat (Chapter 9).

Sporobolomyces Kluyver & van Niel
Twenty-seven species are listed by BARNETT et al. [10], but currently 36 species are known (V. ROBERT, unpubl. observ.).
The species occur mainly in the phyllosphere and on fruits, such as apple, but they have been isolated from a wide range of other substrates. *Sporobolomyces roseus* contains torularhodin as its main pigment, but torulene and β-carotene occur as well [138].

Trichosporon Behrend
Nineteen species are listed by GUÉHO et al. [52], KURTZMAN & FELL [75] and BARNETT et al. [10].

Trichosporon pullulans degrades starch and pullulan due to the production of α-amylase and glucoamylase(s) [28, 29]. *Trichosporon cutaneum*, *T. moniliiforme* and *T. dulcitum* are found to assimilate a wide variety of aromatic compounds [92, 119]. *Trichosporon beigelii* (= *T. cutaneum*) is capable to utilize cheese whey as a carbon and energy source for biomass production. Several species of the genus are of medical importance. *Trichosporon cutaneum* and *T. pullulans* are able to accumulate substantial amounts of lipid [114–117]. *Trichosporon cutaneum* occurs in raw milk and cheese (Chapter 8), salami-type sausages (Chapter 9) and in traditionally fermented doughs (Chapter 17).

1.9 References

[1] ADRIO, J.; LOPEZ, M.; CASQUEIRO, J.; FERNANDEZ, C.; VEIGA, M.: Electrophoretic karyotype of the astaxanthin-producing yeast *Phaffia rhodozyma*. Curr Genet **27** (1995) 447–450.

[2] AINSWORTH, G.C.: Dictionary of fungi, 6th ed. Kew, U.K.: Cab (1971).

[3] AULAKH, H.S.; STRAUS, S.E.; KWON-CHUNG, K.J.: Genetic relatedness of *Filobasidiella neoformans (Cryptococcus neoformans)* and *Filobasidiella bacillispora* (*Cryptococcus bacilisporus*) as determined by deoxyribonucleic acid composition and sequence homology studies. Int. J. Syst. Bacteriol. **31** (1981) 97–103.

[4] AVIS, T.J.; CARON, S.J.; BOEKHOUT, T.; HAMELIN, R.C.; BÉLANGER, R.R.: Molecular and physiological analysis of the powdery mildew antagonist *Pseudozyma flocculosa* and related fungi. Phytopathol. **91** (2001) 249–254.

[5] AZUMI, M.; GOTO-YAMAMOTO, N.: AFLP analysis of type strains and laboratory and industrial strains of *Saccharomyces* sensu stricto and its application to phenetic clustering. Yeast **18** (2001) 1145–1154.

[6] BAI, F.Y.; LIANG, H.Y.; JIA, J.H.: Taxonomic relationships among the taxa in the *Candida guilliermondii* complex, as revealed by comparative electrophoretic karyotyping. Int. J. Syst. Evol. Microbiol. **50** (2000) 417–422.

[7] BANNO, I.: Studies on the sexuality of *Rhodotorula*. J. Gen. Appl. Microbiol. **13** (1967) 167–196.

[8] BARNETT, J.A.: Biochemical differentiation of taxa with special reference to the yeasts. In: The Fungi, Vol. 3 (edited by Ainsworth, G.C.; Sussman, A.S.). London, U.K.: Academic Press (1968) 557–595.

[9] BARNETT, J.A.: The nutritional tests in yeast systematics. J. Gen. Microbiol. **99** (1977) 183–190.

[10] BARNETT, J.A.; PAYNE, R.W.; YARROW, D.: Yeasts: characteristics and identification, 3rd ed. Cambridge, U.K.: Cambridge University Press (2000).

[11] BEGEROW, D.; BAUER, R.; BOEKHOUT, T.: Phylogenetic placements of ustilaginomycetous anamorphs as deduced from nuclear LSU rDNA sequences. Mycol. Res. **104** (2000) 53–60.

[12] BELLOCH, C.; BARRIO, E.; GARCIA, M.D.; QUEROL, A.: Inter- and intraspecific chromosome pattern variation in the yeast genus *Kluyveromyces*. Yeast **14** (1998) 1341–1354.

[13] BOEKHOUT, T.: Systematics of anamorphs of Ustilaginales (smut fungi) – a preliminary survey. Stud. Mycol. **30** (1987) 137–149.

[14] BOEKHOUT, T.; BOSBOOM, R.W.: Karyotyping of *Malassezia* yeasts: taxonomic and epidemiological implications. Syst. Appl. Microbiol. **17** (1994)146–153.

[15] BOEKHOUT, T.; KAMP, M.; GUÉHO, E.: Molecular typing of *Malassezia* species with PFGE and RAPD. Med. Mycol. **36** (1998) 365–372.

[16] BOEKHOUT, T.; RENTING, M.; SCHEFFERS, W.A.; BOSBOOM, R.: The use of karyotyping in the systematics of yeasts. Antonie van Leeuwenhoek **63** (1993) 157–163.

[17] BOEKHOUT, T.; YAMADA, Y.; WEIJMAN, A.C.M.; ROEIJMANS, H.J.; BATENBURG-VAN DER VEGTE, W.H.: The significance of coenzyme Q, carbohydrate composition and septal ultrastructure for the taxonomy of ballistoconidia-forming yeasts and fungi. Syst. Appl. Microbiol. **15** (1992) 1–10.

[18] BOEKHOUT, T.; VAN BELKUM, A.; LEENDERS, A.C.A.P.; VERBRUGH, H.A.; MUKAMURANGWA, P.; SWINNE, D.; SCHEFFERS, W.A.: Molecular typing of *Cryptococcus neoformans*: Taxonomic and epidemiological aspects. Int. J. Syst. Bacteriol. **47** (1997) 432–442.

[19] BOEKHOUT, T.; THEELEN, B.; DIAZ, M.; FELL, J.W.; HOP, W.C.; ABELN, E.C.; DROMER, F.; MEYER, W.: Hybrid genotypes in the pathogenic yeast *Cryptococcus neoformans*. Microbiology **147** (2001) 891–907.

[20] BRITTEN, R.J.; GRAHAM, D.E.; NEUFELD, B.R.: Analysis of repeating DNA sequences by reassociation. In: Methods in Enzymology Vol. 29, Nucleic Acids and protein synthesis (edited by Grossman, L.; Moldave, K.). New York, U.S.A.: Academic Press (1974) 363–418.

[21] BYRON, J.K.; CLEMONS, K.V.; MCCUSKER, J.H.; DAVIS, R.W.; STEVENS, D.A.: Pathogenicity of *Saccharomyces cerevisiae* in complement factor five-deficient mice. Infect. Immun. **63** (1995) 478–485.

[22] CHOUDHURY, S.R.; TRAQUIAR, J.A.; JARVIS, W.R.: 4-Methyl-7,11-heptadecadenal and 4-methyl-7,11-heptadecadienoic acid: new antibiotics from *Sporothrix flocculosa* and *Sporothrix rugulosa*. J. Nat. Prod. 57 (1994) 700–704.

[23] CLAISSE, M.L.; PÉRÉ-AUBERT, G.A.; CLAVILLIER, L.P.; SLONIMSKY, P.: Méthodes d'estimation de la concentration des cytochromes dans les cellules entières de levures. Eur. J. Biochem. **16** (1970) 430–438.

[24] CLEMONS, K.V.; MCCUSKER, J.H.; DAVIS, R.W.; & STEVENS, D.A.: Comparative pathogenesis of clinical and nonclinical isolates of *Saccharomyces cerevisiae*. J. Infect. Dis. **169** (1994) 859–867.

[25] COTTRELL, M.; KOCK, J.L.F.; LATEGAN, P.M.; BRITZ, T.J.: Long-chain fatty acid composition as an aid in the classification of the genus *Saccharomyces*. Syst. Appl. Microbiol. **8** (1986) 166–168.

[26] DE BARROS LOPES, M.; RAINIERI, S.; HENSCHKE, P. A.; LANGRIDGE, P.: AFLP fingerprinting for analysis of yeast genetic variation. Int. J. Syst. Bacteriol. **49** (1999) 915–924.

[27] DE MOT, R.; VERACHTERT, H.: Purification and characterization of extracellular amylolytic enzymes from the yeast *Filobasidium capsuligenum*. Appl. Envir. Microbiol. **50** (1985) 1474–1482.

[28] DE MOT, R.; VERACHTERT, H.: Secretion of α-amylase and multiple forms of glucoamylase by the yeast *Trichosporon pullulans*. Can. J. Microbiol. **32** (1986) 47–51.

[29] DE MOT, R.; DEMEERSMAN, M.; VERACHTERT, H.: Comparative study of starch degradation and amylase production by non-ascomycetous yeast species. Syst. Appl. Microbiol. **5** (1984) 421–432.

[30] DIAZ, M.R.; BOEKHOUT, T.; THEELEN, B.; FELL, J.W.: Molecular sequence analyses of the intergenic spacer (IGS) associated with rDNA of the two varieties of the pathogenic yeast, *Cryptococcus neoformans*. Syst. Appl. Microbiol. **23** (2000) 535–545.

[31] DIDDENS, H.A.; LODDER, J.: Die anaskosporogenen Hefen, zweite Hälfte. Amsterdam, The Netherlands: Noord-Hollandse Uitgevers Maatschappij (1942).

[32] DOBZHANSKY, T.: Organismic molecular aspects of species formation. In: Molecular evolution (edited by Ayala, F.J.). Sunderland, Massachusetts, U.S.A.: Sinauer Associates (1976) 95–105.

[33] EZAKI, T.; HASHIMOTO, Y.; YABUUCHI, E.: Fluorometric deoxyribonucleic acid-deoxyribonucleic acid hybridization in microdilution wells as an alternative to membrane filter hybridization in which radioisotopes are used to determine genetic relatedness among bacterial strains. Int. J. Syst. Bacteriol. **39** (1989) 224–229.

[34] FELL, J.W.: Teliospore-forming yeasts. In: The yeasts, a taxonomic study, 3rd ed. (edited by Kreger-van Rij, N.J.W.). Amsterdam, The Netherlands: Elsevier (1984) 491–495.

[35] FELL, J.W.; STATZELL TALLMAN, A.: *Leucosporidium* Fell, Statzell, Hunter & Phaff. In: The yeasts, a taxonomic study, 3rd ed. (edited by Kreger-van Rij, N.J.W.). Amsterdam, The Netherlands: Elsevier (1984) 496–508.

[36] FELL, J.W.; STATZELL TALLMAN, A.: *Rhodosporidium* Banno. In: The yeasts, a taxonomic study, 3rd ed. (edited by Kreger-van Rij, N.J.W.). Amsterdam, The Netherlands: Elsevier (1984) 509–531.

[37] FELL, J.W.; ROEIJMANS, H.; BOEKHOUT, T.: Cystofilobasidiales, a new order of basdiomycetous yeasts. Int. J. Syst. Bacteriol. **49** (1999) 907–913.

[38] FELL, J.W.; STATZELL, A.C.; HUNTER, I.L.; PHAFF, H.J.: *Leucosporidium* gen. n., the heterobasidiomycetous stage of several yeasts of the genus *Candida*. Antonie van Leeuwenhoek 35 (1969) 433–462.

[39] FELL, J.W.; BOEKHOUT ,T.; FONSECA, A.; SCORZETTI, G.; STATZELL-TALLMAN, A.: Biodiversity and systematics of basidiomycetous yeasts as determined by large-subunit rDNA D1/D2 domain sequence analysis. Int. J. Syst. Evol. Microbiol. **50** (2000) 1351–1371.

[40] FELSENSTEIN, J.: Phylogenies from molecular sequences: inference and reliability. Ann. Rev. Genet. **22** (1988) 521–565.

[41] FIOL, J.B.; CLAISSE, M.: Les cytochromes des *Kluyveromyces*: Détermination et implications systématiques. Mycopathol. **78** (1987) 177–184.

[42] Flegel, T.W.: Let's call a yeast a yeast. Can. J. Microbiol. **23** (1977) 945–946.

[43] FONSECA, A.; SAMPAIO, J.P.: *Rhodosporidium lusitaniae* sp. nov., a novel homothallic basidiomycetous yeast species from Portugal that degrades phenolic compounds. Syst. Appl. Microbiol. **15** (1992) 47–51.

[44] FURUKAWA, T.; OGINO, T.; MATSUYOSHI, T.: Fermentative production of citric acid from n-paraffins by *Saccharomycopsis lipolytica*. J. Ferment. Technol. **60** (1982) 281–286.

[45] GAILLARDIN, C.; DUCHATEAU-NGUYEN, G.; TEKAIA, F.; LLORENTE, B.; CASAREGOLA, S.; TOFFANO-NIOCHE, C.; AIGLE, M.; ARTIGUENAVE, F.; BLANDIN, G.; BOLOTIN-FUKUHARA, M.; BON, E.; BROTTIER, P.; DE MONTIGNY, J.; DUJON, B.; DURRENS, P.; LEPINGLE, A.; MALPERTUY, A.; NEUVEGLISE, C.; OZIER-KALOGEROPOULOS, O.; POTIER, S.; SAURIN, W.; TERMIER, M.; WESOLOWSKI-LOUVEL, M.; WINCKER, P.; SOUCIET, J.; WEISSENBACH, J.: Genomic exploration of the Hemiascomycetous yeasts: 21. Comparatrive functional classification of genes. FEBS Lett. **487** (2000) 134–149.

[46] GILL, C.O.; HALL, M.J.; RATLEDGE, C.: Lipid accumulation in an oleaginous yeast (*Candida* 107) growing on glucose in single-stage continuous culture. Appl. Envir. Microbiol. **33** (1977) 231–239.

[47] GIMÉNEZ-JURADO, G.; VAN UDEN, N.: *Leucosporidium fellii* spec. nov., a basidiomycetous yeast that degrades L(+)-tartaric acid. Antonie van Leeuwenhoek **55** (1989) 133–141.

[48] GOFFEAU, A.; BARRELL, B.G.; BUSSEY, H.; DAVIS, R.W.; DUJON, B.; FELDMANN, H.; GALIBERT, F.; HOHEISEL, J.D.; JACQ, C.; JOHNSTON, M.; LOUIS, E.J.; MEWES, H.W.; MURAKAMI, Y.; PHILIPPSEN, P.; TETTELIN, H.; OLIVER, S.G.: Life with 6000 genes. Science **275** (1997) 1051–1052.

[49] GOLUBEV, W.I.: Catabolism of m-inositol and taxonomic value of D-glucuronate assimilation in yeasts. Mikrobiologiya **58** (1989) 276–283.

[50] GOLUBEV, W.I.: Perfect state of *Rhodomyces dendrorhous* (*Phaffia rhodozyma*) Yeast **11** (1995) 101–110.

[51] GROTH, C.; HANSEN, J.; PISKUR, J.: A natural chimeric yeast containing genetic material from three species. Int. J. Syst. Bacteriol. **49** (1999) 1933–1938.

[52] GUÉHO, E.; SMITH, M.TH.; DE HOOG, G.S.; BILLON-GRAND, G.; CHRISTEN, R.; BATENBURG-VAN DER VEGTE, W.H.: Contributions to a revision of the genus *Trichosporon*. Antonie van Leeuwenhoek **61** (1992) 289–316.

[53] GUILLIERMOND, A.: Les levures. Paris, France: Doin et Fils (1912).

[54] HAJLAOUI, M.R.; BÉLANGER, R.R.: Antagonism of the yeast-like phylloplane fungus *Sporothrix flocculosa* against *Erysiphe graminis* var. *tritici*. Biocontr. Sci. Technol. **3** (1993) 427–434.

[55] HARA, T.; SHIMODA, T.; NONAKA, K.; OGATA, S.: Colorimetric detection of DNA-DNA hybridization in microdilution wells for taxonomic application on bacterial strains. J. Ferment. Bioengin. **72** (1991) 122–124.

[56] HEITMAN, J.; CASADEVALL, A.; LODGE, J.K.; PERFECT, J.R.: The *Cryptococcus neoformans* genome project. Mycopathol. **148** (1999) 1–7.

[57] HENDRIKS, L.; GORIS, A.; VAN DE PEER, Y.; NEEFS, J.-M.; VANCANNEYT, M.; KERSTERS, K.; HENNEBERT, G.; DE WACHTER, R.: Phylogenetic analysis of five medically important *Candida* species as deduced on the basis of small ribosomal subunit RNA sequences. J. Gen. Microbiol. **137** (1992) 1223–1230.

[58] HENDRIKS, L.; GORIS, A.; VAN DE PEER, Y.; NEEFS, J.-M.; VANCANNEYT, M.; KERSTERS, K.; BERNY, J.-F.; HENNEBERT, G.: Phylogenetic relationships among Ascomycetes and Ascomycete-like yeasts as deduced from small ribosomal subunit RNA sequences. Syst. Appl. Microbiol. **15** (1992) 98–104.

[59] HESLOT, H.: Genetics and genetic engineering of the industrial yeast *Yarrowia lipolytica*. In: Advances in Biochemical Engineering/Biotechnology, Vol. 43 (edited by Fiechter, A.). Berlin, Germany: Springer (1990) 43–73.

[60] JAHNKE, K.-D.: Assessing natural relationships by DNA analysis: techniques and applications. Stud. Mycol. **30** (1987) 227–246.

[61] JARVIS, W.R.; SHAW, L.A.; TRAQUIAR, J.A.: Factors affecting antagonism of cucumber powdery mildew by *Stephanosascus flocculosus* and *S. rugulosus*. Mycol. Res. **92** (1989) 162–165.

[62] JENQ, W.J.; SPECKMAN, R.A.; CRANG, R.E.; STEINBERG, M.P.: Enhanced conversion of lactose to glycerol by *Kluyveromyces fragilis* utilizing whey permeate as a substrate. Appl. Envir. Microbiol. **55** (1989) 573–578.

[63] JESPERSEN, L.; VAN DER KUHLE, A.; PETERSEN, K.M.: Phenotypic and genetic diversity of *Saccharomyces* contaminants isolated from lager breweries and their phylogenetic relationship with brewing yeasts. Int. J. Food. Microbiol. **60** (2000) 43–53.

[64] JOHNSON, E.A.; CONKLIN, D.E.; LEWIS, M.J.: The yeast *Phaffia rhodozyma* as a dietary source for salmonids and crustaceans. J. Fish. Res. Board Can. **34** (1977) 2417–2421.

[65] JOHNSON, E.A.; LEWIS, M.J.; GRAU, C.R.: Pigmentation of egg yolks with astaxanthin from the yeast *Phaffia rhodozyma*. Poultry Sci. **59** (1980) 1777–1782.

[66] KANEKO, Y.; BANNO, I.: Reexamination of *Saccharomyces bayanus* strains by DNA-DNA hybridization and electrophoretic karyotyping. IFO Res. Comm. **15** (1991) 30–41.

[67] KENDRICK, B.: Yeasts and yeast-like fungi – new concepts and new techniques. Stud. Mycol. **30** (1987) 479–486.

[68] KITAMOTO, D.; AKIBA, S.; HIOKI, C.; TABUCHI, T.: Extracellular accumulation of mannosylerythritol lipids by a strain of *Candida antarctica*. Agric. Biol. Chem. **54** (1990) 31–36.

[69] KREGER-VAN RIJ, N.J.W.: The yeasts, a taxonomic study, 3rd ed. Amsterdam, The Netherlands: Elsevier.

[70] KREGER-VAN RIJ, N.J.W.; VEENHUIS, M.: A comparative study of the cell wall structure of basidiomycetous and related yeasts. J. Gen. Microbiol. **68** (1971) 87–95.

[71] KURTZMAN, C.P.: Synonymy of the yeast genera *Hansenula* and *Pichia* demonstrated through comparisons of deoxyribonucleic acid relatedness. Antonie van Leeuwenhoek **50** (1984) 209–217.

[72] KURTZMAN, C.P.: Molecular taxonomy of fungi. In: Gene manipulation in fungi (edited by Bennett, J.W.; Lasure, L.L.). New York, U.S.A.: Academic Press (1985) 35–56.

[73] KURTZMAN, C.P.: Prediction of biological relatedness among yeasts from comparison of nuclear DNA complementarity. Stud. Mycol. **30** (1987) 459–468.

[74] KURTZMAN, C.P.: DNA-DNA hybridization approaches to species identification in small genome organisms. Meth. Enzym. **224** (1993) 335–348.

[75] KURTZMAN, C.P.; FELL, J.W.: The yeast, a taxonomic study, 4th ed. Amsterdam: Elsevier (1998).

[76] KURTZMAN, C.P.; PHAFF, H.J.: Molecular taxonomy. In: The yeasts, 2nd ed., Vol. 1 (edited by Rose, A.H.; Harrison, J.S.). London, U.K.: Academic Press (1987) 63–94.

[77] KURTZMAN, C.P.; ROBNETT, C.J.: Phylogenetic relationships among species of *Saccharomyces, Schizosaccharomyces, Debaryomyces* and Schwanniomyces determined from partial ribosomal RNA sequences. Yeast **7** (1991) 61–72.

[78] KURTZMAN, C.P.; ROBNETT, C.J.: Molecular relationships among hyphal ascomycetous yeasts and yeastlike taxa. Can. J. Bot. **73** (1995) S824–S830.

[79] KURTZMAN, C.P.; ROBNETT, C.J.: Synonymy of the yeast genera *Wingea* and *Debaryomyces*. Antonie van Leeuwenhoek **66** (1994) 337–342.

[80] KURTZMAN, C.P.; ROBNETT, C.J.: Identification of clinically important ascomycetous yeasts based on nucleotide divergence in the 5' end of the large-subunit (26S) ribosomal DNA gene. J. Clin. Microbiol. **35** (1997) 1216–23.

[81] KURTZMAN, C.P.; ROBNETT, C.J.: Identification and phylogeny of ascomycetous yeasts from analysis of nuclear large subunit (26S) ribosomal DNA partial sequences. Antonie van Leeuwenhoek. **73** (1998) 331–71.

[82] KURTZMAN, C.P.; SMILEY, M.J.; JOHNSON, C.J.; WICKERHAM, L.J.; FUSON, G.B.: Two new and closely related heterothallic species. *Pichia amylophila* and *Pichia mississippiensis*: characteri-

zation by hybridization and deoxyribonucleic acid reassociation. Int. J. Syst. Bacteriol. **30** (1980) 208–216.

[83] KWON-CHUNG, K.J.: Nuclear genotypes of spore chains in *Filobasidiella neoformans* (*Cryptococcus neoformans*). Mycologia **72** (1980) 418–422.

[84] KWON-CHUNG, K.J.; FELL, J.W.: The genus *Filobasidiella* Kwon-Chung. In: The yeasts, a taxonomic study, 3rd ed. (edited by Kreger-van Rij, N.J.W.). Amsterdam, The Netherlands: Elsevier (1984) 472–482.

[85] LAASER, G.; MÖLLER, E.; JAHNKE, K.-D.; PRILLINGER, H.; PRELL, H.: Ribosomal restriction analysis as a taxonomic tool in separating physiologically similar basidiomycetous yeasts. Syst. Appl. Microbiol. **11** (1987) 170–175.

[86] LACHANCE, M.-A.: Current views on the yeast species. Microbiol. Sci. **2** (1985) 122–126.

[87] LODDER, J.: Die anaskosporogenen Hefen, erste Hälfte. Amsterdam, The Netherlands: Noord-Hollandsche Uitgevers Maatschappij (1934).

[88] LODDER, J.: The yeasts, a taxonomic study, 2nd ed. Amsterdam, The Netherlands: North-Holland Publ. Comp. (1970).

[89] MCARTHUR, C.R.; CLARK-WALKER, G.D.: Mitochondrial DNA size diversity in the *Dekkera/Brettanomyces* yeasts. Curr. Genet. **7** (1983) 29–35.

[90] MCCUSKER, J.H.; CLEMONS, K.V.; STEVENS, D.A.; DAVIS, R.W.: Genetic characterization of pathogenic *Saccharomyces cerevisiae* isolates. Genetics **136** (1994) 1261–1269.

[91] MESSNER, R.; PRILLINGER, H.; ALTMANN, F.; LOPANDIC, K.; WIMMER, K.; MOLNAR, O.; WEIGANG, F.: Molecular characterization and application of random amplified polymorphic DNA analysis of *Mrakia* and *Sterigmatomyces* species. Int. J. Syst. Bacteriol. **44** (1994) 694–703.

[92] MIDDELHOVEN, W.J.; KOOREVAAR, M.; SCHUUR, G.W.: Degradation of benzene compounds by yeasts in acidic soils. Plant and Soil **145** (1992) 37–43.

[93] MILLER, M.W.; YONEYAMA, M.; SONEDA, M.: *Phaffia*, a new yeast genus in the Deuteromycotina (Blastomycetes). Int. J. Syst. Bacteriol. **26** (1976) 286–291.

[94] MILLS, S.C.; CHILD, J.J.; SPENCER, J.F.T.: The utilization of aromatic compounds by yeasts. Antonie van Leeuwenhoek **37** (1971) 281–287.

[95] MOLINA, F.I.; JONG, S.-C.; HUFFMAN, J.L.: PCR amplification of the 3' external transcribed and intergenic spacers of the ribosomal DNA repeat unit in three species of *Saccharomyces*. FEMS Microbiol. Lett. **108** (1993) 259–264.

[96] MOLINA, F.I.; SHEN, P.; JONG, S.-C.: Validation of the species concept in the genus *Dekkera* by restriction analysis of genes coding for rRNA. Int. J. Syst. Bacteriol. **43** (1993) 32–35.

[97] MOLNAR, O.; PRILLINGER, H.; LOPANDIC, K.; WEIGANG, F.; STAUDACHER, E.: Analysis of coenzyme Q systems, monosaccharide patterns of purified cell walls, and RAPD-PCR patterns in the genus *Kluyveromyces*. Antonie van Leeuwenhoek **70** (1996) 67–78.

[98] MONTROCHER, R.; CLAISSE, M.: Spectrophotometric analysis of yeasts: cytochrome spectra of some *Candida* and related taxa. J. Gen. Appl. Microbiol. **34** (1988) 221–232.

[99] NAUMOVA, E.S.; NAUMOV, G.I.; MOLINA, E.I.: Genetic variation among European strains of *Saccharomyces paradoxus*: results from DNA fingerprinting. Syst. Appl. Microbiol. **23** (2000) 86–92.

[100] NAUMOV, G.I ; JAMES, S.A.; NAUMOVA, E.S.; LOUIS, E.J.; ROBERTS, I.N.: Three new species in the *Saccharomyces* sensu stricto complex: *Saccharomyces cariocanus, Saccharomyces kudriavzevii* and *Saccharomyces mikatae*. Int. J. Syst. Evol. Microbiol. **50** (2000) 1931–1942.

[101] NILSSON, R.; ENEBO, L.; LUNDIN, H.; MYRBÄCK, K.: Mikrobielle Fettsynthese unter Verwendung von *Rhodotorula glutinis* nach dem Lufthefeverfahren. Svensk Kemisk Tidskr. **55** (1943) 41–51.

[102] OBERWINKLER, F.: Heterobasidiomycetes with ontogenetic yeast-stages – systematic and phylogenetic aspects. Stud. Mycol. **30** (1987) 61–74.

[103] ODA, Y.; FUJISAWA T.: Intraspecific divergence of *Saccharomyces kluyveri* as revealed by the nucleotide sequences of 18S-28S rRNA spacer regions and alpha-galactosidase MEL genes. Biosci. Biotechnol. Biochem. **65** (2001) 164–166.

[104] PARK, W.S.; MURPHY, P.A.; GLATZ, B.A.: Lipid metabolism and cell composition of the oleaginous yeast *Apiotrichum curvatum* grown at different carbon to nitrogen ratios. Can. J. Microbiol. **36** (1990) 318–326.

[105] PETERSON, S.W.; KURTZMAN, C.P.: Ribosomal RNA sequence divergence among sibling species of yeasts. Syst. Appl. Microbiol. **14** (1991) 124–129.

[106] PHAFF, H.J.: The species concept in yeasts: physiologic, morphologic, genetic, and ecological parameters. In: Current developments in yeast research (edited by Stewart, G.G.; Russell, I.). Oxford, U.K.: Pergamon Press (1981) 635–643.

[107] PHAFF, H.J.: Trends in yeast research. Yeast (spec. issue) **5** (1989) 341–349.

[108] PHAFF, H.J.; MILLER, M.W.; MRAK, E.M.: The life of yeasts, 2nd ed. Cambridge, Massachusetts: U.S.A.: Harvard Univ. Press (1978).

[109] PHAFF, H.J.; STARMER, W.T.; TREDICK-KLINE, J.: *Pichia kluyveri* sensu lato – a proposal for two new varieties and a new anamorph. Stud. Mycol. **30** (1987) 403–414.

[110] PISKUR, J.: 2001. Origin of the duplicated regions in the yeast genomes. Trends Genet. **17** (2001) 302–303.

[111] PRILLINGER, H.; MOLNAR, O.; ELISKASES-LECHNER, F.; LOPANDIC, K.: Phenotypic and genotypic identification of yeasts from cheese. Antonie van Leeuwenhoek **75** (1999) 267–283.

[112] PRILLINGER, H.; OBERWINKLER, F.; UMILE, C.; TLACHAC, K.; BAUER, R.; DÖRFLER, C.; TAUFRATZHOFER, E.: Analysis of cell wall carbohydrates (neutral sugars) from ascomycetous and asidiomycetous yeasts with and without derivatization. J. Gen. Appl. Microbiol. **39** (1993) 1–34.

[113] RAGAN, M.A.: Ribosomal RNA and the major lines of evolution: a perspective. Biosystems **21** (1988) 177–188.

[114] RATLEDGE, C.: Microbial oils and fats: an assessment of their commercial potential. Progr. Industr. Microbiol. **16** (1982) 119–206.

[115] RATLEDGE, C.: Lipids and fatty acids. In: Economic Microbiology, Vol. 2, Primary products of metabolism (edited by Rose, A.H..). London, U.K.: Academic Press (1978) 263–302.

[116] RATLEDGE, C.: Lipids. In: Biotechnology, Vol. 4, 1st ed., Microbial products II (edited by Pape, H.; Rehm, H.-J.). Weinheim, Germany: VCH (1986) 185–213.

[117] RATLEDGE, C.; EVANS, C.T.: Lipids and their metabolism. In: The yeasts, Vol. 3, 2nd ed. (edited by Rose, A.H.; Harrison, J.S.). London, U.K.: Academic Press (1989) 367–455.

[118] RATLEDGE, C.; HALL, M.J.: Accumulation of lipid by *Rhodotorula glutinis* in continuous culture. Biotechn. Lett. **1** (1979) 115–120.

[119] REISER, J.; GLUMOFF, V.; KÄLIN, M.; OCHSNER, U.: Transfer and expression of heterologous genes in yeasts other than *Saccharomyces cerevisiae*. In: Advances in Biochemical Engineering/ Biotechnology, Vol. 43 (edited by Fiechter, A.). Berlin, Germany: Springer Verlag (1990) 75–102.

[120] ROLPH, C.E.; MORETON, R.S.; HARWOOD, J.L.: Acyl lipid metabolism in the oleaginous yeast *Rhodotorula gracilis* (CBS 3043). Lipids **24** (1989) 715–720.

[121] SCHEDA, R.: Merkmalsveränderungen bei Hefen der Gattung *Saccharomyces*. Monatsschr. Brauerei **19** (1966) 256–258.

[122] SCHEDA, R.; YARROW, D.: The instability of physiological properties used as criteria in the taxonomy of yeasts. Arch. Mikrobiol. **55** (1966) 209–225.

[123] SEOIGHE, C.; WOLFE, K.H.: Extent of genomic rearrangement after genome duplication in yeast. Proc. Natl Acad. Sci. USA **95** (1998) 4447–4452.

[124] SHEN, R.; LACHANCE, M.-A.: Phylogenetic study of ribosomal DNA of cactophilic *Pichia* species by restriction mapping. Yeast **9** (1993) 315–330.

[125] SPENCER, J.F.T.; GORIN, P.A.J.: Systematics of the genus *Candida* Berkhout: proton magnetic resonance spectra of the mannans and mannose-containing polysaccharides as an aid in classification. Antonie van Leeuwenhoek **35** (1969) 33–44.

[126] SPENCER, J.F.T.; GORIN, P.A.J.: Systematics of the genus *Torulopsis*: proton magnetic resonance spectra of the mannose-containing polysaccharides as an aid in classification. Antonie van Leeuwenhoek **36** (1970) 509–524.

[127] STELLING-DEKKER, N.M.: Die sporogene Hefen. Verhand. Kon. Akad. Wetensch., Sect. 2, (1930) 1–547.

[128] STREIBLOVÁ, E.: Cell division in yeasts. In: Recent advances in Microbiology. Proc. 10th Int. Congr Microbiol (edited by Pérez-Miravette A.; Peláez, D). Mexico City, D.C., Mexico (1971) 131–140.

[129] SUGITA, T.; NISHIKAWA, A.; IKEDA, R.; SHINODA, T.: Identification of medically relevant *Trichosporon* species based on sequences of internal transcribed spacer regions and construction of a database for *Trichosporon* identification. J. Clin. Microbiol. **37** (1999) 1985–1993.

[130] SUGIYAMA, J.; SUH, S.-O.: Phylogenetic analysis of basidiomycetous yeasts inferred from small subunit ribosomal DNA sequence. J. Gen. Microbiol. **139** (1993) 1595–1598.

[131] SUZUKI, M.; NAKASE, T.: The distribution of xylose in the cells of ballistosporogenous yeasts – application of high performance liquid chromatography without derivatization to the analysis of xylose in whole cell hydrolyzates. J. Gen. Appl. Microbiol. **34** (1988) 95–103.

[132] SWANN, E.; TAYLOR, J.W.: Higher taxa of basidiomycetes: an 18S rRNA gene perspective. Mycologia **85** (1993) 923–936.

[133] SWANN, E.; TAYLOR, J.W.: Toward a phylogenetic systematics of the basidiomycota: integrating yeasts and filamentous Basidiomycetes using 18S rRNA gene sequences. Stud. Mycol. **38** (1995) 147–161.

[134] SWANN, E.; TAYLOR, J.W.: Phylogenetic perspectives on basidiomycete systematics: evidence from the 18S rRNA gene. Can. J. Bot. **73** (suppl. 1) (1995) S862–S868.

[135] TEMPLETON, A.R.: Systematics of basidiomycetes based on 5S rRNA sequences and other data. Nature **303** (1983) 731.

[136] THEELEN, B.; SILVESTRI, M.; GUÉHO, E.; VAN BELKUM, A.; BOEKHOUT, T.: Identification and typing of *Malassezia* yeasts using amplified fragment length polymorphism (AFLP^{Tm}), random amplified polymophic DNA (RAPD) and denaturing gradient gel electrophoresis. FEMS Yeast Res. **1** (2001) 79–86.

[137] TORNAI-LEHOCZKI, J.; DLAUCHY, D.: Delimination of brewing yeast strains using different molecular techniques. Int. J. Fd Microbiol. **62** (2000) 37–45.

[138] VALADON, L.R.G.: Carotenoids as additional taxonomic characters in fungi: a review. Trans. Br. Mycol. Soc. **67** (1976) 1–15.

[139] VAN DE PEER, Y.; HENDRIKS, L.; GORIS, A.; NEEFS, J.-M.; VANCANNEYT, M.; KERSTERS, K.; BERNY, J.-F.; HENNEBERT, G.; DE WACHTER, R.: Evolution of basidiomycetous yeasts as deduced from small ribosomal subunit RNA sequences. Syst. Appl. Microbiol. **15** (1992) 250–258.

[140] VAN DER WALT, J.P.: The yeasts – a conspectus. Stud. Mycol. **30** (1987) 19–31.

[141] VANDAMME, E.J.; DERYCKE, D.G.: Microbial inulases: fermentation process, properties, and applications. Adv. Appl. Microbiol. **29** (1983) 139–175.

[142] VAUGHAN MARTINI, A.; KURTZMAN, C.P.: Deoxyribonucleic acid relatedness among species of the genus *Saccharomyces* sensu stricto. Int. J. Syst. Bacteriol. **35** (1985) 508–511.

[143] VAUGHAN-MARTINI, A.; ANGELINI, P.;& CARDINALI, G.: Use of conventional taxonomy, electrophoretic karyotyping and DNA-DNA hybridization for the classification of fermentative apiculate yeasts. Int. J. Syst. Evol. Microbiol. **50** (2000) 1665–1672.

[144] VILGALYS, R.; HESTER, M.: Rapid genetic identification and mapping of enzymatically amplified ribosomal DNA from several *Cryptococcus* species. J. Bacteriol. **172** (1990) 4238–4246.

[145] VILJOEN, B.C.; KOCK, J.L.F.; LATEGAN, P.M.: Long-chain fatty acid composition of selected genera of yeasts belonging to the Endomycetales. Antonie van Leeuwenhoek **52** (1986) 45–51.

[146] WEIJMAN, A.C.M.; GOLUBEV, W.I.: Carbohydrate patterns and taxonomy of yeasts and yeast-like fungi. Stud. Mycol. **30** (1987) 361–371.

[147] WEIJMAN, A.C.M.; RODRIGUES DE MIRANDA, L.: Xylose distribution within and taxonomy of the genera *Bullera* and *Sporobolomyces*. Antonie van Leeuwenhoek **49** (1983) 559–562.

[148] WESTHUIZEN, J.P.J. VAN DER; KOCK, J.L.F.; SMITH, E.J.: The potential use of cellular long-chain fatty acid composition in the taxonomy of the carotenoid pigment producing genera *Rhodosporidium* Banno and *Rhodotorula* Harrison. Syst. Appl. Microbiol. **14** (1991) 282–290.

[149] WHITE, T.J.; BRUNS, T.; LEE, S.; TAYLOR, J.: Amplification and direct sequencing of fungal ribosomal RNA genes for phylogenetics. In: PCR protocols: a guide to methods and applications (edited by Innis, M.A.; Gelfand, D.H.; Sninsky, J.J.; White, T.J.). New York, U.S.A.: Academic Press (1990) 315–322.

[150] WICKERHAM, L.J.: Taxonomy of yeasts. Techn. Bull. 1029. Washington D.C.: USDA (1951).

[151] WILEY, E.O.: Phylogenetics. The theory and practice of phylogenetics systematics. New York, U.S.A.: Wiley & Sons (1981).

[152] WINGE, Ö.; ROBERTS, C.: Inheritance of enzymatic characters in yeasts, and the phenomenon of long-term adaptation. C. R. Trav. Carlsberg, Sér. Physiol. **24** (1949) 263–315.

[153] WOESE, C.R.: Bacterial evolution. Microbiol. Rev. **51** (1987) 221–271.

[154] WOLFE, K.H.; SHIELDS, D.C.: Molecular evidence for an ancient duplication of the entire yeast genome. Nature **387** (1997) 708–713.

[155] WONG, G.J.: A comparison of the mating system of *Tremella mesenterica* and other modified bifactorial species. Stud. Mycol. **30** (1987) 431–441.

[156] YAMADA, Y.; KONDO, K.: Coenzyme Q system in the classification of the yeast genera *Rhodotorula* and *Cryptococcus*, and the yeast-like genera *Sporobolomyces* and *Rhodosporidium*. J. Gen. Appl. Microbiol. **19** (1973) 59–77.

[157] YAMADA, Y.; BANNO, I.; VON ARX, J.A.; VAN DER WALT, J.P.: Taxonomic significance of the coenzyme Q system in yeasts and yeast-like fungi. Stud. Mycol. **30** (1987) 299–308.

[158] YAMADA, Y.; NAGAHAMA, T.; BANNO, I.: The molecular phylogeny of the Q_9-equipped ascomycetous teleomorphic yeast genus *Debaryomyces* Lodder et Kreger-van Rij based on the partial sequences of 18S and 26S ribosomal ribonucleic acids. J. Gen. Appl. Microbiol. **37** (1991) 277–288.

[159] YAMADA, Y.; OHISHI, T.; KONDO, I.: Coenzyme Q system in the classification of the ascosporogenous yeast genera *Hansenula* and *Pichia*. J. Syst. Appl. Microbiol. **19** (1973) 189–208.

[160] YAMAZAKI, M.; GOTO, S.; KOMAGATA, K.: An electrophoretic comparison of the enzymes of *Saccharomyces* yeasts. J. Gen. Appl. Microbiol. **29** (1983) 305–318.

[161] YAMAZAKI, M.; GOTO, S.; KOMAGATA, K.: Taxonomic studies of the genus *Tilletiopsis* on physiological properties and electrophoretic comparison of enzymes. Trans. Mycol. Soc. Jpn **26** (1985) 13–22.

2 Detection, enumeration and isolation of yeasts

TIBOR DEAK

2.1 Introduction

Mycological investigation of foods often aims only at enumerating colony forming units (CFUs) of "yeast and molds" together. While this information may relate to the general contamination of products, it may be meaningless or even misleading for assessing the mycological safety, quality and stability of foods, considering the fundamental differences between the two groups of fungi. Yeasts are mostly unicellular fungi and multiply vegetatively by budding, whereas molds are filamentous fungi growing at the tip of hyphae and often produce vegetative propagules in a large number. This difference has a great impact on the growth rate in cell mass of yeasts and molds. Moreover, approximately two-third of the yeast species occurring frequently in foods are fermentative, whereas molds in food are, with few exceptions, strictly aerobic organisms. None of the yeast species pathogenic for man is known to be transmitted by foods. On the other hand, a large number of mold species produce various mycotoxins in foods. These and other differences justify the separate assessment of yeasts and molds in food. Both groups of fungi can be grown and separated from most bacteria on media of low pH. One of the first media developed has been the wort agar used early in the brewing industry. Many other acidified media, e. g., malt extract agar and potato dextrose agar, have traditionally been used for the detection, isolation and enumeration of yeasts and molds in foods. In time, several other media have been formulated for isolation and enumeration, but also for the specific and differential detection of yeasts and molds. The range and use of these media in mycological examination is far less developed if compared with bacteriology. However, in the literature, several mycological media and cultivation methods have been described. Earlier references have been well reviewed [15, 17, 45, 46, 69, 70, 91, 92] and these will not be recapitulated here.

Since 1984, a series of workshops have been organized on the standardization of methods for the mycological examination of foods. The proceedings of these workshops have been published [101, 126, 127, 135]. These and other recent reviews and chapters provide comprehensive information to the interested reader [18, 20, 53, 128].

The procedure for the detection and enumeration of yeasts from food usually involves a number of steps (Table 2.1-1) to be followed, after their isolation and purification, the identification and typing of yeasts by testing the morphological, physiological, biochemical and molecular characteristics of the culture (see also Chapters 3 and 4). In the following an overview will be given on the basic methodology updated with the current developments.

Tab. 2.1-1 General flow-chart of the procedures for the detection, enumeration and identification of yeasts

1. Preparation of media and equipments
2. Sample preparation and homogenization
3. Serial dilution
4. Inoculation of media
5. Incubation
6. Enumeration
7. Isolation
8. Purification
9. Microscopic investigation
10. Identification tests

2.2 Sample preparation

Yeasts are usually entrapped and/or embedded in gummy, waxy and mucous materials in their natural habitats, whereby they firmly adhere to the substrate and may not be recovered when a light procedure such as short and mild shaking is employed for their isolation. The same is true for surfaces of food processing equipments and machinery on which yeasts form biofilms together with other microorganisms. Several investigators, in particular MARTINI and co-workers [63, 111, 112, 133] have demonstrated that more vigorous and disruptive pre-isolation treatments allow the recovery of a greater variety of yeast species, and increase the number of counted cells by several orders of magnitude. Investigations on various fresh and frozen fruits and vegetables, as well as chicken and fish sample using treatments of increased forces, such as vortexing, jet-streaming and sonication, result in more effective removal of cells than mild shaking. Interestingly, however, vigorous shaking using a Vortex mixer appears to be the most effective method compared with water jet or sonication exerting higher forces.

Since the invention of stomacher, its use mostly replaces blending and vortexing for sample preparation in food investigation. Comparative studies on the method of homogenization have been made for molds [58, 80, 98] and these show no significant differences between stomaching and blending. Neither has to be applied for more than 2 min. No research data exist for yeasts, though it is probable that the above holds true for them, as well.

After homogenization, further dilution or plating must follow within a few minutes in order to avoid settling of cells. KING [99] and BEUCHAT [16] demonstrated that a settling time longer than 1 minute may decrease significantly the population size of yeasts and molds detected in suspension.

2.3 Dilution

The primary homogenate of a food sample is generally prepared in a 1:10 ratio, and further dilutions are also made in a decimal scale. The 1:5 dilution scheme results in significantly higher numbers of fungi with less variation, as detected in ground pepper [23], but no further consideration is given to its routine use for the mycological analysis of foods.

The composition of the diluent has received far more attention. Peptone water (0.1 %), saline (0.85 %), phosphate buffer (0.1 M, pH 7.0) or the combination of these are most commonly used. A wetting agent (e. g., Tween 80, 0.05 %) may be added to enhance separation of cell clumps and filamentous structures. The use of distilled water or deionized water as solvent is recommended, but the sole use of either as the diluent medium is discouraged. MIAN et al. [113] demonstrated that peptone water is a very effective diluent for the investigation of yeasts in foods, with saline and phosphate buffer giving lower counts than those obtained with distilled water as diluent. Yeast suspensions must not be kept in dilution long before plating in order to avoid cell death. After 1 h in dilution, a 5–72 % decrease in cell viability occurs depending on the composition of the diluent and the species of yeast. NaCl in diluent has an adverse effect, even for the salt tolerant yeast, *Debaryomyces hansenii* [3].

Samples to be tested for the presence of osmotolerant yeasts must be diluted using a diluent adjusted to lower water activity (a_w) values to minimize osmotic shock effects. For the analysis of foods with reduced a_w (e. g., concentrates, syrups etc) a diluent containing at least 20–30 % glucose has to be used [88]. HERNANDEZ and BEUCHAT [84] demonstrated that 18 and 26 % glycerol with the same a_w as 40 and 50 % (w/w) glucose (a_w 0.934 and a_w 0.898, respectively) are effective in recovering *Zygosaccharomyces rouxii.* Yet, in another study [1] it has been shown that glucose-containing diluents generally result in a higher recovery.

In addition to the type of solute used to adjust the a_w of the diluent, the composition of the media used subsequently for plating influences the efficiency of recovering *Z. rouxii* as well. Interestingly, using an appropriately adjusted diluent ($a_w < 0.94$), even tryptone glucose yeast extract agar (TGYA) of high water activity value (a_w 0.98) resulted in a high recovery of this yeast from blueberry syrup (a_w 0.818–0.921). Dichloran 18 % glycerol agar (DG18) and malt extract yeast extract 50 % glucose agar (MY50G) both perform well too. The use of 40–50 % glucose diluent in combination with MY50G is recommended for enumerating *Z. rouxii* from high-sugar foods. 0.1 % Peptone water, a generally used diluent for enumerating yeasts, is not suitable for recovering osmotolerant yeasts [1, 84]. ANDREWS et al. [3] have reported that 30 % (w/w) glycerol diluent in combination with tryptone yeast extract agar containing 10 % glucose (TY10G) gives optimal recovery of xerophilic yeasts, even of sublethally injured cells.

2.4 Plating and other methods of enumeration

Plating with agar media is the most commonly used enumeration method for yeasts. The two basic techniques used are the pour plate and the spread plate method. In the former, 1 ml of suspension is mixed with about 15 ml of molten and tempered (45 °C) agar medium and allowed to solidify, whereas by the latter 0.1 ml aliquot is spread on the surface of a preliminary prepared agar plate. Comparative studies have demonstrated that the recovery of yeasts from foods is significantly enhanced by the spread plate method as compared with pour plating [57, 67]. It has been shown that high temperatures associated with pour plating impose a heat stress on yeast cells [26, 97]. BEUCHAT et al. [29] indicated also that the spread plating technique is superior to pour plating for enumeration of fungi in foods, and SEILER [138] concluded that spread plating is preferable to the pour plate methods because it gives a better recovery of yeasts with lower dilution errors.

One advantage of the pour plate method if compared with the spread plate method is that the 0.1 ml aliquots result in a higher detection limit. A maximum of 0.33 ml inoculum can be distributed on the surface of a standard petri dish, and triplicate samples from the original suspension approximate the same sensitivity as can be obtained with the 1.0 ml samples used in pour plating.

The most probable number (MPN) method allows using a larger volume of inoculum, but this technique has a low level of statistical significance and results in significantly higher counts than do other enumeration methods [57, 80, 103]. Another disadvantage is that further steps are required to obtain isolated colonies from broth medium.

Membrane filtration is an appropriate method when low numbers of cells are to be enumerated from fluid products (soft drinks, beverages, milk). Suspensions made by homogenization from solid products may easily cause clogging the membrane, which is generally of 0.5 μm pore diameter for retaining yeast cells. Prefiltration may solve the problem, however, no study was made to show its influence on counts. Otherwise, the precision of the membrane filtration is comparable to that of other enumeration methods [57].

2.5 Incubation

In general, for enumeration of yeasts, an incubation temperature between 25–28 °C is appropriate, but an ambient room temperature (20–22 °C) may also suffice. Under these conditions yeast colonies develop usually in 2–3 days. However, significantly higher numbers of yeasts and molds are detected after 5 days incubation when compared to 3 days incubation on both conventional media and Petrifilm yeast mold (YM) plates [27]. Hence, 25 °C for 5 days is recommended as a standard incubation regime for general purpose enumeration of yeasts [87].

On selective media such as DG18 or TGYA, yeasts develop slowly and 5 to 7 days may be required to get the highest count. In general, colonies develop faster when incubated at

30 °C instead of 25 °C. On TY10G agar, colonies of *Z. rouxii* appear more rapidly at 30 °C than at 25 °C. This enhances the ease of counting after five days, although the number of colonies does not increase after three days of incubation [19]. A higher incubation temperature may be disadvantageous for recovering psychrotrophic species that may represent a majority of the yeasts occurring in some chilled products. An incubation regime of 25 °C for 7 days turns out to be better than 5 °C for 14 days for the recovery of yeasts from a variety of chilled dairy and meat products. In particular the number of recovered yeasts is found to be higher, although some of these products contain a considerable population of psychrophilic species that grow only at the lower temperature [8].

2.6 Media

2.6.1 General purpose media

For the isolation and enumeration of yeasts from foods the use of general purpose media which allow the recovery of all kinds of yeast, while inhibiting bacterial growth and reducing fungal spreading, is recommended [101]. A number of such media exist, and the history and development of these media has been well documented [46, 69, 92, 101]. Comparative studies have indicated that none of the currently used media is effective for enumerating yeasts in all foods [18, 47]. The requirements for an ideal cultivation medium are multiple (Table 2.6-1) and it has to be accepted that no single, all-purpose medium sufficient for any commodity exists. A few media, however, are more suitable for most isolation purposes. Recipes of the media are provided in Chapter 3.

Tab. 2.6-1 Requirements of an ideal fungal enumeration medium (modified after [125])

1. Suppress bacterial growth completely without affecting growth of fungi
2. Be nutritionally adequate and support the growth of relatively fastidious fungi
3. Restrict spreading growth of fungi to facilitate enumeration
4. Allow recognition and differentiation of fungal colonies
5. Permit maximal recovery of fungal population
6. Give satisfactory and reproducible data
7. Be suitable for most purposes and all foods
8. Be of definite but simple composition, easy to prepare and stable

2.6.1.1 Basal media

Traditionally, malt agar, malt extract agar (MEA), Sabouraud-glucose agar (SGA), tryptone-glucose-yeast extract agar (TGYA), potato dextrose agar (PDA) have been used for the cultivation of yeasts. In recent years, TGYA has emerged as the most commonly used general purpose medium for the detection, enumeration and isolation of yeasts from foods.

However, media with a different composition have been described under this name, especially with respect to the concentration of glucose that may vary from 0.1 % (as in plate count agar (PCA)) to 10 % [101]. TGYA can be used either acidified (ATGY, pH 3.5) or supplemented with chloramphenicol (100 mg/L, TGYC). Yeast extract glucose chloramphenicol agar (YGCA), a simpler medium without tryptone, is recommended as an ISO standard medium [89]. Several comparative studies have been carried out to evaluate TGYA in comparison with other general purpose and selective mycological media [21, 47, 54, 55]. The overall conclusion has been that media consisting of tryptone, yeast extract with glucose concentrations ranging between 0.1 and 10 %, and supplemented with chloramphenicol (0.1 %) can be reliably used for the detection of yeasts from most foods. Colony development may be slower if the medium contains 0.1 % glucose only, and YGCA may be somewhat nutrient depleted for the recovery of stressed cells. However, in international laboratory trials no significant differences have been revealed in the efficacy of enumeration using TGYC if compared with other general purpose media for foodborne yeasts.

2.6.1.2 Acidified media

Acidified media can be made from malt extract agar, potato dextrose agar or another basal medium by adjusting the pH to 3.5. Acidification has to done before pouring the agar medium with an appropriate amount of 10 % tartaric acid (but other acids such as lactic, citric, phosphoric, hydrochloric acids may be used as well). Acidified media have been routinely used over the years to enumerate fungi in various foods. However, acidified media are shown to be less suitable for the enumeration of yeasts than those supplemented with antibiotics [12, 57, 149]. Oxytetracycline glucose yeast extract (OGY) agar yields consistently more psychrotrophic yeasts from chilled foods than acidified media [8]. Acidified PDA (APDA) is found inferior when compared to malt extract yeast extract agar (MYA) or PCA supplemented with 52 % (w/w) sucrose to recover osmotolerant yeasts [39]. The use of acidified media can be suitable in analyzing the population of yeasts consisting of strains adapted to high acid conditions, e. g., in fruit purees, pickles and kefyr [25, 47]. In a broad survey comparing ten media [157] it has been found that antibiotic-amended media (antibiotic-supplemented media) are superior to acidified media for the enumeration of yeasts in dairy products of neutral pH value. No significant difference in performance has been observed between the two types of media with cheeses and yoghurt at low pH values.

2.6.1.3 Antibiotic-supplemented media

Oxytetracycline-glucose-yeast extract (OGY) has been one of the first antibiotic supplemented media [118]. Later, Mossel et al. [116] suggested the use of two different antibiotics, oxytetracycline and gentamycin, each in 100 mg/L concentration, to selectively detect yeasts from meat products and other foods heavily loaded with bacteria. Banks and Board

[8] observed, however, that while neither oxytetracycline nor chloramphenicol are inhibitory to the majority (96 to 99 %) of yeasts even at a concentration of 1000 mg/L, gentamycin, on the other hand, prevents growth of nearly 20 % of yeasts at a concentration as low as 50 mg/L. With some types of foods, e. g., meats, a single antibiotic will not be sufficient to control the growth of bacteria, and the use of two antibiotics is recommended [104, 117].

Chloramphenicol, oxytetracycline, chlortetracycline or some other antibiotics appear to be equally effective in controlling bacteria. The first is heat stable and can be added with other ingredients before autoclaving, hence its use is more convenient. However, being carcinogenic, care has to be taken when handling the compound [11].

2.6.1.4 Control of fungal growth

Various attempts have been made to improve the enumeration of yeasts in the presence of filamentous fungi by reducing the colony diameter of spreading molds. Rose bengal added to general purpose media restricts excessive mycelium formation. Dichloran, alone or in combination with rose bengal, has been shown to limit colony diameter and spreading fungi and facilitates the enumeration of yeasts [82]. KING et al. [100] have described a medium containing chloramphenicol for the inhibition of bacteria as well as dichloran and rose bengal to retard the spreading of molds. This dichloran rose bengal chloramphenicol agar (DRBC) has become one of the most commonly used isolation media. However, some yeast and mold strains may be inhibited completely by rose bengal if the medium is exposed to light. The cytotoxic and photodynamic inactivation of yeasts by rose bengal upon illumination has been repeatedly noted [8, 9, 61, 90, 154], and updated [41].

Various chemicals, fungicides, antibiotics, surfactants and others have been evaluated for their suitability to control the spread of fungal growth [30]. BRAGULAT et al. [35] has investigated a wide range of dyes as mould-spreading inhibitor. Apart from the commonly used dichloran and rose bengal, auremine (25 μg/ml) inhibits colony growth and allows enumeration, whereas gentian violet (5 μg/ml) and malachite green (1 μg/ml) inhibit completely the growth of various fungal species.

2.6.2 Selective media

Possible strategies to design specific media to selectively isolate or differentiate yeasts can be based on the use of inhibitors, dyes and specific growth substrates. No strict distinction can be made between selective and differential media. In this review, a medium permitting the development of one specific group or species while controlling or inhibiting the growth of the rest of yeasts will be considered selective, whereas differential media will permit the growth of several yeast species but these can be recognized through colonies of various and different colour, shape or size.

2.6.2.1 Osmotolerant yeasts

Several yeast species, collectively called osmotolerant yeasts, are able to develop and cause spoilage in intermediate and low-moisture foods and beverages (a_w 0.85–0.65), of which *Z. rouxii* and *Z. mellis* occur most frequently [53, 150]. Various media have been used for the detection of these yeasts, and their performance varies depending on the composition of the food and the type of diluent used. It has been found that for the satisfactory recovery of osmotolerant yeasts both the diluent and the medium must be osmotically balanced in order to protect cells from osmotic shock [22, 88].

HOCKING and PITT [86] developed a medium for the recovery of xerophilic fungi containing 18 % glycerol to reduce water activity (a_w 0.955) and dichloran (2 mg/L) to limit the spread of fungal colonies, as well as chloramphenicol (200 mg/L) to inhibit bacteria. The dichloran 18 % glycerol agar (DG18) appears to be appropriate for the enumeration of yeasts in general [61]. DG18 performs better than OGY or DRBC in recovering *Z. rouxii*, whereas malt extract yeast extract agar with 50 % glucose (MY50G) has been found too selective for osmotolerant yeast species [34]. DG18 has been suggested for the routine use to enumerate yeasts from low a_w products. Plate count agar containing 52 % (w/w) sucrose is more effective to recover *Z. rouxii* from orange juice concentrates than yeast extract malt extract agar with 52 % (w/w) sucrose [39]. However, the authors have applied different plating methods in both studies (surface plating in the former and pour plating in the latter case), which may have influenced the results.

DG18, malt extract yeast extract agar with 30 % glucose (MY30G) and tryptone yeast extract agar with 10 % glucose (TY10G) perform equally well to recover xerophilic (osmotolerant) yeasts from concentrated products, but DRBC without osmotic supplement is unsatisfactory for enumerating these yeasts [3]. DG18 and MY30G are more expensive than TY10G. As a further disadvantage, MY30G is viscous and difficult to handle and may caramelize during autoclaving.

BEUCHAT et al. [24] determined the performance of three diluents (0.1 % peptone water, 40 % glucose and 30 % glycerol), in combination with three media (TY10G, MY50G, DG18) in the recovery of *Z. rouxii* from a wide range (a_w 0.73–0.85) of intermediate moisture foods. With the exception of 0.1 % peptone water, all combinations of diluents and media performed well. For ease of use and economic consideration, the use of 40 % glucose diluent in combination with TY10G agar is recommended.

Commercial brands and batches of DG18 using filamentous fungi may differ in performance [72]. Media prepared freshly in the laboratories are more efficient. DG18 has been originally developed for the detection and enumeration of moderately xerophilic fungi [86], and, due to its successful performance, some workers use it as a general purpose enumeration medium for molds and yeasts [61, 88, 138]. However, several reports have noticed, firstly that the recovery of yeasts on DG18 is lower than on TGYC or DRBC, secondly that this depends on the type of food and the species of yeasts therein [47, 120], thirdly that the rate of growth and the size of colonies formed are retarded [3, 139], and fourthly that DG18 inhibits certain yeasts [11, 120]. In another recent collaborative study [55] it has been clearly demonstrated that compared to DRBC, TGYC and PCAC, the recovery of yeasts on DG18 is significantly lower, the development of colonies slower, and the growth of some yeast species (*Rhodotorula mucilaginosa, Cryptococcus albidus, Brettanomyces anomalus* (= *Dekkera anomala*)) inhibited. Hence, the use of DG18 as a general enumeration medium for foodborne yeasts cannot be recommended, notwithstanding it performs well in detecting and isolating xerotolerant and xerophilic molds and osmotolerant yeasts.

2.6.2.2 Preservative and acid-resistant yeasts

Zygosaccharomyces bailii is the most notorious spoilage yeast capable to grow in low-acid and/or preservative-containing foods. Several other yeast species, such as *Pichia membranifaciens* (including its anamorph *Candida valida*), *Issatchenkia orientalis* (anamorph *C. krusei*), *Schizosaccharomyces pombe,* and *C. parapsilosis* also show notable resistance towards acidic pH and high concentrations of preservatives [56]. Acidified media are generally recommended for detecting and enumerating these yeasts. For this purpose, general media, such as MEA or TGY, molten and cooled to 50 °C, are acidified by adding 5 ml/L glacial acetic acid before pouring into Petri dishes. A medium called *Zygosaccharomyces bailii* agar (ZBA, see chapter 3), containing 0.5 % acetic acid in addition to 0.1 % potassium sorbate, has been developed for the selective enumeration of *Z. bailii* [65]. Acidified malt extract agar is inferior for the recovery of *Z. bailii*, whereas ATGY is a better selective medium than ZBA [108, 109]. ATGY and acidified tryptone fructose (4 %) yeast extract agar (ATFY), both with a reduced (0.3 %) acetic acid content, are judged best among selective media [109]. For recovering heat-stressed cells of *Z. bailii*, non-selective yeast extract malt extract agar (YMA) turned out to be the best medium.

In a collaborative study [85] the most effective medium for the selective isolation and enumeration of preservative resistant yeasts was determined. Malt extract agar and tryptone glucose yeast extract agar with and without 0.5 % acetic acid, as well as selective *Z. bailii* agar (ZBA) have been compared. ATGY is the best medium to recover the highest nummer of preservative resistant yeasts, *Z. bailii, P. membranifaciens,* and *Sch. pombe.* ZBA is highly selective because it permits the growth of *Z. bailii* only, although it even inhibits the growth of this species. This may be due to the use of lyophilized cultures in this study, comprising about 25 % sublethally injured cells.

2.6.2.3 Wild yeasts

Wild yeasts is a collective term used in the fermentation industry to denominate any and all yeast species other than the industrial strains of *S. cerevisiae* (and in a few cases some other industrially used species, such as *C. utilis* (anamorph of *Pichia jadinii*) and *Kluyveromyces lactis*). Wild yeasts are considered contaminants, and create a problem when present in large numbers thus resulting in poor quality products. Contamination of pitching yeast with wild yeasts is a notorious problem in the brewing industry, where serious efforts have long been made for their detection. This is, however, a difficult task because wild yeasts include not only species belonging to various genera (so-called non-*Saccharomyces* wild yeasts, e. g., *Brettanomyces, Candida, Debaryomyces, Pichia, Torulaspora* and *Zygosaccharomyces* spp.) but also *Saccharomyces* species and even strains of *S. cerevisiae* other than, but very similar to the brewing strains [6, 37]. Moreover, detection of wild yeasts has to be done out of a far larger number of brewing yeast. Hence the detection method must not only be selective but very sensitive as well. A variety of selective media have been developed, but none of the currently available media can be used alone to detect all kinds of contaminant yeasts. The applied selective principles include the use of: compounds inhibitory to *S. cerevisiae* (e. g. copper, cycloheximide), substrates not used by this yeast (lysine, nitrate, xylose, dextrin etc), elevated incubation temperature (37 °C) or a combination thereof. Detection of wild yeasts in breweries has been the subject of recent publications [64, 93, 153], and will not be further discussed in detail. Table 2.6-2 gives an overview of various differential and selective media used in breweries (see chapter 13).

Tab. 2.6-2 A compilation of various selective and differential media for detection of wild yeasts in breweries ([93, 153] and references therein)

Medium	Selective principle
Lysine	Nitrogen source
CLEN	Cadaverine, lysine, ethylamine, nitrate
XMACS	Xylose,mannitol,adonitol,cellobiose, sorbitol
Actidione	Inhibitor
Copper	Inhibitor
Lin's LWYM	$CuSO_4$
Schwarz	Fuchsin sulphite
Crystal violet	Dye
Dextrin	Carbon source

A related problem exists in the wine industry, although the species of wild yeast are mostly different. The most frequently encountered wild yeasts during the early stages of must fermentation are the so-called apiculate yeasts (*Hanseniaspora* spp. and *Kloeckera* spp.). Wine strains of *S. cerevisiae* are generally more resistant to ethanol and sulphur dioxide, hence a medium containing 12 % (v/v) ethanol and 150 mg/L bisulfite can suppress the growth of non-*Saccharomyces* wild yeasts [102]. The most frequently occurring species of *Hanseniaspora (Kloeckera)* are resistant to cycloheximide and utilize cellobiose for sole carbon source. A medium based on these properties has been used for their selective isolation from grapes and must (T. Deak, unpublished observ.). In compressed yeast and dough, baker's yeast strains of *S. cerevisiae* occur in very high number (10^9–10^{10} cfu/g). A modified lysine agar containing succinate has been used successfully to detect non-*Saccharomyces* wild yeasts, making up only 7.6–7.8 % of baker's yeasts [155].

2.6.3 Differential media

Organic dyes and indicators have long been used as selective and differential substances in bacteriological media. However, very few data are available for the differential isolation of yeasts using dyes. FUNG and co-workers [76, 106] provided evidence that dyes are valuable for the differential isolation of particular groups of yeasts, and that they can be employed for this purpose in the microbiological investigation of foods. These studies have resulted in the development of two specific media. One of these, containing crystal violet, allows growth of *Candida lipolytica* only. Another medium, containing aniline blue, differentiates *C. albicans* from other yeasts in clinical samples, because only *C. albicans* shows fluorescence under UV light [78]. This medium is successfully used in food samples for the presumptive identification and differentiation of *C. albicans* from *C. tropicalis* [75]. Wallerstein Laboratory nutrient agar, originally developed for the growth of microorganisms from beer, contains bromcresol green and was shown to facilitate the differentiation of wild yeasts by different colour and size of colonies [2].

Dyes have also been used to discriminate among strains within a single species. Different colony types develope on Sabouraud glucose agar containing triphenyltetrazolium chloride and allow subtyping of several *Candida* species of clinical importance [130]. Absorption of nine dyes, added to a basal yeast extract peptone glucose agar, has been used to distinguish distillery and brewing strains of *S. cerevisiae* [79]. Characteristic differences are found between the two groups of strains and also between brewing strains, but not between fuel alcohol producing strains. The dye-absorption patterns agree well with SDS-PAGE and RAPD data.

2.6.4 Media for specific yeasts

Use of fluorogenic or chromogenic enzyme substrates has led to the development of a great number of methods for the identification of bacteria, even in primary isolation media. MANAFI [110] listed 23 different commercially available media for the detection of *Escherichia coli* and coliforms, and many others are available also for *Salmonella, Listeria, Staphylococcus,* enterococci, spore-formers and lactic acid bacteria. Compared to these widely used specific media, very few media have been developed for the selective isolation and direct identification of specific groups or species of yeasts. However, developments in this direction have started in food mycology too. Recently, successful attempts to design specific media primarily for foodborne yeasts have been made.

Rapid presumptive detection of the human pathogenic yeast *Filobasidiella (Cryptococcus) neoformans,* has been made possible since long through the formation of dark-brown melanin pigments from various diphenolic compounds (e. g., catechol, dopamin, dihydroxy-phenylalanin, etc.), which are also abundantly present in extract of niger seed (*Guizotia abyssinica*) [40]. The two varieties of the species, *F. neoformans* var. *neoformans* and var. *bacillispora*, can also be differentiated through the ability of the latter to hydrolize glycine and to grow in the presence of canavanine. A medium containing these compounds and an indicator, bromothymol blue, turns blue only in the presence of *F. neoformans* var. *bacillispora* [105]. Applying a similar principle, CARREIRA and LOUREIRO [38] described a differential medium to detect *Yarrowia lipolytica* based on the ability of this yeast to produce brown pigments from tyrosine, which is a unique property among the yeasts.

A number of media have been developed for direct detection and presumptive identification of another pathogenic yeast, *Candida albicans*, responsible for up to 80 % of various mycoses caused by yeast. An early medium (Pagano agar) contains tetrazolium salt and turns red when reduced by *Candida tropicalis*, but remains pale with *C. albicans*, which is unable to reduce this indicator. BOBEY and EDERER [31] demonstrated that substrates conjugated with chromogenic or fluorogenic groups can be used for monitoring various enzymes in yeasts. Based on this principle, a rapid enzyme test kit is developed for the identification of clinically important yeasts applying 4-methylumbelliferyl- and p-nitrophenyl-conjugated substrates. Unfortunately, an effort to adapt this technique for a broader range of foodborne yeast species remained unsuccessful [51]. However, media based on a chromogenic or fluorogenic substrate have proved to be valuable for the rapid, presumptive identification of *Candida albicans* and for its differentiation from other yeasts. Several kinds of commercial media are available such as Candida ID, Albicans ID, fluoroplate Candida agar, CHROMagar Candida [10, 73, 121, 124, 134, 160]. Chromogenic media are very useful to differentiate a number of commonly occurring yeast species of clinical significance (e. g. *C. tropicalis, C. glabrata, C. krusei*) in addition to *C. albicans.*

CHROMagar Candida also discriminates easily between yeasts used in animal feeds as probiotic additives [32]. TORNAI-LEHOCZKI and PÉTER [151] made a preliminary study on the use of CHROMagar Candida as a differential media for food-borne yeasts (Fig. 2.6-1). In addition to *C. albicans*, several other species, not encountered in clinical samples, develop blue-green colonies as well. Different strains of the same species produce very similar colours, thus indicating the reliability of the medium. In some cases, closely related species show different colours, providing a quick method to separate *Z. bailii* from *Z. rouxii*, as well as *Kluyveromyces lactis* from *K. marxianus* (Table 2.6-3).

Applying various strategies such as differential dyes (eozin, methylene blue), selective inhibitors (acetic acid, tellurite), sole carbon source (2-ketogluconate) and detection of enzyme activities (β-glucosidase, alkaline phosphatase), SILONIZ et al. [142] have developed three media for the presumptive identification of several common osmotolerant food spoilage yeasts. On eosin-methylene blue medium, only colonies of *S. cerevisiae* turn metallic green while the others are black or violet; *Z. bailii* tolerates 1 % acetate whereas *Issatchenkia orientalis*, and some strains of *Torulaspora delbrueckii* grow in the presence of 0.5 % acetate and 0.2 % potassium tellurite. Species of similar growth pattern and strains of vari-

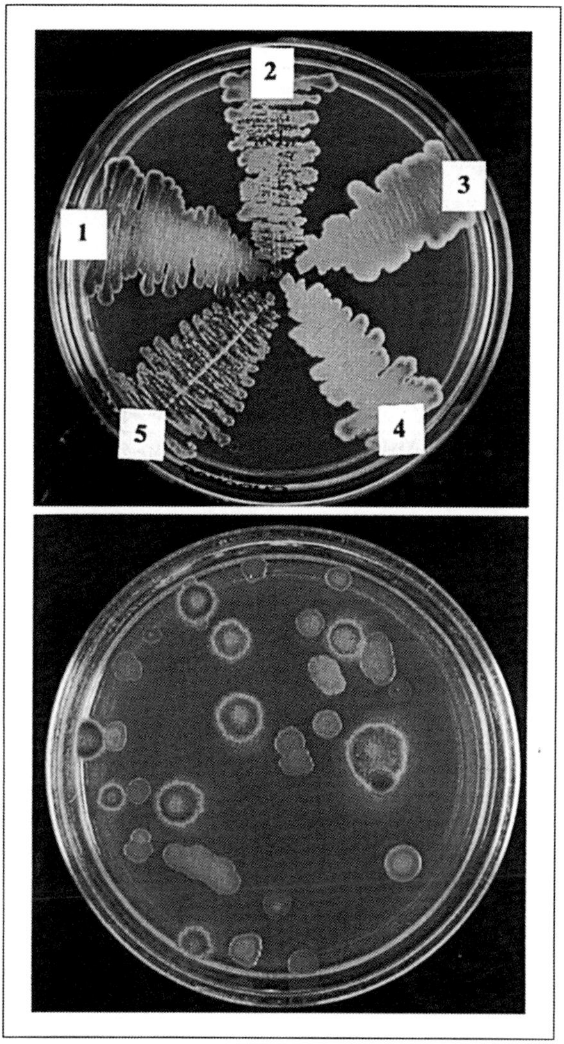

Fig. 2.6-1 Different appearence of yeast colonies on CHROMagar-Candida (top) and Dichloran Rose Bengal Chloramphenicol (DRBC) agar (bottom).
1: *Candida tropicalis*; 2: *Candida glabrata*; 3: *Debaryomyces hansenii*; 4: *Torulaspora delbrueckii*; 5: *Issatchenkia orientalis*.
Note also the restricted growth of molds on DRBC agar.

Tab. 2.6-3 Appearance on CHROMagar Candida of some common foodborne yeasts (Data from [151] and personal communication)

Yeast species	Colony colour	Reverse	Margin	Surface
Candida albicans	Green	Green		Green
C. zeylanoides	Green	Greenish	Gray	
Debaryomyces polymorphus	Green	Greenish		Dull
Cryptococcus laurentii	Bluish green	Deep green		
C. tropicalis	Dark blue	Deep blue	Creamy	Rough
Saccharomyces cerevisiae	Deep purple	Deep purple	Grayish	
S. exiguus	Creamy	Beige		
Kluyveromyces lactis	Beige	Deep beige	White	
K. marxianus	Grayish purple	Grayish purple	Cream	
Torulaspora delbrueckii	Yellow	Yellow	Gray	
Zygosaccharomyces rouxii	White grayish	Grayish		
Z. bailii	Light purple	Light purple	Grayish	Rough

able properties are further discriminated by additional tests: alkaline phosphatase activity in *Z. bailii,* β-glucosidase acivity in *Debaryomyces hansenii,* and growth on 2-ketogluconate of *T. delbrueckii.* In further studies, some of these preliminary results have been exploited to develop more reliable differential media.

Another strategy is to detect a particular enzyme demonstrating its activity on a chromogenic substrate. Using this principle, a medium has been developed for the detection of *D. hansenii* with a conjugated substrate of β-glucosidase [143]. By the same token, the lactose utilizing species *K. marxianus* and *K. lactis* are detected using a chromogenic galactopyranoside substrate (X-gal) upon the induction of the enzyme, β-galactosidase, by a substrate analogue [152]. NGUYEN et al. [119] showed that *K. marxianus* and *K. lactis* can be easily detected by including X-gal into yeast extract peptone glucose agar, as only these species produce blue colonies on this medium. *Dekkera* and *Brettanomyces* species are peculiar in producing acetic acid from ethanol, and their presence can be detected by a color change of an indicator. Moreover, these species form 4-ethylphenol from p-coumaric acid, a product of strong and characteristic odor, further enhancing their recognition [132]. A mineral medium including glucose and formic acid as the only carbon and energy sources, is useful for the selective and differential detection of *Z. bailii* and *Z. bisporus* [136].

2.6.5 Media for specific foods

Media have been developed for enumerating yeasts in specific food products. In particular, the brewing industry has long demanded appropriate media for the differential enumeration of pitching yeast from wild yeasts. The various media, developed for this purpose, among them lysine agar and many others, have been described above. The wine industry is also strongly interested in methods for the specific detection and differentiation of *Saccharomyces* wine yeasts and various non-*Saccharomyces* wild yeasts. A selective medium containing 150 mg/L bisulphite and 12 % (v/v) ethanol is based on the higher tolerance of wine yeasts towards sulphur dioxide and ethylalcohol [102]. HEARD and FLEET [83] found that ethanol-sulphite agar variably supports the growth of wild yeasts. Lysine agar, on the other hand, suppresses the growth of *S. cerevisiae* and enables the enumeration of *Kloeckera apiculata, C. stellata* and other non-*Saccharomyces* yeasts in fermenting must. H_2S-production is a detrimental property of wine yeasts. A bismuth sulphite indicator agar can be used for selecting low or non-H_2S producing wine yeast strains [95].

Various specific media have been devised for different food products. A few examples are mentioned only. A medium containing Schiff's reagent is devised to detect sulphite-binding yeasts in comminuted meat products [62]. The acetaldehyde produced by certain yeast species binds to sulphite and releases basic fuchsin inducing a red coloration of the medium around the colonies. Molybdate (0.187 % phosphomolybdic acid) and Ca-propionate (0.125 %) are used to selectively isolate yeasts from tropical fruits [131]. Malt-yeast extract-sucrose agar is recommended for the enumeration and isolation of molds and yeasts from silage [144]. WELTHAGEN and VILJOEN [157] evaluated ten selective media for their suitability to enumerate yeasts in dairy products. No specific media have been included, and most antibiotic-supplemented media are shown to be superior to acidified media in the recovery of yeasts from dairy products of neutral pH values. However, all media perform equally well in dairy products of low pH.

2.6.6 Performance of media

Since the establishment, under the auspices of the International Committee on Food Microbiology and Hygiene (ICFMH), of the Working Party on Culture Media in 1978, special attention has been focused on the quality assurance and validation of microbiological methods. Several meetings have been held, and reviews, monographs and a book provide details on the subject [7, 42, 43]. Although among the nearly hundred bacteriological media published in monographs only three (DG18, DRBC, OGY) are used for yeasts, the importance of quality control of media and performance testing of laboratories is being increasingly recognized in food mycology.

Monitoring of media is necessary if Good Laboratory Practice is to be maintained. Purpose of the activity is threefold: (i) quality assurance, (ii) validation and standardization, (iii) proficiency testing. Quality assurance of culture media and methods aims to ensure that the methods, media and equipments are all functioning correctly. Validation of methods is carried out in order to demonstrate that the methods are adequate for their intended use. Standardization of methods enables an easier comparison of results from different laboratories. Proficiency testing aims to test the overall ability of laboratories to apply and evaluate the methods [42].

The scope of quality control of culture media is threefold: (i) to assess the quality of commercially available dehydrated media or ready-to-use plates or tubes; (ii) to check the quality of purchased batches of media or their ingredients; (iii) to monitor the procedures of media preparation such as weighing components, heat treatment, and storage conditions. The laboratory quality control of media can be based on five main criteria, namely productivity, selectivity, sensitivity, specificity and accuracy. These can be tested using pure cultures of target and interfering microorganisms, and results can be expressed quantitatively as percentages of recovery related to a reference medium.

A standardized, lyophilized mixture of reference microorganisms must be prepared and used to monitor each new batch arriving in the laboratory. The use of reference species for monitoring media has been suggested [137]. A shortened list of these species is shown in Table 2.6-4.

Tab. 2.6-4 Suggested reference species for monitoring mycological media (after [137])

Spreading molds	*Mucor racemosus, Rhizopus stolonifer*
Other molds	*Eurotium repens, Cladosporium herbarum*
Yeasts	*Saccharomyces cerevisiae, Zygosaccharomyces rouxii*
Bacteria	*Pseudomonas aeruginosa, Bacillus subtilis*

Accuracy of detection and enumeration must also be evaluated regarding repeatability and reproducibility, i.e. testing the quantitative performance within and between laboratories. Methods and media have to be assessed also for their performance using real food samples in order to evaluate the potential influence of the food constituents, the background microbiota and the recovery of stressed microorganisms [44].

Several methods have been proposed to check the quality of media, and these and their statistical evaluation have been reviewed [158]. The majority of the techniques using solid media rely on colony counting by spread plating, the modified Miles-Mishra method or the semi-quantitative ecometric streaking method, whereas for liquid media the serial dilution technique is most widely used. The stab inoculation method is preferable when rapidly spreading test strains are used.

Even if carefully formulated dehydrated media from reputable manufacturers are prepared with due diligence in the laboratory, quality monitoring of produced media is essential if satisfactory and reliable test results are to be obtained. This is particularly the case with selective media such as DRBC, DG18. Significant batch-to-batch variations have been reported in the growth of yeasts and molds on selective media such as DRBC and DG18, containing rose bengal or dichloran [72,137]. The effect of storage time on media quality has been investigated as well [122]. Both colony morphology and antibacterial activity on DRBC and DG18 appear very stable during a long period of storage, but the restrictive capability of mold spreading is lost. It is recommended that storage of DRBC in flasks at 4 °C does not exceed four weeks, and only one week after pouring. DG18 plates have to be made from freshly prepared medium and should be kept at 4 °C for a maximum of one week.

2.7 Toxicity of media on injured cells

Yeast cells surviving in processed foods may undergo sublethal injury and exhibit an increased sensitivity to suboptimal culture conditions. Heat-stressed or frozen and thawed cells as well as those found in low pH, low a_w, or chemically preserved foods may not be recovered using acidified media or media lacking nutrients. This topic has been thoroughly reviewed [13, 14, 53, 69, 148]. Golden and Beuchat [77] reported that the recovery of sublethally heat-injured *Z. rouxii* cells is affected by the concentration and the type of solute in the medium. Thirty three percent glucose (equaling an a_w 0.936) is found to be superior to sucrose, sorbitol or glycerol at concentrations providing the same a_w value. More recently, Fleet and Mian [71] reported that populations of various yeast species suffer 5–85 % sublethal injury induced by freezing and thawing or heating. Injured cells are unable to grow on selective media providing restricted growth conditions. Media containing 3–5 % salt are most inhibitory to injured cells, whereas media deficient in nutrients or with an increased sugar concentration (20 % sucrose) are less inhibitory. While it is generally accepted that antibiotic supplemented media are less selective for growth than acidified media, Fleet and Mian [71] found that in some cases, and depending on the species, 5–20 % of injured cells are not detected on antibiotic-based media. Sublethal damage in cell structure or metabolism can, however, be repaired, upon time and appropriate conditions. When incubated for 3 h at 25 °C, 2 % malt extract broth results in a better recovery of cells than trypticase soy broth. However, even after resuscitation, 5–20 % of the cells remains injured, thus indicating that the viable population may be underestimated if recovery conditions are not optimal. On the other hand, a resuscitation period longer than 3 h allows some cells to multiply, leading to overestimation of the original population [71].

Foods and beverages with high sugar concentrations can also result in stressed yeast cells. Detection of yeasts from concentrated fruit juices consistently results in much lower counts when samples are directly pipetted onto a plate of yeast extract-60 % (w/w) glucose agar [66]. A more effective procedure is dilution of the concentrate from 1:4 to 1:10 in a 30 % glucose-solution followed by plating on low a_w medium. Dilution of the concentrate reduces the sensitivity of the method, which is critical for the detection of osmotolerant yeasts occurring in small numbers. In order to test for the most notorious spoilage organisms of juice concentrates, *Z. bailii*, *Z. rouxii* and *Z. bisporus*, a selective enrichment broth containing 60 % (w/w) glucose and 500 mg/L benzoic acid at pH 4.5 is recommended [66].

The effect of freeze-drying injury and long term storage at freezing, chilling and ambient temperatures has been investigated using DRBC, APDA, DG18 and orange serum agar (OSA) in comparison with TGYC and PCAC media [21]. TGYC and PCAC perform equal and are superior to the other four media for recovering desiccated and stored yeasts.

2.8 Non-traditional and rapid methods

Culturing techniques for detection, enumeration and isolation of yeasts and molds are widely used in food mycology since they are simple, convenient and flexible. However, apart from their low reproducibility, the inherent slowness of culturing methods is an important drawback for their use in quality control. In the last two decades, numerous approaches have been made to develop rapid methods. These non-conventional techniques are often based on novel principles, and many of them can be automated. This subject has been reviewed [48, 49, 53, 74].

2.8.1 Accelerated cultivation methods

Conventional culturing methods can be accelerated using equipments and devices for saving time in routine work. Convenient and automated tools are marketed for media preparation, plate pouring, sample preparation and dilution, and colony counting [68, 74]. The numerous technical improvements can be illustrated by the gravimetric dilutor and spiral plater. These devices facilitate tedious weighting of samples and eliminate dilution error, thus making it possible to count cells over a range of 3–4 $\log_{10}$ units. Both the accuracy and recovery of cells increase, but the sensitivity decreases because of the small volume of sample spread on a plate.

The Petrifilm™ is a series of dry selective media prepared on a support membrane and covered with a film. The medium is rehydrated by adding 1ml of a diluted sample. According to Beuchat and coworkers [26, 27, 28], Petrifilm™ YM plate compares very favorably with conventional agar media (PCAC, APDA) for recovering yeasts and molds from various foods. According to Vlaemynck [156], the Petrifilm™ system produces results similar to traditional plating for enumerating yeasts and moulds in cheese and yoghurt.

The hydrophobic grid membrane filter (HGMF) method, commercialized as the Iso-Grid system, combines advantages of the membrane filter and most probable number (MPN) techniques. The hydrophobic filter confines the growth of colonies to 1600 grid cells, and the positive compartments can be evaluated similarly to MPN statistics. The accuracy is much greater as the large number of grid cells allows counting in 3 to 4 $\log_{10}$ range without dilution, and counts can be obtained after 24 h incubation [36, 65].

2.8.2 Direct counting

Direct microscopy has been used since long for the enumeration of yeasts. Methylene blue staining allows for discrimination of living and dead yeast cells [145]. Viability can be detected more precisely with several fluorescent staining methods, and, in conjunction with membrane filtration, has become an efficient means of enumeration, called direct epifluorescent filter technique (DEFT). Several studies have demonstrated its advantages and limits to enumerate yeasts in various foods. As with other direct microscopic techniques, operator fatigue is one limiting factor in exploiting the full capacity of this otherwise rapid method. An instrument has been developed coupling DEFT with image analysis and turning it into a fully automated but more expensive counting system [123].

Another instrumental counting method is flow cytometry that recently has found diverse applications combined with various fluorescent staining and molecular labeling techniques [4]. Flow cytometry is an efficient enumeration method for rapid (within 15–30 min) determination of yeast counts in milk, grape juice, beer, yoghurt, salads, cheese and other foods [43, 94, 96], and allows real-time monitoring and discrimination between yeasts, moulds and bacterial cells. With fluorescent probes, the method is suitable for determining viability and vitality testing [60, 115]. The large capacity of commercially available instruments with an automated sample processor permits to generate data for predictive modeling [147].

2.8.3 Electrometry

By monitoring changes in impedance, capacitance or conductance, metabolic activity of microorganisms can be estimated in terms of detection time. This in turn, can be used for determining the number of viable cells present when calibrated with colony counts. With the commercially available automated instruments results can be obtained in 6 to 8 h, depending on the size of the initial population in the sample. Impedimetric detection of yeasts has found wide application in the food industry [53, 141]. A modification of the technique, called indirect conductimetry, is an efficient method to study yeasts in beverages [50, 52].

2.8.4 Other non-conventional methods

A diverse range of rapid techniques has been proposed and many of these have potential advantages over traditional culturing methods, but each has limitations as well. The ATP bioluminescence assay, for example, is perhaps the most rapid existing method providing results in about 1 or 2 minutes. Several convenient, portable instruments are available for testing. Yeast population sizes show good correlation with ATP content in beverages and dough [5, 140]. However, yeast populations occur mostly in mixed microbial populations in food, and dead cells and food particles may contain ATP as well, hence the technique is best applicable for hygiene assessment only [81].

Immunoassays are widely used in food bacteriology and detection for mycotoxins. Immunological detection methods have been developed also for yeasts [114]. Because of the lack of commercially available antigens, these methods have not yet found wide practical applications.

Molecular methods based on the detection of biological macromolecules such as cell wall carbohydrates, fatty acids of membrane phospholipids, total proteins and specific enzymes, and, in particular, various types of deoxyribo- and ribonucleic acids (nuclear and mitochondrial DNA, tRNA, mRNA) are common tools in research and routine diagnosis of microorganisms in clinical material and food samples [49, 59, 107, 129, 146, 154]. However, the high specificity and sensitivity of many of these molecular methods predetermine their use for the identification and subtyping of microorganisms rather than the quantitative enumeration and isolation of living cultures (see Chapters 3 and 4).

2.9 Conclusions

In the last 15 years great progress has been made in the standardization of basic cultivation techniques for the detection, enumeration and isolation of yeasts. No single medium can be universally used for testing all yeasts from all foods. A few media (TGYC and DRBC) that support the growth of yeasts, while inhibiting bacteria and retarding the growth of moulds, have emerged as most suitable for general purposes. Other media can be recommended for the detection of specific groups of yeasts, such as DG18 for xerotolerant yeasts, and acetic acid supplemented medium for preservative resistant yeasts. Peptone water can be best used for preparing sample suspensions and dilutions for most purposes. Xerotolerant yeasts, however, require osmotically amended diluents. Recently, a number of selective and differential media has been developed for the isolation and identification of specific yeast species using dyes, chromogenic substrates, inhibitory compounds and unique carbon and nitrogen sources. More work along these lines will bring further improvement of the conventional methodology, together with technical developments and automation of cultivation techniques. Two main lines of non-conventional methods will play increasingly important roles in the future detection and identification of yeasts. Firstly, various rapid and automated methods based on physical and chemical principles have great potential in reducing the time and workload needed to detect and enumerate yeasts by cultivation in foods and beverages. Secondly, molecular methods (see Chapters 3 and 4) hold great promise for practical application in industrial settings to detect, identify and type yeasts.

2.10 Acknowledgement

The author gratefully acknowledges Dr. C. Leao, Dr. V. Loureiro and Dr. J.M. Peinado for kindly providing pre-publication information about media developments, and Dr. L.R. Beuchat for collaboration in media studies.

2.11 References

[1] ABDUL-RAOUF, H.M.; HWANG, C.A.; BEUCHAT, L.R.: Comparison of combinations of diluents and media for enumerating *Zygosaccharomyces rouxii* in intermediate water activity foods. Lett. Appl. Microbiol. **19** (1994) 28–31.

[2] ANDREWS, S.: Comparative study on WL nutrient agar with DRBC and OGY for yeast enumeration in foods. In: Modern methods in food mycology (edited by Samson, R.A.; Hocking, A.D.; Pitt, J.A.; King, A.D.). Amsterdam, The Netherlands: Elsevier (1992) 61–65.

[3] ANDREWS, S.; DE GRAAF, H.; STAMATIN, H.: Optimization of methodology for enumeration of xerophilic yeasts from foods. Int. J. Fd Microbiol. **35** (1997) 109–116.

[4] ATTFIELD, P.V.; KLETSAS, S.; VEAL, D.A.; VAN ROOIJEN, R.; BELL, P.J.L.: Use of flow cytometry to monitor cell damage and predict fermentation activity of dried yeasts. J. Appl. Microbiol. **89** (2000) 207–214.

[5] AUTIO, K.; MATTILA-SANDHOLM, T.: Detection of active yeast cells (*Saccharomyces cerevisiae*) in frozen dough section. Appl. Environ. Microbiol. **58** (1992) 2153–2157.

[6] BACK, W.: Nachweis und Identifizierung von Fremdhefen in der Braurei. Brauwelt **127** (1987) 735–737.

[7] BAIRD, R.M.; BARNES, L.M.; CORRY, J.E.L.; CURTIS, G.D.W.; MACKEY, B.M. (editors): Quality assurance and quality control of microbiological media. Int. J. Fd Microbiol. **2** (1995) 1–136.

[8] BANKS, J.G.; BOARD, R.G.: Some factors influencing the recovery of yeasts and moulds from chilled foods. Int. J. Fd Microbiol. **4** (1987) 197–206.

[9] BANKS, J.G.; BOARD, R.G.; CARTER, J.; DODGE, A.D.: The cytotoxic and photodynamic inactivation of microorganisms by Rose Bengal. J. Appl. Bacteriol. **58** (1985) 391–400.

[10] BAUMGARTNER, C.; FREYDIERE, A.-M.; GILLE, Y.: Direct identification and recognition of yeast species from clinical material by using Albicans ID and CHROMagar Candida plates. J. Clin. Microbiol. **34** (1996) 454–456.

[11] BECKERS, H.J.; DE BOER, E.; DIJKMAN, K.E.; HARTOG, B.J.; VAN KOOIJ, J.A.; KUIK, D.; MOL, N.; NOOITGEDAGT, A.J.; NORTHOLT, .M.D.; SAMSON, R.A.: Comparison of various media for the enumeration of yeasts and molds in food. In: Methods for the mycological examination of food (edited by King, A.D. Jr; Pitt, J.I.; Beuchat, L.R.; Corry, J.E.L.). New York, U.S.A.: Plenum (1986) 66–76.

[12] BEUCHAT, L.R.: Comparison of acidified and antibiotic supplemented potato dextrose agar from three manufacturers for its capacity to recover fungi from foods. J. Fd Prot. **42** (1979) 427–428.

[13] BEUCHAT, L.R.: Injury and repair of yeasts and moulds. In: The Revival of Injured Microbes (edited by Andrews, M.H.E.; Russel, A.P). London, U.K.: Academic Press (1984) 293–308.

[14] BEUCHAT, L.R.: Consideration of media for enumerating injured fungi. In: Methods for the mycological examination of food (edited by King, A.D. Jr; Pitt, J.I.; Beuchat, L.R.; Corry, J.E.L.). New York: Plenum (1986) 168–171.

[15] BEUCHAT, L.R.: Media for detecting and enumerating yeasts and moulds. Int. J. Fd Microbiol. **17** (1992) 145–158.

[16] BEUCHAT, L.R.: Enumeration of fungi in grain flours and meals as influenced by settling time in diluent and by the recovery medium. J. Fd Prot. **55** (1992) 899–901.

[17] BEUCHAT, L.R.: Selective media for detecting and enumerating foodborne yeasts. Int. J. Fd. Microbiol. **19** (1993) 1–14.

[18] BEUCHAT, L.R.: Progress in conventional methods for detection and enumeration of foodborne yeasts. Food Technol. Biotechnol. **36** (1998) 267–272.

[19] BEUCHAT, L.R.: Influence of pH and temperature on performance of tryptone yeast extract (TY10G) agar in supporting colony development by *Zygosaccharomyces rouxii.* J. Fd Mycol. **1** (1998) 181–186.

[20] BEUCHAT, L.R.; COUSIN, M.A.: Yeasts and molds. In: Compendium of methods for the microbiological examination of foods (edited by Pouch-Downes, F.; Ito, K.), Washington, DC, U.S.A.: Am. Publ. Health Assoc. (2001) 209–215.

[21] BEUCHAT, L.R.; FRÄNDBERG, E.; DEAK, T.; ALZAMORA, S.M.; GUERRO, S.; LÓPEZ-MALO, A.; OHLSSON, I.; OLSEN, H.; PEINADO, J.M.; SCHNÜRER, J.; DE SILONIZ, M.I.; TORNAI-LEHOCZKI, J.: Performance of mycological media for supporting colony formation by desiccated food-spoilage yeasts: An inter-laboratory study. Int. J. Fd Microbiol. **70** (2001) 89–96.

[22] BEUCHAT, L.R.; HOCKING, A.D.: Some considerations when analysing foods for the presence of xerophilic fungi. J. Fd Prot. **53** (1990) 984–989.

[23] BEUCHAT, L.R.; HWANG, C.-A.: Comparison of dilution schemes for enumerating fungi in ground pepper: a collaborative study. Int. J. Fd Microbiol. **29** (1996) 157–160.

[24] BEUCHAT, L.R.; JUNG, Y.; DEAK, T.; KEFFLER, T.; GOLDEN, D.A.; PEINADO, J.M.; GONZALO, P.; DE SILONIZ, M.I.; VALDERRAMA, M.J.: An interlaboratory study on the suitability of diluents and recovery media for enumeration of *Zygosaccharomyces rouxii* in high sugar foods. J. Fd Mycol. **1** (1998) 117–130.

[25] BEUCHAT, L.R.; NAIL, B.V.: Evaluation of medium for enumerating yeasts and moulds in fresh and frozen fruit purees. J. Fd Protect. **48** (1985) 312–315.

[26] BEUCHAT, L.R.; NAIL, B.V.; BRACKETT, R.E.; FOX, T.L.: Evaluation of a culture film (Petrifilm™ YM) method for enumerating yeasts and molds in selected dairy and high-acid foods. J. Fd Protect. **53** (1990) 864, 869–874.

[27] BEUCHAT, L.R.; NAIL, B.V.; BRACKETT, R.E.; FOX, T.L.: Comparison of the Petrifilm™ yeast and mold culture film method to conventional methods for enumerating yeast and molds in foods. J. Fd Protect. **54** (1991) 443–447.

[28] BEUCHAT, L.R.; NAIL, B.V.; BRACKETT, R.E.; FOX, T.L.: Evaluation of the Petrifilm™ yeast and mold plate for enumeration of fungi in foods. In: Modern methods in food mycology (edited by Samson, R.A.; Hocking, A.D.; Pitt, J.I.; King, A.D. Jr). Amsterdam, The Netherlands: Elsevier (1992) 21–25.

[29] BEUCHAT, L.R.; NAIL, B.V.; BRACKETT, R.E.; FOX, T.L.: Comparison of pour and spread plating for enumeration of fungi in foods. In: Modern methods in food mycology (edited by Samson, R.A.; Hocking, A.D.; Pitt, J.I.; King, A.D. Jr). Amsterdam, The Netherlands: Elsevier (1992) 27–30.

[30] BEUCHAT, L.R.; TAPIA DE DAZA, M.S.: Evaluation of chemicals for restricting colony spreading by a xerophilic mold, *Eurotium amstelodami,* on dichloran 18 % glycerol agar. Appl. Environ. Microbiol. **58** (1992) 2093–2095.

[31] BOBEY, D.G.; EDERER, G.M.: Rapid detection of yeast enzymes by using 4-methylumbelliferyl substrates. J. Clin. Microbiol. **13** (1981) 393–394.

[32] BOVILL, R.; BEW, J.; ROBINSON, S.: Comparison of selective media for the recovery and enumeration of probiotic yeasts from animal feed. Int. J. Fd Microbiol. **67** (2001) 55–61.

[33] BOUIX, M.; GRABOWSKI, A.; CARPENTIER, M.; LEVEAU, J.Y.; DUTEURTE, B.: Rapid detection of microbial contamination in grape juice by flow cytometry. J. Int. Sci. Vigne Vin **33** (1999) 31–38.

[34] BRAENDLIN, V.: Enumeration of xerophilic yeasts in the presence of xerophilic molds. Int. J. Fd Microbiol. **29** (1996) 185–192.

[35] BRAGULAT, M.R.; ABARCA, M.L.; CASTELLA, G.; CABAÑES, F.J.: Dyes as fungal inhibitors: effect on colony enumeration. J. Appl. Bacteriol. **79** (1995) 578–582.

[36] BRODSKY, M.H.; ENTIS, P.; ENTIS, M.P.; SHARPE, A.N.; JARVIS, G.A.: Determination of aerobic plate and yeast and mold counts in foods using an automated hydrophobic grid-membrane filter technique. J. Fd Protect. **45** (1982) 301–304.

[37] CAMPBELL, I.: Wild yeast in brewing and distillery. In: Brewing microbiology, 2nd ed (edited by Priest, F.G.; Campbell, I.). London, U.K.: Chapman and Hall (1996) 193–208.

[38] CARREIRA, A.; LOUREIRO, V.: A differential medium to detect *Yarrowia lipolytica* within 24 hours. J. Fd Mycol. **1** (1998) 3–12.

[39] CAVA, R.R.; HERNÁNDEZ, P.S.: Comparison of media for enumerating osmotolerant yeasts in orange juice concentrates. Int. J. Fd Microbiol. **22** (1994) 291–295.

[40] CHASKES, S.; TYNDALL, R.L.: Pigment production by *Cryptococcus neoformans* and other *Cryptococcus* species from aminophenols and diaminobenzenes. J. Clin. Microbiol. **7** (1978) 146–152.

[41] CHILVERS, K.F.; REED, R.H.; PERRY, J.D.: Phototoxicity of rose bengal in mycological media – implications for laboratory practice. Lett. Appl. Microbiol. **28** (1999) 103–107.

[42] CORRY, J.E.L. (editor): Laboratory Quality Assurance and Validation of Methods in Food Microbiology. Int. J. Fd Microbiol. **45** (special issue) (1998) 1–81.

[43] CORRY, J.E.L.; CURTIS, G.D.W.; BAIRD, R.M. (editors): Culture Media for Food Microbiology. Progr. Industr. Microbiol. Vol. 34. Amsterdam, The Netherlands: Elsevier (1995).

[44] CURTIS, G.D.W.; BEUCHAT, L.R.: Quality control of culture media – perspectives and problems. Int. J. Fd Microbiol. **45** (1998) 3–6.

[45] DAVENPORT, R.R.: An outline guide to media and methods for studying yeasts and yeast-like organisms. In: Biology and activities of yeasts (edited by Skinner, F.A.; SM Passmore, S.M.; Davenport, R.R.). London, U.K.: Academic Press (1980) 261–278.

[46] DEAK, T,: Foodborne yeasts. Adv. Appl. Microbiol. **36** (1991) 179–278.

[47] DEAK, T.: Media for enumerating spoilage yeasts – a collaborative study. In: Modern methods in food mycology (edited by Samson, R.A.; Hocking, A.D.; Pitt, J.I.; King, A.D. Jr). Amsterdam, The Netherlands: Elsevier (1992) 31–38.

[48] DEAK, T.: Rapid methods in food mycology. In: Rapid methods and automatization in microbiology and immunology (edited by Spencer, R.C.; Wright, E.P.; Newson, S.W.B). Andover, U.K.: Intercept (1994) 327–345.

[49] DEAK, T.: Methods for the rapid detection and identification of yeasts in foods. Trends Fd Sci. Technol. **6** (1995) 287–292.

[50] DEAK, T.; BEUCHAT, L.R.: Comparison of conductimetric and traditional plating techniques for detecting yeasts in fruit juices. J. Appl. Bacteriol. **75** (1993) 546–550.

[51] DEAK, T.; BEUCHAT, L.R.: Modified indirect conductimetric technique for detecting low populations of yeasts in beverage concentrates and carbonated beverages. Fd Microbiol. **12** (1995) 165–172.

[52] DEAK, T.; BEUCHAT, L.R.: Evaluation of the Microscan enzyme based system for the identification of foodborne yeasts. J. Appl. Microbiol. **79** (1995) 439–446.

[53] DEAK, T.; BEUCHAT, L.R.: Handbook of food spoilage yeasts. Boca Raton, U.S.A.: CRC Press (1996).

[54] DEAK, T.; BEUCHAT, L.R.; GUERZONI, M.E.; LILLIE, A.; PETER, G.; ROHM, H.; SCHNÜRER, F.; TABAJDI, P.V.; WESTPHAL, S.: A collaborative study on media for the enumeration of yeasts in foods. Int. J. Fd Microbiol. **43** (1998) 91–95.

[55] DEAK, T.; CHEN, J.; GOLDEN, D.A.; TAPIA, M.S.; TORNAI-LEHOCZKI, J.; VILJOEN, B.L.; BEUCHAT, L.R.: Comparison of dichloran glycerol (DG18) agar with general purpose mycological media for enumerating food-spoilage yeasts. Int. J. Fd Microbiol. **67** (2001) 49–53.

[56] DEAK, T.; REICHART, O.; SZAKMÁR, K.; PÉTER, G.: Spoilage yeasts of unusual sorbate resistance. In: Modern methods in food mycology (edited by Samson, R.A.; Hocking, A.D.; Pitt, J.I.; King, A.D. Jr). Amsterdam, The Netherlands: Elsevier (1992) 55–59.

[57] DEAK, T.; TÖRÖK, T.; LEHOCZKI, J.; REICHART, O.; TABAJDI-PINTÉR, V.; FÁBRI, I.: Comparison of general purpose media for the enumeration of molds and yeasts. In: Methods for the mycological examination of food (edited by King, A.D. Jr; Pitt, J.I.; Beuchat, L.R.; Corry, J.E.L.). New York, U.S.A.: Plenum (1986) 114–119.

[58] DEAK, T.; TABAJDI-PINTÉR, V.; FÁBRI, I.: Evaluation of blending, stomaching and shaking for mold counts in flour. In: Methods for the mycological examination of food (edited by King, A.D. Jr; Pitt, J.I.; Beuchat, L.R.; Corry, J.E.L.). New York, U.S.A.: Plenum (1986) 10–11.

[59] DE BOER, E.; BEUMER, R.R.: Methodology for detection and typing of foodborne microorganisms. Int. J. Fd Microbiol. **50** (1999) 119–130.

[60] DEERE, D.; SHEN, J.; VESEY, G.; BELL, P.; BISSINGER, P.; VEAL, D.: Flow cytometry and cell sorting for yeast viability assessment and selection. Yeast **14** (1998) 147–160.

[61] DIJKMAN, K.E.: Merits and shortcoming of DG18, OGY, OGGY and DRBC media for the examination of raw meats. In: Methods for the mycological examination of food (edited by King, A.D. Jr; Pitt, J.I.; Beuchat, L.R.; Corry, J.E.L.). New York, U.S.A.: Plenum (1986) 120–123.

[62] DILLON, V.M.; BOARD, R.G.: A medium for detecting sulphite-binding yeasts in meat products. Lett. Appl. Microbiol. **8** (1989) 165–167.

[63] DIRIYE, F.V.; SCORZETTI, G.; MARTINI, A.: Methods for the separation of yeast cells from the surfaces of processed, frozen foods. Int. J. Fd Microbiol. **19** (1993) 27–37.

[64] DULL, C.L.; BEALE, H.; BLINDER, K.; DRESSE, J.; GOMES, H.; GONZALEZ, M.; HAYASHI, N.; HUIJBERTS, G.; JOHNSON, B.; MIYAJIMA, K.; MOTOYAMA, Y.; RODRIGUEZ, G.; SMART, K.; SOBCZAK, J.; SPEERS, A.; THOMPSON, A.; TOMPKINS. T.; VAUGHAN, K.; BENDIAK, D.: CLEN medium for the detection of wild yeast. J. Am. Soc. Brew. Chem. **56** (1998) 202–208.

[65] ERICKSON, J.P.: Hydrophobic membrane filtration method for the selective recovery and differentiation of *Zygosaccharomyces bailii* in acidified ingredients. J. Fd Protect. **56** (1993) 234–238.

[66] VAN ESCH, F.: Yeast in soft drinks and concentrated fruit juices. Quest International June (1990).

[67] FERGUSON, R.B.: Comparison of spread and pour plate methods for enumerating yeasts and molds in dry foods using OGY medium. In: Methods for the mycological examination of food (edited by King, A.D. Jr; Pitt, J.I.; Beuchat, L.R.; Corry, J.E.L.). New York, U.S.A.: Plenum (1986) 53–55.

[68] FIRSTENBERG-EDEN, R.; SHARPE, A.N.: Impact of biotechnology and new methodology on microbial testing of foods. In: Biotechnology and food ingredients (edited by Goldberg, I.; Williams, K.). New York, U.S.A.: Van Nostrand Reinhold (1991) 507–535.

[69] FLEET, G.H.: Food spoilage yeasts. In: Yeast technology (edited by Spencer, J.F.T.; Spencer, D.M.). Berlin, Germany, Springer Verlag (1990) 124–166.

[70] FLEET, G.: Spoilage yeasts. Crit. Rev. Biotechnol. **12** (1992) 1–44.

[71] FLEET, G.H.; MIAN, M.A.: Induction and repair of sublethal injury in food spoilage yeasts. J. Fd Mycol. **1** (1998) 85–93.

[72] FRÄNDBERG, E.; OLSEN, M.: Performance of DG18 media, a collaborative study. J. Fd Mycol. **2** (1999) 239–249.

[73] FRICKER-HIDALGO, H.; ORENGA, S.; LEBEAN, B.; PELLOUX, H.; BREMIER-PRICHART, M.P.; AMBROISE-THOMAS, P.; GRILLOT, R.: Evaluation of Candida ID, a new chromogenic medium for fungal isolation and preliminary identification of some yeast species. J. Clin. Microbiol. **39** (2001) 1647–1649.

[74] FUNG, D.Y.C.: Rapid methods and automation for food microbiology. In: Instrumental methods for quality assurance in foods (edited by Fung, D.Y.C.; Matthews, R.F.). New York, U.S.A.: Marcel Dekker (1991) 1–38.

[75] FUNG, D.Y.C.; DEAK, T.; VASAVADA, P.C.; PETER, G.: Evaluation of *Candida* isolation medium in food hygiene in Hungary. J. Rapid Methods **4** (1996) 279–284.

[76] FUNG, D.Y.C.; LIANG, C.: Critical review of isolation, detection, and identification of yeasts from meat products. Crit. Revs. Fd Sci. Nutr. **29** (1990) 341–379.

[77] GOLDEN, D.A.; BEUCHAT, L.R.: Colony formation by sublethally heat-injured *Zygosaccharomyces rouxii* as affected by solutes in the recovery medium and procedure for sterilizing medium. Appl. Environ. Microbiol. **56** (1990) 2319–2326.

[78] GOLDSCHMIDT, M.C.; FUNG, D.Y.C.; GRANT, R.; WHITE, J.; BROWN, I.: New aniline blue dye medium for rapid identification and isolation of *Candida albicans*. J. Clin. Microbiol. **29** (1991) 1095–1099.

[79] GOMES, L.H.; DUARTE, K.M.R.; ARGUESO, J.L.; ECHEVERRIGARAY, S.; TAVARES, F.C.A.: Methods for yeast characterization from industrial products. Fd Microbiol. **17** (2000) 217–223.

[80] HASTINGS, J.W.; TSAI, W.Y.J.; BULLERMAN, L.B.: Comparison of stomaching versus blending in sample preparation for mold enumeration. In: Methods for the mycological examination of food (edited by King, A.D. Jr; Pitt, J.I.; Beuchat, L.R.; Corry, J.E.L) New York, U.S.A.: Plenum (1986) 7–9.

[81] HAWRONSKYJ J.-M.; HOLAH, J.: ATP: a universal hygiene monitor. Trends Food Sci. Technol. **8** (1997) 79–84.

[82] HENSON, O.E.: Dichloran as an inhibitor of mold spreading in fungal plating media: effects on colony diameter and enumeration. Appl. Environ. Microbiol. **42** (1981) 656–660.

[83] HEARD, G.M.; FLEET, G.H.: Evaluation of selective media for enumeration of yeasts during wine fermentation. J. Appl. Bacteriol. **60** (1986) 477–481.

[84] HERNANDEZ, P.; BEUCHAT, L.R.: Evaluation of diluents and media for enumerating *Zygosaccharomyces rouxii* in blueberry syrup. Int. J. Fd Microbiol. **25** (1995) 11–18.

[85] HOCKING, A.D.: Media for preservative resistant yeasts: a collaborative study. Int. J. Fd Microbiol. **29** (1996) 167–175.

[86] HOCKING, A.D.; PITT, J.I.: Dichloran-glycerol medium for enumeration of xerophilic fungi from low-moisture foods. Appl. Environ. Microbiol. **39** (1980) 488–492.

[87] HOCKING, A.D.; PITT, J.I.: Introduction and summary of the first international workshop on standardisation of methods for the mycological examination of foods. In: Modern methods in food mycology (edited by Samson, R.A.; Hocking, A.D.; Pitt, J.I.; King, A.D.). Amsterdam, The Netherlands: Elsevier (1992) 3–7.

[88] HOCKING, A.D.; PITT, J.I.; SAMSON, R.A.; KING, A.D.: Recommendations from the closing session of SMMEF II. In: Modern methods in food mycology (edited by Samson, R.A.; Hocking, A.D.; Pitt, J.I.; King, A.D.). Amsterdam, The Netherlands: Elsevier (1992) 359–368.

[89] ISO TC 34/SC 9N 7954. General guidance for enumeration of yeasts and molds: colony count technique at 25 °C. ISO: International Organization for Standardisation (1987).

[90] JARVIS, B.: Comparison of an improved rose bengal chlortetracycline agar with other media for the selective isolation and enumeration of moulds and yeasts in foods. J. Appl. Bacteriol. **36** (1973) 723–727. 1973.

[91] JARVIS, B.: Methods for detecting fungi in foods and beverages. In: Food and beverage mycology (edited by Beuchat, L.R..). Westport, CT, U.S.A.: AVI Publ. Inc. (1978) 471–504.

[92] JARVIS, B.; WILLIAMS, A.P.: Methods for detecting fungi in foods and beverages. In: Food and Beverage Mycology (edited by Beuchat, L.R.) 2nd ed. New York, U.S.A.: Van Rostrand (1987) 599–636.

[93] JESPERSEN, L.; JAKOBSEN, M.: Specific spoilage organisms in breweries and laboratory media for their detection. Int. J. Fd Microbiol. **33** (1996) 139–155.

[94] JESPERSEN, L.; LASSEN, S.; JAKOBSEN, M.: Flow cytometric detection of wild yeast in lager breweries. Int. J. Fd Microbiol. **17** (1993) 321–328.

[95] JIRANEK, V.; LANGRIDGE, P.; HENSCHKE, A.: Validation of bismuth-containing indicator media for predicting H_2S-producing potential of *Saccharomyces cerevisiae* wine yeasts under enological conditions. Am. J. Enol. Vitic. **46** (1995) 269–273.

[96] JOOSTEN, H.M.L.J.; VAN DICK, W.G.F.M.; SPIKKER, J.W.M.; TRAA, H.M.: Detection of yeasts in quark by flow cytometry. Milchwiss. **51** (1996) 202–204.

[97] KENNEDY, J.E.; PHILIPS, J.R.; OBLINGER, J.R.: Effect of stored pre-pour plates on microbial enumeration using the surface plate method. J. Fd Protect. **43** (1980) 592–594.

[98] KING, A.D. JR: Comparison of fungal counts on foods prepared by blending or stomaching. In: Methods for the mycological examination of food (edited by King, A.D. Jr; Pitt, J.I.; Beuchat, L.R.; Corry, J.E.L.). New York, U.S.A.: Plenum (1986) 11–12.

[99] KING, A.D.: Methodology for routine mycological examination of food – a collaborative study. In: Modern methods in food mycology (edited by Samson, R.A.; Hocking, A.D.; Pitt, J.I.; King, A.D.). Amsterdam, The Netherlands: Elsevier (1992) 11–20.

[100] KING, A.D. JR; HOCKING, A.D.; PITT, J.I.: Dichloran-rose bengal medium for enumeration and isolation of molds from foods. Appl. Environ. Microbiol. **37** (1979) 959–964.

[101] KING, A.D. JR; PITT, J.I.; BEUCHAT, L.R.; CORRY, J.E.L. (editors): Methods for the mycological examination of food. New York, U.S.A.: Plenum (1986).

[102] KISH, S.; SHARF, R.; MARGALITH, P.: A note on a selective medium for wine yeasts. J. Appl. Bacteriol. **55** (1983) 177–179.

[103] KOBURGER, J.A.; NORDEN, A.R.: Fungi in foods VII. A comparison of the surface, pour plate and most probable number methods for the enumeration of yeasts and molds. J. Milk Fd Technol. **38** (1975) 745–756.

[104] KOBURGER, J.A.; RODGERS, M.F.: Single multiple antibiotic amended media to enumerate yeasts and molds. J. Fd Protect **41** (1978) 367–369.

[105] KWON-CHUNG, L.J.; POLACHECK, I.; BENNETT, J.E.: An improved diagnostic medium for separation of *Cryptococcus neoformans* var. *neoformans* (serotypes A and D) and *Cryptococcus neoformans* var. *gattii* (serotypes B and C). J. Clin. Microbiol. **15** (1982) 535–537.

[106] LIN, C.C.S.; FUNG, D.Y.C.: Effect of dyes on the growth of food yeast. J. Fd Sci. **50** (1985) 241–244.

[107] LOUREIRO, V.: Spoilage yeasts in foods and beverages: characterization and ecology for improved diagnosis and control. Fd Res. Internat. **33** (2000) 247–256.

[108] MAKDESI, A.K.; BEUCHAT, L.R.: Performance of selective media for enumerating *Zygosaccharomyces bailii* in acidic foods and beverages. J. Fd Sci. **59** (1996): 652–656.

[109] MAKDESI, A.K.; BEUCHAT, L.R.: Improved selective medium for enumeration of benzoate-resistant, heat-stressed *Zygosaccharomyces bailii*. Fd Microbiol. **13** (1996) 281–290.

[110] MANAFI, M.: New developments in chromogenic and fluorogenic media. Int. J. Fd Microbiol. **60** (2000) 205–218.

[111] MARTINI, A.; CIANI, M.; SCORZETTI, G.: Direct enumeration and isolation of wine yeasts from grape surfaces. Am. J. Enol. Vitic. **47** (1996) 435–440.

[112] MARTINI, A.; FREDERICHI, F.; ROSINI, G.: A new approach to the study of yeast ecology of natural substrates. Can. J. Microbiol. **26** (1980) 856–859.

[113] MIAN, M.A.; FLEET, G.H.; HOCKING, A.D.: Effect of diluent type on viability of yeasts enumerated from foods on pure culture. Int. J. Fd Microbiol. **35** (1997) 103–107.

[114] MIDDELHOVEN, W.J.; NOTERMANS, S.: Immuno-assay techniques for detecting yeasts in foods. Int. J. Fd Microbiol. **19** (1993) 53–62.

[115] MILLARD, P.J.; ROTH, B.L.; TRUONG THI, H.-P.; YUE, S.T.; HAUGLAND, R.P.: Development of the FUN-1 family of fluorescent probes for vacuole labeling and viability testing of yeasts. Appl. Environ. Microbiol. **63** (1997) 2897–2905.

[116] MOSSEL, D.A.A.; VEGA, C.L.; PUT, H.M.C.: Further studies on the suitability of various media containing antibacterial antibiotics for the enumeration of moulds in food and food environments. J. Appl. Bacteriol. **39** (1975) 15–22.

[117] MOSSEL, D.A.A.; DIJKMANN, K.E.; KOOPMANS, M.: Experiments with methods for enumeration and identification of yeasts occurring in foods. In: Biology and activities of yeasts (edited by Skinner, F.A.S.; Passmore, S.; Davenport, R.R.). London, U.K.: Academic Press (1980) 279–288.

[118] MOSSEL, D.A.A.; KLEYNEN-SEMMELING, A.M.C.; VINCENTIE, H.M.; BEERENS, H.; CATSARAS, M.: Oxytetracycline-glucose-yeast extract agar for selective enumeration of moulds and yeasts in foods and clinical material. J. Appl. Bacteriol. **33** (1970) 454–457.

[119] NGUYEN, H.V.; PULVIRENTI, A.; GAILLARDIN, C.: Rapid differentiation of the closely related *Kluyveromyces lactis* var. *lactis* and *Kluyveromyces marxianus* strains isolated from dairy products using selective media and PCR/RFLP of the rDNA non transcribed spacer 2. Can. J. Microbiol. **46** (2000) 1115–1122.

[120] NÚÑEZ, F.; RODRÍGUEZ, M.M.; CÓRDOBA, J.J.; BERMÚDEZ, M.E.; ASENSIÓ, M.A.: Yeast population during ripening of dry-cured Iberian ham. Int. J. Fd Microbiol. **29** (1996) 271–280.

[121] ODDS, F.; BERNAERTS, R.: CHROMagar Candida, a new differential isolation medium for the presumptive identification of clinically important *Candida* species. J. Clin. Microbiol. **32** (1994) 1923–1929.

[122] OLSEN, M.; ANDERSSON, K.; ÅKERSTRAND, K.: Quality control of two rose bengal and modified DRBC and DG18 media. Int. J. Fd Microbiol. **35** (1997) 163–168.

[123] PETTIPHER, G.L.; WATTS, Y.B.; LANGFORD, S.A.; KROLL, R.G.: Preliminary evaluation of COBRA, an automated DEFT instrument, for the rapid enumeration of micro-organisms in cultures, raw milk, meat and fish. Lett. Appl. Microbiol. **14** (1992) 206–209.

[124] PFALLER, M.A.; HOUSTON, A.; COFFMANN, S.: Application of CHROMagar Candida for rapid screening of clinical specimens for *Candida albicans, Candida tropicalis, Candida krusei* and *Candida (Torulopsis) glabrata.* J. Clin. Microbiol. **34** (1996) 58–61.

[125] PITT, J.I.: Properties of the ideal fungal enumeration medium. In: Methods for the mycological examination of food (King, A.D. Jr: Pitt, J.I.; Beuchat, L.R.; Corry, J.E.L.). New York, U.S.A.: Plenum (1986) 63–65.

[126] PITT, J.I.: Collaborative studies on methods in food mycology. Int. J. Fd Microbiol. **29** (1996) 137–139.

[127] PITT, J.I.: Contributions to methods in food mycology. Int. J. Fd Microbiol. **35** (1997) 99–101.

[128] PITT, J.I.; HOCKING, A.D.: Fungi and Food Spoilage, 2nd ed. London, U.K.: Blackie Academic (1997).

[129] QUEROL, A.; RAMON, P.: The application of molecular techniques in wine microbiology. Trends Fd Sci. Technol. **7** (1996) 73–78.

[130] QUINDÓS, G.; FERNÁNDES-RODRIGUEZ, M.; BURGOS, A.; TELLAETEXE, M.; CISTERNA, R.; PONTÓN, J.: Colony morphotype on Sabouraud-triphenyltetrazolium agar: a simple and inexpensive method for *Candida* subspecies discrimination. J. Clin. Microbiol. **30** (1992) 2748–2752.

[131] RALE, V.B.; VAKIL, J.R.: A note on an improved molybdate agar for the selective isolation of yeasts from tropical fruits. J. Appl. Bacteriol. **56** (1984) 409–413.

[132] RODRIGUES, N.; GONCALVES, G.; PEREIRE-DA-SILVA, S.; MALFEITO-FERREIRA, M.; LOUREIRO, V.: Development and use of a new medium to detect yeasts of the genera *Dekkera/Brettanomyces.* J. Appl. Microbiol. **90** (2000) 588–599.

[133] ROSINI, G.; FREDERICHI, F.; MARTINI, A.: Yeast flora of grape berries during ripening. Microb. Ecol. **8** (1982) 83–89.

[134] ROUSELLE, P.; FREYDIERE, A.M.; COUILLEROT, P.J.; DE MONTELOS, H.; GILLE, Y.: Rapid identification of *Candida albicans* by using Albicans ID and fluoroplate agar plates. J. Clin. Microbiol. **32** (1994) 3034–3036.

[135] SAMSON, R.A.; HOCKING, A.D.; PITT, J.I.; KING, A.D. (editors): Modern methods in food mycology. Amsterdam, The Netherlands: Elsevier (1992).

[136] SCHULLER, D.; CORTE-REAL, M.; LEAO, C.: A differential medium for the enumeration of the spoilage yeast *Zygosaccharomyces bailii*. J. Fd Protect. **63** (2000) 1570–1575.

[137] SEILER, D.A.L.: Monitoring mycological media. In: Modern methods in food mycology (edited by Samson, R.A.; Hocking, A.D.; Pitt, J.I.; King, A.D.). Amsterdam, The Netherlands: Elsevier (1992) 73–77.

[138] SEILER, D.A.L.: Report on a collaborative study on the effect of presoaking and mixing time on the recovery of fungi from foods. In: Modern methods in food mycology (edited by Samson, R.A.; Hocking, A.D.; Pitt, J.I.; King, A.D.). Amsterdam, The Netherlands: Elsevier (1992) 79–88.

[139] SEILER, D.A.L.: Effect of diluent and medium water activity on recovery of yeasts from hugh sugar coatings and fillings. In: Methods for the mycological examination of food (edited by King, A.D. Jr; Pitt, J.I.; Beuchat, L.R.; Corry, J.E.L.). New York, U.S.A.: Plenum (1986) 162–163.

[140] SEILER, H.; WENDT, A.: Methodenvergleich für Hefennachweis in Fruchtzubereitungen. Dtsch. Milk Z. **44** (1992) 1348–1360.

[141] SILLEY, P.; FORSITHE, S.: Impedance microbiology – a rapid change for microbiologists. J. Appl. Bacteriol. **80** (1996) 233–243.

[142] SILONIZ, M.I. DE; VALDERRAMA, M.J.; PEINADO, J.M.: Advances in methodology to isolate and identify osmotolerant yeasts in food. Fd Technol. Biotechnol. **37** (1999) 277–280.

[143] SILONIZ, M.I. DE; VALDERRAMA, M.J.; PEINADO, J.M.: A chromogenic medium for the detection of yeasts with β-galactosidase and β-glucosidase activities from intermediate moisture foods. J. Fd Protect. **63** (2000) 651–654.

[144] SKAAR, I.; STENWIG, H.: Malt-yeast extract-sucrose agar, a suitable medium for enumeration and isolation of fungi from silage. Appl. Environ. Microbiol. **62** (1996) 3614–3619.

[145] SMART, K.A.; CHAMBERS, K.M.; LAMBERT, I.; JENKINS, C.: Use of methylene violet staining procedures to determine yeast viability and vitality. J. Am. Soc. Brew. Chem. **57** (1999) 18–23.

[146] SMOLE MOZINA, S.; RASPOR, P.: Molecular techniques for yeast identification in food processing. Fd Technol. Biotechnol. **35** (1997) 55–61.

[147] SORENSEN, B.B.; JAKOBSEN, M.: The combined effects of temperature, pH and NaCl on growth of *Debaryomyces hansenii* analyzed by flow cytometry and predictive microbiology. Int. J. Fd Microbiol. **34** (1997) 209–220.

[148] STEVENSON, K.A.; GRAUMLICH, T.R.: Injury and recovery of yeasts and moulds. Adv. Appl. Microbiol. **23** (1978) 203–217.

[149] THOMSON, G.F.: Enumeration of yeasts and molds: media trial. Fd Microbiol. **1** (1984) 223–227.

[150] TOKOUKA, K.: Sugar- and salt-tolerant yeasts. J. Appl. Bacteriol. **74** (1993) 101–110.

[151] Tornai-Lehoczki, J.; Peter, G.: Typing of foodborne, food spoilage and oportunistic pathogenic yeasts on CHROMagar Candida medium. In: The rising Power of yeasts in science and industry (edited by van Dijken, J.P.; Scheffers, W.A.). Delft, The Netherlands: Delft Univ. Press (2000) 404.

[152] Valderrama, M.J.; de Siloniz, M.I.; Gonzalo, P.; Peinado, J.M.: A differential medium for the isolation of *Kluyveromyces marxianus* and *Kluyveromyces lactis* from dairy products. J. Fd Prot. **62** (1999) 189–193.

[153] Van der Aa Kühle, A.; Jespersen, L.: Detection and identification of wild yeasts in lager breweries. Int. J. Fd Microbiol. **43** (1998) 205–213.

[154] van der Vossen, J.M.B.M.; Hofstra, H.: DNA based typing, identification and detection systems for food spoilage microorganisms. Development and implementation. Int. J. Fd Microbiol. **33** (1996) 35–49.

[155] Viljoen, B.C.; Lues, J.F.R.: The microbial populations associated with post-fermented dough and compressed baker's yeast. Fd Microbiol. **10** (1993) 379–386.

[156] Vlaemynck, G.M.: Comparison of Petrifilm™ and plate count methods for enumerating molds and yeasts in cheese and yoghurt. J. Fd Protect. **57** (1994) 913–914.

[157] Welthagen, J.J.; Viljoen, B.C.: Comparison of ten media for the enumeration of yeasts in dairy products. Fd Res. Internat. **30** (1997) 207–211.

[158] Wenk, G.H.: Microbiological assesment of culture media: comparison and statistical evaluation of methods. Int. J. Fd Microbiol. **17** (1992) 159–181.

[159] Williams, A.P.: A comparison of DRBC, RBC and MEA media for enumeration of molds and yeasts in pure culture and in foods. In: Methods for the mycological examination of food (King, A.D. Jr; Pitt, J.I.; Beuchat, L.R.; Corry, J.E.L.). New York, U.S.A.: Plenum (1986) 85–89.

[160] Willinger, B.; Manafi, M.; Rotter, M.L.: Comparison of rapid methods using fluorogenic-chromogenic assays for detecting *Candida albicans*. Lett. Appl. Microbiol. **18** (1994) 47–49.

3 Methods to identify yeasts

Cletus P. Kurtzman, Teun Boekhout, Vincent Robert, Jack W. Fell and Tibor Deak

3.1 Introduction

Yeasts are commonly identified from either phenotype or, more recently, from diagnostic gene sequences [13]. Methods based on phenotype include fermentation reactions on a select set of sugars and growth responses on various carbon and nitrogen sources or on other diagnostic compounds. Isolates are further characterized phenotypically from the microscopic appearance of vegetative cells as well as sexual states because this information often gives clues to the identity of a strain. Molecular methods for yeast identification include nuclear DNA reassociation as well as the more recent and faster procedure of sequencing species-specific genes or the use of oligonucleotide sequences that react to a species-specific region of genomic DNA. These so-called "molecular probes" may have a marker molecule attached or they may be used in pairs, that when employed in a PCR reaction, amplify a species-specific region of DNA. The DNA-based identifications are far more reliable than those from phenotypic tests, and much faster. Nonetheless, in the absence of facilities for molecular comparisons, satisfactory identifications can often be made from phenotype. For this reason, both phenotype-based methods, as well as molecular techniques are given in this chapter.

3.2 Identification from phenotype – fermentation and growth tests

3.2.1 Fermentation of sugars

Anaerobic utilization of sugars is generally tested by measuring the amount of CO_2 that is trapped in Durham tubes. This method is not very accurate for detecting fermentation by slowly fermenting yeast species because the CO_2 may not evolve rapidly enough to be collected as a gas bubble. Because of this, many yeast species considered to be non-fermentative proved capable of producing ethanol [27]. However, for identification purposes, the use of Durham tubes is recommended because of easy preparation and scoring. An initial fermentation test using only glucose is often done. If no fermentation is detected, other sugars need not be tested. The sugars tested for identification purposes usually include D-glucose, D-galactose, maltose, sucrose, lactose, raffinose, trehalose, and D-xylose.

Sugars are tested at a concentration of 2 % (w/v) (or 4 % for raffinose) in a basal medium of 2 % (w/v) yeast extract, yeast infusion, or 0.3 % yeast extract – 0.5 % peptone [99]. Depending on size of the test tubes used, the volume of medium ranges from 3.0 to 4.5 ml. Each tube contains a small inverted insert tube to collect the CO_2 that may be formed. Sugars are filter-sterilized at a 3X concentration and added aseptically to the autoclaved basal medium concentrate which often contains bromthymol blue as a pH indicator. Insert tubes fill with basal medium during autoclaving. Following addition of the sugar solutions, and again after inoculation, tubes are shaken side-ways by hand to facilitate mixing. Fermentation tubes are inoculated with 0.1–0.2 ml of a heavy aqueous cell suspension prepared from an actively growing slant culture. Tubes are incubated at 25–28 °C and observed every few days for up to four weeks. Psychrophilic species require 12 or 15 °C. The results are scored as follows, depending on the time taken to fill the insert with gas and the amount accumulating:

+ strongly positive, insert filled 70–100 % within 7 days
D delayed positive, insert filled 70–100 % only after more than 7 days
W weakly positive, the insert is less than 30 % filled
– negative, no accumulation of gas in the insert

3.2.2 Growth on carbon compounds

The ability of a yeast to grow aerobically on a particular carbon compound supplied as the sole source of energy (often called assimilation tests) is tested in either liquid media or by the auxanographic method using agar media [1, 6–9, 99–101, 104]. Auxanograms can be read after two to three days, whereas test tubes of liquid media need up to four weeks. However, the latter method is more sensitive. For identification purposes, growth reactions on the following compounds are usually determined:

Hexoses (D-glucose, D-galactose, L-rhamnose, L-sorbose)
Pentoses (D-xylose, D-ribose, L-arabinose, D-arabinose)
Disaccharides (sucrose, maltose, cellobiose, trehalose, lactose, melibiose)
Trisaccharides (raffinose, melezitose)
Polysaccharides (soluble starch, inulin)
Alcohols (erythritol, ribitol (adonitol), D-mannitol, m-inositol, methanol, ethanol, glycerol, glycol, propane 1,2 diol, butane 2,3 diol, galactitol (dulcitol), D-glucitol (sorbitol))
Organic acids (succinate, citrate, DL-lactate, D-gluconate, D-glucuronate, D-galacturonate, 2-keto-D-gluconate, 5-keto-D-gluconate)
Glycosides (α-methyl-D-glucoside, arbutin, salicin)
Other compounds (glucono-δ-lactone, D-glucosamine·HCl, N-acetyl D-glucosamine, decane, hexadecane).

3.2.2.1 Carbon assimilation by auxanogram

Nitrogen-containing agar: 0.5 % w/v $(NH_4)_2SO_4$, 0.1 % KH_2PO_4, 0.05 % $MgSO_4$, 2 % agar (Difco), 0.1 % v/v vitamin solution [99] in distilled water. Commercially available Yeast Nitrogen Base (Difco) with agar is a convenient alternative.

An inoculum of the strain to be tested is suspended in 5 ml sterile distilled water, poured into a petri dish (diam. ca. 9–11 cm), and thoroughly mixed with 10 ml of cooled, molten nitrogen-containing agar. After solidification, crystals of the compounds to be tested are placed on the surface of the agar. Glucose is used as a positive control. Growth is examined after 2–3 days.

3.2.2.2 Assimilation of carbon compounds in liquid medium

Stock 10X solutions of Yeast Nitrogen Base, each with the carbon source to be tested, are sterilized by filtration. Final concentrations of the carbon sources is 0.5 %, except for raffinose which is 1 %. The pH of the 10X stock needs to be readjusted to 5.6 for those carbon compounds that are not neutral, such as organic acids. Water blanks consisting of test tubes with 4.5 ml of distilled water are autoclaved and each tube receives 0.5 ml of the appropriate 10X stock solution.

Soluble starch and inulin are exceptions to this treatment. Each compound is made as a 0.5 % aqueous solution. For soluble starch, 4.5 ml amounts are placed in test tubes and autoclaved for 15 minutes. Because inulin is easily degraded by heat, the solution is filter-sterilized and 4.5 ml is added to sterile test tubes. The soluble starch and inulin tubes each receive 0.5 ml of 10X Yeast Nitrogen Base. Assimilation tubes are inoculated with an aqueous suspension of starved yeast cells as described by WICKERHAM [99]. Turbidity, used as a measurement of growth, is scored after 7, 14 and 21 or 28 days at 25 °C. Growth is faster and more consistent if tubes are shaken during incubation on a rocking shaker (30 x min^{-1}), a reciprocal shaker, or a rotary shaker. Psychrophilic species require lower temperatures, e. g., 12 or 15 °C.

At the CBS Yeast Division the following procedure is practiced. The tubes of test substrates are prepared by dispensing 4.5 ml amounts of a 0.5 % solution of the substrate in deionized water and autoclaved at 110°C for 15 min. D-galacturonic acid is autoclaved for only 10 min. The pH of citric acid, DL-lactic acid, succinic acid, D-gluconic acid, and hemi-saccharate is adjusted to 5.2 with NaOH. The inoculum for the tests is prepared by taking material aseptically from a 24 to 48-hours-old culture and dispersing it in about 3 ml of sterile Yeast Nitrogen Base, taking care to avoid carrying over any of the nutrient medium. Turbidity is assessed by holding a white card with black lines (ca. 0.75 mm wide) behind the tube. The suspension is diluted aseptically until the lines become visible through the tube as dark bands (this is the same as a 2+ reading as in evaluating growth tests) and 2.5 ml of the suspension is added to 25 ml of yeast nitrogen base (10X concentrate). Each tube of test medium is inoculated with 0.5 ml of the final suspension. The tubes are rocked during incubation through as wide an angle as possible without wetting the plugs. Results are read after 1 and 3 weeks. The tubes are shaken to disperse the yeast and the degree of growth is assessed by eye by placing the tubes against a white card with black lines (ca. 0.75 mm thick and 5 mm apart). Using Wickerham's scale, the result is scored as 3+ if the lines are completely obscured, as 2+ if the lines appear as diffuse bands, as 1+ if the lines are slightly blurred, or as negative if the lines are distinct and sharply edged. The test is repeated where the outcome is doubtful.

The results are presented in the descriptions as follows:

+ positive, either a 2+ or a 3+ reading after 1 week
D positive delayed, either a 2+ or 3+ reading after more than 1 week
W weakly positive, a 1+ reading
– negative

3.2.3 Growth on nitrogen compounds

3.2.3.1 Nitrogen assimilation by auxanogram

The following compounds are tested for routine identifications: nitrate, ethylamine, L-lysine and cadaverine. Additional compounds are nitrite, creatine, creatinine and imidazole.

An inoculum of the strain of interest is made in 5 ml sterile distilled water, poured into a petri dish (diam. ca. 9–11 cm), and thoroughly mixed with ca. 10 ml molten, cooled nitrogen-free agar (2 % w/v glucose, 0.1 % w/v KH_2PO_4, 0.05 % w/v $MgSO_4 \cdot 7H_2O$, 2 % w/v agar (Difco), 0.1 % v/v vitamin solution in distilled water [99]). Commercially available Yeast Carbon Base (Difco) with agar is a convenient alternative. After solidification, crystals of the compounds to be tested are placed on the surface of the agar. To avoid over-dosing with nitrite and ethylamine, an inoculating needle is dipped into saturated solutions of the salts and then used as an inoculum. Peptone or $(NH_4)_2SO_4$ are used as positive controls. Growth is recorded after 2–3 days.

3.2.3.2 Assimilation of nitrogen compounds in liquid medium

Testing for assimilation of nitrogen compounds in liquid medium is as described by WICKERHAM [99]. A stock 10X medium is prepared by dissolving 11.7 g of Yeast Carbon Base, together with the required amount of each nitrogen source to be tested, in 100 ml of distilled water. Each of the 10X stocks are filter sterilized. The following amounts of nitrogen sources are used: 0.78 g of potassium nitrate, 0.26 g of sodium nitrite, 0.64 g of ethylamine hydrochloride, 0.68 g of cadaverine dihydrochloride, and 0.56 g of L-lysine. As with the carbon assimilation tests, 0.5 ml of the 10X concentrate is added to 4.5 ml of sterile distilled water. Tubes for negative controls are prepared by adding 0.5 ml of 10X Yeast Carbon Base to 4.5 ml of sterile distilled water. Cells used for inoculum often carry over enough nitrogen to give a weak reaction in nitrogen-free media. Consequently, duplicate tubes must be prepared for nitrogen assimilation tests. One week after inoculation, a small loopful of growth from the first tube is used to inoculate the second tube. Growth in the second tube is used to judge whether a particular nitrogen compound is utilized.

3.2.4 Vitamin requirements

Vitamin requirements are tested by slightly inoculating a test tube containing 5 ml vitamin-free medium (see 3.9–64). Most yeasts grow considerably owing to reserves being carried over within the cells. A second tube of vitamin-free medium, and a complete set of tubes lacking individual vitamins when individual requirements are tested, are inoculated with 1 drop of cell suspension from the first tube. The cultures are incubated with rocking and inspected after 3 and 7 days for growth.

3.2.5 Resistance to cycloheximide

0.5 ml of a filter-sterilized solution containing 0.1 % or 1.0 % cycloheximide in 10X concentrated Yeast Nitrogen Base is added to 4.5 ml 0.5 % w/v glucose in distilled water, giving final concentrations of 0.01 % (100 ppm) or 0.1 % (1000 ppm) cycloheximide, respectively. After inoculation with the strain of interest, growth is scored.

3.2.6 Growth in media at high osmotic pressure

The ability to grow with high concentrations of sugar and salt is tested by growth on agar media containing either 50 % or 60 % (w/w) glucose and in liquid media containing 5 % glucose and either 10 % or 16 % sodium chloride, and in agar media containing 16 % NaCl and 5 % glucose. The tubes are inspected for growth after 7 and 21 days. Growth after 7 days is scored + and growth after 21 days is delayed (D).

3.2.7 Production of acetic acid

A small amount of inoculum is streaked onto slants or plates of YPGA with 0.5 % w/v $CaCO_3$. Production of acetic acid results in a clear zone around the culture.

3.2.8 Urease activity

0.5 ml of Difco Bacto Urea R broth is dispensed into tubes and stored in a freezer. After inoculation, tubes are placed at 37 °C, or the maximum growth temperature for the strain if lower than 37 °C. Change of color of the suspension to deep red after ca. 4–20 hours indicates urease activity.

3.2.9 Extracellular starch production

0.33 % iodine and 0.66 % potassium iodide are dissolved in distilled water (Lugol's solution). A drop of reagent is added to the culture in an assimilation tube containing D-glucose.

Dark blue coloration indicates the presence of extracellular starch-like compounds. It is often helpful to have duplicate glucose assimilation tubes in order to test for starch-like compounds at one and two weeks after inoculation. In some doubtful cases the tests are repeated in Yeast Nitrogen Base medium with 3 % glucose.

3.2.10 Growth at various temperatures

A slant of glucose-peptone-yeast extract agar is inoculated with cells from a young culture and incubated at the chosen temperature for 4 days and then inspected for growth. A longer period of incubation may be required for psychophilic species.

3.2.11 Growth with 1 % acetic acid

A 4-mm loop of the same cell suspension used to inoculate the growth tests is either spotted or streaked onto an agar plate with 1 % acetic acid (several strains can be tested in one petri dish). The plates are incubated and inspected after 6 days for the development of colonies.

3.2.12 Diazonium Blue B reaction

This test is used to differentiate between ascomycetous and basidiomycetous yeasts. The latter organisms show a color reaction in this test. Directly before use, 0.1 % w/v Diazonium Blue B (DBB) salt is dissolved in ice-cold 0.1 M Tris-HCl, pH 7.0. A drop of the freshly prepared ice-cold DBB reagent is placed on cultures earlier inoculated onto YMA plates or slants for periods of ten days to three weeks. Immediate change of colour of the colony to dark red indicates a positive DBB reaction.

3.2.13 Physiological testing using microplate technology

The use of physiological tests in tubes or on plates for the identification of yeasts is time-consuming, labor intensive and requires experienced and highly skilled operators. An alternative to the traditional techniques is the use of microplate technology [75]. Reading of the microplates can be automated and linked to identification software (see Chapter 5). Details for preparation, inoculation, incubation and reading of the microplates as used at the Centraalbureau voor Schimmelculture (CBS) are provided below. The microplates can be obtained from the CBS. Other companies provide microplates (e. g., API, Biolog) for testing physiological properties of yeasts and more information about these alternative systems is given in chapter 5.

3.2.13.1 Preparation of microplates

Microplate wells for assimilation and other growth tests (Nunc, 96 wells, flat bottom) are filled with 100 µl of the media described in section 3.9-42. Microplates are sealed by heat using polypropylene-aluminum sealing foil, and can be stored at –18 °C or lower temperatures for a year or longer. The media are sterilized by heating or filtration prior to their addition to sterile microplates. Alternatively, filled and sealed microplates can be sterilized by gamma irradiation at 4 KGRay. This latter option should be favoured if a gamma radiation facility is available

3.2.13.2 Inoculation and incubation of microplates

Fifty µl of inoculum (MacFarland standard # 2, diluted by a factor of 10) is introduced into each well using a multi-channel (8 or 12 channels) pipette. A loose cellophane cover (do not seal since air should be able to circulate) is placed on the microplate to avoid desiccation of the wells. Replace the cover of the microplate. The microplate is incubated at 25 °C (or other temperature if required) for 3 to 10 days. Agitation of the microplates during incubation is not required.

3.2.13.3 Test reading

Microplates are shaken with a microplate shaker just before automatic reading with a microplate reader. Absorbance values at 405 nm are transferred by cable (RS-232 through a serial port) to the computer and transformed by the BioloMICS software [74] into negative, weak or positive growth reactions (see Chapter 5 for more details). The results of every test are transformed independently.

3.3 Appearance of colonies, cell shape and filamentation

The colour and texture of colonies is recorded after 7 days on GPYA agar.

Cultures are examined microscopically for budding or fission, and for the shape of the cells after being grown for 24–72 hours. Cells may be spherical to elongate in shape and budding may be multilateral, bipolar or polar [39, 51, 70, 87]. The latter is characteristic of some basidiomycetes in which the narrow-based buds form only at the poles (ends) of cells, in contrast to multilateral budding, which occurs essentially all over the cell surface. The bipolar budding of certain ascomycetes is characterized by broad-based budding at cell poles. Filaments are detected by making a single streak on a slide culture with Wickerham's morphology agar and inspecting the culture under low magnification after 10 and 20 days of growth at room temperature. Filaments may be comprised of pseudohyphae, true hyphae, or a mixture of the two types [104].

3.4 Sexual states and mating tests

The initial taxonomic division of yeasts is based on whether they are ascomycetes or basidiomycetes. In the past, finding the sexual state or teleomorph was the only way this decision could be made. Now, it is also possible to decide the taxonomic class from the diazonium blue B test, from ultrastructure of cell walls and from the distinctive nucleotide sequences of ribosomal DNA and other genes. Nonetheless, placement in teleomorphic genera is often aided by the appearance of the sexual state, and for this reason, as well as for others, it is important to examine strains for formation of sexual spores. Examples of the various sexual states are presented by YARROW [104] as well as by many of the other authors of *The Yeasts, A Taxonomic Study*, 4th edition [55].

3.4.1 Ascomycetes

Strains can be grown for two days on YMA and paired by mixing a loopful of growth from each strain on fresh YMA [102, 104]. Alternatively, up to 8 strains may be mixed in one test with the strategy that if mating is detected, strains will then be paired in all combinations. Mixtures should be examined under the microscope at daily intervals for at least a week, although mating usually occurs within the first day or two after mixing. Mated cells are joined by conjugation tubes that are usually long enough to be easily discerned. Occasionally, conjugation tubes are quite short and require careful observation for detection. Conjugating cells may be of equal size or one of the conjugants can be quite small and misinterpreted as a bud if the conjugation tube is short. If ascosporulation does not follow mating, the pairings can be repeated on other media that may be more conducive to ascospore formation such as those listed below. Lack of ascosporulation following mating may also indicate incompatibility that could be due to ploidy differences or genetic divergence sufficiently great to prevent karyogamy [60, 61].

3.4.2 Basidiomycetes

Strains may be paired as described in the preceding section on ascomycetes, or some prefer the following procedure when many strains are to be paired. Strains are mixed in pairs near the margin of an agar plate, and then with a flattened needle, three to five streaks are made across the Petri dish. Mating reactions frequently occur along these streaks. The first indication of the occurrence of a mating reaction is the formation of conjugation tubes, followed by the development of dikaryotic hyphae, which frequently grow submerged [3, 4, 10]. Formation of clamp connections, dikaryotic cells, teliospores and/or basidia can indicate a sexual reaction. Commonly used media include corn meal agar, modified Flegel's conjugation medium, V8 juice agar, 5 % malt extract agar, potato-dextrose agar, and 1.5 % malt extract-0.05 % yeast extract-0.25 % peptone agar. For mating of *Filobasidiella neoformans* the following media have been suggested: corn meal agar, hay infusion agar, V8 juice agar, pigeon manure agar or sunflower seed agar. Germination of teliospores is sometimes difficult, and it can be enhanced by a temporary increase of the temperature to 50–55 °C for 5 or 10 min [4], or soaking in sterile distilled water for 2–10 weeks at 12 °C [33] prior to transfer to appropriate media for germination, e. g., corn meal agar, or 2 % water agar.

3.5 Nuclear staining

The number of nuclei and their behavior play an important role in life cycles of yeasts. A number of rapid and reliable fluorescent nuclear staining techniques have been developed. Nuclear staining of thick-walled and/or pigmented cell, such as teliospores, is sometimes difficult. Some fluorochromes permit quantification of nuclear DNA, such as para-rosanaline (Feulgen), propidium iodide and 4'-6'-diamino-2-phenolindole (DAPI). For non-fluorescent observation, staining using Giemsa is convenient. Cells can be fixed and stained using 1.5 ml microfuge tubes, or adhered to a cover glass using egg albumin as an adhesive.

3.5.1 Staining nuclei using DAPI [19]

Stock solution: 1mg DAPI ml^{-1} distilled water, store at 4 °C in the dark. Working solution: 0.5 µg DAPI ml^{-1} McIlvaine's buffer, pH 4.4 (44.1 ml 0.2 M $Na_2HPO_4 \cdot 2H_2O$ and 55.9 ml 0.1 M citric acid). Fix cells in 70 % ethanol for 60 min in a 1.5 ml centrifuge tube; centrifuge; rinse 5 min in McIlvaine's buffer; centrifuge; resuspend cells in DAPI working solution for ca. 3 h, or overnight; centrifuge; mount cells in 90 % v/v glycerol in McIlvaine's buffer and seal coverglass with nail polish. Slide preparations made in this way may last a year or longer when stored in the dark.

For rapid analysis of the number of nuclei, yeast cells may be suspended directly in DAPI working solution. An additional advantage of DAPI is that mitochondria stain as well. The emission of DAPI depends on the mol % G+C, and therefore the method has to be used with caution when attempting quantitative fluorescence microscopy.

3.5.2 Staining nuclei with propidium iodide [29, 94]

Propidium iodide (PI) stock solution: 150 µg PI ml^{-1} NS buffer (20 mM Tris-HCl pH 7.6, 0.25 M sucrose, 1 mM $MgCl_2$, 0.1 mM $ZnSO_4$, 0.1 mM $CaCl_2$, 0.8 mM phenylmethylsulphonylfluoride (PMSF, toxic), 0.05 % β-mercaptoethanol). Propidium iodide working solution: 1.5–10 µg PI ml^{-1} NS buffer containing 50–500 µg RNase ml^{-1}.

RNase stock solution: 10 mg RNase ml^{-1} in 0.01 M Na-acetate, pH 5.2. Heat 15 min at 100 °C, cool down slowly, and adjust pH to 7.4 by adding 0.1 volume 0.1 M Tris-HCl (pH 7.5). RNase is not deactivated by boiling as is the case for DNases.

Fix cells with 50 % ethanol for 10 min, and subsequently with 70 % ethanol for 12 h. Stain with PI on a microscope slide for 30 min to 16 h (under a coverglass), remove excess dye with tissue, and seal with nail polish. Staining of RNA by PI may cause high background fluorescence. Preparations sometimes improve after a longer incubation time (e. g., cells of *Galactomyces geotrichum* need up to 4 days). Propidium iodide can be used for cytometric quantitation of DNA as the emission is independent of the mol % G+C.

3.5.3 Staining nuclei with mithramycin and ethidium bromide [5]

Mithramycin working solution: 100 µg mithramycine ml^{-1} 15 mM $MgCl_2$ + 30 % ethanol.

Ethidium bromide working solution: 25 µg ethidium bromide ml^{-1} 0.1 M Tris + 0.6 % NaCl, pH 7.4 [18]. Mix equal volumes of mithramycin and ethidium bromide working solutions, and suspend cells this mixture for ca. 30 min. Preparations can be viewed directly. Thick-walled cells may give better results after gentle heating. Mithramycine and ethidium bromide are suspected carcinogens! Acriflavine has been used to stain meiotic chromosomes in filamentous fungi such as *Neurospora crassa* [73], and may be useful for yeasts as well.

3.5.4 Staining nuclei with Giemsa [10]

Giemsa stock solution: 0.76 g Giemsa powder ml^{-1} 50 % (v/v) glycerol in absolute methanol [36]. Phosphate buffer (pH 7.0): Mix 60.8 ml 0.15 M $Na_2HPO_4 \cdot 2H_2O$ with 39.8 ml 0.15M KH_2PO_4.

Cells are dried for 20 min at room temperature and fixed for 30 min in a 3:1 mixture of 92 % ethanol and acetic acid, repeatedly rinsed with water, and hydrolyzed in 1N HCl for 7 min at 60°C. After rinsing once with water and five times with phosphate buffer (pH 7.0), the cells are stained with Giemsa working solution (one volume Giemsa stock solution and nine volumes phosphate buffer, pH 7.0) for 2 h. After rinsing in phosphate buffer (pH 7.0), a coverglass is placed over the cells, excess buffer is blotted away and the coverglass sealed with nail polish.

3.6 DNA based methods for yeast identification

3.6.1 Isolation

Methods used for DNA isolation for taxonomic studies need to be rapid and reliable. Several methods are in use by yeast systematists. Two commonly used large-scale DNA isolation methods are described below. The first method is based on adsorption of DNA on hydroxylapatite columns as described by BRITTEN et al. [15, 71]. The second is based on extraction with various reagents and subsequent precipitation with ethanol or iso-amylalcohol. Small-scale methods for DNA isolation (minipreps) are widely used for PCR amplification. We describe a miniprep method that we have successfully applied to PCR amplification of ribosomal DNA and other genes. Further DNA isolation protocols can be found in CRYER et al. [20], HOLM et al. [42], JOHNSTON [47], MANN and JEFFERY [65], TRECO [92], and TAYLOR and NATVIG [88].

3.6.1.1 DNA isolation using hydroxylapatite [15]

DNA isolated by this method has been used extensively for spectrophotometric analysis of base composition, and DNA reassociation experiments. Grow cells in ca. 400 ml YM, YPG or YPM broth until the late logarithmic or early stationary phase. Harvest by centrifugation or, in the case of extensive hyphal growth, by filtration through a filter paper in a Buchner funnel. Wash with water and then with saline EDTA (0.1 M NaCl, 0.15 M EDTA). Cells can be stored at –20 °C until use. Usually ca. 5 g wet packed cells is sufficient. Add an equal volume of lysing buffer (900 ml 10 M urea, 56.2 ml 5.1 M phosphate buffer (2.4 M $Na_2HPO_4 \cdot 2H_20$ + 2.7 M $NaH_2PO_4 \cdot H_2O$ in distilled water)), 43.8 ml 20 % sodium dodecylsulphate (SDS), break cells at least three times with e. g., a French Press, and check microscopically (the majority of cells should be broken). Repeat if necessary. Centrifuge broken cell debris for 10 min at 10.000 rpm. The supernatant contains the DNA.

Suspend ca. 5 g hydroxylapatite in washing buffer (800 ml 10 M urea, 50 ml 5.1 M phosphate buffer, 150 ml distilled water), and pour column. Pour supernatant containing the DNA on the hydroxylapatite column and let DNA adsorb. Wash with ca. 200 ml washing buffer until $A_{260} = 0.0$. Remove excess urea with ca. 30 ml low phosphate buffer (1 ml 5.1 M phosphate buffer and 480 ml distilled water), elute DNA with eluting buffer (30 ml 5.1 M phosphate buffer and 450 ml distilled water), and collect DNA fractions with a fraction collector at A_{260} (threshold value: $A_{260} = 0.25$, which equals ca. 12.5 µg DNA ml^{-1}). Pool fractions and measure A_{260}. Dialyze overnight against 0.1 X SSC (1 X SSC: 0.15 M NaCl, 15 mM trisodium citrate.$2H_2O$), repeat once and measure absorbance at 260, 230 and 280 nm. Ratios for purified DNA are $A_{230/260} = 0.5$ and $A_{260/280} = 1.85$.

For spectrophotometric DNA reassociations the A_{260} has to be at least 1.5. If necessary, concentrate the sample as follows: Add ca. 2.5 volumes of ice-cold ethanol, let DNA precipitate for 24 h, centrifuge 1 h at 17,000 rpm, remove supernatant and add 0.1 SSC to a final A_{260} of 3.0. The samples can be stored at -20°C until further use.

3.6.1.2 DNA isolation by a modified Marmur method

Isolation of DNA from yeasts by a modification of MARMUR'S [67] method has been used successfully in many laboratories and the procedure given here is similar to that reported by

Price et al. [71]. Strains are usually grown for 3 days at 25 °C on a rotary shaker (200 rpm) in Fernbach flasks containing 1500 ml of YM broth and harvested by centrifugation. Two flasks are usually prepared for each strain. Following harvesting, the cells are suspended in 2X SSE buffer [90] and either broken in a Braun cell homogenizer (B. Braun Biotech, Allentown, PA) with 0.5 mm glass beads or by enzymatic digestion. Following cell breakage, sodium perchlorate and sodium sarcosine are added to the suspension to give concentrations of 1 M and 1 %, respectively, and the mixture is then emulsified by swirling with an equal volume of chloroform: isoamyl alcohol (CIA) (24:1, v/v). The emulsion is maintained for 3 h on a rotary shaker and then separated by centrifugation. The upper DNA-containing aqueous layer is removed with a wide-mouth pipette, and the DNA is precipitated by addition of 1.3 volumes of cold (–20 °C) ethanol. The precipitate is collected by centrifugation, dissolved in 20 ml of 1X SSC containing 2 mg α-amylase and 2 mg of pancreatic RNase, and incubated overnight at room temperature on a rotary shaker. One mg of pronase is then added and the solution is incubated for an additional 4 h. The preparation is placed in a 300-ml Erlenmeyer flask with an equal volume of CIA, emulsified for 30 min on a rotary shaker and then centrifuged. The upper layer is removed with a wide-mouth pipette and placed in a beaker where the DNA is spooled following addition of 1.3 volumes of cold ethanol. The spooled DNA can be dissolved in 20 ml of 0.001 M sodium phosphate buffer for later use.

At this point, the DNA may be further purified by cesium chloride gradient ultracentrifugation or by hydroxylapatite chromatography. If the latter option is chosen, the DNA spool is dissolved in 20 ml of 0.001 M sodium phosphate buffer and treated with a mixture of enzymes containing 2 mg α-amylase, 2 mg pancreatic RNase and 400 units T1 RNase. The DNA-enzyme solution is dialyzed against 0.001 M sodium phosphate buffer overnight at room temperature. The solution is treated with CIA as before and adjusted to 0.2 M with sodium phosphate buffer if the G+C content of the DNA is under 55 % or else to 0.15 M. The solution is now passed through a hydroxylapatite column and the DNA eluted with 0.5 M sodium phosphate buffer. Progress can be monitored by following absorbance at A_{260} in a spectrophotometer flow cell.

3.6.1.3 Miniprep method for isolation of DNA for PCR amplification (modified, after Raeder and Broda [72])

Cells are grown in 25 ml YM broth on a 200 rpm shaker at 25 °C for ca. 48 h, and harvested by centrifugation. After washing with water, the cells are lyophilized overnight. The lyophilized cell mass is pulverized with a pipette tip, and shaken with glass beads (diam 0.5 mm) for ca. 15 min using a wrist-action shaker. Following breaking, 1000 µl extraction buffer (200 mM Tris-HCl, pH 8.4; 200 mM NaCl; 25 mM EDTA; 0.5 % SDS) is added. After pelleting by centrifugation, 600 µl of the supernatant is transferred to a new tube, to which 420 µl phenol is added. After vortexing, 180 µl chloroform is added. Centrifuge for 10 min at 14,000 rpm; 550 µl of the aqueous (upper) phase is transferred to a new tube without disturbing the interface. Add 1 volume of chloroform, and vortex briefly. Centrifuge 5 min at

14,000 rpm, and transfer 400 µl of the aqueous (upper) phase to a new tube. Add 0.54 volume isopropanol, and shake briefly. Centrifuge the precipitated DNA for 10–60 sec at 14,000 rpm. Discard the supernatant, and wash the pellet once with 70 % ethanol. Centrifuge for 3 min at 14,000 rpm, and decant the ethanol carefully. Add 100 µl TE buffer (10 mM Tris-HCl; 1 mM EDTA, pH 8.0), loosen the pellet and solubilize the DNA for 1 h (or overnight) at 55°C. The DNA can be stored at –20 °C.

3.6.1.4 DNA isolation using hexadecyltrimethyl-ammonium bromide (CTAB)

As an alternative to the laboratory use of phenol, the broken cells (see 3.6.1.3) are suspended in 700 µl 2X CTAB buffer (100 mM Tris-HCl [pH 8.4], 1.4 M NaCl, 25 mM EDTA, 2 % hexadecyltrimethyl-ammonium bromide), vortex-mixed with an equal volume of chloroform and centrifuged for 10 min (for references see KURTZMAN and ROBNETT [57, 58]). DNA is precipitated from the aqueous phase by adding 0.54 volume of isopropanol and pelleted for ca. 3 min in an Eppendorf or similar microcentrifuge at 14,000 rpm. The pellet is washed gently with 70 % ethanol, resuspended in 100 µl of TE buffer (10 mM Tris-HCl, 1 mM EDTA [pH 8.0], and dissolved by incubation at 55 °C for 1 to 2 h. Dilute DNA samples for PCR are prepared by adding 4 µl of the genomic stocks to 1 ml of 0.1X TE buffer.

3.6.2 Analysis of base composition

Determination of nuclear DNA base composition as mol % G+C has been the first of the molecular methods to be applied to yeast taxonomy. The taxonomic uses of G+C values are mainly exclusionary because the 1,000 or so known yeast species range in nuclear G+C contents from approximately 27–70 mol %, and overlap between unrelated species is inevitable. Since similar G+C contents do not necessarily mean that strains are conspecific, how dissimilar do they need to be to state that two strains represent different species? This depends to some extent on the method used for determination of base composition. When buoyant density is used, the span of G + C values among strains of a species is usually less than 1 %, but thermal denaturation shows greater variation and strains of a species may exhibit a G + C range of as much as 2 % [56]. The nuclear G+C content of ascomycetous yeasts is about 27–50 %, whereas the span for basidiomycetous yeasts is approximately 50–70 %. Except for the narrow range of 48–52 %, where some overlap occurs, the taxonomic affinity of anamorphs can be reliably determined from their base composition. For species in the doubtful range, placement can be made by examining cell wall structure under the transmission electron microscope or through use of the diazonium blue B test [104]. Although no longer of primary importance for yeast classification, G + C values are sometimes still of interest, and it is for this reason that methodologies for their determination are presented.

3.6.2.1 Spectrophotometric determination of mol % G+C

Dilute the DNA with 0.1X SSC in quartz cuvettes until an A_{260} of 0.3–0.4. *Candida parapsilosis* CBS 604 (mol % G + C = 40.8) is used as a reference strain, and 0.1 x SSC is used as a blank. Increase of temperature is 0.5 °C min^{-1}, and the A_{260} is recorded using a microprocessor-controlled spectrophotometer. T_m values from melting curves can be calculated graphically, or from the first or second derivative [44]. Mol % G+C can be calculated as follows (in 0.1X SSC): mol % G + C= T_m x 2.08 – 106.4. Melting curves of basidiomycetous yeasts frequently show one or two shoulders, which probably represent mtDNA and/or rDNA. Determinations of mol % G + C obtained with HPLC are frequently lower than those obtained from spectrophotometric analyses. This may be due to the presence of mtDNA in the fractions. A short protocol for analyzing the mol % G + C using the high performance liquid chromatography (HPLC) method can be found in HAMAMOTO et al. [38] and NAKASE et al. [68].

3.6.2.2 Determination of mol % G+C content from buoyant density

When a cesium chloride solution is spun in an ultracentrifuge, a density gradient is formed. The position of DNA in the gradient is determined by its G + C content, which can be calculated from the relative position of a second DNA of known density. Determinations are generally made in an analytical ultracentrifuge, but a preparative ultracentrifuge will also serve the purpose. Begin by making a stock cesium chloride solution in the following manner: Add 130 g CsCl to 75 ml of 10 mM Tris buffer, pH 8.5. Dissolve and treat for 20 min with 2 g activated charcoal to remove any material absorbing at 260 nm. Remove the charcoal by filtration through Whatman No. 1 filter paper. Determine the exact concentration of CsCl from the refractive index. Solutions are usually approximately 1.87 g/ml, and about 405 µl is used in a total volume of 500 µl. Combine the required amount of CsCl stock solution, 1 µg of undetermined DNA, 1 µg of *Micrococcus luteus* reference DNA (buoyant density 1.7311 g/ml), and bring to a final volume of 500 µl with distilled water. Load centrifuge cells. Centrifugation is generally for 20 h at 44,000 rpm.

Determine distances of peak centre points for unknown and reference DNAs and determine the G+C content using the following equations [79].

$$\boldsymbol{\rho = \rho_0 + 4.2\ \omega^2(r^2 - r_0^2) \times 10^{-10}\ g/cm^3}$$

where ρ is the density of unknown DNA, ρ_0 the density of known DNA, ω the radians/sec (2π radians/revolution), r the distance of unknown from the center of rotation, and r_0 the distance of standard from the center of rotation.

$$\mathbf{Mol\ \%\ G + C = (\rho - 1.66)/0.098 \times 100}$$

3.6.3 Hybridization of nuclear DNA

Although this is an older method that is being replaced by gene sequencing, hybridization or reassociation of nuclear DNA is a reliable means for estimating the extent of genetic relatedness between strains [56, 71]. DNAs that have approximately 80 % or more nucleotide similarity can form a duplex under appropriate conditions of incubation. Because duplex formation requires considerable nucleotide similarity, genetic resolution from measurements of reassociation extends only to the distance of sibling species.

Protocols commonly used to measure DNA reassociation have been compiled by KURTZMAN [54], and the following account represents only a brief summary of these methods. Methods for measuring DNA relatedness fall into two general categories: (1) the free-solution technique, in which all of the reactants are solubilized, and (2) the filter binding technique, in which the DNA of one strain is immobilized on nitrocellulose or other filter materials and DNA from the other strain is solubilized in the buffer surrounding the membrane. Each method has its strengths and, when properly done, each provides the same measure of relatedness [61, 81].

In order to satisfy reaction kinetics, free-solution hybridizations require shearing of the genomic DNA in 400–500 base pairs fragments. The DNA can be sheared to this length by double passage through a French pressure cell at 10,000 psi or greater. Any fragments that escape shearing can be removed by passage of the solution of sheared DNA through a 0.45 μm membrane filter. The DNA bound to filters for filter hybridization is not sheared, but the probe is of sheared DNA.

3.6.3.1 Spectrophotometric method

The spectrophotometric (optical) method is a convenient non-isotopic procedure that has the advantage of simultaneously providing data for estimates of genome size. The rationale for this method is based on the observation that DNA reassociation is a concentration-dependent, second-order reaction. As a result, if 50 μg/ml of DNA reassociates at a certain rate, 25 μg/ml will take twice as long. Consequently, if a mixture of two DNAs reassociates at the same rate as an equivalent concentration of unmixed DNA, the organisms providing this DNA belong to the same species. If the reaction time of the mixture is the sum of that of the two unmixed DNAs, the organisms are different species. Because the genome sizes of yeasts are relatively small, the midpoint of the reaction, which is often used as a reference, is reached within about 2–12 h, depending on reaction conditions and actual genome sizes.

3.6.3.2 Hydroxylapatite method

Hydroxylapatite (HA) is a form of calcium phosphate that preferentially binds double-stranded DNA when in an appropriate concentration of phosphate buffer. This specificity provides a means for separating renatured DNA duplexes from free-solution reassociation reactions. The technique requires use of radiolabeled probe DNA, but has the potential for relative ease in processing large numbers of samples.

3.6.3.3 S1 nuclease method

DNA hybridization mixtures are freed of single-stranded, unhybridized DNA by hydrolysis with S1 endonuclease. The double-stranded hybrids are then removed from the reaction mix by precipitation with trichloroacetic acid or collected on Whatman DE-81 filters. The probe DNA is radiolabeled.

3.6.3.4 Filter hybridization

Unsheared, single-stranded DNA is attached to nitrocellulose or nylon membrane filters. The filters are incubated in a liquid medium containing a sheared, single-stranded probe that is either labeled with a radioisotope or has another means for detection. Following incubation, the filters are washed and assayed for extent of binding by the probe. The filter method is amenable to easily perform large numbers of comparisons, but leaching of unsheared target DNA from filters during incubation can be a problem.

3.6.3.5 Interpretation of DNA hybridization data

Data from DNA reassociation experiments have been interpreted in the context of the biological species concept. DOBZHANSKY [28] has championed the idea that species can be described in terms of genetics and that, among sexually reproducing and outbreeding organisms, species can be defined as Mendelian populations or arrays of populations that are reproductively isolated from other population arrays. The concept seems apparent for mammals, but its application to yeasts is less straightforward because not all taxa are known to have sexual cycles and some that do may show little outbreeding. Further, species formation does not usually leave a clear separation between groups comprising the new species, and genetically intermediate populations may initially survive [103]. Consequently, in order to use DNA reassociation data to define species, there must be some knowledge of the extent of DNA divergence among members of a yeast species. On the basis of comparisons between the extent of nuclear DNA relatedness and fertility of progeny arising from conventional genetic crosses, conspecific strains generally exhibit DNA relatedness in excess of 70 %. Varietal designations can be accorded to those strains showing 40–70 % DNA relatedness unless genetic crosses demonstrate the absence of interfertility. An exception is the varieties of *Issatchenkia scutulata* which show only 25 % DNA relatedness, but exhibit some intervarietal fertility [53, 56].

3.6.4 Amplification of yeast DNA using polymerase chain reaction (PCR)

DNA isolated by different methods may be used for PCR amplification. We obtained good results with DNA isolated according to the miniprep method as described earlier. DNA in TE/10 buffer (add 4 µl of the DNA solution in TE buffer to 1 ml TE/10 buffer, 10 mM Tris-HCl pH 8.0, 0.1 mM EDTA pH 8.0), is usually used as template in the PCR reactions but the more concentrated DNA in TE buffer is occasionally needed. Visualization of the amplified products is performed by electrophoresis in 1 % agarose minigels in 1X TBE (0.045 M Tris-borate, 0.001 M EDTA pH 8.0), using agarose dye mix (2 % brom-phenol blue; 2 % xylene cyanol; 1 ml ddH_2O, 0.2 g sucrose) and a DNA ladder (10 µl DNA ladder stock (1 $\mu g\ \mu l^{-1}$, Gibco); 10 µl TE buffer; 50 µl ddH_2O; 100 µl agarose dye mix (see above)). Staining is performed with ethidium bromide ($8 \times 10^{-5}\ \mu g\ \mu^{-1}$) and visualized on a UV transilluminator.

3.6.5 DNA methods: protocols for sequencing the D1/D2 domain of the 26S rDNA, 18S rDNA and the internally transcribed spacer (ITS)

3.6.5.1 Analysis of D1/D2 domain of 26S rDNA

The 600–650 nucleotide D1/D2 domain of large subunit (26S) ribosomal DNA (nucleotides 63–642 for *Saccharomyces cerevisiae*) is sufficiently divergent to resolve most yeast species. Furthermore, analysis of these sequences provides predictions of phylogenetic relationships that are similar to those obtained from analysis of small subunit (18S) rDNA sequences.

Methods for nuclear DNA isolation, purification and sequencing vary considerably among laboratories, and any procedures that give high purity DNA and accurate sequences are appropriate. The following protocols have been successfully used in our laboratories and are offered to those who wish to try them. The procedures given include DNA isolation and purification and PCR amplification of the region to be sequenced. Procedures for manual or automated sequencing vary considerably, and the protocols for a particular method need to be consulted. Manufacturers of automated sequencers have protocols based on their particular sequencing reagents and methods.

The D1/D2 domain is amplified with primers NL-1 (See list below at paragraph 3.6.5.5. for the composition of the all primers cited) and NL-4. Amplification is performed for 36 PCR cycles with denaturation at 94 °C for 1 min, annealing at 52 °C, extension at 72 °C for 2 min, except for the final cycle which has a 7 min extension. The amplified DNA is purified with Geneclean II (Bio 101, La Jolla, Calif.) according to the manufacturer's instructions, but a variety of other methods are also available for purification such as the QIAquick kit described below. Visualization of the amplified DNA following purification is by electrophoresis in 1.5 % agarose in 1X TAE buffer (0.04 M Tris-acetate, 0.001 M EDTA [pH 8.0]) with staining in ethidium bromide (8 x 10–5 µg/µl). Both strands of the DNA are sequenced. Primers for these reactions are the external primers NL-1 and NL-4 and the internal primers NL-2A and NL-3A. If long sequencing runs are possible with available equipment, reactions using the internal primers will not be necessary.

3.6.5.2 Alternate method for analysis of the D1/D2 domain of basidiomycetous yeasts

The preceding method for ascomycetous yeasts is often effective for basidiomycetous yeasts, but failures sometimes occur, and the following method provides a good alternative. Cells are grown for 12–14 h in YPG broth (2 % glucose, 0.5 % peptone and 0.1 % yeast extract), centrifuged and washed with distilled water and then converted to spheroplasts by incubating for 2 h at 37 °C in 10 mM citrate buffer, pH 5.8, 1 M sorbitol and 10 mg/ml lysing enzymes from *Trichoderma harzianum* (Sigma), which is freshly prepared for each procedure. DNA is extracted and purified from the spheroplasts using a QIAamp tissue culture kit following standard protocols.

The next step is to prepare a region-specific amplicon for sequence analysis. Problems have been encountered if the same primer is used for production of the amplicon and for sequencing. During the initial amplification, short, incomplete fragments at the primer end of the sequence can be produced, which are not eliminated during the clean up process, unless gel purification is used. Consequently, during cycle sequencing the sequencing primer will amplify these small fragments and the result is multiple banding on the sequencing gels, particularly following the primer front. To avoid this problem, an amplicon can be produced that is longer than the region of interest for sequencing. Sequencing primers are designed that are approximately 100 nucleotides or more inside from the 5′ or 3′ ends of the amplicon. For example, to examine the ITS and D1/D2 regions from a single amplicon, primers were designed for regions in the 18S DNA and well within the 26S DNA. Specifically, these are universal fungal primers ITS5 and LR6. The resulting amplicon is purified with the QIAquick PCR purification kit. Cycle sequencing of the D1/D2 domain employs the forward primer NL-1 (F63) and reverse primer LR3. ITS cycle sequencing primers include the forward strand primer ITS1 and the reverse strand primer ITS4. Preferably, the ITS-D1/D2 region can be sequenced as a single fragment using cycle sequencing primers ITS1 and LR3.

3.6.5.3 Amplification and sequencing of 18S rDNA from ascomycetous yeasts

The 18S rDNA is amplified as two overlapping PCR fragments using the primer combinations P108/M3490 and P1190/M3989 [45, 46]. Using the primers applied for the ascomycetous yeasts ca. 95 % of the 18S rDNA can be determined by double-strand sequencing. The V4 domain [22] is the most promising region of 18S rDNA for species identification and can be sequenced using the amplicon generated with primers P108 and M3490 and the sequencing primers P1190 and M2130. However, the main value of 18S sequences is for phylogenetic comparisons because there is generally insufficient resolution to differentiate closely related species (e. g., *Saccharomyces cerevisiae* and *S. paradoxus*).

3.6.5.4 Amplification and sequencing of 18S rDNA from basidiomycetous yeasts

Primers OLI4 and ITS 4 are used for amplification of basidiomycetous 18S rDNA. For sequencing the following forward primers are being used: NS1, NS26, NS3, Basid3, NS23, NS7, as well as the following reverse primers NS8, NS6, Basid2, NS2, and CNS26. Primers ITS2' replaces primer NS8 in the sequencing of the 18S rDNA of *Malassezia* species.

3.6.5.5 Sequencing primers

3.6.5.5.1 Primers for 26S rDNA

NL-1 (=F63): 5´-GCA TAT CAA TAA GCG GAG GAA AAG
NL-4: 5´-GGT CCG TGT TTC AAG ACG G
NL-2A: 5´-CTT GTT CGC TAT CGG TCT C
NL-3A: 5´-GAG ACC GAT AGC GAA CAA G
LR3: 5´-GGT CCG TGT TTC AAG ACG CC

3.6.5.5.2 Primers for 18S rDNA

P108: 5´-ACC TGG TTG ATC CTG CCA GT
P130: 5´-GTC TCA AAG ATT AAG CCA TG
WIL1 (position 301–318): 5´-ATT TCT GCC CTA TCA ACT
P1190 (position 543–562): 5´-CAA TTG GAG GGC AAG TCT GG
P2130 (position 900–921): 5´-GGT GAA ATT CTT GGA TTT ATT G
P2540 (position 1108–1127): 5´-GGA GTA TGG TCG CAA GGC TG
P3490 (position 1454–1473): 5´-CCG CAC GCG CGC TAC ACT GA
WIL2 (position 318–301): 5´-AGT TGA TAG GGC AGA AAT
M2130 (position 921–900): 5´-CAA TAA ATC CAA GAA TTT CAC C
M2540 (position 1127–1108): 5´-CAG CCT TGC GAC CAT ACT TCC
M3490 (position 1473–1454): 5´-TCA GTG TAG CGC GCG TGC GG
M3989 (position 1775–1754): 5´-CTA CGG AAA CCT TGT TAC GAC T
OLI4: 5´-CTG GTT GAT YCT GCC AGT

ITS4: 5´-TCC TCC GCT TAT TGA TAT GC
NS1: 5´-GTA GTC ATA TGC TTG TCT C
NS26: 5´-CTG CCC TAT CAA CTT TCG A
NS3: 5´-GCA AGT CTG GTG CCA GCA GCC
Basid3: 5´-AGA GTG TTC AAA GCA GGC
NS23: 5´-GAC TCA ACA CGG GGA AAC TC
NS7: 5´-GAG GCA ATA ACA GGT CTG TGA TGC
NS8: 5´-TCC GCA GGT TCA CCT ACG GA
ITS2': 5´-GCT GCG TTC TTC ATC GAT G
NS6: 5´-GCA TCA CAG ACC TGT TAT TGC CTC
Basid2: 5´-CTG TTA AGA CTA CAA CGG
NS2: 5´-GG6.5.5 C TGC TGG CAC CAG ACT TGC
CNS26: 5´-TCG AAA GTT GAT AGG GCA G

3.6.5.5.3 Primers for ITS

ITS1: 5'-TTC GTA GGT GAA CCT GCG G
ITS4: 5'-TCC TCC GCT TAT TGA TAT GC
ITS 5: 5'-GGA AGT AAA GTC GTA ACA AGG
LR6: 5'-CGC CAG TTC TGC TTA CC

3.6.6 Molecular methods for rapid identification of yeasts

Molecular methods provide a means for rapid and accurate identification of yeasts. At present, many of these methods rely on sequencing short (ca. 500–600 nucleotides) species-specific gene sequences or using species-specific oligonucleotide primers derived from these gene sequences in a detection system based on PCR.

The ca. 600-nucleotide D1/D2 domain at the 5´ end of large subunit (26S) rDNA often has sufficient substitutions that individual species can be recognized. Databases of this domain are now available for all known yeast species ([32, 58] and subsequent entries to GenBank), allowing for rapid strain identification when sequencing facilities are available. The ITS domains of rDNA are often more variable than the D1/D2 domain and will prove useful once extensive sequence databases are developed.

Oligonucleotide probes based on unique sequences provide a rapid means for detecting individual strains, species and genera, or even larger taxonomic groups. Such probes are increasingly used for detection of pathogenic species in the clinical laboratory, for detection of food and beverage spoilage organisms and for finding particular species in environmental samples. The limitation at present is the lack of technology for conveniently handling large numbers of species-specific probes, but this is likely to change with advances in methodologies such as molecular array technologies.

There are many examples of oligonucleotide probe technology used for species identification [31]. One of these is from the study of MANNARELLI and KURTZMAN [66] in which species-specific oligonucleotides were developed for detection of human pathogenic species of *Candida*. A forward and reverse primer pair unique to each of several different *Candida* species was developed for domain D1/D2 of the 26S rDNA. The reaction mix included genomic DNA, the species-specific primer pair and a primer pair specific to a conserved region in 18S rDNA, which served as a positive control. The mixture was subjected to PCR and then separation on an agarose gel. If the target DNA reacted, a short band of amplified 26S rDNA was seen on the gel, along with the larger, slower moving control band. If the target and primers did not match, only the control DNA band was visible.

For rapid detection, species-specific oligonucleotide probes can be labeled and this is often done by attaching a fluorescent dye molecule to the oligonucleotide. STENDER et al. [86] developed a fluorescent hybridization probe for *Dekkera bruxellensis*, the cause of "*Brettanomyces*" spoilage of wines. In this case, the oligonucleotide probe had a peptide backbone giving what is known as a peptide nucleic acid (PNA). The fluorescent dye 5 (6)-carboxyfluorescein was attached to the species-specific PNA probe to allow detection of *Brettanomyces* cells by fluorescence microscopy following hybridization with the probe.

Random amplified polymorphic DNA (RAPD) is another methodology that is used for rapid detection of species. The technique is based on amplification of genomic DNA in the presence of one or more short (ca. 10–15-mers) oligonucleotide primers of random sequence. The amplified products are visualized on an agarose gel and strains identified from matching band patterns. HADRYS et al. [37] discussed details of this procedure noting points of technical difficulty. Poor reproducibility of band patterns may occur, and apparently results from small differences in reagent concentrations, from small variations in thermal cycler temperature control, and perhaps from single nucleotide polymorphisms.

Restriction fragment length polymorphisms (RFLPs) have been widely used for separation of yeast species as well as individual strains of a species [16, 59, 62, 63, 97]. Either nuclear or mitochondrial DNA or PCR amplified gene sequences from these DNAs are treated with restriction endonucleases, and the resulting fragments are visualized as band patterns on an agarose gel. With appropriate endonucleases, band patterns can be quite reproducible and therefore useful for strain and species identification. However, single nucleotide substitutions (polymorphisms) can markedly impact band patterns. For this reason, identifications should be based on RFLPs from several DNA sequences.

3.7 Pulsed field electrophoresis (electrophoretic karyotyping)

Electrophoretic patterns of chromosomal DNAs (electrophoretic karyotypes) provide data of potential use in systematic studies of yeasts. The analysis of karyotypes by pulsed-field techniques has received wide application in yeast systematics and genetics. Number and size of individual chromosomes, estimates of total genome sizes, chromosomal rearrangements, and gene assessment are among the characters studied. Taxonomically, these data are mainly used at or below the species level, e. g., to differentiate species, strains, and/or populations.

Due to the presence of chromosomal length polymorphisms, aneuploidy, and comigrating bands, a straightforward taxonomic interpretation of electrophoretic karyotypes is sometimes complicated. Ribosomal DNA-containing chromosomes frequently show considerable size variation [5, 64, 77]. Many species studied show rather variable karyotypes, e. g., *Candida albicans* [43, 77]. In contrast, other species show remarkably stable karyotypes, e. g., those belonging to the basidiomycetous yeast *Malassezia*. All strains of *M. pachydermatis* show a slight length variation of only the smallest chromosomal DNA [11, 12]. In *Saccharomyces cerevisiae*, a correlation has been demonstrated between similarities of DNA as observed in DNA reassociation experiments, and similarities of karyotypes [96].

Chromosomal DNAs of up to ca. 6 Mb can be separated [35], but even 10 Mb has been reported [106]. Separation of chromosomal DNAs larger then ca. 6 Mb is still difficult, and requires long pulse- and run times. Chromosomal DNAs of *Saccharomyces cerevisiae*, *Pichia canadensis* (= *Hansenula wingei*), and *Schizosaccharomyces pombe* are commercially available as size standards.

For each yeast species with an unknown karyotype, the parameters used during electrophoresis have to be optimized. As a rule of thumb, the size of the chromosomal DNAs is inversely proportional to the agarose concentration of the gels and the field strength to be used. The pulse time is direct proportional to the chromosomal size.

3.7.1 Preparation of agar embedded protoplasts using lysing enzymes of *Trichoderma harzianum* (Sigma)

Rotary-shaken (ca. 200 rpm) cultures are grown until late logarithmic phase in 50 ml 1 % yeast extract-0.5 % peptone-4 % glucose (YPG) broth at 25 °C. For basidiomycetous yeasts, 0.05 % yeast extract-0.5 % peptone-0.7 % malt extract (YPM) broth is preferred. Psychrophilic species require lower temperatures, e. g., 12 or 17 °C. Approximately 0.5×10^9 cells are harvested by centrifugation, and washed in 2 ml 0.05 M EDTA, pH 7.5. The pellet is resuspended in ca. 2 ml buffer made up of 0.05 M EDTA, 10 mM Tris-HCl, and 10 mM dithiothreitol (DTT), final pH 7.5. After centrifugation, the pellet is washed with 2 ml CPE buffer (100 ml 40 mM citric acid, 120 mM Na_2PO_4 (pH 6.0) and 4 ml 0.5 M EDTA, pH 7.5). Centrifuge and resuspend in 0.3 ml CPES buffer (CPE buffer containing 1.2 M sorbitol and 5 mM DTT).

Dissolve 2 mg lysing enzyme in 0.5 ml 1 % w/v low-melting agarose in CPE buffer at 38 °C. Mix equal volumes of lysing enzyme/agarose solution and the cell suspension (when preparing protoplasts of hyphal species or large-sized cells, the volumes have to be adjusted). Pipet the cell/lysing enzyme/agarose mixture into a precooled matrix and allow to gel on ice. The agarose/cell blocks are incubated for 1 h at 30 °C in ca. 5 ml CPE buffer. The agarose blocks are rinsed with ca. 5 ml NDS (0.5 M EDTA, pH 7.5, 10 mM Tris-HCl, pH 7.5), 1 % v/v sodium *N*-lauroylsarcosinate, and, subsequently placed for 16 h at 50 °C in 2 ml NDS buffer to which 0.4 mg proteinase K and 1 % v/v sodium *N*-lauroylsarcosinate have been added. After lysis, the agarose blocks are rinsed in NDS buffer containing 1 % v/v sodium *N*-lauroylsarcosinate and stored in NDS buffer containing 1 % v/v sodium *N*-lauroylsarcosinate and 0.2 mg ml^{-1} proteinase K. Sodium *N*-lauroylsarcosinate and proteinase K have to be added just prior to use. The agarose/cell blocks can be stored for several months at 4 °C.

Alternatively, protoplasts of cells can be prepared before embedding in low-melting agarose as follows. Add 2 mg lysing enzyme in 0.5 ml CPE to the resuspended cells in 0.3 ml CPES (see above), and allow degradation of the cell wall for 2–3 h at 30 °C. Check protoplast formation microscopically, and in case of sufficient cell wall degradation, centrifuge for 3 min at 13,000 rpm. Resuspend the protoplasts in 0.3 ml CPES, after which an equal volume of 1 % low-melting agarose in CPE buffer is added. Pipet the protoplast/agarose mixture into a precooled matrix and allow to gel on ice. Rinse the agarose blocks in 2 ml NDS, 1 % v/v *N*-lauroylsarcosinate, and incubate for 16 h at 50 °C in NDS to which 1 % v/v *N*-lauroylsarcosinate and 0.02 % proteinase K are added. Proceed according to the above protocol.

Protoplasts of ascomycetous yeasts usually are prepared using zymolyase [48]. However, this enzyme is not suitable for the preparation of protoplasts of basidiomycetous yeasts. Therefore, we prefer to use lysing enzymes from *Trichoderma harzianum* or *Rhizoctonia solani* (Sigma) for protoplasting both asco- and basidiomycetous yeasts.

3.7.2 Electrophoresis

Depending on the size of the chromosomal DNAs to be separated, 0.5–1.5 % agarose gels (chromosomal grade) in 0.25–0.5X TBE are prepared. The positioning of the agarose/cell blocks in the gel inserts is facilitated by initially flooding the inserts with 0.5X TBE buffer. 0.25–0.5X TBE is used as electrophoretic buffer, and the temperature is kept between 12 and 15 °C. Some electrophoretic parameters, taken from the literature, are presented in Table 3.7-1. After completion, the gels are stained in 0.5 µg ml^{-1} ethidium bromide for 30–120 min, destained with demineralized water, and the DNA visualized with an UV transilluminator. If the DNA banding is obscured by a smear, an RNase treatment is suggested [48] as follows: Incubate the gel for 2 h at 37 °C with gentle shaking in a sealed bag containing 30 ml 0.5 x TBE and 1.5 ml RNase solution (500 µg ml^{-1} pancreatic RNase + 100 units ml^{-1} T1 RNase in 10 mM Tris-HCl, pH 7.5, 15 mM NaCl, heated for 10 min at 100 °C and slowly cooled to room temperature). Rinse twice with electrophoresis buffer during 60 min.

3.8 Maintenance and storage of cultures

Ascomycetous yeasts are often maintained on glucose-peptone-yeast extract agar (GPYA) or on YM agar. Many basidiomycetous yeasts do not survive well on these media, although

Tab. 3.7-1 Electrophoretic parameters for separation of chromosomal DNAs from some yeast species using contour-clamped electric fields (CHEF).

Species	Run time (h)	Pulse time (sec)	Agarose %	Reference
Cryptococcus neoformans	30 40	100–300 400–600	1.0	[5]
Kluyveromyces marxianus and *K. lactis*	32 20	20–120 120	1.5	[84]
Malassezia furfur	36 36	300 300–600	1.0	[11]
Malassezia pachydermatis	48	50–300	1.0	[11]
Saccharomyces cerevisiae	23 7	60 100	1.0	[2]
Schizosaccharomyces pombe	130	3600	0.6	[98]

they grow well on them. Such yeasts are kept on potato-dextrose agar (PDA). Most strains are stored at temperatures between 4 and 12 °C and subcultured at intervals of ca. 6 months. Some yeasts, such as *Arxiozyma* and *Malassezia*, have to be subcultured every month. *Dekkera* and *Brettanomyces* produce excessive amounts of acetic acid, therefore 2 % calcium carbonate is added to the medium to neutralize the acid. Nevertheless, these yeasts still need to be subcultured every 1 to 2 months.

Cultures are frozen in liquid nitrogen, the vapor phase of liquid nitrogen (ca. –170 °C) or in a mechanical freezer at temperatures between –80° and –135 °C for long-term storage. Cultures of all strains held at the Centraalbureau voor Schimmelcultures are frozen and are successfully kept at –80 °C and in liquid nitrogen. The preparation of cultures for freezing is simple and quick. The general method used is as follows: short lengths of polypropylene drinking straws are sealed at one end, labeled with a black felt-tipped pen (e. g., Pentel Permanent Marker), and sterilized in the autoclave at 121 °C for 15 min. The strain to be frozen is grown for about 24 hr in 3.0 ml of liquid medium on a shaker before adding 1.0 ml of a sterile 60 % solution of glycerol in water. The resulting suspension is pipetted into the straws sufficient to half fill them. The straws are then closed by clamping the open ends in the jaws of a sealing machine for plastic packages. The cultures are then either frozen at approximately –30 °C for between 30 and 60 min before being placed in the storage tank under liquid nitrogen, or put directly into a freezer cabinet at –80 °C. Sterile plastic screw-cap ampoules suitable for use in liquid nitrogen are often used but are more expensive and require more storage space. However, cells can be removed multiple times from these ampoules before the freeze-thaw cycles cause significant loss in viability. Storage in the vapor phase of liquid nitrogen generally gives results comparable to immersion storage, which precludes accidental leakage of liquid nitrogen into poorly sealed ampoules and the resulting explosion when ampoules are thawed. Cell viability is often increased if ampoules are thawed quickly in a 37 °C water bath.

3.9 Growth media for yeasts including those for detection, enumeration, and isolation of species from foods and clinical specimens

This listing (in alphabetical order) includes commonly used media as well as those recently developed as selective and differential media for the detection, enumeration and isolation of specific physiological groups, including species of foodborne and clinical yeasts. Comments on use of the media are presented as well as details on preparation. Additional formulae and discussion of media can be found in the literature [23–25, 50, 78] and in manufacturer's manuals (Difco, Oxoid, Merck, BioMerieux and others).

1. Acetic Acid Agar

100 g	glucose
10 g	tryptone
5 g	yeast extract
20 g	agar
1000 ml	deionized water

Cool the molten medium to approximately 45 °C and add 1 ml of glacial acetic acid to each 100 ml, mix rapidly and pour into petri dishes.

Use: Acetic acid agar is a general purpose medium such as TGYC supplemented with 0.5 % acetic acid. It inhibits the majority of yeasts, while it allows the growth of a few acid resistant species including the most notorious spoilage species *Zygosaccharomyces bailii* [40]. According to SILONIZ et al. [82], increasing the acetate concentration to 1 % renders the medium even more selective for acid resistant yeasts. This medium only permits the growth of *Z. bailii, Issatchenkia orientalis* (including its anamorph *C. krusei*) and some strains of *Torulaspora delbrueckii.* Using 0.5 % acetate together with 0.2 % potassium tellurite inhibits *Z. bailii,* whereas the other two species are able to grow. This study has been made, however, on a limited number of species and strains, and more extended experiments are needed to corroborate these findings.

2. Acidified Media

To acidify agar media, a predetermined amount of 1 N HCl is added after autoclaving. Add 7 ml of acid to each litre of medium for a pH of approximately 3.8.
Use: Isolation, to reduce growth of bacteria

3. AF Agar

2.85 g	sucrose
230 mg	aspartic acid
510 mg	glutamic acid

245 mg KH_2PO_4
660 mg $MgSO_4 \cdot 7H_2O$
1700 µg $FeCl_3 \cdot 6H_2O$
510 µg $MnSO_4 \cdot 6H_2O$
4500 µg $ZnSO_4 \cdot 7H_2O$
410 mg KOH
20 g agar
1000 ml distilled water
Use: Sporulation of *Lipomyces* and related species.

4. Canavanine-Glycine-Bromothymol (CGB) Blue Medium
Bromothymol Agar
20 ml bromothymol blue solution
20 g agar
880 ml deionized water
Bromothymol Blue Solution
0.4 g sodium bromothymol blue
100 ml deionized water
Sterilize by filtration.
Stock Solution A
300 mg L-canavanine sulphate
100 g glycine
10 g potassium dihydrogen phosphate
10 g magnesium sulphate heptahydrate
1000 ml deionized water
and either
10 drops Bejectal with vitamin C (Abbott, Chicago)
or
10 mg thiamine hydrochloride
Adjust pH to 5.6 and sterilize by filtration.
Cool 900 ml of bromothymol agar to about 55 °C after autoclaving and add 100 ml of stock solution A. Mix well and pour into tubes or plates.
Use: Identification of varieties of *Filobasidiella (Cryptococcus) neoformans.* A loopful of cells from a young culture is streaked onto CGB agar in either a tube or a petri dish and incubated at 25 °C for up to 5 days. The colour of the medium does not change if the strain is *Filobasidiella neoformans* var. *neoformans*, but turns blue if the strain is the variety *Filobasidiella neoformans* var. *bacillispora.*

5. Carbon Base-Urea Agar
11.7 g Yeast Carbon Base (Difco)
0.2 g acid fuchsine
20 g agar
800 ml deionized water
200 ml filter-sterilized 10 % solution of urea

Add the urea solution to the other ingredients after they have been sterilized and cooled to approximately 50 °C.
Use: Diazonium blue B test.

6. Chalk Agar

20 g	glucose
10 g	finely powdered calcium carbonate
5 g	peptone
20 g	agar
1000 ml	yeast extract

Sterilize by autoclaving and gently agitate to keep the chalk in suspension while the agar is setting.
Use: Isolation and cultivation of *Dekkera (Brettanomyces).*

7. Chloramphenicol

0.1–1.0 g	chloramphenicol
1000 ml	medium

Chloramphenicol can be added to the medium before autoclaving.
Use: Additive to media to inhibit growth of bacteria.

8. Christensen Urea Agar

1 g	peptone
5 g	sodium chloride
2 g	dihydrogen phosphate
0.012 g	phenol red
1000 ml	deionized water

Adjust the pH to 6.8, and then dissolve

20 g	agar

Sterilize by autoclaving at 121°C for 15 minutes, then add 100 ml filter sterilized 20 % solution urea.
Use: Test for hydrolysis of urea.

9. CHROMagar *Candida* (commercial product)

10 g	peptone
22 g	chromogenic mixture
0.5 g	chloramphenicol
15 g	agar
1000 ml	deionized water

Use: Identification of *Candida albicans.* This medium was originally developed for the discrimination and presumptive identification of *Candida albicans* among other yeast species frequently occurring in clinical materials such as *C. tropicalis, C. glabrata, C. krusei* (= *I. orientalis*) and others [69]. Studies by TORNAI-LEHOCZKI and PETER [91] showed that this chromogenic medium can be successfully used to differentiate between foodborne yeasts in mixed population using a single plate. Among the yeasts of clinical significance, only *C. albicans* forms green coloured colonies. However, among foodborne yeasts several additional species can be found showing green and greyish-green colonies. Others can be discriminated by various other colours such as blue, brown to purple, white to grey or cream to yellow (see Chapter 2).

10. Corn Meal Agar

15 g agar
1000 ml maize infusion

Powdered products are available from Oxoid and Difco.

Use: Morphology, cells and filaments, sporulation, ascospores, teliospores, basidiospores, chlamydospores in *Candida albicans* and *C. dubliniensis.*

11. Corn Meal + Tween Agar

1000 ml corn meal agar
10 ml tween 80

Use: Chlamydospores in *Candida albicans* and *C. dubliniensis.*

12. $CuSO_4$ Medium

3 g malt extract
3 g yeast extract
5 g peptone
15 g agar
1000 ml deionized water

Autoclave 2 % (w/v) aqueous solution of $CuSO_4$ and add immediately before pouring the agar to give a final concentration of 200 mg/L (1 ml to 100 ml medium).

Use: Selective media for detecting wild yeasts in breweries. According to comparative studies [21, 52], $CuSO_4$ medium performs best in detecting both *Saccharomyces* and non-*Saccharomyces* wild yeasts with 80 to 91 % efficiency, while lysine agar (see 3.9-35) completely inhibits *Saccharomyces* wild yeasts and is 56–61 % effective in the detection of non-*Saccharomyces* wild yeasts.

13. Custers Agar

50 g glucose
5 g finely powdered calcium carbonate
20 g agar
1000 ml yeast extract

Agitate gently while the agar is setting to keep the chalk in suspension.

Use: Detection of acid production

14. Cycloheximide Medium: 0.01 % and 0.1 %

0.1 g cycloheximide for 0.01 %
1.0 g cycloheximide for 0.1 %
2.5 ml acetone
6.7 g Yeast Nitrogen Base (Difco)
10 g glucose
100 ml deionized water

Dissolve the cycloheximide in the acetone and the other ingredients in the water. Mix the solutions and filter sterilize. To use, add 0.5 ml aseptically to 4.5 ml of sterile water.
Use: Identification character. Resistance to cycloheximide. This medium can also be used for selective isolation.

15. *Cyniclomyces* Medium

15.1 Agar

10 g yeast autolysate
40 g glucose
10 g proteose peptone
20 g agar
1000 ml deionized water

Sterilize 100 ml amounts in bottles by autoclaving. Melt the agar, cool to approximately 45 °C, then adjust the pH with 1N HCl (add approx. 2.8–3.0 ml to 100 ml of agar) to between 3.5 and 4.5 and pour into petri dishes.
Use: Isolation and cultivation of *Cyniclomyces guttulatus.*

15.2 Broth

10 g yeast autolysate
40 g glucose
10 g proteose peptone
1000 ml deionized water

Adjust the pH with 1N HCl to approximately 4.0.
Use: Isolation and cultivation of *Cyniclomyces guttulatus.*

16. D-20 Medium

6.7 g Yeast Nitrogen Base (Difco)
200 g glucose
1 g yeast extract
1 g malt extract
20 g agar
1000 ml deionized water

Use: Isolation.

17. *Debaryomyces hansenii* Differential Medium

5 g yeast extract
3 g malt extract
3 g proteose peptone
10 g glucose
20 g agar
1000 ml deionized water

After autoclaving add filter sterilized solutions of Salmon-Glc (6Cl-3-indoxyl-β-D-glucopyranoside), O-Me-β-Glc (1-O-methyl-D-glucopyranoside), Xgal (5-bromo-4-chloro-3-indolyl-β-D-galactopyranoside), IPTG (isopropyl-β-D-thiogalactopyranoside) and chloramphenicol, in order to make the final amounts per 1000 ml medium as follows:

0.15 g	Salmon-Gluc (6Cl-3-indoxyl-β-D-glucopyranoside, Biosynth, Switzerland)
0.1 g	O-Me-β-Glc (1-O-methyl-β-D-glucopyranoside, Biosynth, Switzerland)
0.08 g	XGal (5-bromo-4-chloro-3-indolyl-β-D-galactopyranoside, Sigma),
0.1 g	IPTG (isopropyl-β-D-thiogalactopyranoside, Sigma) and
0.5 g	Chloramphenicol.

Use: Detectin of *Debryomyces hansenii.*

D. *hansenii* produces salmon coloured colonies when β-glucosidase is induced in the presence of chromogenic substrates [83]. β-Glucosidase can be detected in several yeast species that show pink colonies of different shades, such as *Hanseniaspora guilliermondii, H. uvarum,* and *Pichia anomala.* These species can be distinguished from *D. hansenii* by a different colony and cellular morphology. *Kluyveromyces marxianus* and *K. lactis* have both β-glucosidase and β-galactosidase activities and produce deep blue colour hydrolysing galactopyranosides, XGal and IPTG. When this chromogenic medium is used for the analysis of intermediate moisture foods, only *D. hansenii* shows a salmon colony colour.

18. *Dekkera/Brettanomyces* Differential Medium (DBDM)

6.7 g	Yeast Nitrogen Base (Difco)
6 ml	ethanol (v/v)
0.01 g	cycloheximide
0.1 g	p-coumaric acid
0.022 g	bromocresol green
20 g	agar
1000 ml	deionized water

All ingredients, except the agar, are sterilized by filtration, and the pH adjusted to 5.4.

Use: Isolation, identification of *Dekkera* and *Brettanomyces.* DBDM was developed by RODRIGUES et al. [76] and is based on the principle that *Dekkera* and *Brettanomyces* species, in addition to being strong acid producers, produce 4-ethylphenol from p-coumaric acid that can be detected by its phenolic odor. *Dekkera* and *Brettanomyces* yeasts develop slowly, and even after 8–12 days of incubation, only pinpoint colonies appear, which are yellow to cream coloured, but often turn to green after longer incubation times. Several other yeast species may grow on DBDM. Among these are *C. tropicalis, D. hansenii, K. marxianus,* and *S. unisporus,* which are infrequently found in beverages and juices, as well as *Z. fermentati, Z. florentinus* and *Z. microellipsoides* that occur more frequently in wines. These species, however, grow faster and their colonies appear in less than eight days, but none of these species produce acetic acid and a phenolic smell.

19. Dichloran 18 % Glycerol (DG18) Agar for Selective Isolation of Xerotolerant Yeasts.

5 g	peptone
10 g	glucose
1 g	KH_2PO_4
0.5 g	$MgSO_4 \cdot 7H_2O$
0.002 g	dichloran
0.1 g	chloramphenicol

220 ml glycerol
15 g agar

All ingredients except glycerol are dissolved by heating, volume made to 1000 ml with water and glycerol is then added giving a final concentration of 18 % (w/w). Dichloran (2,6-dichloro-4-nitroaniline) is added as 1 ml of a 0.2 % solution in ethanol. DG18 agar is also available commercially.

Use: DG18 agar provides decreased water activity (a_w 0.95) and has been developed for the enumeration of xerotolerant and moderately xerophilic moulds in commodities such as grains, flours, nuts and spices [41]. As dichloran limits the colony diameter of molds, this medium has become popular for the detection of xerotolerant (osmophilic) yeasts from high sugar foods such as fruit concentrates, syrups, beverage bases. For a more complete recovery of xerotolerant yeasts adapted to high sugar commodities, sample suspension and dilutions need to be prepared in a 10 % glucose-containing diluent before plating on DG18.

20. Dichloran Rose Bengal Chloramphenicol (DRBC) Agar

10 g glucose
5 g peptone
1 g KH_2PO_4
0.5 g $MgSO_4 \cdot 7H_2O$
0.025 g rose bengal
0.002 g dichloran
0.1 g chloramphenicol
15 g agar
1000 ml deionized water

Chloramphenicol can be added before autoclaving. Dichloran (2,6-dichloro-4-nitroaniline) is added in 1 ml of 0.2 % solution in ethanol, and rose bengal as 0.5 ml of 5 % solution in water. Prepared medium should be kept in dark.

Use: DRBC agar, described by KING et al. [49], is one of the most commonly used general purpose media for the enumeration and isolation of yeasts from foods. This medium contains both antibacterial and fungistatic agents. The heat labile chlortetracycline in the original formula has been replaced with chloramphenicol. This method does not eliminate moulds completely, but it does restrict their spread and allows the counting of yeast colonies. The culture medium also provides some differentiation of mixed populations since various yeast species develop different coloured colonies in the presence of rose bengal.

21. Dilute V8 Agar

1 can V8 Vegetable Juice (Campbell Soup Co.)

Dilute juice with equal volume of water and add sodium hydroxide to bring pH to 5.5. Filter through Whatman No. 1 paper then dilute 1:9 and 1:19 before adding 2 % agar.

Use: Sporulation of *Metschnikowia* spp.

22. Diphenyl Solution

1 g diphenyl
100 ml 95 % ethanol

Add 10 ml of the solution aseptically to each litre of molten medium after it has cooled to approximately 45 °C.
Use: Suppresses growth of moulds.

23. Fermentation Medium

1000 ml yeast extract (see 3.9-69)
20 g test sugar (40 g in the case of raffinose)

Some laboratories add 7.5 g peptone and use bromothymol blue as pH indicator. Dispense medium in tubes or bottles containing a small inverted tube (Durham insert). The insert should become full of medium after autoclaving.
Use: Fermentation tests for identification.

24. Fowell's Acetate Agar

5 g sodium acetate (trihydrate)
20 g agar
1000 ml deionized water

Dissolve the acetate in water and adjust pH to between 6.5 and 7 before adding the agar.
Use: Formation of ascospores.

25. Glucose 50 % Agar

500 g glucose
500 ml yeast extract (see 3.9-69)
13 g agar

Sterilize by autoclaving at 115 °C (10 psi) for 10 min. If the medium is over-heated during autoclaving, it turns brown and must be discarded.
Use: Identification, isolation and cultivation of osmophilic and osmotolerant strains.

26. Glucose 60 % Agar

600 g glucose
400 ml yeast extract (see 3.9-69)
22.5 g agar

Sterilize by autoclaving at 115 °C (10 psi) for 10 min. If the medium is over-heated during autoclaving, it turns brown and must be discarded.
Use: Identification, isolation and cultivation of osmophilic and osmotolerant strains.

27. Glucose-Peptone-Yeast Extract Broth (GPYB)

20 g glucose
5 g peptone
1000 ml yeast extract (see 3.9-69)

pH is not adjusted.
Use: Isolation and cultivation.

28. Glucose-Peptone-Yeast Extract Agar (GPYA)

20 g glucose
5 g peptone
20 g agar
1000 ml yeast extract (see 3.9-69)

pH is not adjusted. Dispense 200 ml amounts in bottles (enough for 10–12 plates) or 5 ml in 16 ml diameter test tubes for slants.

Use: Isolation and cultivation.

29. Glucose 2 %-Yeast Extract Agar

20 g glucose
20 g agar
1000 ml yeast extract (see 3.9-69)

Use: Sporulation of *Debaryomyces (Schwanniomyces)* spp.

30. Glucose 1 %-Yeast Extract Agar

10 g glucose
15 g agar
1000 ml yeast extract (see 3.9-69)

Use: Sporulation of *Metschnikowia* and *Wickerhamiella* spp.

31. Gorodkowa Agar

1 g glucose
5 g sodium chloride
10 g peptone
20 g agar
1000 ml tap water

A modified version of this medium contains 2.5 g of glucose, and 10 g of meat extract is substituted for the peptone.

Use: Ascospores (particularly *Debaryomyces* spp.).

32. Hay-Infusion Agar

50 g decomposing hay
1000 ml deionized water
2 g potassium monohydrogen phosphate
15 g agar

Autoclave the hay in the water for 30 min at 121 °C and then filter. Dissolve the phosphate and the agar in the filtrate and adjust the pH to 6.2.

Use: Sporulation of basidiomycetes.

33. Killer-Test Medium

YM agar

Adjust to pH 4.2–4.7 with 0.05 M citrate buffer

After autoclaving add 0.003 % w/v methylene blue (= 0.03 g per 1000 ml medium).

The sensitive strain is spread as a lawn over the agar and the test strains dotted around the perimeter. A clear or dark blue area around a strain after 48 to 72 hrs of incubation denotes a killer activity [105]. The initial pH is 6.5.
Use: Testing strains for killer activity.

34. *Kluyveromyces* Differential Medium (KDM)
Basal medium:

5 g	yeast extract
3 g	malt extract
3 g	proteose peptone
10 g	glucose
20 g	agar
1000 ml	deionized water

After autoclaving add filter sterilized solutions of Xgal (5-bromo-4-chloro-3-indolyl-β-D-galactopyranoside), IPTG (isopropyl-β-D-thiogalactopyranoside) and chloramphenicol, in order to make the final amounts per 1000 ml medium as follows: 0.08 g Xgal (5-bromo-4-chloro-3-indolyl-β-D-galactopyranoside, Sigma), 0.1 g IPTG (isopropyl-β-D-thiogalacto-pyranoside, Sigma) and 0.5 g chloramphenicol. The medium is poured into Petri dishes, and after solidification 5–10 μl of the sample is streaked, and incubated at 28 °C for 7 days.
Use: KDM has been used with success to selectively isolate *Kluyveromyces marxianus* and *K. lactis* from dairy products. KDM is based on the principle that *K. marxianus* and *K. lactis* produce dark blue colonies when β-galactosidase is induced in the presence of the chromogenic substrate [95].

35. Lysine Medium (also commercially available)

11.75 g	Yeast Carbon Base (Difco)
2.3 g	L-lysine · HCl
20 g	agar
1000 ml	deionized water

Use: Selective media for detecting wild yeasts in breweries.

36. Maize Infusion (see also Corn Meal Agar)

42 g	maize
1000 ml	deionized water

Heat the maize in water to 60 °C for 1 hr, filter through paper, then add water to restore the volume to 1 litre.

37. Malt Agar (MA)

1000 ml	malt extract (10° Brix) (brewer's wort)
30 g	agar

Alternatively the following may be used

100 g	malt extract powder
30 g	agar
1000 ml	tap water

Use: Cultivation, isolation, morphology.

38. 5 % Malt Agar
50 g   malt extract powder (Difco)
30 g   agar
1000 ml   deionized water
Use: Ascospores, basidiospores, teliospores.

39. 2 % Malt Agar
20 g   malt extract powder (Difco)
20 g   agar
1000 ml   deionized water
Use: Ascospores, basidiospores, teliospores.

40. Malt Extract (ME)
Unhopped brewer's wort, if this is not obtainable prepare wort as follows:
1000 g   malt
2600 ml   tap water
Stir the malt in the water at 45 °C for 3 hrs, then raise the temperature to 63 °C for 1 hr and filter the infusion through a cheese cloth. Autoclave the filtrate at 110 °C for 15 min then filter through paper. The wort is diluted to a density of 15° Brix (determined with a saccharometer as with degree Balling) and the pH is adjusted to 5.4
Use: Cultivation, isolation, morphology.

41. McClary Acetate Agar
1 g   glucose
1.8 g   potassium chloride
8.2 g   sodium acetate (trihydrate)
2.5 g   yeast extract
15 g   agar
1000 ml   deionized water
Use: Ascospore formation (particularly of *Saccharomyces cerevisiae*).

42. Preparation of Microplates (media composition and location of positions in the assimilation microplates used at the CBS).

42.1 Basal Medium for Carbohydrates Assimilation Tests (BMC)
100 ml   demineralized water
1 g   Yeast Nitrogen Base (Difco)

42.2 Basal Medium for Nitrogen Compounds Assimilation Tests (BMN)
100 ml   demineralized water
1.77 g   Yeast Carbon Base (Difco)

42.3 Basal Medium for Testing Growth with Different Vitamin Combinations (BMV)
100 ml   demineralized water
2.52 g   Vitamin-Free Yeast Base (Difco)
Use: Identification.

Growth media for yeasts

Test	Amount	Position in microplate
N Control	BMN	A1
C Control	BMC	A2
C1 D-Glucose	BMC + 0.78 g	A3
C2 D-Galactose	BMC + 0.78 g	A4
C3 L-Sorbose	BMC + 0.78 g	A5
C4 D-Glucosamine	BMC + 0.78 g	A6
C5 D-Ribose	BMC + 0.78 g	A7
C6 D-Xylose	BMC + 0.78 g	A8
C7 L-Arabinose	BMC + 0.78 g	A9
C8 D-Arabinose	BMC + 0.78 g	A10
C9 L-Rhamnose	BMC + 0.78 g	A11
C10 Sucrose	BMC + 0.78 g	A12
C11 Maltose	BMC + 0.78 g	B1
C12 a,a-Trehalose	BMC + 0.78 g	B2
C13 Me a-D-Glucoside	BMC + 0.78 g	B3
C14 Cellobiose	BMC + 0.78 g	B4
C15 Salicin	BMC + 0.78 g	B5
C16 Arbutin	BMC + 0.78 g	B6
C17 Melibiose	BMC + 0.78 g	B7
C18 Lactose	BMC + 0.78 g	B8
C19 Raffinose	BMC + 0.78 g	B9
C20 Melezitose	BMC + 0.78 g	B10
C21 Inulin	BMC + 0.78 g	B11
C22 Starch	BMC + 0.78 g	B12
C23 Glycerol	BMC + 0.78 ml	C1
C24 Erythritol	BMC + 0.78 g	C2
C25 Ribitol	BMC + 0.78 g	C3
C26 Xylitol	BMC + 0.78 g	C4
C27 L-Arabinitol	BMC + 0.78 g	C5
C28 D-Glucitol	BMC + 0.78 g	C6
C29 D-Mannitol	BMC + 0.78 g	C7
C30 Galactitol	BMC + 0.78 g	C8
C31 myo-Inositol	BMC + 0.78 g	C9
C32 D-Glucono-1,5-lactone	BMC + 0.78 g	C10
C33 2-Keto-D-Gluconate	BMC + 0.78 g	C11
C34 5-Keto-D-Gluconate	BMC + 0.78 g	C12

Test	Amount	Position in microplate
C35 D-Gluconate	BMC + 0.78 g	D1
C36 D-Glucuronate	BMC + 0.78 g	D2
C37 D-Galacturonate	BMC + 0.78 g	D3
C38 DL-Lactate	BMC + 0.78 g	D4
C39 Succinate	BMC + 0.78 g	D5
C40 Citrate	BMC + 0.78 g	D6
C43 Propane 1,2 diol	BMC + 0.78 ml	D7
C44 Butane 2,3 diol	BMC + 0.78 ml	D8
C45 Quinic acid	BMC + 0.78 g	D9
C46 D-glucarate	BMC + 0.78 g	D10
C47 D-Galactonate	BMC + 0.78 g	D11
C48 Palatinose	BMC + 0.78 g	D12
C49 Levulinate	BMC + 0.78 g	E1
C50 L-Malic acid	BMC + 0.78 g	E2
C51 L-Tartaric acid	BMC + 0.78 g	E3
C52 D-Tartaric acid	BMC + 0.78 g	E4
C53 meso-Tartaric acid	BMC + 0.78 g	E5
C54 Galactaric acid	BMC + 0.78 g	E6
C55 Uric acid	BMC + 0.78 g	E7
C56 Gentobiose	BMC + 0.78 g	E8
C57 Ethylene glycol	BMC + 0.78 ml	E9
C58 Tween 40	BMC + 0.78 ml	E10
C59 Tween 60	BMC + 0.78 ml	E11
C60 Tween 80	BMC + 0.78 ml	E12
N1 Nitrate	BMN + 0.11 g	F1
N2 Nitrite	BMN + 0.04 g	F2
N3 Ethylamine	BMN + 0.10 g	F3
N4 L-Lysine	BMN + 0.10 g	F4
N5 Cadaverine	BMN + 0.10 g	F5
N6 Creatine	BMN + 0.11 ml	F6
N7 Creatinine	BMN + 0.087 g	F7
N8 Glucosamine	BMN + 0.11 g	F8
N9 Imidazole	BMN + 0.047 g	F9
N10 D-Tryptophan	BMN + 0.10 g	F10
N11 D-Proline	BMN + 0.10 g	F11
N12 Putrescine	BMN + 0.10 g	F12

Test	Amount	Position in microplate
V1 w/o vitamins	BMV	G1
V2 w/o myo-Inositol	BMV + 300 µg calcium pantothenate, 3.0 µg biotin, 60 µg thiamine hydrochloride, 60 µg pyridoxine hydrochloride, 60 µg niacin, 30 µg p-aminobenzoic acid, 0.3 µg folic acid, 30 µg riboflavin	G2
V3 w/o Pantothenate	BMV + 1500 µg myo-inositol, 3.0 µg biotin, 60 µg thiamine hydrochloride, 60 µg pyridoxine hydrochloride, 60 µg niacin, 30 µg p-aminobenzoic acid, 0.3 µg folic acid, 30 µg riboflavin	G3
V4 w/o Biotin	BMV + 1500 µg myo-inositol, 300 µg calcium pantothenate, 60 µg thiamine hydrochloride, 60 µg pyridoxine hydrochloride, 60 µg niacin, 30 µg p-aminobenzoic acid, 0.3 µg folic acid, 30 µg riboflavin	G4
V5 w/o Thiamin	BMV + 1500 µg myo-inositol, 300 µg calcium pantothenate, 3.0 µg biotin, 60 µg pyridoxine hydrochloride, 60 µg niacin, 30 µg p-aminobenzoic acid, 0.3 µg folic acid, 30 µg riboflavin	G5
V6 w/o Biotin & Thiamin	BMV + 1500 µg myo-inositol, 300 µg calcium pantothenate, 60 µg pyridoxine hydrochloride, 60 µg niacin, 30 µg p-aminobenzoic acid, 0.3 µg folic acid, 30 µg riboflavin	G6
V7 w/o Pyridoxine	BMV + 1500 µg myo-inositol, 300 µg calcium pantothenate, 3.0 µg biotin, 60 µg thiamine hydrochloride, 60 µg niacin, 30 µg p-aminobenzoic acid, 0.3 µg folic acid, 30 µg riboflavin	G7
V8 w/o Pyridoxine & Thiamin	BMV + 1500 µg myo-inositol, 300 µg calcium pantothenate, 3.0 µg biotin, 60 µg niacin, 30 µg p-aminobenzoic acid, 0.3 µg folic acid, 30 µg riboflavin	G8
V9 w/o Niacin	BMV + 1500 µg myo-inositol, 300 µg calcium pantothenate, 3.0 µg biotin, 60 µg thiamine hydrochloride, 60 µg pyridoxine hydrochloride, 30 µg p-aminobenzoic acid, 0.3 µg folic acid, 30 µg riboflavin	G9

Test	Amount	Position in microplate
V10 w/o PABA	BMV + 1500 µg myo-inositol, 300 µg calcium pantothenate, 3.0 µg biotin, 60 µg thiamine hydrochloride, 60 µg pyridoxine hydrochloride, 60 µg niacin, 0.3 µg folic acid, 30 µg riboflavin	G10
O1 Cycloheximide 0.01 %	Demineralized water 100 ml, D-glucose 1.5 g, yeast nitrogen base (Difco) 1.0 g, cycloheximide 0.0075 g	G11
O2 Cycloheximide 0.1 %	Demineralized water 100 ml, D-glucose 1.5 g, yeast nitrogen base (Difco) 1.0 g, cycloheximide 0.075 g	G12
O3 Acetic acid 1 %	100 g glucose, 10 g tryptone, 5 g yeast extract, 1 l deionized water. Cool the molten medium to approximately 45 °C and add 1 ml of glacial acetic acid to each 100 ml	H1
O6 10 % NaCl	10 g NaCl, 5 g glucose, 0.7 g Difco Yeast Nitrogen base, 100 ml deionized water	H2
O7 16 % NaCl	16 g NaCl, 5 g glucose, 0.7 g Difco Yeast Nitrogen base, 100 ml deionized water	H3
O8 Growth at pH = 3	Demineralized water 100 ml, D-glucose 1.5 g, yeast nitrogen base (Difco) 1.0 g, ajust pH with HCl 1N	H4
O9 Growth at pH = 9.5	a. 10 g Peptone, 6 g Yeast extract, 10 g Glucose, 500 ml deionized water b. 400 mmol KCl, 40 mmol NaCl, 1800 mmol Na_2CO_3, 500 ml deionized water. Sterilize a and b seperately by autoclaving. Mix aseptically when cool and dispense in microplate.	H5

pH of tests C 35, C 38, C 39, C 40 and C 46 is adjusted to 5.2 with NAOH

43. Morphology Agar (MoA)

35 g Yeast Morphology Agar (Difco)
1000 ml deionized water
Use: Cellular morphology.

44. Niger-Seed Agar (1)

1 g glucose
20 g agar
200 ml niger-seed infusion (see 3.9-46)
800 ml deionized water

Chloramphenicol and diphenyl are sometimes added to inhibit growth of bacteria and moulds.
Use: Isolation of *Filobasidiella (Cryptococcus) neoformans.*

45. Niger-Seed Agar (2)
50 g pulverized Niger seed
1000 ml deionized water
Boil the seed in the water for 30 minutes. Filter through paper and restore final volume to 1 litre, and then add the following:
10 g glucose
1 g KH_2PO_4
1 g creatinine
15 g agar
Streptomycin 40 units/ml, Penicillin 20 units/ml, and diphenyl solution are added when the medium has cooled to about 50 °C.
Filobasidiella neoformans produces brown, usually mucoid, colonies after 3 to 8 days at 25 °C [85].
Use: Isolation of *Filobasidiella (Cryptococcus) neoformans.*

46. Niger-Seed Infusion
40 g ground or pulverized Niger seed (*Guizotia abyssinica*)
200 ml deionized water
Autoclave the seed in the water for 10 minutes at 115 °C (10 psi) and filter the infusion through gauze.
Use: Isolation of *Filobasidiella (Cryptococcus) neoformans.*

47. Oatmeal Agar
40 g oatmeal
20 g agar
1000 ml tap water
Simmer the oatmeal in the water for 1 hour then filter through cheese cloth. Restore volume to 1 litre and dissolve the agar.
Use: Ascospore formation.

48. pH 10 Medium
Part A
10 g peptone
6 g yeast extract
10 g glucose
500 ml deionized water
Part B
14.9g KCl
1.16g NaCl
95.4g Na_2CO_3
500 ml deionized water

Sterilize parts A and B separately by autoclaving. Mix aseptically when cool and dispense 4 ml amounts into test tubes.
Use: Growth at pH 10.

49. Polyol Agar

6.7 g Yeast Nitrogen Base (Difco)
5 g either ribitol or glucitol
15 g agar
1000 ml deionized water
Use: Sporulation of *Xanthophyllomyces dendrorhous* (= *Phaffia rhodozyma*).

50. Potato-Dextrose Agar (PDA)

20 g glucose (dextrose)
20 g agar
230 ml potato infusion (see 3.9-51)
770 ml deionized water
Use: Cultivation, isolation, morphology (filamentation), sporulation (ascospores, basidiospores, teliospores).

51. Potato Infusion

300 g well washed, grated or homogenized potato
900 ml tap water
Soak the potato in water overnight in a refrigerator. After filtering, autoclave the infusion for 1 hr at 110 °C.

52. Rapid Urea Broth

Prepare according to the instructions on the container and dispense 0.5 ml amounts into sterile tubes. Store frozen until required.
Use: Hydrolysis of urea.

53. Ribitol Agar

5 g ribitol
20 g agar
1000 ml deionized water
Use: Sporulation of *Xanthophyllomyces dendrorhous* (= *Phaffia rhodozyma*).

54. Rice Agar

20 g agar
1000 ml rice infusion
Commercial products are available but the results obtained with them and with an infusion of either polished or 'instant' rice are considerably inferior to those obtained with the medium prepared with an infusion of unpolished rice.
Use: Chlamydospores of *Candida albicans* and *C. dubliniensis*.

55. Rice Infusion

20 g unpolished rice
1000 ml water

Simmer the rice in the water for 45 min, filter, and add water to restore volume to 1 litre. For good results, unpolished rice must be used, because the results obtained with white polished rice are inferior.

Use: Chlamydospores of *Candida albicans* and *C. dubliniensis.*

56. Sabouraud's Glucose Agar (SabG)

20 g glucose
10 g peptone
20 g agar
1000 ml water

The pH is adjusted to 7.0 before adding the agar.

Use: Cultivation, isolation.

57. Salt Medium

16 g or 10 g NaCl
5 g glucose
0.7 g Yeast Nitrogen Base (Difco)
100 ml deionized water

Dispense 4 ml amounts in 16 mm diameter tubes.

Incubate tests with agitation and read after 1 week.

Use: Identification, particularly for discriminating *Zygosaccharomyces rouxii* from *Z. mellis.*

58. Selective Medium for Detecting Spoilage Yeasts in Wine

11.7 g Yeast Carbon Base (Difco)
0.4 g urea
2 g fructose
120 ml ethanol (95 %)

Adjust pH to 4.5 with HCl, and filter sterilize.

Use: Selective media for detecting spoilage yeasts in wine. This medium is based on the principle that alcohol tolerant yeasts can grow in the presence of 12 % (v/v) ethanol. THOMAS and ACKERMAN [89] developed a broth medium deficient in nutrients but permitting the growth of yeasts capable of spoiling wine, such as *S. cerevisiae, Z. bailii* and some strains of *Pichia anomala* and *Debaryomyces hansenii.*

59. Sucrose-Yeast Extract Agar

1.0 g potassium dihydrogen phosphate
0.5 g magnesium sulphate (heptahydrate)
0.1 g calcium chloride
0.1 g sodium chloride
0.5 g yeast extract
20 g sucrose

40 g agar
5 ug biotin
1000 ml deionized water
Use: Sporulation of *Filobasidiella neoformans* (= *Cryptococcus neoformans*).

60. Tryptone Glucose Yeast Extract Chloramphenicol (TGYC) Agar

5 g tryptone
100 g glucose
5 g yeast extract
0.1 g chloramphenicol
15 g agar
1000 ml deionized water
Use: TGYC agar is recommended for general purpose use if there is no risk that moulds overgrow yeast colonies. The medium can be prepared from several commercial basal media, such as plate count agar with 1 % glucose, with the addition of the antibiotic. For the enumeration of yeasts, an elevated (10 %) concentration of glucose is recommended [26].

61. V8 Agar

350 ml V8 Vegetable Juice (Campbell Soup Co.)
5 g compressed yeast suspended in 10 ml of water
14 g agar
350 ml deionized water
Mix the yeast and V8 juice, adjust the pH to 6.8 and steam for 10 minutes. Adjust the pH of the cool mixture to 6.8 again. Add the mixture to the agar which has been dissolved in the 350 ml of water.
Use: Ascospore formation.

62. Vegetable Wedges

Cylinders about 1 cm in diameter are cut from washed vegetables with a cork borer or apple corer. The wedges are cut obliquely from these cylinders and put into test tubes with enough water to cover the bottom half of the wedge. Vegetables commonly used are: carrot, beet, cucumber and turnip.
Use: Ascospore formation.

63. Vitamin Solutions for Growth Tests

2000 μg inositol
400 μg calcium pantothenate
2 μg biotin
400 μg thiamine hydrochloride
400 μg pyridoxine hydrochloride
400 μg niacin
200 μg para-aminobenzoic acid
2 μg folic acid
100 ml distilled water

Solutions are prepared with the following omissions:
1. all vitamins
2. inositol
3. calcium pantothenate
4. biotin
5. thiamine
6. biotin and thiamine
7. pyridoxine
8. thiamine and pyridoxine
9. niacin
10. *p*-aminobenzoic acid (PABA)
11. no omissions (complete medium for positive controls)

To use, add 0.5 ml aseptically to 4.5 ml of Difco vitamin-free basal medium in 16 mm diameter tubes. Difco vitamin-free basal medium is not being sold anymore; as an alternative, vitamin-free medium according to WICKERHAM [99] may be prepared (see 3.9-64).
All glassware used in preparing these media must be thoroughly cleaned with acid.
Use: Vitamin requirements, identification.

64. Vitamin Free Medium (after [99])

5 g $(NH_4)_2SO_4$
10 g glucose
10 mg L-histidine monohydrochloride$\cdot H_2O$
20 mg DL-methionine
20 mg DL-tryptophane
500 µ H_3BO_3
40 µg $CuSO_4 \cdot 5H_2O$
100 µg KI
200 µg $FeCl_3 \cdot 6H_2O$
400 µg $MnSO_4 \cdot H_2O$
200 µg $Na_2MoO_4 \cdot 2H_2O$
400 µg $ZnSO_4 \cdot 7H_2O$
1 g KH_2PO_4
0.5 g $MgSO_4 \cdot 7H_2O$
0.1 g NaCl
0.1 g $CaCl_2 \cdot 2H_2O$
1000 ml distilled water

65. Water Agar

20 g agar
1000 ml deionized water
Use: Germination of teliospores.

66. *Yarrowia lipolytica* Differential Medium

66.1 YL1 Medium

5 g peptone
5 g yeast extract
1.8 g L-tyrosine
0.28 g $MnSO_4 \cdot 7H_2O$
5 g lactate (90 % sol)
20 g agar
1000 ml deionized water

Adjust pH at 6.2.

Use: Isolation, identification of *Yarrowia lipolytica.* YL1 is based on the principle that only *Y. lipolytica* produces brown pigments from tyrosine in the presence of Mn^{2+} ions [17]. After incubation at 25 °C for 24 h, colonies of *Y. lipolytica* develop a unique deep brown colour.

66.2 YL2 Medium

2.5 g yeast extract
10 g malt extract
10 g glucose
0.1 g crystal violet
20 g agar
1000 ml deionized water

Use: Isolation, identification of *Yarrowia lipolytica.* This differential medium is based on the fact that only *Y. lipolytica* can grow in the presence of 0.01 % crystal violet [34]. After incubation at 21 °C for 24 h, *Y. lipolytica* forms white colonies.

67. Yeast Extract-Malt Extract Broth (YM broth)

3 g yeast extract
3 g malt extract
5 g peptone
10 g glucose
1000 ml deionized water

The pH is not adjusted and ranges between 6 and 7. This medium is available from Difco.

Use: Sporulation, cultivation, isolation.

68. Yeast Extract-Malt Extract Agar (YM agar)

3 g yeast extract
3 g malt extract
5 g peptone
10 g glucose
20 g agar
1000 ml deionized water

The pH is not adjusted and ranges between 6 and 7. This medium is available from Difco.

Use: Sporulation, cultivation, isolation. YMA + 2 % NaCl can be used for sporulation of *Zygosaccharomyces rouxii.*

69. Yeast Extract (liquid)

1000 g compressed baker's yeast
5000 ml deionized water

Mix ingredients and keep at 50 °C for 24 hours, add the whites of 2 eggs to clarify the extract, shake well and filter.

Alternatively:

5 g yeast extract powder
1000 ml deionized water

Use: Sporulation, cultivation, isolation.

70. *Zygosaccharomyces bailii* Medium (ZBM) for Acidified Foods and Fruit Juice Concentrates

30 g Sabouraud dextrose agar
20 g fructose
2.5 g yeast extract
0.25 g trypan blue
5 ml acetic acid glacial
0.1 g potassium sorbate

Use: ZBM has been developed by ERICKSON [30] for the detection of *Z. bailii* in acidified foods and fruit juice concentrates. Selectivity is obtained by the joint application of acetate and sorbate. Enrichment in nonselective medium is recommended for 48 h, and 0.1 ml of the enrichment is filtered through an Iso-Grid membrane that is then placed on the surface of the ZBM. Growth within 48 h is considered as a positive result. A few other yeast species, such as *Pichia membranifaciens* and *P. anomala,* may grow but show a different appearance (rough colonies filling the grids in contrast to shiny, smooth, round colonies of *Z. bailii*).

71. *Zygosaccharomyces bailii* Selective Medium (ZBSM)

1000 ml mineral medium (see 3.9-71.1)
10 g glucose
40 ml formic acid
0.05 g bromocresol green

71.1 Mineral Medium [after 93]

1700 ml autoclaved basal medium (see 3.9-71.2) to which is added
8.5 ml filter sterilized vitamin solution see 3.9-71.3)
8.5 ml filter sterilized biotin solution (see 3.9-71.4)
8.5 ml filter sterilized trace element solution A (see 3.9-71.5)
8.5 ml filter sterilized trace element solution B (see 3.9-71.6)

71.2. Basal Medium

8.5 g $(NH_4)_2SO_4$
8.5 g $MgSO_4 \cdot 7H_2O$

0.225 g $CaCl_2 \cdot 2H_2O$
1700 ml demineralized water

71.3 Vitamin Solution
80 mg calcium pantothenate
200 mg inositol
160 mg niacin
160 mg pyridoxin·HCl
160 mg thiamine·HCl
100 ml demineralized water

71.4 Biotin Solution
2 mg biotin
100 ml demineralized water

71.5 Trace Element Solution A
100 mg H_3BO_3
20 mg KI
40 mg $NaMoO_4 \cdot 2H_2O$
100 ml demineralized water

71.6 Trace Element Solution B
8 mg $CuSO_4 \cdot 5H_2O$
40 mg $FeCl_3 \cdot 6H_2O$
80 mg $MnSO_4 \cdot 4H_2O$
80 mg $ZnSO_4 \cdot 7H_2O$
100 ml 0.001 N HCl

Instead of the above mineral medium, commercially available Bacto Yeast Nitrogen Base Without Amino Acids (Difco) can be used (6.7 g/L).

Use: Isolation of *Zygosaccharomyces bailii* from wine. *Z. bailii* is able to grow in a mineral medium with glucose and formic acid as the only carbon and energy sources and gives rise to alkalinization detected with an acid-base indicator [80]. The sample is membrane filtered and the membrane is placed on the plate, incubated for 96 h at 30 °C. A change in the colony colour from green to blue is considered as a positive result. A formic acid concentration of 0.3 % (v/v) increases the recovery of *Z. bailii*. However, *Z. bisporus* also exhibits positive results.

3.10 References

[1] AHEARN, D.G.; ROTH, F.J.; FELL, J.W.; MEYERS, S.P.: Use of shaken cultures in the assimilation test for yeast identification. J. Bacteriol. **79** (1960) 369–371.

[2] BAKALINSKY, A.T.; SNOW, R.: The chromosome constitution of wine strains of *Saccharomyces cerevisiae*. Yeast **6** (1990) 367–382.

[3] BANDONI, R.J.: Conjugation in *Tremella mesenterica*. Can. J. Bot. **41** (1963) 468–474.

[4] BANDONI, R.J.; LOBO, K.J.; BREZDAN, S.A.: Conjugation and chlamydospores in *Sporobolomyces odorus*. Can. J. Bot. **49** (1971) 683–686.

[5] BARLOGIE, B.; SPITZER, G.; HART, J.S.; JOHNSTON, D.A.; BUCHNER, T.; SCHUMANN, J.; DREWINKO, B.: DNA histogram analysis of human hemopoietic cells. Blood **48** (1976) 245–258.

[6] BARNETT, J.A.: Biochemical differentiation of taxa with special reference to the yeasts. In: The Fungi, Vol. 3 (edited by Ainsworth, G.C.; Sussman, A.S.). London, U.K.: Academic Press (1968) 557–595.

[7] BARNETT, J.A.: The nutritional tests in yeast systematics. J. Gen. Microbiol. **99** (1977) 183–190.

[8] BARNETT, J.A.; INGRAM, M.: Technique in the study of yeast assimilation reactions. J. Appl. Bacteriol. **18** (1955) 131–149.

[9] BARNETT, J.A.; PAYNE, R.W.; YARROW, D.: Yeasts: characteristics and identification, 3rd ed. Cambridge, U.K.: Cambridge University Press (2000).

[10] BAUER, R.: Uredinales – germination of basidiospores and pycnospores. Stud. Mycol. **30** (1987) 111–125.

[11] BOEKHOUT, T.; BOSBOOM, R.W.: Karyotyping of *Malassezia* yeasts: taxonomic and epidemiological implications. Syst. Appl. Microbiol. **17** (1994) 146–153.

[12] BOEKHOUT, T.; KAMP, M.; GUÉHO, E.: Molecular typing of *Malassezia* species with PFGE and RAPD. Med. Mycol. **36** (1998) 365–372.

[13] BOEKHOUT, T.; ROBERT, V.; SMITH, M.TH.; STALPERS, J.; YARROW, D.; BOER, P.; GIJSWIJT, G.; KURTZMAN, C.P.; FELL, J.W.; GUÉHO, E.; GUILLOT, J.; ROBERTS, I.: Yeasts of the World 2.0. Amsterdam, The Netherlands: ETI Biodiversity Center (2002).

[14] BOEKHOUT, T.; VAN BELKUM, A.; LEENDERS, A.C.A.P.; VERBRUGH, H.A.; MUKAMURANGWA, P.; SWINNE, D.; SCHEFFERS, W.A.: Molecular typing of *Cryptococcus neoformans*: taxonomic and epidemiological aspects. Int. J. Syst. Bacteriol. **47** (1997) 432–442.

[15] BRITTEN, R.J.; GRAHAM, D.E.; NEUFELD, B.R.: Analysis of repeating DNA sequences by reassociation. In: Methods in Enzymology Vol. 29, Nucleic Acids and protein synthesis (edited by Grossman, L.; Moldave, K.). New York, U.S.A.: Academic Press (1974) 363–418.

[16] CADEZ, N.; RASPOR, P.; DE COCK, W.A.M.; BOEKHOUT, T.; SMITH, M.TH.: Molecular identification and genetic diversity within species of the genera *Hanseniaspora* and *Kloeckera*. FEMS Yeast Res. **1** (2002) 279–289.

[17] CARREIRA, A.; LOUREIRO, V.: A differential medium to detect *Yarrowia lipolytica* within 24 hours. J. Fd Mycol. **1** (1998) 3–12.

[18] COLE, G.T.: *Graphiola phoenicis*: a taxonomic enigma. Mycologia **75** (1983) 93–116.

[19] COLEMAN, A.W.; MAGUIRE, M.J.; COLEMAN, J.R.: Mithramycin- and 4'-6'diamidino-2-phenylindole (DAPI)-DNA staining for fluorescence microspectrophotometric measurement of DNA in nuclei, plastids, and virus particles. J. Histochem. Cytochem. **29** (1981) 959–968.

[20] CRYER, D.R.; ECCLESHALL, R.; MARMUR, J.: Isolation of DNA. In: Methods in cell biology, Vol. 12 (edited by Prescott, D.M.). New York, U.S.A.: Academic Press (1975) 39–44.

[21] DE ANGELO, J.; SIEBERT, K.J.: A new medium for the detection of wild yeast in brewing culture yeast. Am. Soc. Brew. Chem. J. **45** (1987)135–140.

[22] DE RIJK, P.; NEEFS, J.M.; VAN DE PEER, Y.; DE WACHTER, R.: Compilation of small ribosomal subunit RNA sequences. Nucl. Acids Res. **20** (1992) 2075–2089.

[23] DEAK, T.: 1992. Simplified Techniques for Identifying Foodborne Yeasts. Abstr. of 8th Int. Symp. Yeasts. Atlanta, U.S.A. (1992) 45.

[24] DEAK, T.; BEUCHAT, L.R.: Comparison of the SIM, API 20C, and ID 32C systems for identification of yeasts isolated from fruit juice concentrates and beverages. J. Fd Protect. **56** (1993) 583–592.

[25] DEAK, T.: Beuchat, L.R.: Evaluation of the microscan enzyme-based system for the identification of foodborne yeasts. J. Appl. Bactriol. **79** (1995) 439–446.

[26] DEAK, T.; BEUCHAT, L.R.; GUERZONI, M.E.; LILLIE, A.; PETER, G.; ROHM, H.; SCHNÜRER, F.; TABAJDI, P.V.; WESTPHAL, S.: A collaborative study on media for the enumeration of yeasts in foods. Int. J. Fd. Microbiol. **43** (1998) 91–95.

[27] DIJKEN, J.P. VAN; BOSCH, E. VAN DEN; HERMANS, J.J.; RODRIGUES DE MIRANDA, L.; SCHEFFERS, W.A.: Alcoholic fermentation by 'non-fermentative' yeasts. Yeast **2** (1986) 123–127.

[28] DOBZHANSKY, T.: Organismic molecular aspects of species formation. In: Molecular evolution (edited by Ayala, F.J.). Sunderland, Massachusetts, U.S.A.: Sinauer Associates (1976) 95–105.

[29] EILAM, T.; BUSHNELL, W.R.; ANIKSTER, Y.; MCLAUGHLIN, D.J.: Nuclear DNA content of basidiospores of selected rust fungi as estimated from fluorescence of propidium iodide-stained nuclei. Phytopathol. **82** (1992) 705–712.

[30] ERICKSON, J.P.: Hydrophobic membrane filtration method for the selective recovery and differentiation of *Zygosaccharomyces bailii* in acidified ingredients. J. Fd Protect. **56** (1993) 234–238.

[31] FELL, J.W.: Rapid identification of yeast species using three primers in a polymerase chain reaction. Mol. Mar. Biol. Biotechnol. **2** (1993) 174–180.

[32] FELL, J.W.; BOEKHOUT ,T.; FONSECA, A.; SCORZETTI, G.;& STATZELL-TALLMAN, A.: Biodiversity and systematics of basidiomycetous yeasts as determined by large-subunit rDNA D1/D2 domain sequence analysis. Int. J. Syst. Evol. Microbiol. **50** (2000) 1351–1371.

[33] FELL, J.W.; STATZELL, A.C.; HUNTER, I.L.; PHAFF, H.J.: *Leucosporidium* gen. n., the heterobasidiomycetous stage of several yeasts of the genus *Candida*. Antonie van Leeuwenhoek **35** (1969) 433–462.

[34] FUNG, D.Y.C.; LIANG, C.: Critical review of isolation, detection and identification of yeasts from meat products. CRC Crit. Rev. Fd Sci. Nutr. **29** (1990) 341–379.

[35] GUNDERSON, K.; CHU, G.: Pulsed-field electrophoresis of megabase-sized DNA. Mol. Cell. Biol. **11** (1991) 3348–3354.

[36] GURR, E.: The rational use of dyes in biology. London, U.K.: Leonard Hill (1965).

[37] HADRYS, H.; BALICK, M.; SCHIERWATER, B.: Applications of random amplified polymorphic DNA (RAPD) in molecular ecology. Mol. Ecol. **1** (1992) 55–63.

[38] HAMAMOTO, M.; SUGIYAMA, J.; KOMAGATA, K.: DNA base composition of strains in the genera *Rhodosporidium*, *Cystofilobasidium*, and *Rhodotorula* determined by reverse-phase high-performance liquid chromatography. J. Gen. Appl. Microbiol. **32** (1986) 215–223.

[39] HAWKSWORTH, D.L.; KIRK, P.L.; SUTTON, B.C.; PEGLER, D.N.: 1995. Dictionary of Fungi, 8th ed., Wallingford, U.K.: CAB International (1995).

[40] HOCKING, A.D.: Media for preservative resistant yeasts: a collaborative study. Int. J. Fd Microbiol. **29** (1996) 167–175.

[41] HOCKING, A.D.; PITT, J.A.: Dichloran-glycerol medium for enumeration of xerophilic fungi from low-moisture foods. Appl. Envir. Microbiol. **39** (1980) 488–492.

[42] HOLM, C.; MEEKS-WAGNER, D.W.; FANGMAN, W.L.; BOTSTEIN, D.: A rapid, efficient method for isolating DNA from yeast. Gene **42** (1986) 169–173.

[43] IWAGUCHI, S.; HOMMA, M.; TANAKA, K.: Variations in the electrophoretic karyotype analysed by the assignment of DNA probes in *Candida albicans*. J. Gen. Microbiol. **136** (1990) 2433–2442.

[44] JAHNKE, K.-D.: Assessing natural relationships by DNA analysis: techniques and applications. Stud. Mycol. **30** (1987) 227–246.

[45] JAMES, S.A.; COLLINS, M.D.; ROBERTS, I.N.: Genetic interrelationship among species of the genus *Zygosaccharomyces* as revealed by small-subunit rRNA gene sequences. Yeast **10** (1994) 871–881.

[46] JAMES, S.A.; ROBERTS, I.N.; COLLINS, M.D.: Phylogenetic heterogeneity of the genus *Williopsis* as revealed by 18S rRNA gene sequences. Int. J. Syst. Bacteriol. **48** (1998) 591–596.

[47] JOHNSTON, J.R.: Yeast genetics, molecular aspects. In: Yeast, a practical approach (edited by Campbell, I.; Duffus, J.H.). Oxford, U.K.: IRL Press (1988) 107–123.

[48] JONGE, P. DE; DE JONGH, F.C.M.; MEIJERS, R.; STEENSMA, H.Y.; SCHEFFERS, W.A.: Orthogonal-field-alternation gel electrophoresis banding patterns of DNA from yeasts. Yeast **2** (1986) 193–204.

[49] KING, D.A.; HOCKING, A.D.; PITT, J.A.: Dichloran rose bengal medium for enumeration and isolation of molds from foods. Appl. Envir. Microbiol. **37** (1979) 959–964.

[50] KING, A.D.; PITT, J.A.; BEUCHAT, L.R.; CORRY, J.E.L.: Methods for the mycological examination of food. New York, U.S.A.: Plenum Press (1986).

[51] KREGER-VAN RIJ, N.J.W.: Classification of Yeasts. In: The yeasts, Vol. 1, 2nd ed. (edited by Rose, A.H.; Harrison, J.S.). London, U.K.: Academic Press (1987) 5–61.

[52] KÜHLE, VAN DER, A.; JESPERSEN, L.: Detection and identification of wild yeasts in lager breweries. Int. J. Fd Microbiol. **43** (1998) 205–213.

[53] KURTZMAN, C.P.: Prediction of biological relatedness among yeasts from comparison of nuclear DNA complementarity. Stud. Mycol. **30** (1987) 459–468.

[54] KURTZMAN, C.P.: DNA-DNA hybridization approaches to species identification in small genome organisms. Meth. Enzymol. **224** (1993) 335–348.

[55] KURTZMAN, C.P.; FELL, J.: The Yeasts, a taxonomic study, 4th Ed. Amsterdam, The Netherlands: Elsevier (1998).

[56] KURTZMAN, C.P.; PHAFF, H.J.: Molecular taxonomy. In: The yeasts, 2nd ed., Vol. 1 (edited by Rose, A.H.; Harrison, J.S.). London, U.K.: Academic Press (1987) 63–94.

[57] KURTZMAN, C.P.; ROBNETT, C.J.: Identification of clinically important ascomycetous yeasts based on nucleotide divergence in the 5' end of the large-subunit (26S) ribosomal DNA gene. J. Clin. Microbiol. **35** (1997) 1216–1223.

[58] KURTZMAN, C.P.; ROBNETT, C.J.: Identification and phylogeny of ascomycetous yeasts from analysis of nuclear large subunit (26S) ribosomal DNA partial sequences. Antonie van Leeuwenhoek. **73** (1998) 331–371.

[59] KURTZMAN, C.P.; ROBNETT, C.J.; BASEHOAR-POWERS, E.: *Zygosaccharomyces kombuchaensis*, a new ascosporogenous yeast from 'Kombucha tea'. FEMS Yeast Res. **1** (2001) 133–138.

[60] KURTZMAN, C.P.; SMILEY, M.J.; JOHNSON, C.J.: Emendation of the genus *Issatchenkia* Kudriavzev and comparison of species by deoxyribonucleic acid reassociation, mating reaction, and ascospore ultrastructure. Int. J. Syst. Bacteriol. **30** (1980) 503–513.

[61] KURTZMAN, C.P.; SMILEY, M.J.; JOHNSON, C.J.; WICKERHAM, L.J.; FUSON, G.B.: Two new and closely related heterothallic species. *Pichia amylophila* and *Pichia mississippiensis*: characterization by hybridization and deoxyribonucleic acid reassociation. Int. J. Syst. Bacteriol. **30** (1980) 208–216.

[62] LACHANCE, M.A.: Ribosomal DNA spacer variation in the cactophilic yeast *Clavispora opuntiae*. Mol. Biol. Evol. **7** (1990) 178–193.

[63] MAGEE, B.B.; D'SOUZA, T.M.; MAGEE, P.T.: Strain and species identification by restriction fragment length polymorphisms in the ribosomal DNA repeat of *Candida* species. J. Bacteriol. **169** (1987) 1639–1643.

[64] MALESZKA, R.: Clark-Walker, G.D.: A petite positive strain of *Kluyveromyces lactis* has a 300 kb deletion in the rDNA cluster. Curr. Genet. **16** (1989) 429–432.

[65] MANN, W.; JEFFERY, J.: Isolation of DNA from yeasts. Anal. Biochem. **178** (1989) 82–87.

[66] MANNARELLI, B.M.; KURTZMAN, C.P.: Rapid identification of *Candida albicans* and other human pathogenic yeasts by using short oligonucleotides in a PCR. J. Clin. Microbiol. **36** (1998) 1634–1641.

[67] MARMUR, J.: A procedure for the isolation of DNA from microorganisms. J. Mol. Biol. **3** (1961) 208–218.

[68] NAKASE, T.; OKADA, G.; SUGIYAMA, J.; ITOH, M.; SUZUKI, M.: *Ballistosporomyces*, a new ballistospore-forming anamorphic yeast genus. J. Gen. Appl. Microbiol. **35** (1989) 289–309.

[69] ODDS, F.; BERNAERTS, R.: CHROMagar Candida, a new differential isolation medium for the presumptive identification of clinically important *Candida* species. J. Clin. Microbiol. **32** (1994) 1923–1929.

[70] OGLETREE, F.F.; ABDELAL, A.T.; AHEARN, D.G.: Germ-tube formation by atypical strains of *Candida albicans*. Antonie van Leeuwenhoek **44** (1978) 15–24.

[71] PRICE, C.W.; FUSON, G.B.; PHAFF, H.J.: Genome comparison in yeast systematics: delimitation of species within the genera *Schwanniomyces, Saccharomyces, Debaryomyces,* and *Pichia*. Microbiol Rev. **42** (1978) 161–193.

[72] RAEDER, U.; BRODA, P.: Rapid preparation of DNA from filamentous fungi. Lett. Appl. Microbiol. **1** (1985) 17–20.

[73] RAJU, N.B.: A simple fluorescent staining method for meiotic chromosomes of *Neurospora*. Mycologia **78** (1986) 901–906.

[74] ROBERT, V.: BioloMICS, a general system for identification, classification and archiving of strains and species data. Http:://www.bio-aware.com (2000).

[75] ROBERT, V.; EVRARD, P.; HENNEBERT, G.L.: BCCM/Allev 2.00 an automated system for the identification of yeasts. Mycotaxon **64** (1997) 455–463.

[76] RODRIGUES, N.; GONCALVES, G.; PEREIRA-DA-SILVA, S.; MALFEITO-FERREIRA, M.; LOUREIRO, V.: Development and use of a new medium to detect yeasts of the genera *Dekkera /Brettanomyces*. J. Appl. Microbiol. **90** (2001) 588–599.

[77] RUSTCHENKO-BULGAC, E.P.: Variations of *Candida albicans* electrophoretic karyotypes. J. Bacteriol. **173** (1991) 6586–6596.

[78] SAMSON, R.A.; HOCKING, A.D.; PITT, J.I.; KING, A.D.: Modern Methods in Food Mycology. Amsterdam, The Netherlands: Elsevier (1992).

[79] SCHILDKRAUT, C.L.; MARMUR, J.; DOTY, P.: Determination of the base composition of deoxyribonucleic acid from its buoyant density in CsCl. J. Mol. Biol. **4** (1962) 430–433.

[80] SCHULLER, D.; CORTE-REAL, M.; LEAO, C.: A differential medium for the enumeration of the spoilage yeast *Zygosaccharomyces bailii* in wine. J. Fd Protect. **63** (2000)1570–1575.

[81] SEIDLER, R.J.: Mandel, M.: Quantitative aspects of DNA renaturation: DNA base composition, state of chromosome replication, and polynucleotide homologies. J. Bacteriol. **106** (1971) 608–614.

[82] SILÓNIZ, DE, M.I.; VALDERRAMA, M.-J.; PAYO, E.; PEINADO, J.M.: Advances in the development of a methodology to identify common yeast contaminants of high sugar food products. Fd Technol. Biotechnol. **37** (1999) 277–280.

[83] SILÓNIZ, DE, M.I.; VALDERRAMA, M.-J.; PEINADO, J.M.: A chromogenic medium for the detection of yeasts with β-galactosidase and β-glucosidase activities from intermediate moisture foods. J. Fd Protect. **63** (2000) 651–654.

[84] SOR, F.; FUKUHARA, H.: Analysis of chromosomal DNA patterns of the genus *Kluyveromyces*. Yeast **5** (1989) 1–10.

[85] STAIB, F.: Membranfiltration und Negersaat (*Guizotia abyssinica*) – Nährboden für den *Cryptococcus neoformans*-Nachweis (Braunfarbeffekt). Z. Hyg. Infekt. **149** (1963) 329–335.

[86] STENDER, H.; KURTZMAN, C.; HYLDIG-NIELSEN, J.J.; SORENSEN, D.; BROOMER, A.; OLIVEIRA, K.; PERRY-O'KEEFE, H.; SAGE, A.; YOUNG, B.; COULL, J.: Identification of *Dekkera bruxellensis* (*Brettanomyces*) from wine by fluorescence in situ hybridization using peptide nucleic acid probes. Appl. Envir. Microbiol. **67** (2001) 938–941.

[87] STREIBLOVÁ, E.: Cell division in yeasts. In: Recent advances in Microbiology. Proc. 10th Int. Congr. Microbiol. (edited by Pérez-Miravette A.; Peláez, D). Mexico City, D.C., Mexico (1971) 131–140.

[88] TAYLOR, J.W.; NATVIG, D.O.: Isolation of fungal DNA. In: Zoosporic fungi in teaching and research (edited by Fuller, M.S.; Jaworski, A.). Athens, Georgia, U.S.A.: Southeastern Publ. Corp. (1987) 252–258.

[89] THOMAS, D.S.; ACKERMAN, J.C.: A selective medium for detecting yeasts capable of spoiling wine. J. Appl. Bacteriol. **65** (1988) 299–308.

[90] TIMBERLAKE, W.E.: Low repetitive DNA content in *Aspergillus nidulans*. Science **202** (1978) 973–975.

[91] TORNAI-LEHOCZKI, J.; PETER, G.: Typing of foodborne, food spoilage and oportunistic pathogenic yeasts on CHROMagar Candida medium. In: The Rising Power of Yeasts in Science and Industry (edited by van Dijken, J.P.; Scheffers, W.A.). Delft, The Netherlands: Delft Univ. Press (2000) 404.

[92] TRECO, D.A.: Preparation of yeast DNA. In: Current protocols in molecular biology (edited by Ausubel, F.M.). New York, U.S.A.: Wiley & Sons (1989) 13.11.1–13.11.5.

[93] UDEN, N. VAN: Transport-limited fermentation and growth of *S. cerevisiae* and its competitive inhibition. Arch. Mikrobiol. **58** (1967) 155–168.

[94] UNO, I.; MATSUMOTO, K.; ADACHI, K.; ISHIHANA, T.: Characterization of cyclic AMP-requiring yeast mutants altered in the catalytic subunit of proteinkinase. J. Biol. Chem. **259** (1984) 12508–12513.

[95] VALDERRAMA, M.-J.; SILÓNIZ, DE, M.I.; GONZALO, P.; PEINADO, J.M.: A differential medium for the isolation of *Kluyveromyces marxianus* and *Kluyveromyces lactis* from dairy products. J. Fd Protect. **62** (1999) 189–193.

[96] VAUGHAN MARTINI, A.; MARTINI, A.; CARDINALE, G.: Electrophoretic karyotyping as a taxonomic tool in the genus *Saccharomyces*. Antonie van Leeuwenhoek **63** (1993) 145–156.

[97] VILGALYS, R.; HESTER, M.: Rapid genetic identification and mapping of enzymatically amplified ribosomal DNA from several *Cryptococcus* species. J. Bacteriol. **172** (1990) 4238–4246.

[98] VOLLRATH, D.; DAVIS, R.W.: Resolution of DNA molecules greater than 5 megabases by contour-clamped homogeneous electric fields. Nucl. Acids Res. **15** (1987) 7865–7876.

[99] WICKERHAM, L.J.: Taxonomy of yeasts. Techn. Bull. 1029. Washington D.C., U.S.A. (1951).

[100] WICKERHAM, L.J.: Recent advances in the taxonomy of yeasts. Ann. Rev. Microbiol. **6** (1952) 317–332.

[101] WICKERHAM, L.J.; BURTON, K.A.: Carbon assimilation tests for the classification of yeasts. J. Bacteriol. **56** (1948) 363–371.

[102] WICKERHAM, L.J.; KURTZMAN, C.P.; HERMAN, A.I.: 1970. Sexual reproduction in *Candida lipolytica*. Science **167** (1970) 1141.

[103] WINGE, Ö.; ROBERTS, C.: Inheritance of enzymatic characters in yeasts, and the phenomenon of long-term adaptation. Compt. Rendus Trav. Lab. Carlsberg, Sér. Physiologique **24** (1948) 263–315.

[104] YARROW, D.: Methods for the isolation, maintenance and identification of yeasts. In: The yeasts, a taxonomic study, 4th ed. (edited by Kurtzman, C.P.; Fell, J.W.). Amsterdam, The Netherlands: Elsevier (1998) 77–100.

[105] YOUNG, T.W.: Yagiu, M.: A comparison of the killer character in different yeasts and its classification. Antonie van Leeuwenhoek **44** (1978) 59–77.

[106] ZHANG, T.Y.; SMITH, C.L.; CANTOR, C.R.: Secondary pulsed field gel electrophoresis: a new method for faster separation of larger DNA molecules. Nucl. Acids Res. **19** (1991) 1291–1296.

4 PCR methods for tracing and detection of yeasts in the food chain

Jos M.B.M. van der Vossen, Hakim Rahaoui, Monique W.C.M. de Nijs and Bob J. Hartog

4.1 Introduction

Yeasts are important microorganisms for the food industry, and contribute in a positive way in the processing and/or ripening of wine, beer, bread, certain cheeses, kefyr drinks, and whey fermentation [5]. This also demonstrates that many food products are important habitats for yeasts. Yeasts metabolize and proliferate better than bacteria under extreme environmental conditions in terms of pH, water activity and low temperature, and consequently are often involved in spoilage of food products. The low pH of beverages, dairy products, salad dressings and mayonnaise is by no means an obstruction for the growth of yeasts. Some yeast species can utilize organic acids such as lactic, citric and acetic acids. Particularly important for the food industry is that some yeast species can utilize weak acid preservatives such as benzoic acid, propionic acid and sorbic acid [17, 23, 24].

At present, control of microbial spoilage is becoming an increasing challenge for the food industry. The scale at which food products are produced is enormous, and food ingredients represent a considerable economical value. Therefore, spoilage of foods has a huge financial impact. In addition, the consumer demand for milder processing, preservation and storage conditions add to the increased impact of spoilage in the food industry [8].

Proliferation of spoilage yeasts depends highly on four groups of parameters: 1. intrinsic parameters (such as water activity, pH, redox potential, nutrients, antimicrobial compounds); 2. extrinsic parameters (temperature, humidity, atmosphere); 3. modes of processing and preservation, and 4. implicit parameters (e. g., direct and indirect microbial interactions) [5, 19]. The composition of the typical spoilage flora is determined by the above listed parameters. To control microbial food spoilage, it is important to know the specific types of spoilage microorganisms that are able to proliferate in a food product and their ingredients. This offers the opportunity to design measures to circumvent spoilage by the microorganisms identified by tuning the conditions of processing and storage. If an altered product formulation is undesirable, the other possible way to prevent spoilage is by blocking the entrance of specific spoilage organisms into the production chain. This requires the recognition where these microorganisms enter the production chain, which can only be established by tracing the spoilage organism. For this reason an adequate high resolution typing system is needed to allow the discrimination of the spoilage from the non-spoilage organisms. For monitoring the presence of specific spoilage yeasts, polymerase chain reaction (PCR) based detection systems are of great help. Rapid methods based on PCR allow a quick response in terms of adequate measures to prevent further damage.

For identification, typing and detection of spoilage yeast's, to trace routes and sources of contamination in the food production chain, it is necessary to have adequate tools. Traditional identification methods based on morphological and physiological tests are often inadequate in this respect, because they lack discriminatory power and are influenced by environmental conditions [7, 27]. As a consequence, misidentification can occur easily. Traditional methods do not allow fine typing of yeasts at the subspecies level, which is essential for tracing routes of contamination [27]. DNA-based methods have advantages over the traditional phenotypic methods, since they are not influenced by environmental conditions and allow differentiation at various levels ranging from species to strain [27].

This chapter provides an overview of typing techniques for spoilage yeasts based on the polymerase chain reaction (PCR) and how the implementation of these techniques is helpful in designing measures to prevent food spoilage by yeasts. In addition, this chapter will show how the PCR methods can be implemented for the rapid and sensitive detection of specific yeasts in the food area.

4.2 Typing of yeasts by PCR-mediated methods

4.2.1 Basic methodology

PCR has become the state of the art method for the detection and characterization of nucleic acid sequences originating from various organisms. Saiki et al. [21] showed how DNA can be amplified exponentially by using Taq DNA polymerase, and two oligonucleotides for priming DNA synthesis, after repetitions of a temperature cycle. A temperature cycle is necessary to melt the DNA at high temperature, followed by annealing of the primers at lower temperatures and subsequent primer extension by polymerase activity at its optimal temperature for activity. Williams et al. [30] demonstrated that PCR can also be used to generate DNA patterns in the so-called random amplified polymorphic DNA (RAPD) analysis. More recently, restriction fragment length polymorphism analysis of PCR products (PCR-RFLP) and PCR-fingerprinting methods have been developed [1, 2, 3, 4, 12, 13, 14, 15, 22]. Amplified fragment length polymorphism (AFLP™), developed by KeyGene (Wageningen, The Netherlands), is a fingerprinting technique based on the selective PCR amplification of restriction fragments from a total digest of genomic DNA [28].

4.2.2 Prerequisites for yeast typing

In order to type yeasts with PCR-based methods, there is a couple of important requirements. A pure culture of yeast cells is essential for unambiguous typing results. The other prerequisite is the release of DNA from the cells in order to make it available to the PCR reaction. For the rapid release of typing data a quick DNA isolation method has been developed that precedes the PCR reaction. This DNA isolation method is based on milling the yeast cells in the presence of zirconium beads [3], and yields high quantities of DNA with an average size of 5 Kb, which appears appropriate for application in PCR-based typing methods. For AFLP typing, however, DNA of higher quality is required. DNA used for this purpose is isolated by grinding the yeast cells in the presence of liquid nitrogen followed by a subsequent treatment with CTAB buffer. This method originally has been developed for the isolation of plant DNA, and is described by WOLFF et al. [31].

4.2.3 PCR-Restriction Fragment Length Polymorphism (PCR-RFLP)

PCR-RFLP is a PCR mediated approach in which a specific nucleic acid sequence is amplified by PCR, and the amplicon is digested subsequently by a four base pair recognizing restriction endonuclease enzyme to generate a restriction pattern. By using this approach on the small subunit ribosomal RNA encoding region (SS rDNA or 18S rDNA) of yeasts, discrimination at genus and species levels is possible. In order to amplify the 18S rDNA, a yeast universal primer pair has been developed based on the conserved regions in this gene [4] (Fig. 4.2-1). The primer sequences are 5′– GTC TCA AAG ATT AAG CCA TG –3′ (forward primer) and 5′– TAA GAA CGG CCA TGC ACC AC –3′ (reverse primer). The 18S rDNA specific primer set allows the amplification of a c. 1230 bp fragment of the 18S rDNA of all yeast strains tested, which indicates the broad application potential of the defined primer set. Patterns are generated in this 18S rDNA targeted PCR-RFLP approach by using different four base pair recognizing restriction endonucleases. The PCR conditions are 4 min at 94 °C followed by 30 cycles of 1 min at 94 °C, 1 min at 48 °C and 2 min at 72 °C. Finally the mixture is heated at 72 °C for 5 min and subsequently cooled to 4 °C. Ten microliter of the PCR product is directly digested by using different restriction enzymes (e. g., *Mse*I, *Ava*I, *Taq*I, *Hha*I, *Cfo*I) in a final volume of 20 µl.

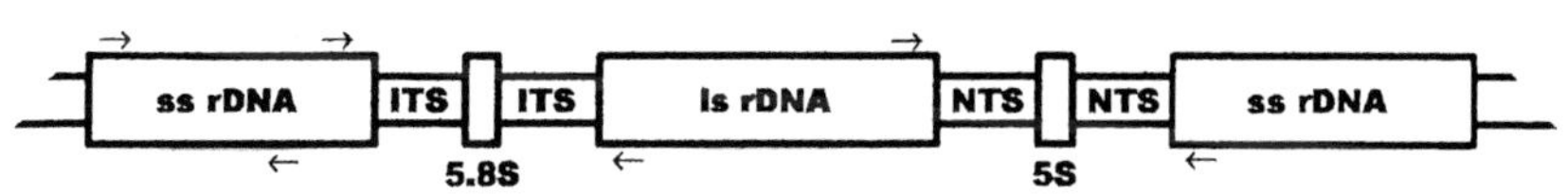

Fig. 4.2-1 Schematic presentation of the ribosomal DNA repeat unit containing the small subunit ribosomal rDNA (ss rDNA) sequence, the internal transcribed spacer (ITS) region with the 5.8S rRNA coding sequence, the large subunit rDNA (ls rDNA) sequence, and the non-transcribed spacer (NTS = intergenic spacer (IGS)) region containing the 5S rRNA coding sequence, respectively.

PCR-RFLP of 18S rDNA appears useful to make a first discrimination between groups of yeast strains. The approach is a low resolution typing method and is therefore suited for the discrimination at genus and species level because distant relationships are always recognized by similarity in some band positions. High-resolution typing methods, such as RAPD, are usually not applicable for the identification of species [16]. Genus and species specific bands are generally absent in RAPD patterns. The restriction enzyme of choice determines the level of discrimination of the rDNA PCR-RFLP typing approach [4]. As an example, the restriction enzyme *Mse*I allows the differentiation of the species *Pichia membranifaciens* (= *Candida valida*), *Saccharomyces cerevisiae*, *Yarrowia lipolyitca* (= *C. lipolytica*), *Zygosaccharomyces bailii and Z. rouxii.* Each of these species shows a unique profile in which all strains belonging to the same species have an identical pattern. Digestion with the restriction endonuclease *Ava*II results in identical patterns for *Z. bailii* and *Z. rouxii,* and digestion with *Cfo*I allows differentiation between *P. membranifaciens* and the other four species only. Some strains obtained from culture collections, which belonged to a certain species based on traditional identification systems, do give atypical patterns. This suggests that the 18S rDNA targeted typing PCR approach is generally a useful identification method. However, some species can not be separated in this PCR-RFLP technique. This is the case for *Z. bailii, Z. bisporus* and *Z. lentus,* which is in agreement with their similar 18S rDNA nucleotide sequences [10, 25]. Since the 18S-rDNA PCR-RFLP technique is not always discriminatory, additional typing data are required to discriminate between the three *Zygosaccharomyces* species. The 18S-rDNA PCR-RFLP analysis using different restriction enzymes is useful for discrimination at the genus to species level.

4.2.4 PCR-RFLP analyses of ribosomal spacer sequences

PCR-RFLP analysis of ribosomal spacer sequences is another approach applicable for yeast typing. Two spacer regions, the internal transcribed spacer (ITS) and the intergenic spacer (IGS), are present in and around the tandemly repeated genes coding for rRNA, respectively (Figure 4.2-1). The ITS region can be amplified with primers ITS1: 5′– TCC GTA GGT GAA CCT GCG G –3′ and NL2: 5′– CTC TCT TTT CAA AGT GCT TTT CAT CT –3′ [29]. The IGS region can be amplified with primers JV51ET: 5’– TGA ACG CCT CTA AGY CAG AAT C –3′, and JV52ET: 5′– TTA TAC TTA GAC ATG CAT GGC-3′ [2].

The length of the ITS amplification products may differ from species to species [6]. Therefore, species can be differentiated more or less based on size polymorphisms of the amplicons. ITS PCR-RFLP provides a higher level of discrimination than the 18S rDNA PCR-RFLP approach. By combining the 18S rDNA PCR-RFLP and the ITS PCR-RFLP approaches, yeast species can be easily differentiated. For some species, like *Z. bailii,* ITS PCR-RFLP allows discrimination at the subspecies level (Fig. 4.2-2). This applies also to *S. cerevisiae* as can be concluded from Table 4.2-1, in which each letter represents a different banding pattern.

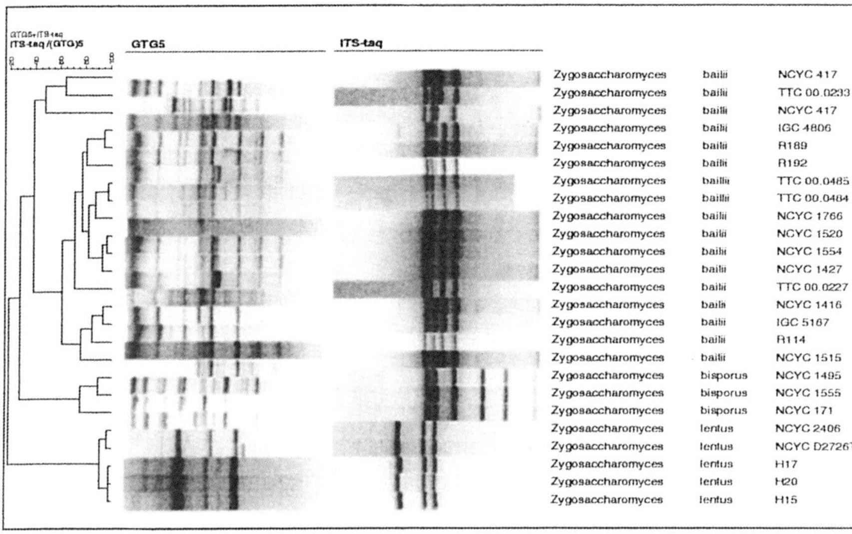

Fig. 4.2-2 Cluster analysis of GTG5 PCR-fingerprint patterns and ITS PCR-RFLP patterns of acid tolerant *Zygosaccharomyces* species.

The PCR-RFLP approach using the complete IGS region is also useful for differentiation at the subspecies level. Amplification of the complete IGS region with a newly developed IGS primer set (Fig. 4.2-1) and subsequent restriction enzyme digestion analysis enables the discrimination of various *S. cerevisiae* strains [2]. The level of discrimination obtained in *S. cerevisiae* is comparable to that observed with some RAPD approaches and PCR fingerprinting using the primer $(GTG)_5$ [2]. The resolution of the different PCR-RFLP approaches using ITS and IGS regions within *S. cerevisiae* is indicated in Table 4.2-1. In this table each pattern type derived from one typing approach is represented by a letter code. The more different letters in a column, the more polymorphisms have been observed. Polymorphism in *S. cerevisiae* has also been observed by MOLINA et al. [18] for the part of the IGS region in between the 26S and the 5S rDNA sequences. The complete IGS region provides possibilities for typing yeast species that do not contain the 5S rDNA sequence in the IGS region. Although some species such as *P. membranifaciens*, *Rhodotorula mucilaginosa* and *Z. bailii* fail to produce an amplification product with the IGS primer set used [2], the IGS PCR-RFLP analysis is useful for differentiation at the intra-species level of those yeasts that allow amplification of the IGS region.

Tab. 4.2-1 Pattern types of *S. cerevisiae* strains obtained from RAPD, PCR-fingerprinting, and ITS- and IGS-targeted RFLP. Identical pattern types are represented by an identical letter code in one column. The last column, overall pattern type, was obtained from the combination of the results from the different typing approaches. A different letter was given to a new combination of pattern types.

	RAPD with primers			**PCR-fingerprinting**		**ITS RFLP**		**IGS RFLP**	**Overall**
	24	28	OPA11	GAC_5	GTG_5	Taq	Mse	Taq	**pattern**
Strain									
W5	A	A	A	A	A	A	A	A	A
W7	A	A	A	A	A	B	B	A	B
W8	A	A	A	A	B	B	B	A	C
W11	B	A	A	A	C	B	B	A	D
W13	A	A	A	A	B	B	C	A	E
B15	A	B	A	A	F	B	C	A	F
B22	C	C	B	B	D	A	D	B	G
B23	C	C	B	B	D	A	D	B	G
B32	C	C	B	B	D	A	D	B	G
B33	C	C	B	B	D	A	D	B	G
B34	D	C	B	B	D	A	D	B	H
B35	B	A	C	A	B	B	D	B	I
B36	D	C	nd	B	D	A	D	B	J
B38	D	C	D	B	E	C	C	C	K
B45	E	A	E	A	B	A	C	A	L
IGC4455	F	A	F	A	F	A	C	nd	M

nd = not determined

4.2.5 PCR-fingerprinting

PCR-fingerprinting using the simple repeat primers $(GAC)_5$, $(GTG)_5$ and bacteriophage M13 core sequence (GAGGGTGGXGGXTCT) is useful to discriminate at the species level [2, 3, 14]. In addition, this typing approach allows discrimination at the subspecies level. The method has been developed on the basis of previous results obtained from DNA fingerprinting analysis [14], in which the simple repeat primers were used as DNA probes in Southern hybridization, to generate patterns with simple repeated DNA motifs that occur at multiple genomic sites in Eukaryotes. Although $(GACA)_4$ has also been reported to be a useful primer for yeast typing by DNA fingerprinting [14], the primer fails to give good PCR fingerprinting patterns with many yeasts. The other simple repeat DNA sequences, in particular the GTG and GAC repeats, appear to be generally present in Eukaryotes, including the yeasts. PCR-fingerprinting appears to be a robust system for generating banding patterns, and enables the discrimination between three closely related species *Z. bailii*, *Z. bisporus* and *Z. lentus* (Fig.

4.2-2). Since this approach has been successful to differentiate between closely related species, the $(GTG)_5$ fingerprints are used to build a database for the rapid identification of new isolates. At present, this TNO-database contains the PCR fingerprint patterns of 650 yeast species, in particular *Issatchenkia orientalis* (including the anamorph *C. krusei*), *Pichia anomala* (including the anamorph *C. pelliculosa*), *Pichia fermentans* (including the anamorph *C. lambica*), *P. membranifaciens* (including the anamorph *C. valida*), *Saccharomyces cerevisiae, S. pastorianus, S. bayanus, Zygosaccharomyces bailii, Z. lentus, Z. rouxii, Z. bisporus, Z. fermentati,* and *Yarrowia lipolytica* (including the anamorph *C. lipolytica*).

The fact that PCR-fingerprinting provides information at the subspecies level renders this approach applicable for tracing routes of contamination in the production chain [3]. Table 4.2-1 shows to what extent this method allows discrimination within *S. cerevisiae.* The primer $(GTG)_5$ provides a higher resolution than $(GAC)_5$ in PCR-fingerprinting, and this observation is true for many other species. Therefore, PCR-fingerprinting with primer $(GTG)_5$ is implemented in our analysis of yeasts from the food production chain both for the identification and to trace the routes of contamination [3].

4.2.6 Random amplified polymorphic DNA (RAPD)

RAPD assays, using selected 10-mer oligonucleotides, allow the discrimination at the species and subspecies levels since minor differences between strains belonging to the same species are revealed [1, 4]. Moreover, discrimination close to the strain level has been demonstrated with selected 10-mer primers (Table 4.2-2) in RAPD analysis. Unfortunately, no primers have been found to enable discrimination at the strain level.

Tab. 4.2-2 Oligonucleotide primer sequences used in RAPD assay

Primer number	Nucleotide sequence (5´–3´)
13	CCG CCA CTG T
15	CGG CCC CGG T
18	GCA AGT AGC T
20	AGG AGA ACG G
21	GCT CGT CGC T
24	GCG TGA CTT G
25	TGG TCC TGC G
26	TGC TGG GCG G
28	AGG AGG AGG A
OPA-01	CAG GCC CTT C
OPA-11	CAA TCG CCG T
OPA-18	AGG TGA CCG T

The reproducibility of RAPD has been a matter of discussion. There is no doubt that RAPD patterns differ among laboratories. Nevertheless, the problem of reproducibility within a single laboratory can be solved by obeying to very strict rules concerning the overall temperature profiles, especially the time involved in heating the reaction mixture from the annealing temperature to the temperature of primer-extension during the amplification procedures [20]. The discriminatory power of the RAPD typing approach is determined by the 10-mer primer used. Table 4.2-1 shows how the resolution varies within *S. cerevisiae* strains depending on the RAPD primer. Each letter in a column represents a different pattern. The more different letters are used in a column, the more polymorphism is visualized by the typing approach used. The table also shows that the combination of RAPD and other typing methods enhances the resolution up to the strain level [2]. This is represented by an overall pattern letter code in the last column of Table 4.2-1. Interestingly, the four strains B22, B23, B32, and B33 of *S. cerevisiae*, that showed an identical overall RAPD type, are isolated from the same production chain. This observation provides support for the use of high-resolution typing techniques to trace routes of contamination of a particular yeast strain.

4.2.7 Amplified Fragment Length Polymorphism (AFLP)

AFLP typing, developed by Keygene (Wageningen, The Netherlands), is a PCR-based method that includes restriction enzyme digestion of the genome of interest. The method was developed to screen plant cultivars differing in agronomic traits [28]. The typing system was also shown to be useful for the typing of microorganisms including yeasts [11]. AFLP typing is based on the principle of the selective amplification of a subset of restriction fragments from a mixture of DNA fragments originating from a digestion of genomic DNA. The number of resulting amplicons depends on the organism and the oligonucleotide primers used in PCR, which have selective overhangs at the 3′ prime-ends (Fig. 4.2-3). The resolution can be influenced by the choice of the restriction enzymes and the selectivity of the oligonucleotide primers used for amplification.

The AFLP technique allows a better discrimination between *Z. bailii* isolates than PCR fingerprinting with $(GTG)_5$ (Fig. 4.2-4, B. Mayo and H. Rahaoui, unpubl. observ.). The technique enables the differentiation of highly related bacterial strains as well. The good reproducibility of the AFLP technique is an advantage over RAPD and PCR-fingerprinting. However, to fully profit from this good reproducibility we recommend the use of a commercial AFLP kit and the standardized protocol of the supplier (e. g., Applied Biosystems). Fragment analysis can be routinely performed on automated sequence analysis systems in combination with specific software such as Bionumerics (Applied Maths, Kortrijk, Belgium) or AFLP-Quantar (Keygene, Wageningen, The Netherlands).

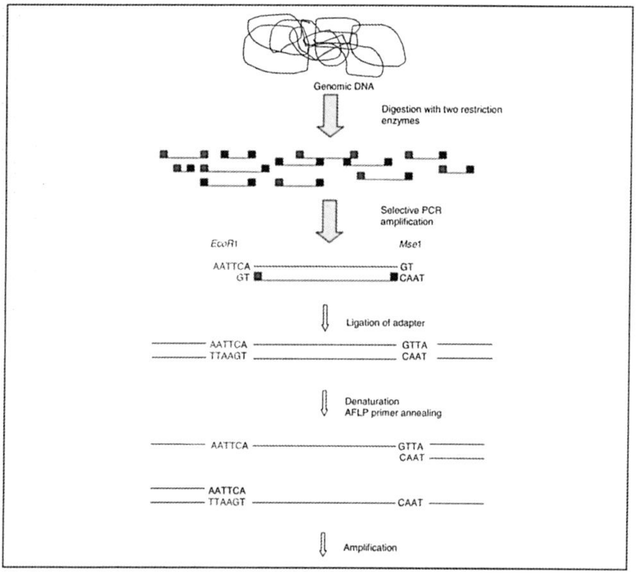

Fig. 4.2-3 Schematic presentation of the AFLP technique. In the first step total cellular DNA is digested with two restriction enzymes (*Eco*R1 and *Mse*1 for example). Subsequently, adapters are ligated to both ends of the fragments. After ligation of the adapters, a portion of the extended fragments is amplified in a selective way by the use of selective primers. These primers are selective at their 3' prime end. If there is no full match of the primer with the extended fragment, such a fragment will not be amplified. On the contrary, if there is a full match at both ends, the extended fragment will be amplified. By separating the extended and amplified fragments by gel electrophoresis a banding pattern (AFLP pattern) will be the result.

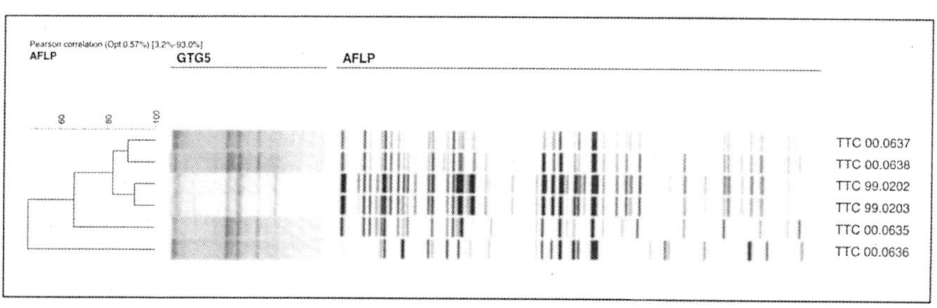

Fig. 4.2-4 Cluster analysis of GTG5 PCR fingerprint patterns and AFLP patterns of *Zygosaccharomyces bailii* strains. The GTG5 patterns display a high degree of similarity. The AFLP patterns on the other hand do show more heterogeneity.

4.3 Implementation of PCR based methods in food production lines

PCR typing of spoilage yeasts is used to monitor production processes and to trace routes of contamination in food production lines. PCR fingerprinting with $(GTG)_5$ and PCR-RFLP of 18S rDNA and ITS are used to identify yeast isolates from food production plants. These identification approaches are successful, when combined with a database containing well-identified pattern types (e. g., the database present at TNO, Zeist, The Netherlands). Implementation of these PCR based typing techniques in the food production chain is performed in several industries.

4.3.1 Sampling and culture conditions

In order to find the origin and transmission route of a contaminant, it is important to find identical genotypes of the spoilage yeast upstream in the production plant and the end product. Immediately after the observation of spoilage, samples need to be taken from various points in the production line, including ingredients and environmental samples. Isolation of yeasts from the spoiled sample needs to be carried out with caution to reduce the risk of isolation of ubiquitous environmental yeasts. On the other hand, the spoilage yeast of interest may refuse to grow if culture conditions are not chosen properly. We give some examples to illustrate the pitfalls of tracing spoilage yeasts in the food chain.

The first example is the spoilage of a fruit juice stored in tanks at low temperature. The spoilage yeast encountered turned out to be a psychrophilic species of the genus *Mrakia*, which is only able to grow below 15 °C. Consequently, culturing of the juice samples is required at low temperatures. PCR-RFLP of the ITS and PCR-fingerprinting allow easy discrimination between *Mrakia* and other spoilage yeasts such as *Zygosaccharomyces* species

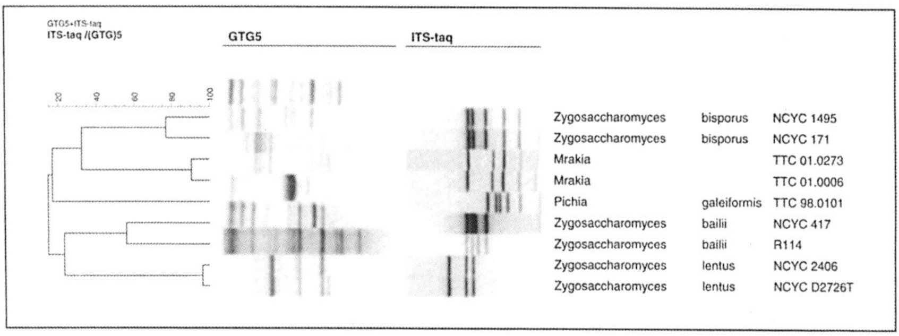

Fig. 4.3-1 Cluster analysis of GTG5 PCR-fingerprinting and ITS PCR RFLP patterns of acid tolerant yeasts.

(Fig. 4.3-1). In this case, the colony morphology of the *Mrakia* species differs from those of *Zygosaccharomyces* species as well (Fig. 4.3-2).

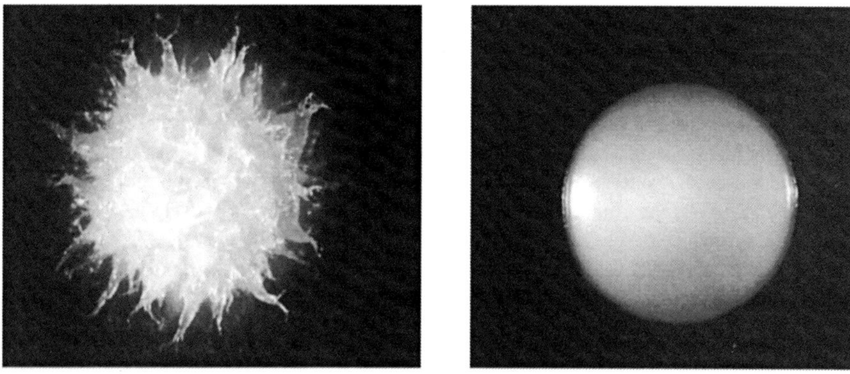

Fig. 4.3-2 Colony morphology of *Mrakia* species (left panel) and *Zygosaccharomyces* species (right panel).

Another example is the search for spoilage yeasts in the production environment of sauces with a low pH. The species *Z. bailii, Z. bisporus* or *Z. lentus* are known to occasionally spoil sauces with a low pH. Sampling and subsequent culturing in the presence of 1 % acetic acid, as is normally recommended for the enrichment of these species, resulted in the isolation of an extreme acid tolerant isolate of *Pichia galeiformis*, as shown by nucleotide sequence analysis of the ribosomal RNA encoding region. Strains of this later species were abundantly found in the production environment and identified by ITS PCR-RFLP and PCR fingerprinting. This species was never found responsible for spoilage of the end product. Because of the overgrowth by *P. galeiformis,* the acid tolerant *Zygosaccharomyces* species, finally found to cause spoilage of the end product, seemed to be absent. This observation has led to the design of a new medium for the isolation of *Zygosaccharomyces* species, containing weak acid preservatives, and incubation under anaerobic conditions (J.M.B.M. VAN DER VOSSEN, unpubl. observ.). The use of this medium and culturing conditions resulted in the enrichment and isolation of *Zygosaccharomyces* species from the production chain. This allowed us to compare their genotype with that of the spoilage strain with ITS PCR-RFLP and PCR fingerprinting. From these examples it can be concluded that sampling and culture conditions are as important as PCR fine typing methods to successfully trace routes and sources of contamination by yeasts. The comparison of isolate-specific molecular typing data, obtained from the yeasts isolated from the end product and samples taken upstream in the production line, can be conclusive in understanding the source of yeast contamination.

4.3.2 Examples of tracing spoilage yeast

PCR typing methods are successfully used to trace contamination routes of yeasts. The following examples illustrate the application of PCR based monitoring techniques in production chains. In one occasion, iso-glucose syrup contaminated with *Z. bailii* from a particular supplier has been identified as the source of spoiled mayonnaise by showing that isolates both from the syrup and mayonnaise displayed identical genotypic patterns after PCR fingerprinting with GTG and ITS PCR-RFLP typing. In another case, a contract packer has been responsible for the spoilage of mayonnaise filled in plastic bags by *Z. bailii.* By using PCR typing it was shown that *Z. bailii* strains present in the mayonnaise in the plastic bags contained the same genotype as the strains isolated from the filling machine, which was not properly cleaned, of the contract packer. In addition, the spoilage type could not be isolated from the mayonnaise in the container from the mother plant. In another example, a spoilage outbreak of *I. orientalis* in mayonnaise has been demonstrated to originate from supplied egg yolk, using PCR typing techniques. Pots of mayonnaise contained occasionally a yeast colony on top of the surface of the product. A rapid monitoring of samples taken upstream from the finished product resulted in the isolation of the spoilage type in the egg yolk. Further inspection of the egg yolk container indicated that the tap of the container was dirty. In this case, the tap of the egg yolk container was found difficult to clean, and the design of the tap was adapted to improve cleanability. In a fourth case of yeast spoilage, an in-house spoilage yeast, *Z. bailii,* remained present after *in situ* cleaning and sterilization of the production line. A valve in the line appeared to contain an area that was not cleaned in the sanitation process. At this particular spot in the process line, *Z. bailii* strains were isolated that had the same PCR fingerprinting and ITS PCR-RFLP pattern as observed in the spoilage strains from the end product. Based on this outcome the hygienic situation of the processline had been improved accordingly.

4.4 Methods for yeast detection

For several yeasts it is known that if they are present, they will spoil the product. This is in particular true for *Z. bailii, Z. bisporus* and *Z. lentus* in low pH products such as mayonnaise, salad dressings, fruit juices, and for *Brettanomyces* isolates in wine and fruit juices. The contamination level of these spoilage yeasts in the production environment of the products mentioned must be kept as low as possible. In order to get an estimate of the risks of a spoilage incident, a rapid monitoring system can be of great advantage. Rapid monitoring systems allow rapid interventions, which cannot be established by methods that take a week before the outcome is known. Slow methods only allow a retrospective view on the yeast problems in a food production plant.

Some rapid PCR based methods have been developed for the detection of yeast species. However, since components of the food may inhibit the PCR reaction and/or contamination levels may be low, additional efforts are needed to guarantee successful PCR detection of the yeast of interest in the sample. False negative results need to be eliminated, and, on the other hand, only the species of interest should give a positive PCR result. False positive signals, due to a PCR product derived from another species, have to be excluded as well. Thus far, only a limited number of specific yeast detection systems has been developed.

STUBBS et al. [26] developed a PCR coupled ligase detection reaction that allows the specific detection of *Z. bailii*. Since *Z. bailii* is an important yeast responsible for the spoilage of mayonnaise and salad dressings, a *Z. bailii* specific PCR detection system is useful for monitoring purposes. Demonstrating the absence of *Z. bailii* at critical points in the process, ensures to a higher extent the spoilage free production of mayonnaise and salad dressings. In the wine area, IBEAS et al. [9] have developed a detection system for *Dekkera-Brettanomyces* strains by isolating and sequencing a specific randomly isolated fragment from the species, which did not cross-hybridize with the yeasts of the flor in sherry. In a nested PCR detection system based on this *Dekkera* specific fragment, the target DNA is efficiently amplified, and less then 10 yeast cells can be detected in PCR without the need for DNA purification. The detection method allowed the detection of *Dekkera* cells in suspected sherry samples (1,5 ml). Therefore, specific detection systems based on the PCR methodology have been developed for the assessment of the microbiological quality of raw materials and end products. These detection systems are highly sensitive and allow the detection of specific yeasts at low contamination levels.

4.5 Conclusions

The typing methods described are successfully implemented in the food industry (Fig. 4.5-1). PCR mediated methods are useful for the rapid identification of yeast strains involved in spoilage and those isolated from the production environment. PCR-mediated typing allows discrimination below the species level, and is useful to trace sources and routes of contamination. Based on the outcome of high-resolution typing methods, appropriate intervention measures have been conducted in the food industries involved. Most of the examples originate from the beverage and the sauce industry. Nevertheless, these methods are generally applicable where yeasts and other microorganisms are involved in spoilage.

The typing methods presented can also be helpful in strain selection procedures. Since PCR typing data can be coupled to physiological data, predictions can be made about both spoilage capacity (negative microbiology) and fermentation potential (positive microbiology). This can be of great help to formulate or to select appropriate spoilage indicators as well as typical strains with a potential for food fermentations.

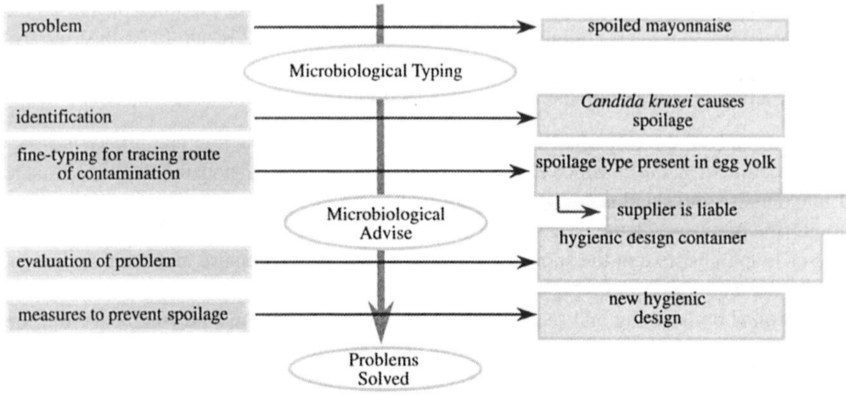

Fig. 4.5-1 The result of the implementation of microbiological typing in troubleshooting in case of a spoilage incident in the food industry

Molecular genetic techniques are progressing rapidly. At present, we are entering the decade of the micro-array technology. This technology has a great potential for the analysis of complex microbiological ecosystems, as well as to collect information on gene expression under various process conditions to monitor the physiological state of microbial cells. The micro-array technology can provide an insight in the *in-situ* microbial situation of ingredients, processes and end products, and is therefore adding to the quality and safety of food products (see also Chapter 7).

4.6 References

[1] BALEIRAS COUTO, M.M.; VAN DER VOSSEN, J.M.B.M.; HOFSTRA, H.; HUIS IN 'T VELD, J.H.J.: RAPD analysis: a rapid technique for differentiation of spoilage yeasts. Int. J. Fd Microbiol. **24** (1994) 249–260.

[2] BALEIRAS COUTO, M.M.; EIJSMA, B.; HOFSTRA, H.; HUIS IN 'T VELD, J.H.J.; VAN DER VOSSEN, J.M.B.M.: Evaluation of molecular typing techniques to assign genetic diversity among strains of *Saccharomyces cerevisiae*. Appl. Environ. Microbiol. **62** (1996) 41–46.

[3] BALEIRAS COUTO, M.M.; HARTOG, B.J.; HUIS IN 'T VELD, J.H.J.; HOFSTRA, H.; VAN DER VOSSEN, J.M.B.M.: Identification of spoilage yeasts in a food production chain by microsatellite PCR fingerprinting. Fd Microbiol. **13** (1996) 59–67.

[4] BALEIRAS COUTO, M.M.; VOGELS, J.T.W.E.; HOFSTRA, H.; HUIS IN 'T VELD, J.H.J.; VAN DER VOSSEN, J.M.B.M.: Random amplified polymorphic DNA and restriction enzyme analysis of PCR amplified rDNA in taxonomy: two identification techniques for food-borne yeasts. J. Appl. Bacteriol. **79** (1995) 525–535.

[5] DEÁK, T.: Food borne yeasts. Adv. Appl. Microbiol. **39** (1991) 179–278.

[6] ESTEVE-ZARZOSO, B.; BELLOCH,; URUBURU, F.; QUEROL, A.: Identification of yeasts by RFLP analysis of the 5.8S rRNA gene and the two ribosomal transcribed spacers. Int. J. Syst. Bacteriol. **49** (1999) 329–337.

[7] HOFSTRA, H.; VAN DER VOSSEN, J.M.B.M.; VAN DER PLAS, J.: Microbes in food processing technology. FEMS Microbiol. Rev. **15** (1994) 175–183.

[8] HUIS IN 'T VELD, J.H.J.: Microbial and biochemical spoilage of foods: an overview. Int. J. Fd Microbiol. **33** (1996) 1–18.

[9] IBEAS, J. I.; LOZANO, I.; PERDIGONES, F.; JIMENEZ, J.: 1996. Detection of *Dekkera-Brettanomyces* strains in sherry by a nested PCR method. Appl. Envir. Microbiol. 62:998–1003.

[10] JAMES, S.A.; COLLINGS, M.D.; ROBERTS, I.N.: Genetic interrelationship among species of the genus *Zygosaccharomyces* as revealed by small-subunit rRNA gene sequences. Yeast **10** (1994) 871–881.

[11] JANSSEN, P.; COOPMAN, R.; HUYS, G.; SWINGS, J.; BLEEKER, M.; VOS, P.; ZABEAU, M.; KERSTERS, K.: Evaluation of the DNA fingerprinting method AFLP as a new tool in bacterial taxonomy. Microbiology **142** (1996) 1881–1893.

[12] KUNZE, G.; KUNZE, I.; BARNER, A.; SCHULZ, R.: Classification of *Saccharomyces cerevisiae* strains by genetical and biochemical methods. Monatsschr. Brauwissensch. **46** (1993) 132–136.

[13] LAVALLEE, F.; SALVA, Y.; LAMY, S.; THOMAS, D.Y.; DEGRE, R.; DULAU, L.: PCR and DNA fingerprinting used as quality control in the production of wine yeast strains. Am. J. Enol. Vitic. **45** (1994) 86–91.

[14] MEYER, W.; MITCHELL, T.G.; FREEDMAN, E.; VILGALYS, R.: Hybridization probes for conventional DNA fingerprinting used as single primers in the polymerase chain reaction to distinguish strains of *Crytococcus neoformans*. J. Clin. Microbiol. **31** (1993) 2274–2280.

[15] MAIWALD, M.; KAPPE, R.; SONNTAG, H.G.: Rapid presumptive identification of medically relevant yeasts to the species level by polymerase chain reaction. J. Med. Vet. Mycol. **32** (1994) 115–122.

[16] MESSNER, R.; PRILLINGER, H.: *Saccharomyces* species assignment by long range ribotyping. Antonie van Leeuwenhoek **67** (1995) 363–370.

[17] MILLER, M.W.: Yeasts in food spoilage: an update. Fd Technol. **33** (1979) 76–80.

[18] MOLINA, F.I.; JONG, S.-C.; HUFFMAN, J.L.: PCR amplification of the 3' external transcribed and intergenic spacer of the ribosomal DNA repeat unit in three species of *Saccharomyces*. FEMS Microbiol. Lett. **108** (1993) 259–264.

[19] MOSSEL, D.A.A.: Microbial deterioration of foods. In: Microbiology of foods. The ecological essentials of assurance and assessment of safety and quality, 3rd ed. Utrecht, The Netherlands: Utrecht University (1982) 29–49.

[20] PENNER, G.A.; BUSH, A.; WISE, R.; KIM, W.; DOMIER, L.; KASHA, K.; LAROCHE, A.; SCOLES, G.; MOLNAR, S.J.; FEDA, G.: Reproducibility of random amplified polymorphic DNA (RAPD) analysis among laboratories. PCR protocols and applications **2** (1993) 341–345.

[21] SAIKI, R.K.; GELFAND, D.H.; STOFFEL, S.; SCHARF, S.J.; HIGUCHI, R.; HORN, G.T.; MULLIS, K.B.; EHRLICH, H.A.: Primer-directed enzymatic amplification of DNA with a thermostable DNA polymerase. Science **239** (1988) 487–491.

[22] SHEN, P.; JONG, S.-C.; MOLINA, F.I.: Analysis of ribosomal DNA restriction patterns in the genus *Kluyveromyces*. Antonie van Leeuwenhoek **65** (1994) 99–105.

[23] STEELS H.; JAMES S.A.; ROBERTS I.N.; STRATFORD M.: *Zygosaccharomyces lentus*: a significant new osmophilic, preservative-resistant spoilage yeast, capable of growth at low temperature. J. Appl. Microbiol. **87** (1999) 520–527.

[24] STEELS H.; JAMES S.A.; ROBERTS I.N.; STRATFORD M.: Sorbic acid resistance: the inoculum effect. Yeast **16** (2000) 1173–1183.

[25] STEELS, H.; BOND, C.J.; COLLINS, M.D.; ROBERTS, I.N.; STRATFORD, M.; JAMES, S.A.: *Zygosaccharomyces lentus* sp. nov., a new member of the yeast genus *Zygosaccharomyces*. Int. J. Syst. Bacteriol. **49** (1999) 319–327.

[26] STUBBS, S.; HUTSON, R.; JAMES, S.; COLLINS, M.D.: Differentiation of the spoilage yeast *Zygosaccharomyces bailii* from other *Zygosaccharomyces* species using 18S rDNA as target for a non-radioactive ligase detection reaction. Lett. Appl. Microbiol. **19** (1994) 268–272.

[27] VAN DER VOSSEN, J.M.B.M.; HOFSTRA, H.: DNA based typing, identification and detection systems for food spoilage microorganisms: development and implementation. Int. J. Fd Microbiol. **33** (1996) 35–49.

[28] VOS, P.; HOGERS, R.; BLEEKER, M.; REIJANS, M.; VAN DE LEE, T.; HORNES, M.; FRIJTERS, A.; POT, J.; PELEMAN, J.; KUIPER, M.; ZABEAU, M.: AFLP: a new technique for DNA fingerprinting. Nucl. Acids Res. **23** (1995) 4407–4414.

[29] WHITE, T.J.; BRUNS, T.; LEE, S.; TAYLOR, J.W.: Amplification and direct sequencing of fungal ribosomal RNA genes for phylogenetics. In: PCR protocols: a guide to methods and applications (edited by Innis, M.A; Gelfand, D.H.; Sninsky, J.J.; White, T.J.). San Diego, California, U.S.A.: Academic Press Inc. (1990) 315–322.

[30] WILLIAMS, J.G.K.; KUBELIK, A.E.; LEVAK, K.J.; RAFALSKI, J.A.; TINGEY, S.C.: DNA polymorphisms amplified by arbitrary primers are useful as genetic markers. Nucl. Acids Res. **18** (1990) 6531–6335.

[31] WOLFF, K.; PETERS – VAN RIJN, J.; HOFSTRA, H.: RFLP analysis in *Chrysanthemum*. 1. Probe and primer development. Theor. Appl. Genet. **88** (1994) 472–478.

5 Data processing

VINCENT ROBERT

5.1 Introduction

The procedures and methods described in the previous chapters produce different types of data that can be used for a wide range of goals. Some users want to isolate and count the number of colony forming units (cfu) per gram of substrate, and others want to identify strains at the genus or the species level. Others need to track particular strains or are interested in the properties of the strains they are dealing with. Therefore, there is a need to store and handle data in proper and different ways.

In this chapter, we present and briefly discuss a non-exhaustive selection of techniques and methods that are used to identify and classify organisms or their properties. The problems and difficulties that can be encountered in data management are discussed, together with the differences between identification based on strain and species databases.

Data are the building blocks used to construct a classification and, therefore, identification procedures depend directly on them. The nature and the combination (e. g., physiology, sequences, morphology, quantitative versus qualitative), the amount (e. g., number of tests performed and accounted, repetitions) and the quality (e. g., reliability, subjectivity of data retrieval methods) of the data to be included in a database as well as the goals underlying the creation of data banks have a profound impact on the ways the data have to be stored, processed and summarized. For example, sequence data can be used with different objectives in mind than physiological ones. While no data transformation procedures are required for sequences, physiological information can be observed in a variety of ways and may be interpreted and, therefore, transformed before being stored in databases. Comparisons of sequences can be performed using algorithms that have properties, which are completely different than those used for physiological features.

All the steps involved in the handling of data, including data storing, comparing, processing and summarizing, result in a reduction of fit between the original information and the obtained summary (Fig. 5.1-1). This loss can be reduced by a clear understanding of the methods used by the available data retrieval systems, often linked to – commercial – identification systems, their processing methods, advantages and limitations.

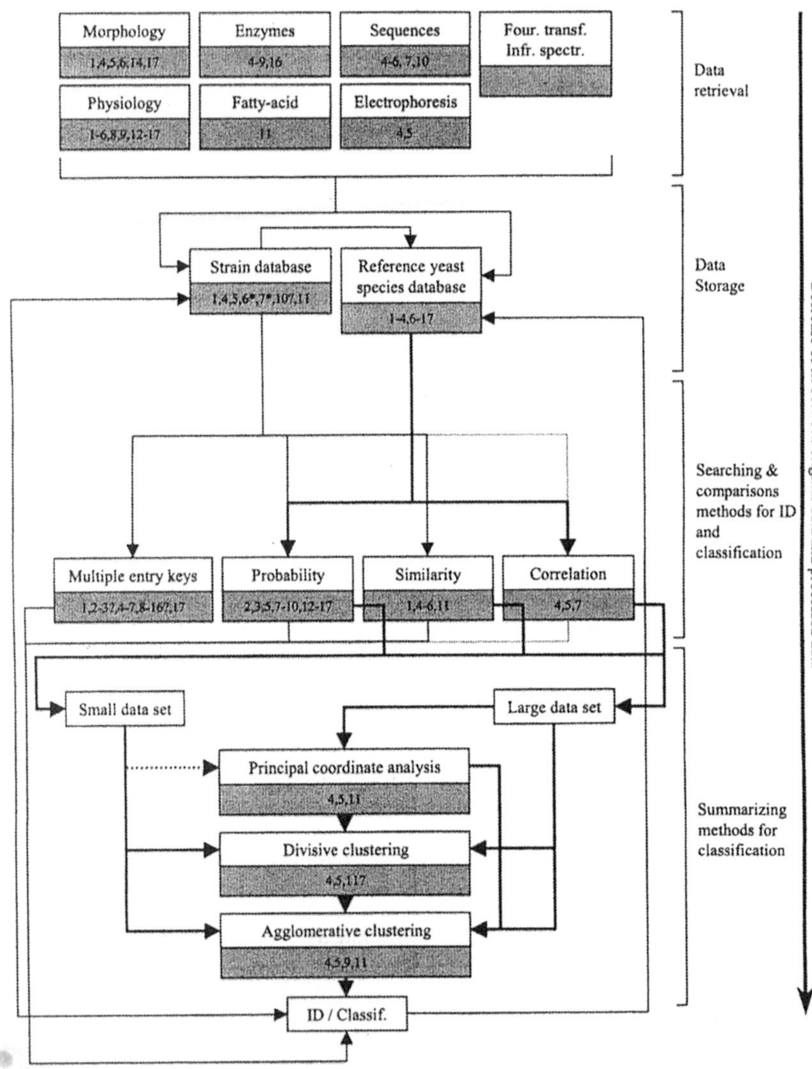

Fig. 5.1-1 Scheme showing the main steps involved in data processing and management. Each step represent losses of fit between the original and processed data. The numbers in the lower grey part of the boxes are related to yeast identification system described in Table 5.3-1. Question marks attached to the number means 'uncertain information', * means partly only. Identification and classification processes are represented by thin and bold arrows respectively. Dotted arrows means rarely applied, not necessary or not suitable.

5.2 Identification and classification

5.2.1 Basic principles

Identification of organisms is an important step in understanding and analyzing biological processes. The ability to recognize and to name the organisms that are responsible for a given spoilage event or those that produce a good wine, is of key importance for the food industry. Recognition and naming are two different steps. Recognition is the process where similarities and differences between an unknown isolate and a set of previously known or recorded organisms are analyzed in order to find an exact or an almost exact match (reference system). When the recognition is completed and if the identical reference organism turns out to be a species description, the unknown can be named and identified at the species level. Most identification systems for yeasts, which are based on morphological, physiological and molecular data, have been developed to provide such a species level identification.

The definition of a yeast species is a summary made by systematicians and is based on the observation of taxonomically informative characteristics of one or several strains, thought to belong to the same biological, phenetic and/or phylogenetic entity. These entities can be large like in *Saccharomyces cerevisiae* or small when the species is known from a single strain only. The size or the volume of such a cluster does not only depend on the number of strains used for its description, but also on the variability of the set of characteristics that have been selected and observed.

To illustrate these concepts that have a huge impact on the way the identification or the comparative results are interpreted, we have plotted a series of species in a three dimensional space (Fig. 5.2-1). The three dimensions of the space represent three different characters and each dot or volume, a strain. The position of the strains is obtained by their states for the three characteristics. When the three states are single values, a discrete point represents the strains (see strain A in Fig. 5.2-1a), whereas if the states for one or several characteristics are ranges, the strains are described as surfaces or volumes (see strain B in Fig. 5.2-1a). Strains are grouped into four clusters (A to D) and each cluster represents a species. The identification of the two unknown strains (U1 and U2), represented by the strikethrough balls, will be different depending on the reference system employed. If the references are known strains (Fig. 5.2-1a), none of the unknown strains shows a perfect match with any reference. On the other hand, in a species (or any taxonomic level above the strain) reference system (Fig. 5.2-1b), strain U1 remains unidentified, while U2 belongs to species C. The species reference system allows a generalization and an easier extrapolation of the potential of the identified strains by using the published species-related data. In such a system, basic information like the number of strains used to produce the species description or definition, their relative positions and volume (i. e., their intrinsic variability) are rarely, if ever, available. New characters added to the system (i. e., introduction of new dimensions) change the spatial position of the clusters. Figures 5.2-1c and 5.2-1d illustrate this principle. Four species or groups

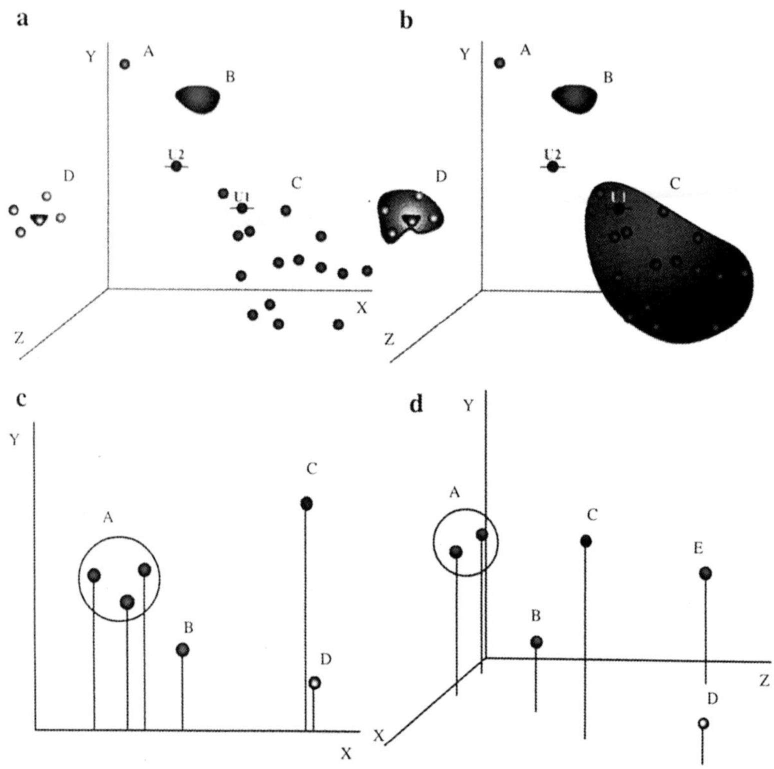

Fig. 5.2-1 Schematic representation of the relative positions of different strains and species in a simplified three-dimensional space. On the left of the figure (Fig. 5.2-1a), the reference system is based on strains, while on the right (Fig. 5.2-1b), the reference system is based on groups of single (clusters A and B) or multiple strains (clusters C and D) representing species (or any taxonomic level above the strain). Both unknown strains U1 and U2 can not be identified with an exact match in a strain based system (Fig. 5.2-1a), while U1 can be attributed to cluster C in a species based system (Fig. 5.2-1b). New characters added to a system (i. e., introduction of new dimensions), change the spatial position of the clusters (Fig. 5.2-1c). The addition of a new dimension or criterion (Fig. 5.2-1d) leads to the splitting of group A into two groups, A and E.

(A to D) are found when two criteria are accounted (Fig. 5.2-1c). The addition of a new dimension or criterion (Fig. 5.2-1d) leads to the splitting of group A into two groups, A and E.

The addition of new and relevant data is useful and contributes to a better understanding of the real positioning of the cluster, but the use of previously published data related to the species name is reduced. Splitting and clumping of species concepts, although useful for a better understanding of the system, lead to the same type of problems. In a species concept based reference system, the volume of the different clusters can be quite different. In our example (Figs. 5.2-1a and 5.2-1b), clusters A and B have small volumes, while C and D are larger. This is mainly due to the number of strains used to define the clusters. Therefore, well-documented species will attract more additional strains by a kind of a gravitational effect. This can result in the creation of very large clusters with a potentially high internal variability (e. g., *Candida albicans* and *Saccharomyces cerevisiae*). In such clusters, the extrapolation of the potential properties of a given strain on the basis of the species profile can be hazardous and misleading.

Taxonomic concepts of species in yeasts are based on a mixture of biological, phenetic and phylogenetic principles. In the past, groupings were mainly based on phenetic observations and on the sexual characteristics of the strains. The introduction of molecular techniques, such as DNA sequencing, enforced the use of the phylogenetic species concept. Therefore, the earlier phenetic descriptions, based on the morphology and the physiology of the species, need to be constantly reviewed and, consequently, the related databases have to be reconstructed. For this reason it is important to use up-to-date databases or reference books. Several identification systems for species are based on outdated data and should be used with care. We will shortly review the existing identification systems, their relevance and how the identity can be determined.

These principles have to be clearly understood by the users of any taxonomical system and their derived products, such as publications and identification systems. This is not only true in order to take full advantage of them, but also to recognize their limitations.

Until now, most identification systems are based on the species reference system. Some allow for the creation of strain-based databases. In general, systems that allow the identification of species and genera, as well as matching with one's own data on strains, are preferred. The possibility to tune the system by the addition, the modification or the subtraction of either features or characteristics or tests is a major advantage allowed by modern software. The latter also permit the creation of user-designed databases that fit to particular and local needs or questions. See below for more detailed discussions and the descriptions of existing identification and classification systems.

5.2.2 Searching and comparisons methods

5.2.2.1 Dichotomous and multiple entry keys

Traditionally, the results obtained from the identification tests and observations are used when following a dichotomous key. The use of such keys has several drawbacks. With some keys, one first determines the genus by characteristics such as the presence and form of ascospores before proceeding to a key for the species level [55]. KREGER-VAN RIJ [45], KURTZMAN and FELL [47] and BARNETT et al. [8] present keys based on physiological tests only or physiological tests together with morphological and sexual characters, which lead directly to the species. Some keys include only a selected set of species that is considered the most likely to occur in a given situation. An isolate, which does not belong to this set, may either be misidentified or be unidentifiable with such a key. On the other hand, a key that includes all the yeast species is very long and requires many tests to be done (see BARNETT et al. [8], where the key to all 704 species involves more than 100 tests). If the results required by a key are not all available, the identification can not be completed until they are done. Moreover, if an erroneous or an unexpected result has been recorded for one of the tests, then either an incorrect identification is made or the organism is found to be unidentifiable. In non-computerized keys, the order of the questions is important since the tests or the observations have to be done accordingly.

Computerized keys are usually multiple entry keys (MEK), and allow the user to ask any question in any order. MEK are superior to printed or dichotomous keys. Some of the problems highlighted above for predefined printed dichotomous keys can be avoided. Using MEK, identifications may be quick and easy, but a single mistake or difference in the observed results may lead to errors. Therefore, identification keys should be used with care. However, MEK are very useful to search for a set of properties in a given database.

5.2.2.2 Probabilistic methods

The use of computer programs, such as BARNETT et al. [9], avoids many of the shortcomings of dichotomous keys. These programs match the results obtained for a strain to be identified against the properties of species in a data matrix. A list of possible species can usually be obtained, even when insufficient tests have been done to allow a definite identification. A list of tests needed to complete the identification can be requested from the program. Moreover, the program permits the user to stipulate that allowance is made during the matching procedure for one or more errors made while performing the tests or typing the results.

The principle used to compare the profile of an unknown (R) with the reference database is based on the use of the Bayes's theorem for the computation of a probabilistic coefficient [86]:

$$P(t_i|R)=\frac{P(R|t_i)}{\sum_{i=1}^{n} P(R|t_i)}$$

where,

$P(t_i|R)$ is the probability that an unknown giving the set of results or profile R is a member of taxon t_i;

$P(ti|R)$ is the probability that a member of taxon *ti* as results or profile R;

$P(ti|R) = P(r_1, r_2 \ldots r_n|ti) = P(r_1|ti)\,P(r_2|ti) \ldots P(r_n|ti)$;

(note that this formula is valid if $r_i|ti$ are independent from one another);

where, $r_1, r_2 \ldots r_n$ are the results obtained for the n tests, criteria or characters of a given taxon;

and, $P(r_1|ti)\,P(r_2|ti) \ldots P(r_n|ti)$ are the probabilities of the occurrence of results $r_1, r_2 \ldots r_n$ for a given taxon *ti*.

The following example shows how this method can be applied.

Tab. 5.2-1 Example of the computation of species probabilistic coefficients associated with a given set of tests.

Species description	Strains used to describe the species	Test 1	Test 2	Test 3	Test 4
	Strain 1	+	–	+	+
	Strain 2	+	–	–	+
Species 1	Based on 2 strains	0.99	0.01	0.50	0.99
	Strain 3	–	–	+	+
Species 2	Based on 1 strain	0.01	0.01	0.99	0.99
	Strain 4	+	+	+	+
	Strain 5	+	+	+	+
	Strain 6	–	+	+	+
Species 3	Based on 3 strains	0.66	0.99	0.99	0.99

Suppose 3 species described on the basis of 4 physiological or biochemical tests and a given number of strains (see Table 5.2-1). For each species and test, a probabilistic coefficient is computed for the occurrence of a positive result. For example, when a strain is positive for a test, the probability coefficient for this test is set to 0.99. A value of 1.00 is usually not attributed in order to allow some degree of flexibility, to avoid problems of small samples or to allow for mistakes in using tests [44, 52, 62]. For a negative result, the probability is set to 0.01. All strains thought to belong to one species are used to compute average probabilistic coefficients for each test.

If, for the 4 tests described in Table 5.2-1, an unknown strain has the following profile (for tests 1 to 4): +, -, +, + then the identification probability for the 3 species will be computed as follows:

Species 1: 0.99 x (1-0.01) x 0.50 x 0.99 = 0.4851

Species 2: 0.01 x (1-0.01) x 0.99 x 0.99 = 0.0097

Species 3: 0.66 x (1-0.99) x 0.99 x 0.99 = 0.0064

To calculate the normalized likelihood for a species, the likelihood is divided by the sum of the likelihood of all species in the system and multiplied by 100. The results are expressed as percentages.

In our example, the sum of the likelihood's for the 3 species is 0.4851 + 0.0097 + 0.064 = 0.5012, and the normalized likelihood is therefore:

Species 1: (0.4851 / 0.5012) x 100 = 96.8

Species 2: (0.0097 / 0.5012) x 100 = 1.9

Species 3: (0.0064 / 0.5012) x 100 = 1.3.

The unknown strain is supposed to be the same species as the one with the highest normalized likelihood in the system. In our example the strain is identified as species 1 with a likelihood of 96.8 %.

This probabilistic method has the advantage of being easy to implement, it is quick and does not require fast computers. In addition, identifications obtained with this technique are usually clear-cut and users tend to like it.

However, probabilistic methods can only be applied to discrete ordered monotonous characters. Continuous or unordered non-monotonous characters can only be analyzed after transformations or reductions. This can lead to many additional problems that will not be discussed here. The method used to create probabilistic profiles for species (see Table 5.2-1) is another major issue since the number and the diversity of strains taken into account can have a profound impact on the basic probabilities and, therefore, on the resulting identifications. It is important to notice that a single mistake or discrepancy in the reading of a test can lead to major mistakes in the identification. In our example, even if species 1 and 3 differ by only one test (#2), the unknown strain has a likelihood of 96.8 % of being species 1 and only 1.3 % likelihood of being species 3. This can be misleading in an identification procedure, especially when characters (tests) are not reliable or if they are variable, which is often the case in the morphological and physiological features used in yeast taxonomy. In Table 5.2-2, a strain profile is being compared to the same database of yeast species using two different methods. The first one is probabilistic and the second is based on similarity methods. Both methods provide the same name in the first position, and most of the proposed names are the same, but their order is different. In addition, results obtained with the probabilistic method suggest that only *Candida ernobii* is possible, even if few additional differences were found with the following species. This is a dangerous practice, especially with tests that are subject to variation or are unevenly reliable such as the physiological ones. One should not only consider the first name proposed by the software, but also carefully evaluate the result using additional criteria (e. g., morphology, ecology, molecular data) when available.

Tab. 5.2-2 Results obtained from the comparison of the physiological results of a strain compared to a database containing more than 700 yeast species using two different methods of comparison, namely probabilistic (from [9]) and similarity (from [69]).

Position	Probabilistic method	%	Similarity method	%
1	*Candida ernobii*	100	*Candida ernobii*	100
2	*Hanseniaspora guilliermondii*	0	*Sporobolomyces roseus*	98.5
3	*Hanseniaspora uvarum*	0	*Hanseniaspora guilliermondii*	97
4	*Sporobolomyces roseus*	0	*Hanseniaspora uvarum*	97
5	*Hanseniaspora valbyensis*	0	*Hanseniaspora valbyensis*	95.5
6	*Hanseniaspora osmophila*	0	*Trichosporon dulcitum*	95.4
7	*Kluyveromyces lactis*	0	*Hanseniaspora osmophila*	93.9
8	*Trichosporon dulcitum*	0	*Sporidiobolus johnsonii*	93
9	*Dekkera bruxellensis*	0	*Hanseniaspora occidentalis*	92.4
10	*Hanseniaspora occidentalis*	0	*Hanseniaspora vineae*	92.4

Probabilistic methods are also used to search sequence databases to identify an unknown from a reference database using pairwise alignment algorithms such as BLASTN [2], FASTA or SSEARCH [63]. Although, BLASTN and FASTA use different scoring methods, they both apply two main criteria. The first one is the length of the sequences under comparison. The second is the similarity or homology between the pairs of sequences. Sequences that have an identical sequence of nucleotides but have different lengths may not be ranked in the top positions. This property can result in misleading conclusions in sequence identification procedures. It is our experience that sequences that were identical but of different lengths, e. g., 563 bp for the query and 329 bp for the reference, were ranked in position 318 while the first sequences showed similarities close to 96% only (and lower)!

Increasing the Expect (E) value from 10 (i. e., default option in BLASTN) to 1,000 or 10,000 may result in more short sequences to be reported with high similarities. These short sequences would otherwise not be reported because of low statistical significance.

BLASTN uses by default a filter for low complexity regions such as repeats like "TTATAAAAAAATTT". With the low complexity option selected, this sequence query may be replaced by "TTATNNNNNNNTTT" leading to misleading conclusions. In Figure 5.2-2a, the use of the low complexity option on a pairwise alignment between the exact same sequence (GenBank accession #: BD000079) results in two local alignments. A first portion of 17 bases was correctly aligned but the second part showed a low similarity (89%)

a

```
Score = 33.4 bits (17), Expect = 222
Identities = 17/17 (100%), Strand = Plus / Plus

Query: 1 taattcaacaagttgta 17
         |||||||||||||||||
Sbjct: 1 taattcaacaagttgta 17


Score =646 bits (336), Expect = 0.0
Identities = 379/422 (89%), Strand = Plus / Plus

Query: 97  ataattcttgatatgatatgatatgatatgatttagttctttgtctgnnnnnnnnnnnnn 156
            |||||||||||||||||||||||||||||||||||||||||||||||
Sbjct: 97  ataattcttgatatgatatgatatgatatgatttagttctttgtctgttttttttttttt 156

Query: 157 nnnnncaaactttcttttaatgactttataccaaaaatttcaaaaatttccnnnnnnn 216
                |||||||||||||||||||||||||||||||||||||||||||||||
Sbjct: 157 tttttcaaactttcttttaatgactttataccaaaaatttcaaaaatttccaaaaaaa 216

Query: 217 nnnncaataatgttctttttacgtctctttcctttacaaaatatatttattgcctgcct 276
                |||||||||||||||||||||||||||||||||||||||||||||||||||||||
Sbjct: 217 aaaacaataatgttctttttacgtctctttcctttacaaaatatatttattgcctgcct 276

Query: 277 catttttttcaaatacnnnnnnnnnccctgtaacagtaattagtaaattgnnnnnnntaat 336
            ||||||||||||||||          ||||||||||||||||| ||||||||       ||||
Sbjct: 277 catttttttcaaatacttttttttccctgtaacagtaattagtaaattgaaaaaaataat 336

Query: 337 tattaatttaagtaaatagcagcaacattgaggtttacaaatatagtaataatagtctat 396
            ||||||||||||||||||||||||||||||||||||||||||||||||||||||||||||
Sbjct: 337 tattaatttaagtaaatagcagcaacattgaggtttacaaatatagtaataatagtctat 396

Query: 397 ctacaaccaatattaaataatttgatcatttaaaacaacattaaatttgaaatttagtac 456
            ||||||||||||||||||||||||||||||||||||||||||||||||||||||||||||
Sbjct: 397 ctacaaccaatattaaataatttgatcatttaaaacaacattaaatttgaaatttagtac 456

Query: 457 ataattaataaaagaaaagaggagaaacaaaagcataataaatcattaaaatttgagtat 516
            ||||||||||||||||||||||||||||||||||||||||||||||||||||||||||||
Sbjct: 457 ataattaataaaagaaaagaggagaaacaaaagcataataaatcattaaaatttgagtat 516

Query: 517 ag 518
            ||
Sbjct: 517 ag 518
```

b

```
Score =996 bits (518), Expect = 0.0
Identities = 518/518 (100%), Strand = Plus / Plus

Query: 1   taattcaacaagttgtatctttttttactgctctttttttaatgatctctctttattttt 60
            ||||||||||||||||||||||||||||||||||||||||||||||||||||||||||||
Sbjct: 1   taattcaacaagttgtatctttttttactgctctttttttaatgatctctctttattttt 60

Query: 61  tttcaatcaatttatttatttaatttttcactttataattcttgatatgatatgatat 120
            ||||||||||||||||||||||||||||||||||||||||||||||||||||||||||||
Sbjct: 61  tttcaatcaatttatttatttaatttttcactttataattcttgatatgatatgatat 120

Query: 121 gatatgatttagttctttgtctgtttttttttttttttttcaaactttcttttaatga 180
            ||||||||||||||||||||||||||||||||||||||||||||||||||||||||||||
Sbjct: 121 gatatgatttagttctttgtctgtttttttttttttttttcaaactttcttttaatga 180

Query: 181 ctttataccaaaaatttcaaaaatttccaaaaaaaaaaacaataatgttctttttacgt 240
            ||||||||||||||||||||||||||||||||||||||||||||||||||||||||||||
Sbjct: 181 ctttataccaaaaatttcaaaaatttccaaaaaaaaaaacaataatgttctttttacgt 240

Query: 241 ctctttcctttacaaaatatatttattgcctgcctcatttttttcaaatacttttttttt 300
            ||||||||||||||||||||||||||||||||||||||||||||||||||||||||||||
Sbjct: 241 ctctttcctttacaaaatatatttattgcctgcctcatttttttcaaatacttttttttt 300

Query: 301 ccctgtaacagtaattagtaaattgaaaaaaataattattaatttaagtaaatagcagca 360
            ||||||||||||||||||||||||||||||||||||||||||||||||||||||||||||
Sbjct: 301 ccctgtaacagtaattagtaaattgaaaaaaataattattaatttaagtaaatagcagca 360

Query: 361 acattgaggtttacaaatatagtaataatagtctatctacaaccaatattaaataatttg 420
            ||||||||||||||||||||||||||||||||||||||||||||||||||||||||||||
Sbjct: 361 acattgaggtttacaaatatagtaataatagtctatctacaaccaatattaaataatttg 420

Query: 421 atcatttaaaacaacattaaatttgaaatttagtacataattaataaaagaaaagaggag 480
            ||||||||||||||||||||||||||||||||||||||||||||||||||||||||||||
Sbjct: 421 atcatttaaaacaacattaaatttgaaatttagtacataattaataaaagaaaagaggag 480

Query: 481 aaacaaaagcataataaatcattaaaatttgagtatag 518
            ||||||||||||||||||||||||||||||||||||||
Sbjct: 481 aaacaaaagcataataaatcattaaaatttgagtatag 518
```

Fig. 5.2-2 BLASTN (www.ncbi.nlm.nih.gov/BLAST/) self pairwise alignment (GenBank accession #: BD000079), with selecting (Fig. 5.2-2a, left) and without selecting the "Low complexity" option (Fig. 5.2-2b, right).

because 43 bases out of 422 were replaced by "N". Without the low complexity option, the same alignment shows a perfect match between both sequences (Fig. 5.2-2b). For taxonomic identification purposes, it is therefore essential to remove this option.

Therefore, and even with the best parameters selected, users must treat the results provided by these programs with great caution. When a good match is not found it is advised to try other settings for the program (see for BLASTN, www.ncbi.nlm.nih.gov/BLAST/; for FASTA, Fasta.bioch.virginia.edu/Fasta/) and to try other alignment software.

Some versions of the Blastn and Fasta packages were modified to sort the results according to similarities only. This allows better identification results, but has the potential disadvantage of presenting more irrelevant results based on short regions that are identical. The ETI [15] and the BioloMICS software packages [69, 70] have incorporated such modifications. In the first one, only the Blastn algorithm can be used. In the second, in addition to two new and original alignment algorithms especially designed for taxonomic identification, both FASTA (including SSEARCH) and BLASTN produce ordered alignment results by similarity or probability.

In conclusion, probabilistic methods are not really suitable and can be misleading when applied in taxonomic identification procedures. Therefore, users should avoid systems using these techniques or should be aware of the limitations and the potential problems arising from them.

5.2.2.3 Similarity or distance methods

Similarity or distance methods are among the most natural techniques for the comparison of biological profiles. Human beings are constantly comparing objects based on some sort of similarity procedures. A very wide range of algorithms and formulas can be used to compute similarity coefficients [53, 79]. Data of all types can be processed with similarity methods with or without transformation of the original data. The choice of the algorithm(s) to be used has an important impact on the final identification or classification results. To illustrate this, we will compare two numerical values (ranges) using three different algorithms. Sizes of vegetative cells of a strain to be identified range from 5 to 6 µm, and this strain has to be compared to a species with cells varying between 4 and 8 µm. One algorithm may lead to the conclusion that sizes of the strain are within the range ($4 < 5$ and $6 < 8$) of the given species and the similarity coefficient is 100%. Another, based on overlaps of ranges could give a coefficient of (6-5)/(8-4) = 25%. A third method based on the comparisons of medians would give the following distance coefficient: [(4+8)/2]-[(5+6)/2] = 6-5.5 = 0.5. This distance value can be transformed into a similarity value by dividing the obtained distance by the highest median. In the example, this would give a similarity value of 1-(0.5/6) = 91.7 %. The conclusion of this simple example is that the algorithms selected have a profound impact on the results obtained. Algorithms must be selected on the basis of the goals to be reached (simple searches, identification, classification, etc.) and on the data to be compared (reliable versus approximate, discrete versus continuous, phenotypic versus molecular, etc.). Methods used to compare sizes are different from those used to compare colours, electrophoretic data or sequences. Some algorithms do provide symmetrical similarity coefficients ($S(X_1, X_2) = S(X_2, X_1)$), which are suitable for classification but not necessarily for identification. In some cases, the triangle rule is not respected ($D(X_1, X_2) + D(X_2, X_3) \geq D(X_1, X_3)$), making interpretation of similarity or distance matrices potentially more difficult. The choice of the most appropriate comparison method must be based on all the parameters mentioned before and has to be reviewed when the objectives change. Ideally, a global similarity coefficient is computed as a summary of an array of local similarity values. Gower [32] has described the general similarity coefficient as:

$$S(X_1, X_2) = \frac{\sum_{i=1}^{n} W_i S_{i12}}{\sum_{i=1}^{n} W_i}$$

where,

S_{i12} is the similarity value for character i (e. g., a physiological test or a morphological feature or sequence region) between two objects, 1 and 2 (e. g., a strain and a species);

n is the number of characters accounted or compared;

W_i is the weight attributed to character i.

An example of the results obtained in an identification using the general similarity coefficient of Gower [32] is given in Table 5.2-2.

Similarity methods are widely used to compare objects or records and have several advantages over dichotomous keys and probabilistic methods: 1) they can be applied to nearly all types of data; 2) data transformations (a source of information loss) are usually not necessary; 3) they are flexible and can be quite easily adapted to a wide range of problems and objectives; 4) comparison reports are not artificially over-discriminant (see Table 5.2-2) and keying out is less likely or impossible; 5) characters can be weighted according to their relative importance; 6) similarity matrices can be summarized easily by clustering techniques.

Similarity methods have some disadvantages as well: 1) comparison reports may be difficult to interpret for non-trained users; 2) similarity coefficients based on a given set of data may vary greatly as a function of the methods used for the comparisons; 3) the complexity of some algorithms may lead to heavy computations and the time to obtain results may become an issue when very large numbers of records and characteristics have to be compared; 4) ideally, the algorithms used in identification should not (always) be the same as the ones used for classification. Therefore, different settings should be selected, resulting in a more complex software management; 5) although weighting of characters can be useful, it remains difficult to attribute the right value; 6) missing data and non-equivalent sets of data can be an issue. For example, if 5 criteria out of 15 are identical, the similarity is equal to 5/15 = 30%. The same coefficient will be obtained when 300 criteria out of 1000 are identical. In the later case, more data are included and the set of characteristics employed is obviously larger and hence different. Consequently, identical similarity coefficients may cover different realities when incomplete data sets are being used.

In conclusion, similarity methods are suitable for identification and classification purposes, but algorithms have to be selected with care and interpretation must be adapted to the properties of the latter.

5.2.2.4 Correlation methods

While similarity methods are ideal for the identification and classification of records or species (or strains), correlation methods have to be used when one wants to compare properties, characteristics of records, species or strains (see [53] for more details). With correlation methods, provided that the required transformations are being done, it is possible to compare "apples and pears". This can be useful when one studies the effects of a heat treatment on the viability of the tested strains or to know whether there is a correlation between the presence of a given gene and a physiological property. Correlation analyses are useful because they allow a better understanding of the relationships between the characteristics of a given set of records (e. g., species or strains).

The formula of the Pearson product-moment correlation coefficient is the following for the comparison of characters j and k on n records [53]:

$$r_{jk} = \frac{\sum_{i=1}^{n}(X_{ij} - \overline{X}_j)(X_{ik} - \overline{X}_k)}{\sqrt{\sum_{n=1}^{n}(X_{ij} - \overline{X}_j)^2 \sum_{i=1}^{n}(X_{ik} - \overline{X}_k)^2}}$$

where,

X_{ij} and X_{ik} are the values of record i for characters j and k, respectively;

$\overline{X}_j$ and $\overline{X}_k$ are the average values for characters j and k for the n records accounted.

From the formula, it is obvious that both characters' data (j and k) must have the same format and scale, which is rarely the case. Therefore, data transformations are usually necessary and include normalization (or standardization) and reduction procedures. Using normalization, values of all characters are subtracted from the mean of the given character for the selected objects or records and then divided by the standard deviation. The new means of all characters are equal to 0 and their variance is equal to 1. Then, all data are reduced to fit within a given range, usually 0 and 1.

When Pearson's correlation is equal to 0, then the two characters under comparison are independent. The sign of the correlation coefficient indicates the direction of the correlation, while the magnitude of the coefficient denotes how closely values adhere to the trend, and hence the strength of the correlation. The closer the correlation coefficient is to 1 or −1, the stronger the correlation. While the coefficient of correlation is a measure of the intensity of the dependence between two characters, significance or statistical tests can be applied to the obtained coefficient of correlation in order to demonstrate the presence or the absence of the relation [53]. However, SNEATH and SOKAL [79] argue that a bivariate normal distribution, which is an important requirement for statistical inference, is unlikely in taxonomic data due to the heterogeneity and the non-independence of the column or the character vectors.

This method has also been used to compare records, strains and taxa by MICHENER and SOKAL [58], BOYCE [16], MOSS [61], SNEATH and SOKAL [79], but it is certainly more suited to the comparison of characters [26].

The absence of correlation (value close to 0) means that both characters are not linearly related. Non-linear relations may not be assessed by the Pearson's formula, and other similarity methods should be applied in this latter case.

One should take into consideration that the reliability and the confidence attached to the obtained coefficient are directly and positively related to the number of "observations" (records, strains or species in our case).

5.2.2.5 Summarizing methods

When comparing a strain with a species database, the results are usually ordered as presented in Table 5.2-2. This is probably the best reporting method in identification, when one strain is compared with a set of records for a selection of characters. Similarity or distance matrices are obtained, when all records are compared to one another (see Table 5.2-3). These matrices are usually information rich, but are difficult to analyze and to interpret. Several groups of clustering techniques can be applied to summarize the information. Hierarchical methods are the most commonly used and usually result in a tree-like representation where clusters are linked hierarchically or related (Fig. 5.2-3a and 5.2-3b). In non-hierarchical techniques, objects are clustered, but no connexion between clusters is graphically displayed. Clustering can also be either divisive or agglomerative (see [40, 41, 53, 59, 66, 79] for more details).

In divisive clustering (K-means [56] or Virtual Centre Analysis [13, 68]) all objects initially belong to the same single cluster, which is gradually broken down into smaller clusters. The division process is stopped when an *a priori* decided number of groups is obtained (note: other methods of stopping can be applied). This *a priori* decision is rather subjective and uneasy. Groupings are never definitive and objects can be clustered together at a certain level, then be separated and (possibly) later re-grouped again with some members of the original group. This latter property is a major advantage of divisive over agglomerative meth-

Tab. 5.2-3. Results obtained from the comparison of morphological, physiological and molecular data of one strain (CBS 14) and 28 species belonging to the genera *Rhodosporidium* and *Saccharomyces*. Only a portion of the distance matrix is displayed. Note that CBS 14 is identical to *Rhodosporidium toruloides* (see zero distance values in the rectangle). Figures 5.2-3a, 5.2-3b and 5.2-3c are based on these data. Results obtained from [69].

CBS 14	**0**						
Rhodosporidium toruloides	0	0					
Rhodosporidium sphaerocarpum	0.06	0.065	0				
Rhodosporidium paludigenum	0.119	0.053	0.093	0			
Rhodosporidium lusitaniae	0.219	0.054	0.115	0.095	0		
Rhodosporidium kratochvilovae	0.09	0.049	0.079	0.095	0.146	0	
Rhodosporidium fluviale	0.091	0.03	0.065	0.054	0.119	0.08	...
Rhodosporidium diobovatum	0.045	0.022	0.054	0.03	0.114	0.043	...
Rhodosporidium babjevae	0.154	0.053	0.113	0.041	0.171	0.099	...
Saccharomyces unisporus	0.448	0.188	0.271	0.349	0.349	0.361	...
Saccharomyces turicensis	0.469	0.183	0.343	0.429	0.433	0.457	...
Saccharomyces transvaalensis	0.418	0.206	0.308	0.386	0.387	0.371	...
Saccharomyces spencerorum	0.373	0.151	0.252	0.321	0.32	0.315	...
Saccharomyces servazzii	0.433	0.225	0.289	0.386	0.378	0.398	...
Saccharomyces rosinii	0.455	0.21	0.311	0.398	0.381	0.383	...
Saccharomyces pastorianus	0.328	0.142	0.243	0.302	0.33	0.297	...
Saccharomyces paradoxus	0.328	0.16	0.261	0.302	0.339	0.297	...
Saccharomyces mikatae	0.421	0.141	0.319	0.332	0.396	0.39	...
Saccharomyces martiniae	0.448	0.188	0.252	0.349	0.33	0.352	...
Saccharomyces kunashirensis	0.418	0.156	0.265	0.35	0.319	0.344	...
Saccharomyces kudriavzevii	0.509	0.212	0.356	0.437	0.442	0.424	...
Saccharomyces kluyveri	0.224	0.133	0.215	0.219	0.244	0.195	...
Saccharomyces exiguus	0.358	0.142	0.224	0.284	0.282	0.269	...
Saccharomyces dairenensis	0.463	0.215	0.298	0.404	0.378	0.389	...
Saccharomyces cerevisiae	0.299	0.095	0.188	0.227	0.234	0.233	...
Saccharomyces castellii	0.418	0.17	0.243	0.349	0.311	0.334	...
Saccharomyces cariocanus	0.343	0.168	0.243	0.35	0.383	0.359	...
Saccharomyces bayanus	0.313	0.114	0.216	0.255	0.281	0.261	...
Saccharomyces barnettii	0.403	0.188	0.298	0.367	0.358	0.343	...

ods. This is particularly true when the number of objects to be clustered is becoming more important (see below). Divisive clustering is computationally intensive and is not widely used, because no graphical representation of the relative position of and within the clusters is proposed (see Table 5.2-4).

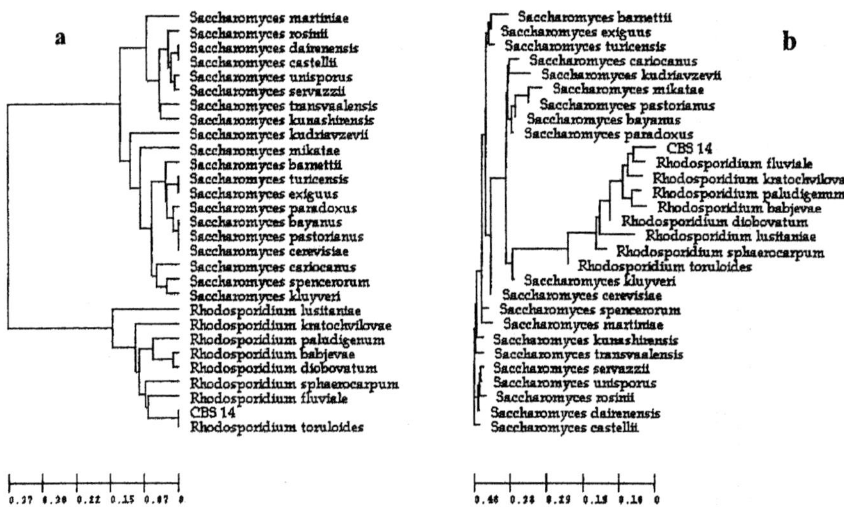

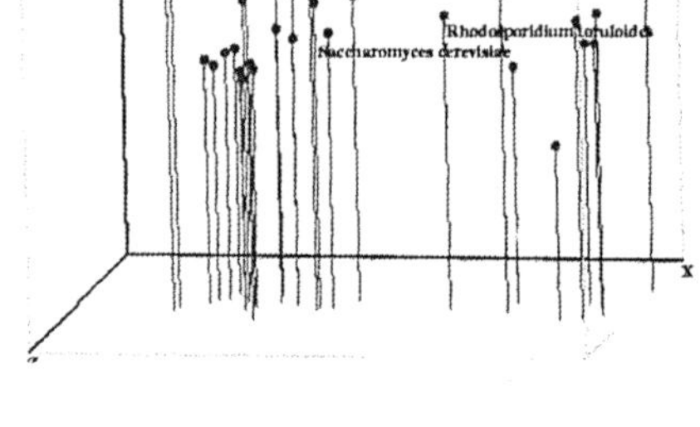

Fig. 5.2-3 Graphical summary/representation of data given in part in Tab. 5.2-4, using two different agglomerative clustering methods: UPGMA (Fig. 5.2-3a) and Neighbor-Joining (Fig. 5.2-3b), and an ordination method, Principal Coordinate Analysis (Fig. 5.2-3c). Results obtained from [69].

Tab. 5.2-4 Results obtained from a divisive clustering method (Virtual Centre Analysis [69]) when 2 groups are selected *a priori*. These results are based on the data displayed partially in Table 5.2-3. Distance between species and the virtual centre of the their group is given in column A, while the distance with the virtual centre of the other group is given in column B. For example, *S. cerevisiae* is the closest species to the centre of group 1 (0.35% distance). The distance between the centre of group 2 and *S. cerevisiae* is 25.43%. Results obtained from [69].

Group 1	A	B
Saccharomyces barnettii	6.28%	40.01%
Saccharomyces bayanus	4.64%	28.19%
Saccharomyces cariocanus	6.01%	37.62%
Saccharomyces castellii	1.99%	37.26%
Saccharomyces cerevisiae	0.35%	25.43%
Saccharomyces dairenensis	4.54%	43.68%
Saccharomyces exiguus	1.12%	32.67%
Saccharomyces kluyveri	6.28%	25.33%
Saccharomyces kudriavzevii	11.2%	47.67%
Saccharomyces kunashirensis	4.02%	39%
Saccharomyces martiniae	4.54%	39.09%
Saccharomyces mikatae	11.1%	36.18%
Saccharomyces paradoxus	5.47%	32.67%
Saccharomyces pastorianus	8.33%	30.21%
Saccharomyces rosinii	6.39%	46.37%
Saccharomyces servazzii	5.41%	43.68%
Saccharomyces spencerorum	1.93%	36.34%
Saccharomyces transvaalensis	7.15%	43.68%
Saccharomyces turicensis	4.23%	47.89%
Saccharomyces unisporus	4.35%	40.93%
Group 2		
CBS 14	7.35%	44.78%
Rhodosporidium babjevae	8.74%	42.19%
Rhodosporidium diobovatum	2.33%	29.17%
Rhodosporidium fluviale	3.03%	39.49%
Rhodosporidium kratochvilovae	8.26%	37.37%
Rhodosporidium lusitaniae	11.11%	35.93%
Rhodosporidium paludigenum	5.46%	38.63%
Rhodosporidium sphaerocarpum	5.83%	27.65%
Rhodosporidium toruloides	3.29%	18.91%

In agglomerative clustering, agglomeration starts with the most closely related pairs of objects and gradually links to less closely related objects. When objects are associated in one cluster, no further dislocation is possible. This is a disadvantage of this method, which is compensated by the widely used and appreciated graphical tree-like representation. Several agglomerative clustering algorithms are available. The most commonly used are the Unweighted Pairgroup Method using Arithmetic mean (UPGMA, Figure 5.2-3a), the Weighted Pairgroup Method using Arithmetic mean (WPGMA), the complete-linkage and the single-linkage methods. These methods differ in the way the distances are calculated when new members of a growing cluster are included. In single linkage, when an object or a group of objects *i* are to be linked with another cluster *j*, the smallest distance between a member of group *i* and a member of group *j*, is the distance at which both groups are connected. In complete linkage, it is the distance between the most distant members of both groups and in UPGMA it is the average distances between all members of both groups. In UPGMA, no weighting related to the relative size of the groups to be joined is performed, contrarily to WPGMA. Neighbor-Joining (NJ, Figure 5.2-3b) is another agglomerative-clustering method aimed at the construction of phylogenetic trees [73]. The NJ method tries to find the shortest possible tree. It is usually used as an alternative to parsimony methods, which can be slow and computationally intensive.

Agglomerative clustering methods are used for a wide range of applications, including the identification and the classification of objects (strains, species, etc.) or characters. It is important to note that agglomerative clustering may be increasingly misleading when the number of objects or characters to be grouped increases. To demonstrate this problem, the evolution of the cophenetic coefficient of correlation between the original distance matrix, on which the four clustering methods are based, and the cophenetic matrix, representing the trees resulting from the clustering algorithms, is plotted (Fig. 5.2-4), with an increasing number of objects. This cophenetic coefficient (r) is a measure of the fit between the original distance matrix and the tree representing it. Results from four different agglomerative clustering methods (UPGMA, WPGMA, single and complete linkages) are plotted, as well as the average r value of the four methods. ROHLF [72] subjectively interpreted the degree of fit as shown in Table 5.2-5. Using his assumptions, the trees based on morphological, physiological and molecular data of yeasts represents the original matrix very poorly when more than 30 objects or characters are to be clustered. The UPGMA showed the best performance while single linkage performed poorly. These conclusions, however, are not universal and depend on the structure of the data. Caution should be paid to tree-like and other graphical representations since they represent 2 or 3-dimensional views and summaries of n-dimensional realities (i. e., the original matrix).

Principal Component Analysis (PCA) is a member of a larger family of techniques, called ordination methods, that include Principal Coordinate Analysis (PCoA), Seriation, Multidimensional Scaling and Factorial Analysis as well. The latter techniques will not be treated in detail here (for further details, see [53, 79]).

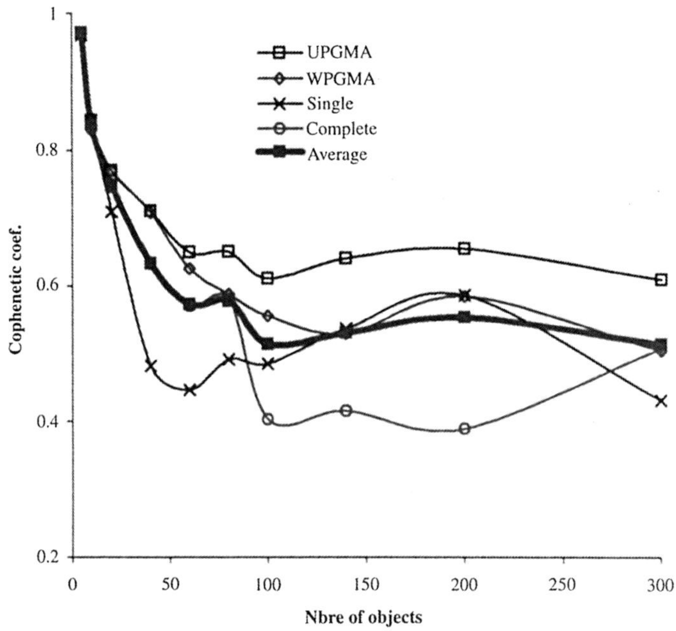

Fig. 5.2-4. Evolution of the cophenetic coefficient of correlation between the original distance matrix, on which the clustering methods have been based, and the cophenetic matrix, representing the tree resulting from the various clustering algorithms, when the number of objects to be grouped is increasing. Results of four different methods (UPGMA, WPGMA, single and complete linkages) are plotted, as well as the average (obtained from [69, 72]).

Tab. 5.2-5 Subjective interpretation by ROLF [72] of the cophenetic coefficient of correlation (r) between the original distance matrix and the cophenetic matrix representing the tree obtained from agglomerative clustering methods.

Level	Interpretation
$0.9 \leq r$	Very good fit.
$0.8 \leq r < 0.9$	Good fit.
$0.7 \leq r < 0.8$	Poor fit.
$r < 0.7$	Very poor fit.

In PCA (and in PcoA as well) the space is reduced to fewer dimensions, usually 2 or 3. In fact, PCA is just a rotation in a multi-dimensional space of the original system of axes. The original axes, representing the characters of the objects, are replaced by new orthogonal axes, which carry most of the variance. Two or three dimensional graphical representations can be proposed (see PCoA in Fig. 5.2-3c). PCA and PCoA are usually superior to agglomerative-clustering methods in the global representation or positioning of groups of objects. However, these ordination algorithms are less efficient in the representation of closely related objects (see example below).

From our example (Table 5.2-4), one can see that various methods of summarizing the data, result in different classifications. Since we know that strain CBS 14 is identical to *Rhodosporidium toruloides*, one would expect these two entities to be clustered together and be identically positioned (*i*). Secondly, all *Saccharomyces* and *Rhodosporidium* species are expected to cluster in two separate groups (*j*). The UPGMA method (Fig. 5.2-3a) does fulfil both conditions *i* and *j*. In NJ (Fig. 5.2-3b), *i* is not respected, while *j* is more or less fulfilled. In divisive clustering (Table 5.2-4) both conditions are fulfilled, but the distance between CBS 14 and *Rh. toruloides* with the virtual centre of group 2 is not identical (7.35% and 8.74% respectively), despite that both objects are identical when compared pairwise. These differences can be explained by the structure of the data and by the algorithms used for the comparisons. There are no algorithmic abnormalities here. In PCoA (Fig. 5.2-3c), *i* is not respected, while *j* is fulfilled. One might conclude from these observations that agglomerative clustering (in our example, UPGMA) is the best method and, to a lesser extent, the divisive clustering technique. Corollary, the other methods should not be used. The validity of these conclusions is directly related to the objectives to be reached and to the data to be analyzed (i. e., the number of objects, the type of characters, identification versus classification, etc.) and it should not be forgotten that summarizing implies loss of information and, therefore, leads to bias. Therefore, the use of several techniques in parallel or successively is advisable with large and complex datasets. As a strategy, one may apply firstly PCA, PCoA or a divisive clustering method to isolate well-defined groups, followed by an agglomerative clustering technique, such as UPGMA, on these smaller groups.

In conclusion, summarizing methods can be useful but should be used with care. Their limitations should be well understood in view of the objectives to be reached.

5.3 Yeasts data management and identification systems

As discussed previously in this chapter, the identification of isolates at the species level, although very useful, may not be sufficient for many purposes. There is a need for users' data to be handled and compared with adequate algorithms that has to be addressed by companies and academic institutions. In this part, we will briefly and non-exhaustively review some of the systems available for yeast identification, data-storage and -analysis.

Originally, yeast identification was based on morphological and sexual features (see Chapters 1-3). It soon became obvious that these were insufficient to perform a reliable identification. Therefore, BEIJERINCK [11] introduced the auxanogram growth tests for yeast characterization, identification and classification. He was soon followed by HANSEN [34, 35, 36, 37], and the physiological methods were further developed by WICKERHAM and BURTON [85], WICKERHAM [84], BARNETT and INGRAM [7], BARNETT [6], AHEARN et al. [1] and many others (see Chapter 3). The problem with these techniques is that their implementation is time consuming, labour intensive and requires experienced and highly skilled operators. Therefore, several companies have developed simpler and alternative identification systems. BioMérieux (e. g., API, VITEK) and Biolog (Microlog system) are probably commercially the most successful, but other systems are available as well.

The API system (API 20C or API/ATB ID 32C) is the most commonly used system for the identification of yeasts. API kits consist of plastic strips in which 20 or 32 growth tests are done in small wells (see also Table 5.3-1 for reviews). Results can be read manually or automatically. Reading is done at 18 to 24 hours or after 4 hours for the Rapid ID 32. Identification is performed using a coding system and by comparing the results to a database of 42 (API 20C) or 63 species (API ID 32C). Identification is based on the computation of probabilities and the species database can not be modified by the user. Although these commercial identification systems are widely used, their scope is limited to the clinical field. Results obtained from the strips are usually quite reliable (see Table 5.3-1) and can be used to construct a user designed database. However, it must be clearly stated that identifications performed on the basis of physiological features only or in combination with morphological observations are often approximate. Many species of interest to the food industry can not be reliably distinguished on the basis of physiological and morphological features (see *Saccharomyces* sensu stricto and Chapter 4).

The MicroLog YT Station from Biolog can identify 267 yeast species using a panel of 94 tests in a microplate [14] (see also Table 5.3-1). The microplate is divided in two sections. In the first one, growth tests with carbohydrates are performed. In the second section, oxidation tests are done using the redox dye tetrazolium violet. The dye is reduced to a purple formazan if the yeast strain oxidizes the substrate. Since 29 substrates are tested in both sections of the microplate, the observed results can be redundant. It has been observed by HEARD et al. [38] and PRAPHAILONG et al. [65] that no difference occurred between the two sets of tests, thus suggesting that the actual number of tests is 63 and not 94. Microplates can be read manually or automatically. Results are compared to the databases and a probabilistic identification is provided. An agglomerative clustering procedure can be performed to produce a tree-like representation of the position of the unknown strain with closely related species. The system can be used in the food industry since most food-related species are present. However, the reliability of the system seems rather variable and questionable depending on the groups of species to be identified (see [38, 57, 65] for more details). Strain records can be appended to the database, making it a valuable system for ecological surveys or in quality control.

Tab. 5.3-1 Comparison of different yeast identification and data related management systems. See also FREYDIERE et al. [31] for an extensive review of identification systems for yeasts.

#	System	Characters	Method	No. of tests	Species in database	Time (hr) for results	Identification method	User's update of strains/ species database	Database modifiable Add/delete characters	Use in food-borne yeast identification	Accuracy % (references)	Internet
1	Allev 2.00	Growth, morphology	Automated & manual	96	693	240	Similarity	Yes/yes	No	Yes	98.2 [71]	www.belspo.be/bccm/ allev2.htm
2	API 20C strip	Growth	Manual	20	42	72	Probability	No/no	No	No	68 [17], 72.1 [21], 77 [12], 78-98.9 [80], 86 [22], 88 [24], 88.5-96 [18], 90 [77], 98 [51], 99.3 [30]	www.biomerieux.com
3	ATB ID 32C	Growth	Automated & manual	32	63	48	Probability	No/no	No	Limited	76 [24], 94 [25], 97 [33], 98 [17]	www.biomerieux.com
4	BioloMICS	Growth, morphology, sequences, gels	Automated & manual	152	754	24-240	Similarity & correlation	Yes/yes	Yes	Yes	98.2 [71]	www.cbs.knaw.nl/yeast/ webc.asp www.bio-aware.com
5	BioNumerics	Growth, morphology, sequences, gels			0		Similarity, correlation & probability	Yes/yes	Yes	Yes, if user is building database		www.applied-maths.com
6	ETI	Growth, morphology sequences	Manual	152	754	24-240	Similarity	Yes/no	No	Yes		www.eti.uva.nl
7	Genbank	Sequences					Probability	Yes/yes	No	Yes		www.ncbi.nlm.nih.gov
8	Micronaut-C	Growth, oxydase	Automated	44	50	24-48	Correlation & probability	No/no	No	?		www.merlin-diagnostika.de
9	MicroLog YT (Biolog)	Growth, oxydase	Automated	94	267	24-72	Probability	No/no	No	Yes	48.8 [57], 68-70.8 [65], 74 [73], 0-100 [38*]	www.biolog.com
10	MicroSeq	Sequences	Automated		?		Probability	No ?/no ?	No	Yes		home.appliedbiosystems.com
11	MIS	Cell fatty acid content	Automated		203	72	Similarity	Yes/Yes	No	Yes	68 [19], 75 [43]	www.midi-inc.com
12	Minitek	Growth	Manual	12	28	72	Probability	No/no	No	No	97 [54]	www.bd.com

Tab. 5.3-1 Continued

#	System	Characters	Method	No. of tests	Species in database	Time (hr) for results	Identification method	User's update of strains/ species database	Database modifiable Add/delete characters	Use in food-borne yeast identification	Accuracy % (references)	Internet
13	Quantum II	Growth	Automated	20	34	24	Probability	No/no	No	No	82 [64], 86 [75]	abbottdiagnostics.com/ systems_tests/index.cfm
14	Reva	Growth, morphology		30	121				No	Yes		www.imv.kiev.ua/key/yeasts/ default.htm
15	Vitek YBC	Growth	Automated	26	36	24-48	Probability	No/no	No	No	83 [64], 86–96 [83], 97.2 [27], 89.7 [30], 98 [33]	www.biomerieux.com
16	Vitek 2 ID-YST	Growth & enzyme	Automated	47	51	15	Probability	No/no	No	No		www.biomerieux.com
17	Yeast ID PC version 5 Barnett *et al* [9].	Growth, morphology		106	669		Probability	No/no	No	Yes		

* depending on genera tested

Vitek 2 ID-YST is another yeast identification system developed by BioMérieux. It allows 47 biochemical tests to be performed, including 29 conventional (carbohydrate growth, urea and nitrate tests) and 18 enzymatic tests. The database contains 51 species that can be identified within 15 hours. Again, VITEK 2 ID-YST is mainly dedicated towards the identification of clinically important yeasts and is therefore not suitable for food-borne yeasts. Vitek YBC is a reduced sister system of the Vitek 2 ID-YST (Table 5.3-1).

The Micronaut C system from Merlin-Diagnostika has a database of 50 species that can be identified in 24 to 48 hours by assessing the growth of the strains on 44 different media. Results are read by a microplate reader connected to a computer and correlation analyses are performed to provide an identification. Strain data can be recorded. The system has been mainly designed for clinically important species and not for food-borne yeasts. Merlin-Diagnostika also provides other ready-to-use microplates (384 wells) that include more than 800 distinct biochemical and physiological tests (see taxa profiles at www.merlin-diagnostika.de), which can be useful in taxonomic and ecological studies and in quality control.

DEÁK [23] has proposed a simplified identification scheme for yeasts associated with food. His scheme encompasses 76 species and uses less than 20 physiological tests and morphological observations. An Internet web site (www.imv.kiev.ua/key/yeasts/default.htm) with an identification module has been designed by REVA [67] in order to identify yeast strains using a modified version of this scheme. Thirty physiological and morphological features are employed to identify 121 species, including all major food-related species. Identification is based on the computation of a similarity coefficient (O.N. REVA, pers. comm.), although it is presented as a probabilistic one. ROBERT et al. [71] developed an automated system (Allev 2.00) for the characterization and the identification of yeasts using a large panel of physiological properties (in a 96 well microplate which is automatically read by a microplate reader, see Table 5.3-1). This system is not available anymore, but has been replaced by a more advanced program called BioloMICS (see below and Table 5.3-1). SEILER and BUSSE [78] have also developed a yeast identification system based on 46 physiological tests in microplates.

Several other systems (see Table 5.3-1 for a partial list), mainly based on growth or enzymatic properties, are available but their reliability is often poor and the number of species included limited (e. g., Minitek, Quantum II). Most of them are medically oriented and are not very useful in food-related studies.

The Sherlock Microbial Identification System (MIS) from MIDI uses gas chromatography to assess the whole-cell fatty-acid content. The yeast library contains 203 species, including some food related species. However, important species and genera such as *Debaryomyces* are lacking. Identification is based on similarity and users can create their own libraries. In addition, several statistical options such as PCA and clustering are available. The reliability of this system has been reviewed [19, 43], and seems to depend on the quality of the Sabouraud glucose agar used for the preparation of the cultures [42].

Fourier transformation infrared (FTIR) spectroscopy seems to have an interesting potential in identification procedures down to the strain level [39, 61, 76, 81]. KUMMERLE et al. [46] have observed that FITR correctly identified 97.5% of 332 food-borne fermentative yeasts.

Many different electrophoretic techniques (RAPD, PCR-finger-printing, real time PCR, RFLP, AFLP, PFGE; see chapters 3 and 4 for further discussions) are available and are widely used, especially by researchers interested in the recognition and characterization of strains [3, 4, 5, 28, 82]. Several polyphasic programs (see below) can be used to compare and store electrophoretic profiles and to link these data with other types of data. Speed, efficiency and sensitivity are the major advantages of some of these electrophoretic techniques. However, reproducibility and exportability of results can be major issues with some electrophoretic methods.

Sequencing databases are widely available (eg: Genbank) and the D1/D2 domain of the large subunit of the ribosomal DNA (26S rDNA) has been sequenced for almost all accepted yeast species [29, 48, 49, 50]. This part of the genome can be used to identify species and genera. It can also be used at the strain level, but the absence of differences is not a definitive indication of identity. Sequences of the small subunit (18S rDNA) and the internal transcribed spacers (ITS1, ITS2) of the rDNA are also available, but not all species have been sequenced yet. This work is expected to happen in the near future. While 18S sequences are useful at higher taxonomic levels (sometimes species, but usually genus and above), ITS sequences can be used for the distinction of strains and closely related species. The actin gene seems to have the potential to discriminate both strains and species [20]. When searching public sequence databases, such as Genbank, users should be aware that anybody can submit sequence data with a name attached to it, and there is no guarantee on the reliability of either those data. Therefore, care must be taken when using Genbank or any other public sequence database. To avoid these problems, the Microbial Identification System (MicroSeq) from Applied Biosystems will provide, in the near future, a comprehensive and curated fungal species database (~2000 species, including the most important yeasts) with all the necessary algorithmic tools [10]. However, and as far as we know, the algorithmic alignment problems described above are still present in this system. This database will only be available on payment of a fee for each query or alignment (see www.appliedbiosystems.com). Another alternative for yeast identification, is to use the CBS web site (<http://www.cbs.knaw.nl/yeast/webc.asp>), which is free of charge and frequently updated. Here, potential alignment problems are solved by using the option for similarity alignment.

Polyphasic software for identification and management allows the handling of many different types of data, such as administrative, morphological, physiological, ecological, molecular, geographical, bibliographic and taxonomic information. Presently, two polyphasic systems are available for yeasts. The first one is a CD-ROM named Yeasts of the World, which is produced by the Expert center for Taxonomic Identification (ETI) of the University of Amsterdam and CBS [15], and contains a database of all accepted yeast species. Type strain data are available as well. Similarity-based identification can be done using a wide range of morphological, physiological and molecular (26S, 18S and ITS) data. Images of macro- and microscopic features of yeast species can be viewed. Users can store their own records but multiple or pairwise comparisons between those self-recorded strains are not possible. The second polyphasic software is called BioloMICS [69, 70], which contains the same species databases as the ETI CD-ROM, but allows the user to add, modify and delete all records (strains and species) and characteristics with a large choice of similarity and correlation based algorithms for multi-directional identification, classification, phenetic and phylogenetic comparisons. Users can design their own databases, including administrative, morphological, physiological (including automated reading of microplates designed by the user), ecological, sequence, gel, molecular, bibliographic and taxonomic data. Data can be published, used and compared on the Internet using the web version of the software (see CBS web site). BioNumerics (www.applied-maths.com) is another interesting polyphasic software with many advanced statistical, identification and classification features allowing the user to create self-designed databases. A database of yeast species is not provided but it should not be too difficult to create one. This software is, however, quite complex and certainly requires some training.

5.4 Conclusion and future

As surveyed in this chapter, the processes of identification and classification are not always easy and straightforward. Selection of algorithms and systems for identification, classification and data management is very important, and one has to consider the objectives to be reached and the quantity, the "quality" and the type of data to be handled.

In the future, classification and identification procedures at the species level, will be complemented with identification and comparisons based on strain data. It is therefore important to use data storage, handling and comparison software that are able to perform all the tasks at the strain level and in a polyphasic way.

The emergence of new types of data, such as microarrays and 2D-gel data, will bring new dimensions to food microbiology. The amount of data to be handled will become much greater and difficult, if not impossible, to manage by the currently available systems. Manual data management will certainly not longer be possible.

New techniques, such as microarrays, have great potential for identification purposes. It is possible to compare complete genomes in a single experiment and to get a very precise identification within a few hours. The information provided will not be limited to taxonomic information, but will be extended to functional data as well. Several commercial companies, as well as public scientific institutions, are already developing projects that will result in the availability of such microarrays.

5.5 References

[1] AHEARN, D.G.; ROTH, F.J.; FELL, J.W.; MEYERS, S.P.: Use of shaken cultures in the assimilation test for yeast identification. J. Bacteriol. **79** (1960) 369–371.

[2] ALTSCHUL, S.F.; GISH, W.; MILLER, W.; MYERS, E.W.; LIPMAN, D.J.: Basic local alignment search tool. J. Mol. Biol. **215** (1990) 403–410.

[3] BALEIRAS COUTO, M.M.; EIJSMA, B.; HOFSTRA, H.; HUIS IN 'T VELD, J.H.J.; VAN DER VOSSEN, J.M.B.M.: Evaluation of molecular typing techniques to assign genetic diversity among strains of *Saccharomyces cerevisiae*. Appl. Envir. Microbiol. **62** (1996) 41–46.

[4] BALEIRAS COUTO, M.M.; HARTOG, B.J.; HUIS IN 'T VELD, J.H.J.; HOFSTRA, H.; VAN DER VOSSEN, J.M.B.M.: Identification of spoilage yeasts in a food production chain by microsatellite PCR fingerprinting. Fd Microbiol. **13** (1996) 59–67.

[5] BALEIRAS COUTO, M.M.; VOGELS, J.T.W.E.; HOFSTRA, H.; HUIS IN 'T VELD, J.H.J.; VAN DER VOSSEN, J.M.B.M.: Random amplified polymorphic DNA and restriction enzyme analysis of PCR amplified rDNA in taxonomy: two identification techniques for food-borne yeasts. J. Appl. Bacteriol. **79** (1995) 525–535.

[6] BARNETT, J.A.: Biochemical differentiation of taxa with special reference to the yeasts. In: The Fungi. An Advanced Treatise, Vol. 3 (edited by Ainsworth, G.G.; Sussman, A.S.) New York, U.S.A.: Academic Press (1968) 557–595.

[7] BARNETT, J.A.; INGRAM, M.: Techniques in the study of yeast assimilation reactions. J. Appl. Bacteriol. **18** (1955) 131–149.

[8] BARNETT, J.A.; PAYNE, R.W.; YARROW, D.: Yeasts: Characteristics and Identification, 3rd ed. Cambridge, U.K.: Cambridge University Press (2000).

[9] BARNETT, J.A.; PAYNE, R.W.; YARROW, D.: Yeasts identification PC program, Version 5. Norwich, U.K.: J. Barnett (2000)

[10] BASSAM, B.J.: The universal fungal identification system using the D2 region of the ribosomal gene. In: Abstr. 5th Int. Conf. *Cryptococcus* and Cryptococcosis, Adelaide, Australia (2002) 22.

[11] BEIJERINCK, W.M.: L'auxanographie, ou la méthode de l'hydrodiffusion dans la gelatine appliquée aux recherches biologiques. Arch. Neerl. Sc. Ex. Nat. **23** (1889) 367–372.

[12] BERGAN, T.; HOLLUM, A.B.; VANGDAL, M.: Evaluation of four commercial biochemical test systems for identification of yeasts. Eur. J. Clin. Microbiol. **1** (1982) 217–222.

[13] BERTHET, J.F.; FEYTMANS, E.; STEVENS, D.; GENETTE, A.: New divisive method of classification illustrated by its applications to ecological problems. In: Proc. 9th Int. Biometric Conf. Boston, Vol. **2** (1976) 366–382.

[14] BOCHNER, B.R.: Sleuthing out bacterial identities. Nature **339** (1989) 157–158.

[15] BOEKHOUT, T.; ROBERT, V.; SMITH, M.TH.; STALPERS, J.; YARROW, D.; BOER, P.; GIJSWIJT, G.; KURTZMAN, C.P.; FELL, J.W.; GUÉHO, E.; GUILLOT, J.; ROBERTS, I.: Yeasts of the world, CD-ROM. Amsterdam, The Netherlands: Expert center for Taxonomic Identification (2002).

[16] BOYCE, A.J.: Values of some methods in numerical taxonomy with reference to hominoid classification. In: Phenetic and phylogenetic classification (edited by Heywood, V.H. ; McNeill, J.). London, U.K.: Systematics Association Publications **6** (1964) 47–65.

[17] BUCHAILLE, L.; FREYDIÈRE, A.M.; GUINET, R.; GILLE, Y.: Evaluation of six commercial systems for identification of medically important yeasts. Eur. J. Clin. Microbiol. Inf. Dis. **17** (1998) 479–488.

[18] BUESCHING, W.J.; KUREK, K.; ROBERTS, G.D.: Evaluation of modified API 20C system for identification of clinically important yeasts. J. Clin. Microbiol. **9** (1979) 565–569.

[19] CRIST, A.E.JR.; JOHNSON, L.M.; BURKE, P.J.: Evaluation of clinically isolated yeasts. J. Clin. Microbiol. **34** (1996) 2408–2410.

[20] DANIEL, H.M.; SORRELL, T.C.; MEYER, W.: Partial sequence analysis of the actin gene and its potential for studying the phylogeny of *Candida* species and their teleomorphs. Int. J. Syst. Evol. Microbiol. **51** (2001) 1593–1606.

[21] DAVEY, K.G.; CHANT, P.M.; DOWNER, C.S.; CAMPBELL, C.K.; WARNOCK, D.W.: Evaluation of the auxacolor system, a new method of clinical yeast identification. J. Clin. Path. **48** (1995) 807–809.

[22] DE LOUVOIS, J.; MULHALL, A.; HURLEY, R.: Biochemical identification of clinically important yeasts. J. Clin. Microbiol. **32** (1979) 715–718.

[23] DEÁK, T.: Simplified techniques for identifying foodborne yeasts. Int. J. Fd Microbiol. **19** (1993) 15–26.

[24] DEÁK, T.; BEUCHAT, L.R.: Comparison of the SIM, API 20C, and ID 32C systems for identification of yeasts isolated from fruit juice concentrates and beverages. J. Fd Prot. **56** (1993) 583–592.

[25] DOUCET, M.; CANIAUX, I.; MONGET, D.: ATB 32C: a new yeast identification system. In: 88th Ann. Meeting Am. Soc. Microbiol. (1988) Abstr. C-303.

[26] EADES, D.C.: The inappropriateness of the correlation coefficient as a measure of taxonomic resemblance. Syst. Zool. **14** (1965) 98–100.

[27] EL-ZAATARI, M.; PASARELL, L.; MCGINNIS, M.R.; BUCKNER, J.; LAND, G.A.; SALKIN, I.F.: Evaluation of the updated Vitek yeast identification data base. J. Clin. Microbiol. **28** (1990) 1938–1941.

[28] ESTEVE-ZARZOSO, B.; BELLOCH, C.; URUBURU, F.; QUEROL, A.: Identification of yeasts by RFLP analysis of the 5.8S rRNA gene and the two ribosomal transcribed spacers. Int. J. Syst. Bacteriol. **49** (1999) 329–337.

[29] FELL, J.W.; BOEKHOUT, T.; FONSECA, A.; SCORZETTI, G.; STATZELL-TALLMAN, A.: Biodiversity and systematics of basidiomycetous yeasts as determined by large-subunit rDNA D1/D2 domain sequence analysis. Int. J. Syst. Evol. Microbiol. **50** (2000) 1351–1371.

[30] FENN, J.P.; SEGAL, H.; BARLAND, B.; DENTON, D.; WHISENANT, J.; CHUN, H.; CHRISTOFFERSON, K.; HAMILTON, L.; CARROLL, K.: Comparison of updated Vitek yeast biochemical card and API 20C yeast identification systems J. Clin. Microbiol. **32** (1994) 1184–1187.

[31] FREYDIERE, A.-M.; GUINET, R.; BOIRON, P.: Yeast identification in clinical microbiology laboratory: phenotypical methods. Med. Mycol. **39** (2001) 9–33.

[32] GOWER, J.C.: A general coefficient of similarity and some of its properties. Biometrics **27** (1971) 857–871.

[33] GUTIERREZ, J.; MARTIN, E.; LOZANO, C.; CORONILLA, J.; NOGALES, C.: Evaluation of the ATB 32C, Automicrobic system and API 20C using clinical isolates. Ann. Biol. Clin. **50** (1994) 443–446.

[34] HANSEN, E.C.: Recherches sur la physiologie et la morphologie des ferments alcooliques. VII. Action des ferments alcooliques sur les diverses espèces de sucre. Compt. Rend. Trav. Lab. Carlsberg **2** (1888) 143–192.

[35] HANSEN, E.C.: Recherches sur la physiologie et la morphologie des ferments alcooliques. Compt. Rend. Trav. Lab. Carlsberg **3** (1891) 44–66.

[36] HANSEN, E.C.: Recherches sur la physiologie et la morphologie des ferments alcooliques. IX. Sur la vitalité des ferments alcooliques et leur variation dans les milieux nutritifs et a l'état sec. Compt. Rend. Trav. Lab. Carlsberg **4** (1898) 93–121.

[37] HANSEN, E.C.: Recherches comparatives sur les conditions de la croissance végétative et le développement des organes de reproduction des levures et des moisissures de la fermentation alcoolique. Compt. Rend. Trav. Lab. Carlsberg **5** (1902) 68–107.

[38] HEARD, G.M.; FLEET, G.H.; PRAPHAILONG, W.; ADDIS, E.: Evaluation of the Biolog system for identification of yeasts from cheese. Int. Dairy Fed. (2000) 117–124.

[39] HELM, D.H.; LABISCHINSKI, H.; SCHALLEHN, G.; NAUMANN, D.: Classification and identification of bacteria by Fourier transformation infrared spectroscopy. J. Gen. Microbiol. **137** (1991) 69–79.

[40] JAIN, A.K.; DUBES,R.C.: Algorithms for clustering data. New Jersey, U.S.A.: Prentice-Hall (1988).

[41] KAUFMAN, L.; ROUSSEEUW, P.J.: Finding groups in data. An introduction to cluster analysis. New York, U.S.A: Wiley (1990).

[42] KELLOG, J.A.; BANKERT, D.A.: Variation in microbial identification system accuracy in identifying clinical isolates of yeasts depending on commercial source of Sabouraud dextrose agar. In: 98th Meeting Am. Soc. Microbiol. Atlanta, U.S.A. (1998) Abstr. C293.

[43] KELLOG, J.A.; BANKERT, D.A.; CHATURVEDI, V.: Limitations of current microbial identification system for identification of clinical yeast isolates. J. Clin. Microbiol. **36** (1998) 1197–1200.

[44] KIMURA, E.; MIBUKURA, Y.; MIURA, S.: Statistical diagnosis of electrocardiogram by theorem of Bayes – a preliminary report. Jap. Heart J. **4** (1963) 469–488.

[45] KREGER-VAN RIJ, N.J.W.: The yeasts: a taxonomic study, 3rd edition. Amsterdam, The Netherlands: Elsevier (1984).

[46] KUMMERLE, M.; SCHERER, S.; SEILER, H.: Rapid and reliable identification of food-borne yeasts by Fourier transform infrared spectroscopy. Appl. Envir. Microbiol. **64** (1998) 2207–2214.

[47] Kurtzman, C.P.; Fell, J.W.: The yeasts: a taxonomic study, 4th edition. Amsterdam, The Netherlands: Elsevier (1998).

[48] KURTZMAN, C.P.; ROBNETT, C.J.: Molecular relationships among hyphal ascomycetous yeasts and yeast-like taxa. Can. J. Bot. **73** (1995) S824–S830.

[49] KURTZMAN, C.P.; ROBNETT, C.J.: Identification of clinically important ascomycetous yeasts based on nucleotide divergence in the 5' end of the large-subunit (26S) ribosomal DNA gene. J. Clin. Microbiol. **35** (1997) 1216–1223.

[50] KURTZMAN, C.P.; ROBNETT, C.J.: Identification and phylogeny of ascomycetous yeasts from analysis of nuclear large subunit (26S) ribosomal DNA partial sequences. Antonie van Leeuwenhoek **73** (1998) 331–371.

[51] LAND, G.A.; HARRISON, B.A.; HULME, K.L.; COOPER, B.H.; BYRD, J.C.: Evaluation of the new API 20C strip for yeast identification against conventional method. J. Clin. Microbiol. **10** (1979) 357–364.

[52] LAPAGE, S.; BASCOMB, S.; WILLCOX, W.R.; CURTIS, M.A.: Computer identification of bacteria. In: Automation, mechanization and data handling in microbiology (edited by Bailli, A.; Gilbert, R.J.). Soc. Appl. Bacteriol. Techn. Ser. **4**. London, U.K.: Academic Press (1970) 1–22.

[53] LEGENDRE, P.; LEGENDRE, L.: Numerical Ecology. Amsterdam, The Netherlands: Elsevier (1998).

[54] LIN, C.C.S.; FUNG, D.Y.C.: Comparative biochemical reactions and identification of food yeasts by the conventional methods, Fung's minimethod, Minitek, and the Automicrobic System. CRC Crit. Rev. Biotechnol. **7** (1987) 1–16.

[55] LODDER, J.; KREGER-VAN RIJ, N.J.W.: The yeasts. A taxonomic study, 2nd ed. Amsterdam, The Netherlands: North-Holland Publisher (1952).

[56] MACQUEEN, J.: Some methods for classification and analysis of multivariate observations. In: Proc. 5th Berkeley Symp. Math. Stat. Prob. **1** (edited by Le Cam, L.M.; Neyman, J.). Berkeley, U.S.A.: Univ. California Press (1967) 281–297.

[57] MCGINNIS, M.R.; MOLINA, T.C.; PIERSON, D.L.; MISHRA, S.K.: Evaluation of the Biolog Microstation system for yeast identification. J. Med. Vet. Mycol. **34** (1996) 349–352.

[58] MICHENER, C.D.; SOKAL, R.R.: A quantitative approach into a problem of classification. Evolution **11** (1957) 130–162.

[59] MIRKIN, B.: Mathematical classification and clustering. Dordrecht, The Netherlands: Kluwer Academic Publishing (1996).

[60] MOSS, W.W.: Some new analytic and graphic approaches to numerical taxonomy with an example from the *Dermanyssidae* (Acari). Syst. Zool. **16** (1967) 177–207.

[61] NAUMANN, D.; HELM, D.; LABISCHINSKI, H.; GIESBRECHT, P.: The characterization of microorganisms by Fourier transform infrared spectroscopy (FT-IR). In: Modern techniques for rapid microbiological analysis (edited by Nelson, W.H.). New York, U.S.A.: VCH Publishers (1991) 43–96.

[62] PAYNE, R.W.; PREECE, D.A.: Identification keys and diagnostic tables: a review. J. Roy. Statist. Soc. A. **143** (1980) 253–292.

[63] PEARSON, W.R.: Flexible sequence similarity searching with the FASTA3 program package. Methods Mol. Biol. **132** (2000) 185–219.

[64] PFALLER, M.A.; PRESTON, T.; BALE, M.; KOONTZ, F.P.; BODY, B.A.: Comparison of the Quantum II, API Yeast Ident, and AutoMicrobic system for identification of clinical yeast isolates. J. Clin. Microbiol. **26** (1988) 2054–2058.

[65] PRAPHAILONG, W.; VAN GESTEL, M.; FLEET, G.H.; HEARD, G.M.: Evaluation of the Biolog system for the identification of food and beverage yeasts. Lett. Appl. Microbiol. **24** (1997) 455–459.

[66] QUACKENBUSH, J.: Computational analysis of microarray data. Nature Rev. Genet. **2** (2001) 418 –427.

[67] REVA, O. N.; SOROKULOVA, I. B.; SMIRNOV, V.V.: Simplified technique for identification of the aerobic spore-forming bacteria by phenotype. Int. J. Syst. Evol. Microbiol. **51** (2001) 1361–1371.

[68] ROBERT, V.: Informatics and treatment of taxonomical data. Applications to Mycology. Ph.D. thesis. Catholic University of Louvain, Louvain-la-Neuve, Belgium (1997).

[69] ROBERT, V.: BioloMICS a general system for identification, classification and archiving of strains and species data. www.bio-aware.com (2000).

[70] ROBERT, V.: The BioloMICS (Biological Manager for Identification, Classification and Statistics) software package as a response to some classification and identification issues. In: Abstr. 5th Int. Conf. *Cryptococcus* and Cryptococcosis, Adelaide, Australia (2002) 23.

[71] ROBERT, V.; EVRARD, P.; HENNEBERT, G.L.: BCCM/Allev 2.00 an automated system for the identification of yeasts. Mycotaxon **64** (1997) 433–439.

[72] ROHLF, F.J.: NTSYSpc 2.00 software. Applied Biostatistics Inc., New York, U.S.A. (1998).

[73] ROMAGNOLI, M.; OSTERHOUT, J.; MERZ, W.G.: Evaluation of the Biolog system for the identification of medically important yeasts. In: Abstr. 95th Ann. Meet. Am. Soc. Microbiol. City, U.S.A. (1995) Abstr. F116.

[74] SAITOU, N.; NEI, M.: The neighbor-joining method: a new method for reconstructing phylogenetic trees. Mol. Biol. Evol. **4(4)** (1987) 406–425.

[75] SALKIN, I.F.; SCHADOW, K.H.; BANKAITIS, L.E.; MCGINNIS, M.R.; KEMMA, M.E.: Evaluation of Abbott Quantum II yeast identification system. J. Clin. Microbiol. **22** (1985) 442–444.

[76] SCHMALRECK, A.F.; TRANKLE, P.; VANCA, E.; BLASCKLE-HELLMESSEN, R.: Differentiation and characterization of yeast pathogenic for humans (*Candida albicans, Exophiala dermatitidis*) and algae pathogenic for animals (*Prototheca* spp.) using Fourier transform infrared spectroscopy (FT-IR) in comparison with conventional methods. Mycoses **41** (1998) 71–77.

[77] SCHUFFENECKER, I.; FREYDIERE, A.-M., DE MONTCLOS, H.; GILLE, Y.: Evaluation of four commercial systems for identification of medically important yeasts. Eur. J. Microbiol. Infect. Dis. **12** (1993) 255–260.

[78] SEILER, H.; BUSSE, M.: Identifizierung von Hefen mit Mikrotiterplatten. Forum Mikrobiol. **11** (1988) 505–509.

[79] SNEATH, P.H.A.; SOKAL, R.R.: Numerical Taxonomy. San Francisco, U.S.A.: W.H. Freeman (1973).

[80] ST.-GERMAIN, G.; BEAUCHESNE, D.: Evaluation of the Microscan rapid yeast identification panel. J. Clin. Microbiol. **29** (1991) 2296–2299.

[81] TIMMINS, E.M.; HOWELL, B.K.; ALSBERG, B.K.; NOBLE, W.C.; GOODACRE, R.: Rapid differenciation of closely related *Candida* species and strains by pyrolysis-mass spectrometry and Fourier transform infrared spectroscopy. J. Clin. Microbiol. **36** (1998) 367–374.

[82] VOS, P.; HOGERS, R.; BLEEKER, M.; REIJANS, M.; VAN DE LEE, T.; HORNES, M.; FRIJTERS, A.; POT, J.; PELEMAN, J.; KUIPER, M.; ZABEAU, M.: AFLP: a new technique for DNA fingerprinting. Nucl. Acids Res. **23** (1995) 4407–4414.

[83] WADLIN, J.K.; HANKO, G.; STEWART, R.; PAPE, J.; NACHAMKIN, J.: Comparison of three commercial systems for identification of yeast commonly isolated in the clinical microbiology laboratory. J. Clin. Microbiol. **37** (1999) 1967–1970.

[84] WICKERHAM, L.J.: Taxonomy of Yeasts. Technical Bulletin No. 1029. Washington, U.S.A.: USDA (1951).

[85] WICKERHAM, L.J.; BURTON,K.A.: Carbon assimilation tests for the classification of yeasts. J. Bacteriol. **56** (1948) 363–371.

[86] WILLCOX, W.R.; LAPAGE, S.P.: Methods used in a program for computer-aided identification of bacteria. In: Biological Identification with Computers (edited by R.J. Pankhurst). London, U.K.: Academic Press (1975) 103–119.

6 Spoilage yeasts with emphasis on the genus *Zygosaccharomyces*

STEPHEN A. JAMES and MALCOLM STRATFORD

6.1 Introduction

Yeasts, as a group of microorganisms, are well known for their positive contributions in the production of a variety of different foods and beverages, including beer, bread, chocolate, cider, saké, soy sauce and wine. Indeed, *Saccharomyces cerevisiae* is perhaps the best-recognised yeast species for such purposes, and is often referred to simply as either baker's or brewer's yeast, depending in which process it is being used (i. e., baking or brewing). However, as well as having a positive role in food and beverage production, yeasts can also have a negative role, with a limited number of species (including *S. cerevisiae*) noted for their ability to cause spoilage.

In nature, yeasts are found in a variety of different habitats, including soil, freshwater, salt water and air [43]. Likewise, these predominantly unicellular fungi can also be found associated with many different animals, insects (e. g., *Drosophila* spp.) and plants (especially fruit and vegetables) [6]. Consequently, due to their widespread distribution, and presence on the surfaces of many fruit and vegetables [13], yeasts are routinely encountered as contaminants in a variety of different foods and beverages, including alcoholic and soft drink beverages, bakery products, dairy products (e. g., yoghurt and cheese), fresh and processed fruits and vegetables, fresh and processed meats and seafood [6, 19, 44, 56, 78]. Indeed in a recent survey by BARNETT et al. [6], of the 678 listed yeast species, more than 150 species from over 40 different genera (representing approx. half of all currently accepted genera) were noted as having a food/beverage association. Typically yeasts become the dominant contaminants of foods and beverages when competition from other microorganisms, particularly bacteria and moulds, is restricted by low pH, presence of preservatives (e. g., benzoic and sorbic acids), high sugar and/or alcohol concentrations [56]. Yeasts have the ability to grow or cause spoilage in foods at near neutral pH, but rarely do so, following the thermal treatments given to "low acid foods" to destroy bacterial pathogens.

However, despite the number of yeast species recorded as being food/beverage contaminants, very few are in fact viewed as significant spoilage species. Indeed, in their recent review of fungi associated with food spoilage, PITT and HOCKING [56] listed a total of only 12 yeast species (Table 6.1-1) which they saw as being responsible for causing the vast majority of spoilage incidents by yeasts of foods and beverages prepared according to normal standards of good manufacturing practice (GMP). In a 'forensic approach' to spoilage of soft drinks, DAVENPORT [10, 11] noted that most yeast contaminants encountered could be categorised as either opportunistic spoilage species (Group 2) or indicators of poor factory hygiene (Group 3). Unlike many of the significant (Group 1) spoilage yeasts listed in Tab-

Table 6.1-1 Yeast species most commonly associated with food and beverage spoilage (adapted from [56]).

Species	Spoilage properties
Brettanomyces bruxellensis (= ***Dekkera bruxellensis****)*	production of off-odours*
Candida holmii (= ***Saccharomyces exiguus****)*	moderately preservative^{R}
Candida krusei (= ***Issatchenkia orientalis****)*	preservative^{R}, film forming
Candida parapsilosis	lipolytic, fermentative
Debaryomyces hansenii	osmotolerant
Kloeckera apiculata (= ***Hanseniaspora uvarum****)*	fermentative
Pichia membranifaciens	preservative^{R}, film forming
Saccharomyces cerevisiae	fermentative, preservative^{R**}
Schizosaccharomyces pombe	preservative^{R}
Zygosaccharomyces bailii	fermentative, osmotolerant, preservative^{R}
Zygosaccharomyces bisporus	fermentative, osmotolerant, preservative^{R}
Zygosaccharomyces rouxii	fermentative, osmotolerant preservative^{R}

Teleomorphic (i. e., sexual) species names are highlighted in bold type.
* Due to production of acetic acid
** Some strains only
^{R} resistance

le 6.1-1, opportunistic spoilage species generally only give rise to spoilage if processing errors occur, such as under-dosing or omission of preservatives, gross errors in hygiene or use of inadequate pasteurisation temperatures. Key characteristics noted by DAVENPORT [10, 11, 12] that distinguished the major spoilage species, such as *Z. bailii*, from opportunistic spoilage species included preservative resistance, osmotolerance, and high fermentative ability. Group 2 opportunistic spoilage yeasts of soft drinks included *Candida parapsilosis*, *Pichia membranifaciens* and *Debaryomyces hansenii* (= *Candida famata*) [71]. In contrast, poor-hygiene-indicator species (e. g., *Aureobasidium pullulans* and *Rhodotorula* spp., indicators of dust contamination when found in soft drinks) are generally incapable of causing spoilage, even when processing errors do arise and allow significant infection.

6.2 Detrimental aspects of *Zygosaccharomyces*

To the food and drinks industries, the most problematic spoilage yeasts encountered are undoubtedly those belonging to the genus *Zygosaccharomyces*. Indeed, this genus is often re-

Table 6.2-1 Species belonging to the genus *Zygosaccharomyces* and typical food products from which they have been recovered [adapted from refs. 6, 33, 42, 64, and 75].

Species	Foods isolated from
Z. bailii	Apple juice, confectionery, fitness drinks, grape and blackcurrant juice, grape juice, honey, marzipan, mayonnaise, orange-juice concentrate, pickles, salad cream, sorghum-brandy mash, wine, Worcester sauce
Z. bisporus	Fermenting cucumbers, orange drink
Z. cidri	Cider
Z. fermentati	Cola drink, orange drink, strawberry drink
Z. florentinus	Cola drink, lemonade, orangeade, soda water
Z. kombuchaensis	Kombucha tea
Z. lentus	Orange squash, tomato ketchup, whole-orange juice, wine
Z. mellis	Honey, strawberry juice, sugar,
Z. microellipsoides	Apple juice, lemonade
Z. mrakii	None recorded
Z. rouxii	Candied fruit, cane sugar, chocolate filling, fruit juice concentrates (e. g., apple, orange), honey, jam, maple syrup, marmalade, marzipan, molasses, orange syrup, soft drinks, soy sauce, sugar syrup, wine

garded as being synonymous with food spoilage. Currently the genus comprises of 11 species (Table 6.2-1), three of which, namely *Z. bailii*, *Z. rouxii* and to a lesser extent *Z. bisporus* are well recognised as representing significant food and beverage spoilage organisms (Table 6.1-1). Key physiological characteristics that make these yeasts such problematic spoilage agents include significant resistance to weak acid preservatives (notably *Z. bailii*), extreme osmotolerance (notably *Z. rouxii* and *Z. mellis*), and ability to vigorously ferment hexose sugars such as glucose and fructose (i. e., most members of the genus). Recently, Steels et al. [64] using ribosomal DNA (rDNA) sequences identified a new *Zygosaccharomyces* species, *Z. lentus*, which was found to have similar spoilage characteristics to those possessed by *Z. bailii*, *Z. bisporus* and *Z. rouxii* (i. e., osmotolerance and preservative resistance) [65]. Furthermore, unlike *Z. bailii*, *Z. lentus* was also shown to be able to grow (albeit slowly) at low temperature (i. e., 4 °C), suggesting that this species might pose a threat as a potential spoilage organism of chilled food products.

Foods at particular risk to spoilage by these yeasts tend to be acidic (pH 2.5 to 5.0) and contain high concentrations of fermentable sugars [75]. Foods commonly spoiled include fruit juices, juice concentrates, soft drinks, sugar syrups, jams and preserves, honey, tomato sauce, mayonnaise and wines [6, 38, 75].

With regard to the types of spoilage that yeasts cause, this is very much dependent upon the food or beverage that they are found in. Typical spoilage includes generation of taints, odours and off-flavours, development of hazes, and excessive gas production. Clouds and hazes generally refer to the appearance of substantive yeast growth, which is barely visible at 10^5 cells/ml, but may reach 10^7 cells/ml. Such growth may not be visible in cloudy food-products, such as fruit juices, but 40 % of commercial fruit juices in Tetrapaks were found to contain viable yeasts including *Z rouxii* and *Z. bailii* [14]. On occasion yeast growth has the appearance of particulates, usually as a result of pseudohyphal formation or flocculation. Spoilage of ciders by *Saccharomycodes ludwigii* often results in 1–3 mm sized particles. Alternatively, surface-films can be produced, notably by *Pichia membranifaciens* or *Issatchenkia orientalis* (= *Candida krusei*) [56, 80], and on occasion by *Zygosaccharomyces* [35, 72]. In more solid foods, yeast spoilage usually takes the form of unsightly colonial surface growth, although gas bubbles can form within the food by fermentation.

Where substantial yeast growth has occurred, it is likely that other attributes of yeast spoilage will also become evident. It is rare for spoilage flavours, odours or taints to occur without yeast growth to at least 10^5 cells/ml. Growth of fermentative yeast may result in gas production and also ethanol, imparting a sweet alcoholic taste. Off flavours and taints may also be caused by microbial secondary metabolite formation, following growth of substantial yeast populations (e. g., ethyl acetate [44]). Alternatively yeast enzymes acting on components of the food may also result in unpleasant taints. As an example, certain spoilage yeasts and moulds are able to decarboxylate sub-inhibitory levels of the preservative sorbic acid (2,4-hexadienoic acid) to form 3,5-pentadiene with a characteristic kerosene smell.

Excessive gas production is a direct and noticeable consequence of high fermentative ability in spoilage yeasts. It may be speculated that a significantly higher proportion of foods spoiled by fermentative yeasts result in consumer complaints, as compared with spoilage by non-fermentative yeasts. Indeed, if left unchecked, yeast fermentation in a food or beverage (i. e., fermentative spoilage) can generate very significant amounts of CO_2, which in turn can lead to the distortion or explosion of either the product (e. g., fondant-filled chocolate cream eggs) or the product packaging. While none of the spoilage-associated *Zygosaccharomyces* species are known to be pathogenic to man, TODD [77] nevertheless reported several cases where yeast ingestion was believed to have caused gastrointestinal disorders. Similarly, product spoilage by *Zygosaccharomyces* species can lead to serious injury (e. g., eye damage), particularly if glass bottles are caused to explode [21].

6.3 Physiological background of spoilage by *Zygosaccharomyces*

The key physiological characteristic that make the *Zygosaccharomyces* such notorious spoilage yeasts are their resistance to the preservatives commonly used in the food industry. Preservative resistance has been widely studied in yeast but in recent years this has predominantly involved the moderately preservative-resistant yeast *S. cerevisiae,* due to its role in the pioneering of genetic manipulation and subsequent genome sequencing [20]. It is commonly assumed that many of the mechanisms of resistance identified in *S. cerevisiae* will have similar counterparts in the more resistant *Zygosaccharomyces* yeasts. The preservatives examined are those most commonly used in foods: sorbic and benzoic acids used in beverages, processed fruit, fish, vegetables, spreads, sauces and confectionery; acetic acid in pickles, sauces, dressings and vinegars; propionic acid in bread; sulphite (SO_2) in wines and cider. These are often collectively termed the "weak-acid preservatives". *Zygosaccharomyces spp.* yeasts are known to be phenomenally resistant to sorbic, benzoic, propionic and acetic acids, and have the capacity to adapt to higher concentrations [26, 49, 82, 83, 84]. KABARA and EKLUND [34] have reviewed the antimicrobial nature of these acids and their use in foods.

Weak acid preservatives are fungistatic rather than fungicidal and are commonly thought to inhibit growth of yeasts by acidification of the cytoplasm. Weak acids in aqueous solution form dynamic equilibria between undissociated acid molecules and their respective charged anions, with acid molecules predominating at low pH. It is known that uncharged weak acid molecules dissolve easily in lipids and are able to pass rapidly through the plasma membrane by Fick-type diffusion [9, 27, 62, 69, 85]. The classical "weak acid theory" proposes that uncharged acid molecules, on passing into the neutral pH of the cytoplasm, are forced to dissociate, thus accumulating as charged anions within the cytoplasm, and simultaneously releasing protons. Excessive proton release overcomes cytoplasmic buffering and causes acidification of the cell, inhibiting key enzymes and preventing active transport. This theory was independently proposed to account for the action of 2,4-dinitrophenol [36], acetic acid [48], sulphite [67], sorbic acid (A. WARTH, pers. commun.) and benzoic acid [37]. Cytoplasmic acidification has been experimentally verified in *S. cerevisiae* for acetic acid [48], benzoic acid [37], and in both *S. cerevisiae* and *Z. bailii* for sulphite [53]. Other mechanisms of action for preservatives have been proposed and are reviewed by KABARA and EKLUND [34]. More recent work has suggested a membrane disrupting action for sorbic acid [68] and involvement of benzoic acid in glycolysis [51].

Microbial mechanisms of resistance can, in broad terms, be summarized as three possible strategies: i) destruction of the inhibiting agent; ii) prevention of entry, or removal, of the inhibitor from the cell; iii) alteration of the inhibitor target, or amelioration of the damage caused. All three strategies have been suggested in accounting for yeast resistance to weak-acid preservatives.

Sulphite removal by production of binding agents was shown to be a key factor prior to growth of *S. cerevisiae* and *Saccharomycodes ludwigii* [67, 70]. The ability of *Z. bailii* strains to metabolize acetate in the presence of glucose may contribute to the resistance of this yeast to acetic acid [63], by removing acetic acid and thereby raising the pH of the cell cytoplasm [75]. Metabolism of benzoic acid by *Z. bailii* via p-hydroxybenzoic acid and catechol, was reported by INGRAM [26], and recently the molecular mechanism for this transformation was elucidated, with the *ZbyME2* gene shown to confer the ability to metabolize both sorbate and benzoate [46]. However, other studies have concluded that metabolism of weak acids is of limited benefit and insufficient to account for the resistance of *Z. bailii* to preservatives [26, 66, 82].

Prevention of entry, or removal of acidic preservatives was suggested by observations that resistant yeasts showed slower permeation by benzoic or propionic acids [85]. WARTH proposed that in *Z. bailii,* an active "sorbate pump" conferred resistance by extrusion of preservatives [82, 83, 84, 85]. This view was disputed by observations that the distribution of weak acids was exactly as predicted from the pH of media and cytoplasm [7, 8]. However, in *S. cerevisiae* active efflux of benzoate was suggested as a resistance mechanism [22], a view strengthened by the discovery that deletion of any of a number of major facilitator superfamily transporters, increased sensitivity to preservatives. These included *PDR12* for sorbate, benzoate and acetate [25, 55], *SSU1* for sulphite [50], and *AZR1* for acetic acid [73].

Amelioration of the damage caused by preservatives (i. e., low pH cytoplasm), involves removal of acid from the cell. The plasma membrane H^+-ATPase ejects protons from the cell in an energy-dependent manner, and has been shown to aid resistance to acetic and sorbic acids [24, 79]. Hsp30 may also affect weak-acid resistance via regulation of the H^+-ATPase [54].

In light of these reports, these characteristics will be discussed in more detail in context of each of the three main *Zygosaccharomyces* spoilage species (i. e., *Z. bailii*, *Z. bisporus* and *Z.rouxii*), as well as the recently described *Z. lentus* [64].

6.3.1 *Zygosaccharomyces bailii*

Among the genus *Zygosaccharomyces*, *Z. bailii* undoubtedly represents the most problematic spoilage yeast to the food and beverage industries. The reason for this is that one of the most distinctive physiological characteristic of this species, from a spoilage perspective, is its exceptional resistance to weak acid preservatives such as acetic, benzoic, propionic and sorbic acids, which are commonly used in the preservation of foods and beverages. Individual cells in populations of *Z. bailii* have been shown to vary considerably in their resistance to sorbic acid, with rare "super cells" able to grow in levels of preservative double that of the average population [66]. Often, *Z. bailii* strains are recovered (from spoiled food products) whose resistance to these types of preservatives far exceeds the levels legally permitted in Europe (i. e., 300 mg l^{-1} in soft drinks [2]). Besides displaying phenomenal resistance to these types of preservative, *Z. bailii* is also remarkable for the fact that exposure to low, sub-inhibitory levels of this type of preservative can often lead to adaptation. As a result, strains of this species can gain the ability to both survive and proliferate in concentrations of (weak acid) preservative much higher than prior to adaptation.

Besides being preservative resistant, other features that contribute to the spoilage capacity of *Z. bailii* are: i) its ability to vigorously ferment glucose, ii) ability to cause product spoilage from an initial inoculum of as little as one viable cell per litre or package [10, 17] and iii) moderate osmotolerance (in comparison to *Z. rouxii*). Like *S. cerevisiae*, *Z. bailii* can continue to ferment glucose under significant atmospheric pressure [28], but unlike *S. cerevisiae*, *Z. bailii* can do so in the presence of high levels of preservative. Indeed, sugar fermentation by *Z. bailii* and the *Zygosaccharomyces* species in general is quite unique, as unlike other yeasts, these species appear to metabolize fructose in preference to glucose (a phenomenon referred to as fructophily) [15]. Consequently, the growth rate of *Z. bailii* in a food product is often enhanced if the level of fructose present exceeds 1 % of the total product composition. Although not as osmotolerant as *Z. rouxii* (see following section), *Z. bailii* is nevertheless still capable of growing at a water activity (a_w) as low as 0.80 at 25 °C [56].

6.3.2 *Zygosaccharomyces bisporus*

Genealogically, *Z. bisporus* has been shown by ribosomal DNA (rDNA) sequence analysis to be closely related to *Z. bailii* [29]. Consequently, this species shares many of the physiological characteristics possessed by *Z. bailii* [6], including preservative resistance and osmotolerance, which can sometimes hamper accurate identification based on conventional physiological methods between isolates of these two species [88]. However, in contrast to *Z. bailii*, *Z. bisporus* is reported to be marginally more osmotolerant [76]. While individual strains show considerable variation, *Z. bisporus* is on the whole slightly less resistant to weak acid preservatives than *Z. bailii* (H. Steels, pers. commun.). From a spoilage perspective, despite having similar physiological characteristics to *Z. bailii*, *Z. bisporus* probably does not represent the same sort of (spoilage) threat as *Z. bailii*, as it does not appear to be as widely distributed as *Z. bailii* [6].

6.3.3 *Zygosaccharomyces lentus*

Recently, a spoilage yeast was isolated from spoiled whole-orange drink, and initially identified by conventional methods [6, 89], as *Z. bailii*. However, closer physiological examination of this isolate revealed it to possess atypical *Z. bailii* characteristics, displaying extremely poor growth when grown aerobically in shake culture, and failing to grow in the presence of 1 % acetic acid (a diagnostic test used to distinguish *Z. bailii* and *Z. bisporus* from other *Zygosaccharomyces* species [6, 38]). Subsequent 18S rDNA sequence analysis of this isolate, as well as four additional physiologically atypical *Z. bailii* strains, revealed the existence of a phylogenetically distinct taxon, closely related to, but nevertheless separate from, *Z. bailii* and *Z. bisporus* [64]. Based on its slow growth under aerobic conditions, Steels *et al.* [64] named this novel *Zygosaccharomyces* species *Z. lentus*. Figure 6.3-1 shows a scanning electron micrograph of *Z. lentus* IGC 5316, a strain isolated from spoiled wine [64].

As a consequence of the close genealogical relationship exhibited between these three *Zygosaccharomyces* spp. (Fig. 6.3-2), *Z. lentus* was found to share many of the spoilage characteristics possessed by both *Z. bailii* and *Z. bisporus*, including osmotolerance (i. e., ability to grow in the presence of 60 % [w/v] sugar), resistance to weak acid preservatives (e. g., benzoic and sorbic acids), as well as resistance to dimethyldicarbonate (DMDC; Vel-

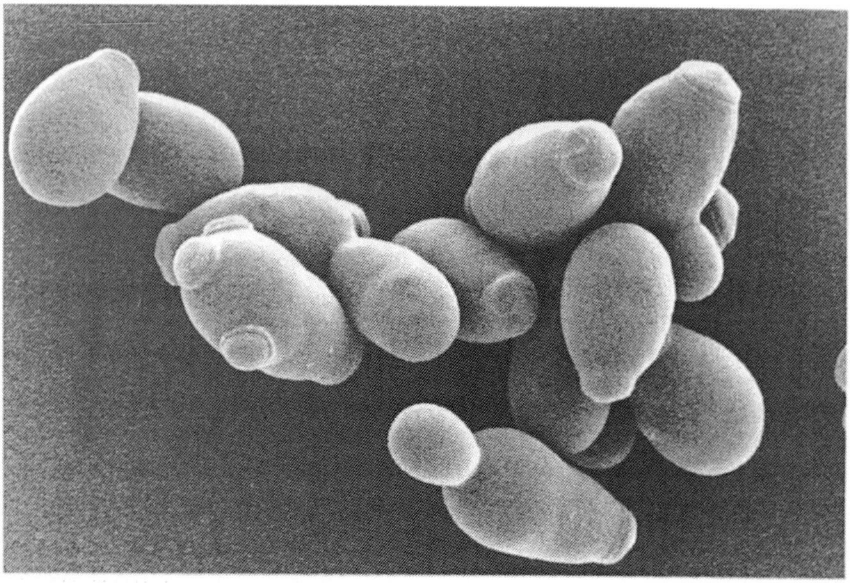

Fig. 6.3-1 Scanning electromicrograph of *Z. lentus* IGC 5316, isolated from spoiled wine [64] (c/o Mark Kirkland, Unilever Research Colworth Laboratory)

corin [Tm]) a sterilant occasionally used in the beverage industries [64, 65]. To date, nine isolates of *Z. lentus* have been identified and characterized, of which five are known to have originated from spoiled food products (i. e., orange squash, orange juice, tomato ketchup and wine [64]). However, unlike both *Z. bailii* and *Z. bisporus*, *Z. lentus* will grow (albeit slowly) at low temperature (i. e., 4 °C). Such ability therefore raises the possibility that this recently described *Zygosaccharomyces* species could pose a real threat to the food and drinks industries as a spoilage agent of chilled products [64].

6.3.4 *Zygosaccharomyces rouxii*

As a spoilage organism, *Z. rouxii* is physiologically distinctive as being the most osmotolerant yeast species known. Indeed, of all known organisms only the filamentous fungus *Xeromyces bisporus* displays greater osmotolerance than *Z. rouxii* [56]. This yeast has been recorded as being able to grow at a water activity (a_w) of 0.62 in fructose and 0.65 in sucrose/glycerol [76]. Consequently, *Z. rouxii* is particularly noted for its spoilage of raw cane

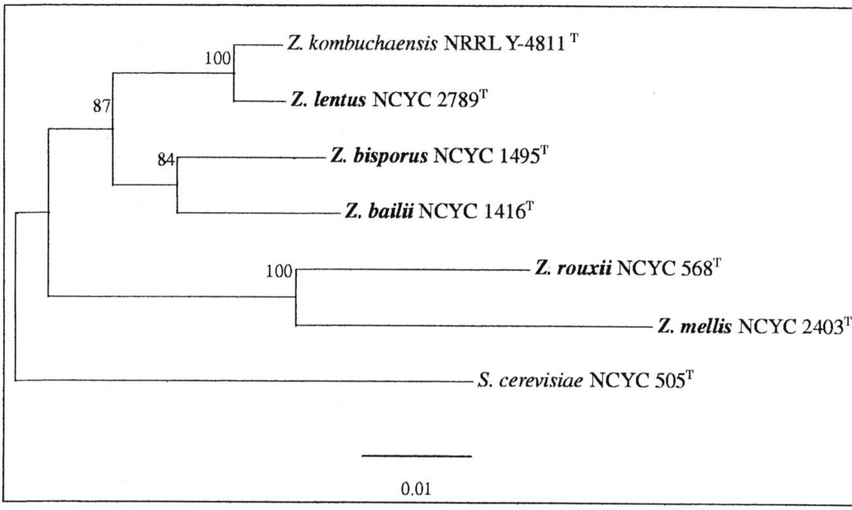

Fig. 6.3-2 Dendrogram showing the phylogenetic relationship between members of the *Zygosaccharomyces* sensu stricto species group [42] based on 26S rDNA D1/D2 gene sequences. Those *Zygosaccharomyces* species noted for their ability to cause spoilage are highlighted in bold font. The tree was constructed by using the neighbour-joining method [58]. Bootstrap values, expressed as percentages of 100 replications, are given at branch nodes (only values > 50 % are shown). Scale bar, 1-estimated base substitution per 100 nucleotides.

sugar and food products of high sugar content, such as fruit juice concentrates, honey, jam, marzipan and sugar syrups [6, 33, 76]. Besides being osmotolerant, *Z. rouxii* is also preservative resistant; though not to the same degree as *Z. bailii*, and like all *Zygosaccharomyces* will vigorously ferment hexose sugars (e. g., glucose). Thus, as a causative agent of fermentative food spoilage, *Z. rouxii* ranks second only to *Z. bailii* [56].

6.3.5 Other *Zygosaccharomyces* spoilage species

Other, less well-known members of the genus *Zygosaccharomyces* can also cause spoilage of foods. These include *Z. cidri*, *Z. fermentati*, *Z. florentinus*, *Z. mellis*, and *Z. microellipsoides* (Table 6.2-1). *Z. mellis*, frequently confused with and mistaken for *Z. rouxii*, is perhaps best known for causing spoilage in honey. *Z. microellipsoides* has been described as "less dangerous than *S. cerevisiae*" [59] and is closely related to *Torulaspora delbrueckii* [30]. Both *T. delbrueckii* and *Z. microellipsoides* are fermentative [6], osmotolerant [60], and widely distributed in soft drinks plants. Both species are capable of causing spoilage, but are relatively sensitive to preservatives. *Z. cidri*, *Z. fermentati* and *Z. florentinus* are all moderately osmotolerant, fermentative species, capable of causing spoilage. They are generally more sensitive to preservatives than *S. cerevisiae* (H. STEELS, pers. commun.) and less prone to cause spoilage.

6.4 Specific methods to study spoilage by *Zygosaccharomyces*

In order to prevent, or at the very least reduce the likelihood of product spoilage by yeasts, and particularly by osmotolerant and preservative resistant species such as *Z. bailii*, *Z. bisporus*, *Z. lentus* and *Z. rouxii*, it is essential to firstly, be able to detect their presence in a food or beverage, and secondly, be able to accurately identify them, preferably to the species level. Accurate identification of a food-borne yeast can, in many instances, help to establish whether or not the contaminating species represents a spoilage threat to the food product in which it is found. By comparing the physiological characteristics of a yeast contaminant to those of a particular food or beverage, food microbiologists can often determine the spoilage potential of the yeast. For example, if the contaminant is identified as being a highly fermentative yeast species (e. g., *S. cerevisiae, T. delbrueckii* or *Z. bailii*), product spoilage becomes a real possibility, if there are fermentable sugars present that the yeast can utilize. Furthermore, knowledge of the yeasts most commonly associated with a particular food product, can also provide insights into the best means of eradicating them from the food or beverage (e. g., through addition of weak acid preservatives, or their elimination from raw materials), reducing the likelihood of future incidences of product contamination and/or spoilage.

Standard methods routinely used to detect for the presence of food-borne yeasts are generally based upon spread plating (see Chapter 2). Typical plating media used include Malt Extract Agar (MEA) and Tryptone Glucose Yeast extract agar (TGY), on to which the food product is either spread directly, or indirectly, having first been homogenised with a suitable aqueous diluent (e. g., 0.1 % peptone [19, 56]). If however, the product being tested has a high sugar content (e. g., sugar syrups), then it is often better to use a diluent that has sucrose added to it (to a final concentration of 10 % or higher [19]). Use of buffered diluents in such circumstances, will often help reduce the death and loss of yeast cells from osmotic shock (as well as aid in the recovery of sub-lethally injured cells), and thus aid in the detection of osmotolerant species such as *Z. rouxii*. Similarly, diluents containing added salt are recommended for the isolation of yeasts from high salt foods [19]. Once inoculated, spread plates are typically incubated at either 25 °C or 30 °C, and checked daily for signs of yeast growth (i. e., appearance of discrete colonies). In the case of the *Zygosaccharomyces* spp. (e. g., *Z. bailii*), HOCKING [23] found that incubating agar plates at 30 °C, rather than at 25 °C, reduced the time needed to detect the presence of these yeast species, as they grew quicker at the elevated temperature.

For preservative-resistant species such as *Z. bailii* (and indeed other such species e. g., *Pichia membranifaciens* and *Schizosaccharomyces pombe*), two media routinely used to selectively screen (also referred to as 'target isolation' [19]) for their presence in products are acidified MEA (MAA [i. e., Malt Acetic Agar]) and acidified TGY (TGYA) [56]. In both cases, the basal medium (MEA or TGY) is supplemented with 0.5 % acetic acid. One % acetic acid is used as a key diagnostic test for differentiating *Z. bailii* and *Z. bisporus* from other *Zygosaccharomyces* species [6]. As well as these two media, Erickson [16] also developed a medium, ZBM (i. e., *Zygosaccharomyces bailii* medium), to use for the specific detection of *Z. bailii*. Like both MAA and TGYA, ZBM medium contains 0.5 % acetic acid, but is additionally supplemented with 2.5 % NaCl and 0.01 % potassium sorbate. While this medium appears highly selective for detecting the presence of *Z. bailii* in food products, results from a study by HOCKING [23] indicated that this medium might fail to detect the presence of other preservative resistant species (e. g., *Schizosaccharomyces pombe*). While such yeasts are not as preservative resistant as *Z. bailii*, they still nevertheless pose a significant spoilage threat. For the detection and isolation of osmotolerant species such as *Z. rouxii*, typical plating media used generally contain 40–60 % glucose (e. g., MY50G, i. e., Malt Yeast Extract 50 % Glucose Agar) which lowers the a_w of the medium and provides an optimum osmotic pressure to specifically select for these yeasts.

However, if initially present in low numbers, yeasts may go undetected in a food or beverage when spread plating methods are used. If the contaminating species is *Z. bailii*, such an oversight can prove both damaging and costly. Recently, PITT and HOCKING [56] reported that an inoculum of only 5 viable cells of a preservative-adapted *Z. bailii* strain was sufficient to cause spoilage of a canned carbonated soft drink. Indeed, the authors went on to caution that, in their opinion, presence of even a single healthy cell of *Z. bailii* would, given sufficient time, lead to product spoilage, confirming the earlier opinions of VAN ESCH [17] and DAVENPORT [10]. In such circumstances, two techniques routinely used (prior to spread plating) to help monitor products for the presence of yeasts in low cell numbers are membrane filtration and inclusion of an enrichment step [56]. Membrane filtration (e. g., hydrophobic grid membrane filtration) is particularly suited for detection of low cell numbers in beverages (i. e., 1 cell/100mls). This technique physically separates viable yeast cells from product ingredients, which may affect yeast growth, and so can improve overall levels of detection.

For the reasons given earlier in this section, it is essential that once a yeast is detected in a product, it is reliably and accurately identified, ideally to the species level. Traditionally, this has been achieved on the basis of morphological, physiological and biochemical characterization [6, 89]. Typically, such identification requires the use of between 50 and 100 diagnostic tests to reliably identify most yeasts to species level, and routinely takes 1 to 3 weeks to obtain a result (i. e., a species identification). Such a method is both labour intensive and time consuming, and interpretation of resulting data can require considerable expertise.

Furthermore, accurate identification, to species level, can sometimes be hampered by ambiguous test results due to strain variability [6]. This latter problem can make differentiating between some species difficult, particularly those possessing similar overall physiologies. For example, amongst the *Zygosaccharomyces*, isolates of *Z. bailii* and *Z. bisporus* can only reliably be differentiated from one another on the basis of their differing responses to a single growth test, namely trehalose assimilation (of the two species, only *Z. bailii* is capable of assimilating this disaccharide sugar [6, 38]). Nevertheless, for practical purposes, several manual and automated yeast identification systems (e. g., API 20C system [BioMérieux], API ID 32C strips [BioMérieux], Biolog MicroStation system [Biolog Inc.]) have been developed based on this phenotypic approach and are routinely used in many food microbiology laboratories. The degree of success with such techniques varies widely, depending on the skill and experience of the individuals concerned (see also Chapter 5).

However, in an effort to improve the overall accuracy and reliability of yeast identification, attention has, in recent years, shifted to the use of DNA-based technologies, such as the polymerase chain reaction (PCR) [57], for the development of alternative methods of identification. PCR-mediated methods currently in use for identifying and typing yeasts include DNA sequence analysis of the small- and large-subunit ribosomal RNA encoding genes (i. e., 18S rDNA and 26S rDNA respectively [29, 40, 41]), random amplified polymorphic DNA (RAPD) analysis [3, 86], and microsatellite PCR fingerprinting [5].

To date, ribosomal DNA (rDNA) sequence analysis has proved by far the most useful method for identifying and characterising yeasts to the species level. This method has gained in popularity since the advent of PCR [57] and the subsequent development of protocols for directly sequencing PCR-amplified DNA products. Development of such protocols has made the amplification and sequencing of yeast ribosomal DNA, either from genomic DNA or directly from individual colonies [29], extremely straightforward. Due to the fact that different regions of the ribosomal DNA display differing levels of sequence divergence has meant that these genes (i. e., 18S and 26S rDNA) and their associated spacer regions (i. e., ITS1 and ITS2) have proved extremely useful for differentiating between yeasts at a variety of different taxonomic levels (e. g., genus, species and subspecies levels) [18, 29, 30, 41].

With regard to the *Zygosaccharomyces* species, ribosomal DNA sequence analysis has proved a powerful method for identifying these yeasts [29, 30, 42, 64], and particularly those species difficult to discriminate between using conventional physiological-based approaches (e. g., *Z. bailii* and *Z. bisporus*; *Z. mellis* and *Z. rouxii* [38]). Furthermore, such sequences have also proved useful in the identification of new *Zygosaccharomyces* species. As discussed in section 6.3.3, Steels and co-workers [64] used 18S rDNA sequences to establish the fact that a number of spoilage-associated *Zygosaccharomyces* isolates, originally characterised as physiologically atypical *Z. bailii* strains, were in fact representatives of a quite separate and phylogenetically distinct species, which the authors named *Z. lentus*. More recently, KURTZMAN and colleagues [42], described another new food-associated *Zygosaccharomyces*, *Z. kombuchaensis* (isolated from kombucha tea), based on sequence analysis of the variable D1 and D2 domains of the 26S rDNA. In comparison to the 18S rDNA, the D1/D2 region of the 26S rDNA displays far more sequence divergence, to the extent that it can be used to differentiate between the vast majority of known yeast species, both ascomycetous and basidiomycetous [18, 41].

Besides their importance in yeast identification, rDNA sequences have also proved useful for studying yeast evolution, and assessing the genealogical relationships (i. e., phylogeny) between different yeast species and genera. In a number of studies, 18S rDNA, 26S rDNA D1/D2 and ITS sequences have all been used to investigate the genealogical relationships between species within the genus *Zygosaccharomyces* [29, 30], and between the *Zygosaccharomyces* species and those of other related genera [30, 32, 41, 42, 64]. Results from these studies have demonstrated that the genus *Zygosaccharomyces*, as currently described [6, 39], is in fact genealogically intermixed with species from other genera (i. e., *Kluyveromyces*, *Saccharomyces* and *Torulaspora*). However despite this phylogenetic intermixing, *Z. bailii*, *Z. bisporus* and *Z. rouxii*, the three members of the genus most noted for their ability to cause spoilage, are nevertheless genealogically closely related to one another (Fig. 6.3-2). Indeed these three species, along with *Z. kombuchaensis* [42], *Z. lentus* [64] and the recently re-established *Z. mellis* [38], form a distinct species group, which is quite separate from other *Zygosaccharomyces* and non-*Zygosaccharomyces* yeasts [32, 41, 64]. In their recent description of *Z. kombuchaensis*, KURTZMAN and co-workers [42] refer to this group of species as the *Zygosaccharomyces* sensu stricto.

For identification purposes, ribosomal DNA sequencing is an extremely useful (and accurate) method for identifying and characterizing unknown yeast isolates, as well as differentiating between different yeast species [40, 41]. Generally this is more than adequate for most purposes. However, there may be occasions, particularly if a product is contaminated with more than one yeast strain (either of the same species or of different species), for the need to use alternative typing methods with the ability to resolve below the species level. Two such typing methods that appear to offer this level of (taxonomic) resolution are random amplified polymorphic DNA (RAPD) analysis and microsatellite PCR fingerprinting.

In RAPD analysis, DNA fingerprints are generated by the amplification of random DNA fragments by low stringency PCR, using a single small (typically 10-base) oligonucleotide primer of arbitrary nucleotide sequence, and analyzed by subsequent gel electrophoresis [86]. Microsatellite PCR fingerprinting is a related technique, which uses oligonucleotide primers of defined sequence (e. g., $[GTG]_5$ and $[GAC]_5$) to amplify simple repetitive DNA sequences and generate a DNA fingerprint pattern. Although similar in principle, the two molecular typing methods differ in the fact that RAPD analysis relies on using a primer annealing temperature that is much lower than that typically used for microsatellite PCR fingerprinting (i. e., 37 °C as opposed to 55 °C) [81]. Consequently, problems can arise regarding the reproducibility of DNA fingerprint patterns generated by RAPD analysis, particularly between different typing laboratories. Nevertheless, both PCR-based techniques are quick and easy to execute, require only small amounts of template DNA (not necessarily of high purity), and can be used for screening/typing large numbers of yeast isolates (for further details see Chapter 4).

To date, the two typing methods have been tested on a variety of different yeasts, including a number of spoilage-associated species (e. g., *S. cerevisiae*, *Z. bailii*, *Z. bisporus* and *Z. rouxii*) and found useful for both species and strain discrimination [3, 4, 5, 47]. In a recent study, Baleiras Couto et al. [5] compared the use of microsatellite PCR fingerprinting with that of a standard API identification system for typing 126 yeast strains recovered, over a 14-month period, from a mayonnaise and salad dressing production facility. Results from the study showed that microsatellite PCR fingerprinting was a far superior method for identifying all the yeast species recovered during the survey. Indeed, the authors noted that the API system failed to distinguish between the spoilage species *Z. bailii* and *Z. bisporus*, and could only identify isolates of these two species to the genus level (i. e., as *Zygosaccharomyces* spp.). In contrast, the microsatellite PCR fingerprinting not only distinguished between these two species, but also between individual strain types as well. In fact, the level of discrimination observed at the sub-species level, led the authors of the study to conclude that this typing method could prove useful for tracking individual spoilage isolates and identifying their source of origin.

6.5 Quality control

Until recently, the main method of quality control (QC) used by the food and beverage industries was that of end-product testing. As the name suggests, this method involves the random sampling of the final product, and determining the total number of viable microbes (referred to as the total viable cell count) present in each sample. Typically for yeasts, this is achieved by plating out sample aliquots onto a suitable detection medium (e. g., MEA or TGY agars), incubating the plates at a set temperature (e. g., 25 or 30 °C), and counting the number of yeast colonies that develop after a specific period of time (e. g., 2 to 5 days). However, as a means of assessing the microbial stability of a food or beverage, total viable cell counts are of limited use. For instance, detection of a high number of yeast cells in a product does not necessarily indicate that spoilage is imminent. In the case of beverages of low nutrient content (e. g. wine), the vast majority of yeast contaminants typically present at the time of bottling will fail to survive for more than a few days, and so spoilage of such beverages is unlikely to arise [74]. Likewise, detection of low numbers of yeasts in a product does not guarantee that it will be safe from spoilage. This is particularly true in the case of *Z. bailii*, a yeast species more than capable of causing soft drink spoilage from an initial inoculum of only one viable cell per container [10, 17, 56].

An alternative approach to end-product testing as a method of quality control, is the use of quality management systems, such as the Hazard Analysis Critical Control Point (HACCP) system [45]. The aim of the HACCP system is to identify and control the (microbial) hazards associated with the manufacture of a food or beverage, removing the need to test the end product. Hazard analysis consists of analyzing all the procedures and raw ingredients used in the production of a food/beverage. The analysis identifies the critical control points, such as locations, processing equipment or raw ingredients, where hazards (i. e., introduction of microbial contamination) are most likely to occur. Once identified, these points can then be monitored on a regular basis as part of the overall quality control programme, and systems put in place to minimise the risk of microbial infection (e. g., pasteurization of raw ingredients). Recent EC legislation necessitates identification of hazards using the principles of HACCP for all producers of food and drink [1].

6.6 Future prospects and conclusions

At present, the vast majority of yeast spoilage of foods and beverages that have been processed and packaged according to normal standards of good manufacturing practice can be attributed to a limited number of species [56]. Amongst this list of species (Table 6.1-1), *Z. bailii* is without doubt the most problematic spoilage yeast currently encountered by the food and beverage industries, due to its exceptional resistance to preservatives. However, with continual demand for the development of new products, the possibility exists for other yeast species to emerge as spoilage threats, some of which may not have been previously encountered or characterized, such as *Z. lentus* [64]. One potential source for introducing new yeasts (both spoilage and non-spoilage species) into a product or processing facility is through the use of novel ingredients, such as tropical fruits in the manufacture of soft drinks. Furthermore, as well as being potential sources of yeast infection, addition of such ingredients to a product may also alter its overall chemical composition, thereby rendering it susceptible to spoilage by other commonly encountered species.

Likewise, changes in preservation strategies may also bring about an increase in product spoilage. Growing consumer concern regarding the presence of chemical residues (e. g., pesticides) in foods and beverages may, in time, lead to a widespread reduction in the levels of chemical preservatives added to foods and beverages. This in turn, may result in an increase in the incidences of spoilage, particularly by yeasts of moderate preservative resistance, such as *T. delbrueckii* and *Z. microellipsoides*, two species widely distributed in soft drinks factories. At present, the levels of preservative added are sufficient to prevent these species from causing significant product spoilage. However, if in the future preservative levels are reduced, yeasts such as *T. delbrueckii* and *Z. microellipsoides* could be encountered much more frequently as spoilage agents.

Reliable and effective methods of detection and identification are therefore key to safeguarding foods and beverages from spoilage by yeasts and other microbes, such as bacteria and moulds. Indeed, recent advances in DNA-based technologies, are undoubtedly paving the way towards a more accurate and robust means of identifying yeasts to the species level. Methods such as ribosomal DNA sequencing [40, 41], have already proved useful in the identification of many new yeasts, including two new species of *Zygosaccharomyces* [42, 64], one of which (*Z. lentus*) has been shown to possess significant spoilage potential [65].

With the amount of yeast sequence data ever increasing, and now including the complete genome sequences for *S. cerevisiae* [20] and *Sch. pombe* [87], as well as partial genome sequences for a further 13 species [61], there is real scope for the design of species-specific detection systems. While still in their infancy, PCR-based methods have already been developed for detecting the presence of a number of spoilage-associated yeasts, (e. g., *Z. len-*

tus [31]; *S. cerevisiae, Z. bailii* and *Z. bisporus* [52]). Furthermore, with the introduction of PCR-based typing techniques (e. g., microsatellite PCR fingerprinting [81] and RAPD analysis [86]), there is also now the means available for differentiating between individual spoilage strains of either the same or different species, and identifying their sources of origin [5]. With adaptation and development, such methods clearly lend themselves for use in food microbiology laboratories, and inclusion in quality assurance (e. g., HACCP) and quality control programmes as effective microbial monitoring systems.

6.7 References

[1] ANON.: EC Directive 93/94/EEC on Hygiene of Food Stuffs. European Council Official Journal No.L 175/1 (19th July 1993).

[2] ANON.: EC Directive 95/2/EC, Food additives other than colours and sweeteners. European Council Official Journal No L. 61 (18th March 1995).

[3] BALEIRAS COUTO, M.M.; VAN DER VOSSEN, J.M.B.M.; HOFSTRA, H.; HUIS IN'T VELD, J.H.J.: RAPD analysis: a rapid technique for differentiation of spoilage yeasts. Int. J. Fd Microbiol. **24** (1994) 249–260.

[4] BALEIRAS COUTO, M.M.; VOGELS, J.T.W.E.; HOFSTRA, H.; HUIS IN'T VELD, J.H.J.; VAN DER VOSSEN, J.M.B.M.: Random amplified polymorphic DNA and restriction enzyme analysis of PCR amplified rDNA in taxonomy: two identification techniques for food-borne yeasts. J. Appl. Bacteriol. **79** (1995), 525–535.

[5] BALEIRAS COUTO, M.M.; HARTOG, B.J.; HUIS IN'T VELD, J.H.J.; HOFSTRA, H.; VAN DER VOSSEN, J.M.B.M.: Identification of spoilage yeasts in a food-production chain by microsatellite polymerase chain reaction fingerprinting. Fd Microbiol. **13** (1996), 59–67.

[6] BARNETT, J.A.; PAYNE, R.W.; YARROW, D.: Yeasts: characteristics and identification, 3rd ed. Cambridge, U.K.: Cambridge University Press (2000).

[7] COLE, M.B.: The effect of weak acids and pH on *Zygosaccharomyces bailii*. Ph.D. thesis: University of East Anglia, Norwich, U.K. (1987).

[8] COLE, M.B.; KEENAN M.H.J.: Effects of weak acids and external pH on the intracellular pH of *Zygosaccharomyces bailii*, and its implications in weak acid resistance. Yeast **3** (1987) 23–32.

[9] CONWAY, E.J.; DOWNEY, M.: An outer metabolic region of the yeast cell. Biochem. J. **47** (1950) 347–355.

[10] DAVENPORT, R.R.: Forensic microbiology for the soft drinks business. Soft Drinks Management International April (1996) 34–35.

[11] DAVENPORT, R.R.: Forensic microbiology II, case book investigations. Soft Drinks International April (1997) 26–30.

[12] DAVENPORT, R.R.: Microbiology of soft drinks. In: Chemistry and technology of soft drinks and fruit juices (edited by Ashurst P.R.). Sheffield, U.K.: Sheffield Academic Press & CRC Press: Boca Raton (1998) 197–216.

[13] DO CARMO-SOUSA, L.: Distribution of yeasts in nature. In: The Yeasts, Vol. I (edited by Rose, A.H.: Harrison J.S.). London, U.K.: Academic Press (1969) 79–105.

[14] DRAGONI, I.; COMI, G.: Prensenza di muffe e lieviti in succhi di frutta prodotti industrialmente. Ind. Bevande **14** (1985) 599–606.

[15] EMMERICH, W.; RADLER F.: The anaerobic metabolism of glucose and fructose by *Saccharomyces bailii*. J. Gen. Microbiol. **129** (1983) 3311–3318.

[16] ERICKSON, J.P.: Hydrophobic membrane filtration method for the selective recovery and differentiation of *Zygosaccharomyces bailii* in acidified ingredients. J. Fd Protect. **56** (1993) 234–238.

[17] VAN ESCH, F.: Yeasts in soft drinks and fruit juice concentrates. De Ware (n) Chemicus **17** (1987) 20–31.

[18] FELL, J.W.; BOEKHOUT, T.; FONSECA, A.; SCORZETTI, G.; STATZELL-TALLMAN, A.: Biodiversity and systematics of basidiomycetous yeasts as determined by large-subunit rDNA D1/D2 domain sequence analysis. Int. J. Syst. Evol. Microbiol. **50** (2000) 1351–1371.

[19] FLEET, G.: Spoilage yeasts. Crit. Rev. Biotechnol. **12** (1992) 1–44.

[20] GOFFEAU, A.; BARREL, B.G.; BUSSEY, H.; DAVIS, R.W.; DUJON, B.; FELDMANN, H.; GALIBERT, F.; HOHEISEL, J.D.; JACQ, C.; JOHNSTON, M.; LOUIS, E.J.; MEWES, H.W.; MURAKAMI, Y.; PHILIPPSEN, P.; TETTELIN, H.; OLIVER, S.G.: Life with 6000 genes. Science **274** (1996) 543–567.

[21] GRINBAUM, A.; ASHKENAZI, I.; TREISTER, G.; GOLDSCHMIED-REOUVEN, A.; BLOCK, C.S.: Exploding bottles: eye injury due to yeast fermentation of an uncarbonated soft drink. Br. J. Ophthalmol. **78** (1994) 883.

[22] HENRIQUES, M.; QUINTAS, C.; LOUREIRO-DIAS, M.C.: Extrusion of benzoic acid in *Saccharomyces cerevisiae* by an energy-dependent mechanism. Microbiology **143** (1997) 1877–1883.

[23] HOCKING, A. D.: Media for preservative resistant yeasts: a collaborative study. Int. J. Fd Microbiol. **29** (1996) 167–175.

[24] HOLYOAK, C.D.; STRATFORD, M.; MCMULLIN, Z.; COLE, M.B.; CRIMMINS, K.; BROWN, A.J.; COOTE, P.J.: Activity of the plasma membrane H^+-ATPase and optimal glycolytic flux are required for rapid adaptation and growth of *Saccharomyces cerevisiae* in the presence of the weak-acid preservative, sorbic acid. Appl. Envir. Microbiol. **62** (1996) 3158–3164.

[25] HOLYOAK, C.D.; BRACEY, D.; PIPER, P. W.; KUCHLER, K.; COOTE, P.J.: The *Saccharomyces cerevisiae* weak-acid inducible ABC transporter Pdr12 transports fluorescein and preservative anions from the cytosol by an energy-dependant mechanism. J. Bacteriol. **181** (1999) 4644–4652.

[26] INGRAM, M.: Studies on benzoate-resistant yeasts. Acta Microbiol. **7** (1960) 95–105.

[27] INGRAM, M.; OTTAWAY, F.J.H.; COPPOCK, J.B.M.: The preservative action of acidic substances in food. Chem. Ind. (London) **75** (1956) 1154–1163.

[28] ISON, R.W.; GUTTERIDGE, C S.: Determination of the carbonation tolerance of yeasts. Lett. Appl. Microbiol. **5** (1987) 11–13.

[29] JAMES, S.A.; COLLINS, M.D.; ROBERTS, I.N.: Genetic interrelationship among species of the genus *Zygosaccharomyces* as revealed by small-subunit rRNA gene sequences. Yeast **10** (1994) 871–881.

[30] JAMES, S.A.; COLLINS, M.D., ROBERTS, I.N.: Use of an rRNA internal transcribed spacer region to distinguish phylogenetically closely related species of the genera *Zygosaccharomyces* and *Torulaspora*. Int. J. Syst. Bacteriol. **46** (1996) 189–194.

[31] JAMES, S.A.; COLLINS, M.D.; ROBERTS, I.N.: Genetic analysis of food spoilage yeasts. In: Methods in Biotechnology, Vol. 14, Food Microbiology Protocols (edited by Spencer, J.F.T.: Ragout de Spencer, A.L.). Totowa, New Jersey, U.S.A.: Humana Press (2001) 37–51.

[32] JAMES, S.A.; CAI, J.; ROBERTS, I.N.; COLLINS, M.D.: A phylogenetic analysis of the genus *Saccharomyces* based on 18S rRNA gene sequences: description of *Saccharomyces kunashirensis* sp. nov. and *Saccharomyces martiniae* sp. nov. Int. J. Syst. Bacteriol. **47** (1997) 453–460.

[33] JERMINI, M.F.G.; GEIGES, O.; SCHMIDT-LORENZ, W.: Detection, isolation and identification of osmotolerant yeasts from high-sugar products. J. Fd Protect. **50** (1987) 468–472.

[34] KABARA, J.J.; EKLUND, T.: Organic acids and esters. In: Food Preservatives (edited by Russell, N.J.: Gould, G.W.). Glasgow and London, U.K.: Blackie Academic & Professional (1991) 45–71.

[35] KATO, N.: Antimicrobial activity of fatty acids and their esters against a film-forming yeast in soy sauce. J. Fd Safety **3** (1981) 121–126.

[36] KOTYK, A.: Uptake of 2,4- dinitrophenol by the yeast cell. Folia Microbiol. (Prague) **7** (1962) 109–114.

[37] KREBS, H.A.; WIGGINS, D.; STUBS, M.; SOLS, A.; BEDOYA, F.: Studies on the mechanism of the antifungal action of benzoate. Biochem. J. **214** (1983) 657–663.

[38] KURTZMAN, C.P.: DNA relatedness among species of the genus *Zygosaccharomyces*. Yeast **6** (1990) 213–219.

[39] KURTZMAN, C.P.: *Zygosaccharomyces* Barker. In: The yeasts: a taxonomic study, 4th ed. (edited by Kurtzman, C. P.: Fell, J. W.). Amsterdam, the Netherlands: Elsevier (1998) 424–432.

[40] KURTZMAN, C.P.; BLANZ, P.A.: Ribosomal RNA/DNA sequence comparisons for assessing phylogenetic relationships. In: The yeasts: a taxonomic study, 4th ed. (edited by Kurtzman, C.P.: Fell, J.W.). Amsterdam, the Netherlands: Elsevier (1998) 69–74.

[41] KURTZMAN, C.P.; ROBNETT, C.J.: Identification and phylogeny of ascomycetous yeasts from analysis of nuclear large subunit (26S) ribosomal DNA partial sequences. Antonie van Leeuwenhoek **73** (1998) 331–371.

[42] KURTZMAN, C.P.; ROBNETT, C.J.; BASEHOAR-POWERS, E.: *Zygosaccharomyces kombuchaensis*, a new ascosporogenous yeast from 'Kombucha tea'. FEMS Yeast Research **1** (2001) 133–138.

[43] LACHANCE, M-A.; STARMER, W.T.: Ecology and yeasts. In: The yeasts: a taxonomic study, 4th ed. (edited by Kurtzman, C.P.: Fell, J.W.). Amsterdam, the Netherlands: Elsevier (1998) 21–30.

[44] LEGAN, J.D.; VOYSEY, P.A.: Yeast spoilage of bakery products and ingredients. J. Appl. Bacteriol. **70** (1991) 361–371.

[45] MAYES, T.: HACCP training. Fd Control **5** (1994) 190–195.

[46] MOLLAPOUR, M.; PIPER, P.W.: The *ZbYME2* gene from the food spoilage yeast *Zygosaccharomyces bailii* confers not just *YME2* functions in *Saccharomyces cerevisiae*, but also capacity for catabolism of sorbate and benzoate, two major weak organic acid preservatives. Mol. Microbiol. **42** (2001) 919–930.

[47] MOLNÁR, O.; MESSNER, R.; PRILLINGER, H.; STAHL, U.; SLAVIKOVA, E.: Genotypic identification of *Saccharomyces* species using random amplified polymorphic DNA analysis. Syst. Appl. Microbiol. **18** (1995) 136–145.

[48] NEAL, A.L.; WEINSTOCK, J.O.; LAMPEN, J.O.: Mechanisms of fatty acid toxicity for yeast. J. Bacteriol. **90** (1965) 126–131.

[49] NEVES, L.; PAMPULHA, M.E.; LOUREIRO-DIAS, M.C.: Resistance of food spoilage yeasts to sorbic acid. Lett. Appl. Microbiol. **19** (1996) 8–11.

[50] PARK, H.; BAKALINSKY, A.T.: *SSU1* mediates sulphite efflux in *Saccharomyces cerevisiae*. Yeast **16** (2000) 881–888.

[51] PEARCE, A.K.; BOOTH, I.R.; BROWN, A.J.P.: Genetic manipulation of 6-phosphofructo-1-kinase and fructose 2,6-bisphosphate levels affects the extent to which benzoic acid inhibits growth of *Saccharomyces cerevisiae*. Microbiology **147** (2001) 403–410.

[52] PEARSON, B.M.; MCKEE, R.A.: Rapid identification of *Saccharomyces cerevisiae*, *Zygosaccharomyces bailii* and *Zygosaccharomyces rouxii*. Int. J. Fd Microbiol. **16** (1992) 63–67.

[53] PILKINGTON, B.J.; ROSE, A.H.: Reactions of *Saccharomyces cerevisiae* and *Zygosaccharomyces bailii* to sulphite. J. Gen. Microbiol. **134** (1988) 2923–2830.

[54] PIPER, P.W.; ORTIZ-CALDERON, C.; HOLYOAK, C.D.; COOTE, P.J.; AND COLE, M.B.: Hsp30, the integral plasma membrane heat shock protein of *Saccharomyces cerevisiae*, is a stress-inducible regulator of plasma membrane ATPase. Cell Stress Chaparones **2** (1997) 12–24.

[55] PIPER, P.W.; MAHE, Y.; THOMPSON, S.; PANDJAITAN, R.; HOLYOAK, C.D.; EGNER, R.; MUHLBAUER, M.; COOTE, P.J.; KUCHLER, K.: The Pdr12 ABC transporter is required for the development of weak organic acid resistance in yeast. EMBO J. **17** (1998) 4257–4265.

[56] PITT, J.I.; HOCKING, A.D.: Fungi and food spoilage, 2nd ed. London, U.K.: Blackie Academic & Professional (1997).

[57] SAIKI, R.K.; GELFAND, D.H.; STOFFEL, S.; SCHARF, S.J.; HIGUCHI, R.; HORN, G.T.; MULLIS, K.B.; EHRLICH, H.A.: Primer-directed enzymatic amplification of DNA with a thermostable DNA polymerase. Science **239** (1988) 487–491.

[58] SAITOU, N.; NEI, M.: The neighbor-joining method: a new method for reconstructing phylogenetic trees. Mol. Biol. Evol. **4** (1987) 406–425.

[59] SAND, F.E.M.J.; VAN GRINSVEN, A.M.: Comparison between the yeast flora of Middle Eastern and Western European soft drinks. Antonie van Leeuwenhoek **42** (1976) 523–532.

[60] SAND, F.E.M.J.; VAN DEN BROEK, W.C.M.; VAN GRINSVEN, A.M.: Yeasts isolated from concentrated orange juice. Proc. 5th Int. Spec. Symp. Yeasts (1977) 121–122.

[61] SOUCIET, J.-L.; AIGLE, M.; ARTIGUENAVE, F.; BLANDIN, G.; BOLOTIN-FUKUHARA, M.; *et al.*: Génolevures: genomic exploration of the hemiascomycetous yeasts. FEBS Letters **487** (2000) 1–149.

[62] SOUSA, M.J.; MIRANDA, L.; CORTE-REAL, M.; LEAO, C.: Transport of acetic acid in *Zygosaccharomyces bailii* – effects of ethanol and their implications on the resistance of the yeast to acidic environments. Appl. Envir. Microbiol. **62** (1996) 3152–3157.

[63] SOUSA, M.J.; RODRIGUES, F.; CORTE-REAL, M.; LEAO, C.: Mechanisms underlying the transport and intracellular metabolism of acetic acid in the presence of glucose in the yeast *Zygosaccharomyces bailii*. Microbiology **144** (1998) 665–670.

[64] STEELS, H.; BOND, C.J.; COLLINS, M.D.; ROBERTS, I.N.; STRATFORD, M.; JAMES, S.A.: *Zygosaccharomyces lentus* sp. nov., a new member of the yeast genus *Zygosaccharomyces* Barker. Int. J. Syst. Bacteriol. **49** (1999) 319–327.

[65] STEELS, H.; JAMES, S.A.; ROBERTS, I.N.; STRATFORD, M.: *Zygosaccharomyces lentus*: a significant new osmophilic, preservative-resistant spoilage yeast, capable of growth at low temperature. J. Appl. Microbiol. **87** (1999) 520–527.

[66] STEELS, H.; JAMES, S A.; ROBERTS, I.N.; STRATFORD, M.: Sorbic acid resistance: the inoculum effect. Yeast **16** (2000) 1173–1183.

[67] STRATFORD, M.: Sulphite metabolism and toxicity in *Saccharomyces cerevisiae* and *Saccharomycodes ludwigii*. Bath, U.K.: Ph.D thesis, University of Bath (1983).

[68] STRATFORD, M.; ANSLOW P.A.: Evidence that sorbic acid does not inhibit yeast as a classic "weak-acid preservative". Lett. Appl. Microbiol. **27** (1998) 203–206.

[69] STRATFORD, M.; ROSE, A. H.: Transport of sulphur dioxide by *Saccharomyces cerevisiae*. J. Gen. Microbiol. **132** (1986) 1–6.

[70] STRATFORD, M.; MORGAN, P.; ROSE, A.H.: Sulphur dioxide resistance in *Saccharomyces cerevisiae* and *Saccharomycodes ludwigii*. J. Gen. Microbiol. **133** (1987) 2173–2179.

[71] STRATFORD, M.; HOFMAN, P.D; COLE, M.B.: Fruit juices, fruit drinks and soft drinks. In: The Microbiological Safety and Quality of Food (edited by Lund, B. M.; Baird-Parker, T. C.; Gould, G. W.). Gaithersburg, Maryland, U.S.A.: Aspen Publishers (2000) 836–869.

[72] SUZZI, G.; ROMANO, P.: Flocculazione in *Zygosaccharomyces*. Ind. Bevande **19** (1990) 306–308.

[73] TENREIRO, S.; ROSA, P.C.; VIEGAS, C.A.; SA-CORREIA, I.: Expression of the *AZR1* gene (ORF YGR224w), encoding a plasma membrane transporter of the major facilitator superfamily, is required for adaptation to acetic acid and resistance to azoles in *Saccharomyces cerevisiae*. Yeast **16** (2000) 1469–1481.

[74] THOMAS, D.S.: Yeasts as spoilage organisms in beverages. In: The Yeasts: Yeast technology, Vol. 5, 2nd ed. (edited by Rose, A. H.: Harrison, J. S.). London, U.K.: Academic Press (1993) 517–561.

[75] THOMAS, D.S.; DAVENPORT, R.R.: *Zygosaccharomyces bailii* – a profile of characteristics and spoilage activities. Fd Microbiol. **2** (1985) 157–169.

[76] TILBURY, R.H.: Xerotolerant yeasts at high sugar concentrations. In: Microbial growth and survival in extremes of environment (edited by Gould, G. W.; Corry, J. E. L.). London, U.K.: Academic Press (1980) 103–128.

[77] TODD, E.C.: Foodborne disease in Canada – a 5-year summary. J. Fd Protect. **46** (1983) 650–657.

[78] TUDOR, E.A.; BOARD, R.G.: Food-spoilage yeasts. In: The Yeasts: Yeast technology, Vol. 5, 2nd ed. (edited by Rose, A.H.: Harrison, J. S.). London, U.K.: Academic Press (1993) 435–516.

[79] VALLEJO, C.G.; SERRANO, R.: Physiology of mutants with reduced expression of plasma membrane H^+-ATPase. Yeast **5** (1987) 307–319.

[80] VOLLEKOVA, A.; MALIK, F.; VOLLEK, V.; LINCZENYIOVA, K.: Characterization of yeasts isolated from red wine surface film. Folia Microbiol. **41** (1996) 347–352.

[81] VAN DER VOSSEN, J.M.B.M.; HOFSTRA, H.: DNA based typing, identification and detection systems for food spoilage microorganisms: development and implementation. Int. J. Fd Microbiol. **3** (1996) 35–49.

[82] WARTH, A.D.: Mechanism of resistance of *Saccharomyces bailii* to benzoic, sorbic and other weak acids used as food preservatives. J. Appl. Bacteriol. **43** (1977) 215–230.

[83] WARTH, A.D.: Effect of benzoic acid on growth yield of yeasts differing in resistance to preservatives. Appl. Envir. Microbiol. **54** (1988) 2091–2095.

[84] WARTH, A.D.: Relationships between the resistance of yeasts to acetic, propionic and benzoic acids and to methyl paraben and pH. Int. J. Fd Microbiol. **8** (1989) 343–349.

[85] WARTH, A.D.: Transport of benzoic acid propionic acids by *Zygosaccharomyces bailii*. J. Gen. Microbiol. **135** (1989) 1383–1390.

[86] WILLIAMS, J.G.K.; KUBELIK, A.E.; LEVAK, K.J.; RAFALSKI, J.A.; TINGEY, S.C.: DNA polymorphisms amplified by arbitrary primers are useful as genetic markers. Nucl. Acid Res. **18** (1990) 6531–6335.

[87] WOOD, V.; GWILLIAM, R.; RAJANDREAM, M.-A.; LYNE, M.; LYNE, R.; *et al.*: The genome sequence of *Schizosaccharomyces pombe*. Nature **415** (2002) 871–880.

[88] YARROW, D.: *Zygosaccharomyces* Barker. In: The yeasts: a taxonomic study, 3rd ed. (edited by Kreger-van Rij, N. J. W.). Amsterdam, The Netherlands: Elsevier (1984) 449–465.

[89] YARROW, D.: Methods for the isolation, maintenance and identification of yeasts. In: The yeasts: a taxonomic study, 4th ed. (edited by Kurtzman, C.P.: Fell, J.W.). Amsterdam, the Netherlands: Elsevier (1998) 77–100.

7 Yeast stress response to food preservations systems

STANLEY BRUL, FRANS M. KLIS, HANS DE NOBEL, SUUS J. C. M. OOMES, PETER COOTE and KLAAS. J. HELLINGWERF

7.1 Introduction

In food manufacturing the challenge is more and more to come to a proper balance between product quality and product stability and safety. In particular mildly acidic products containing significant amounts of sugar are often prone to spoilage by various yeast species [e. g., 6, 24, 35]. Acid-tolerant *Zygosaccharomyces bailii, Z. lentus,* and the baker's yeast *Saccharomyces cerevisiae* are microorganisms of prime concern [e. g., 18, 38, 41]. To ensure the tenability of foods (i. e., closed and sufficient open shelf-life) food manufacturers often include in their products preservatives such as weak organic acids to prevent product spoilage. More natural alternatives, appealing to the current consumer trend, include cell wall lytic systems and membrane permeabilizing biomolecules [3, 4, 6, 11, 22, 31, 40]. However, the application of these is still limited and in most cases their exact mode of action is still under investigation.

As the consumer trend is to move away from harsh preservation systems to obtain better tasting and healthy foods, there is a strong interest in the application of mild physical treatments that may substitute for a severe heat treatment. These include high-pressure processing and the application of pulsed electric fields [1, 6, 13, 14].

In many instances food manufacturers notice that the microorganisms under study can elicit stress responses and survival strategies against the indicated treatments. To effectively circumvent and combat this resistance development it is needed to understand the mechanisms that govern it [e. g., 1, 3, 6, 7, 37, 45]. The current developments in the field of molecular microbiology, including genome-wide screening of cellular response at the transcript, protein, and metabolite levels, open the possibility for a full and comprehensive study of yeast stress response against food preservation systems [e. g., 20, 46; also reviewed in 8]. In this chapter we will show the strength of this approach by illustrating the stress response of yeasts against classical weak organic acids and against a number of novel systems including cell wall lytic enzymes, oligosaccharides that interfere with normal cell wall biogenesis, and membrane permeabilizing small biomolecules. At the outlook of the chapter we will discuss the development of a new 'systems biology' view of microbial cells by integrating the information provided by the various modern genomics technologies.

7.2 Classical food preservatives

In order to understand the mode-of-action of classical preservatives, such as weak organic acids, many experiments have been done. The acids studied included sorbic, benzoic and acetic acid. The classical notion that a weak organic acid is active at its pK and thus in its protonated form, led to many speculations on the actual mechanism of microbial (yeast) growth inhibition [discussed in Chapter 6, see also 18, 26, 42]. Prominent in this respect was the idea that the acid would affect the intracellular pH as it dissociates in a proton and anion in the cytosol. It was subsequently shown that the cell counteracts the lowering of the intracellular pH by activating a proton-pumping ATP-ase and a plasmamembrane-bound Multidrug resistance (Mdr) pump Pdr12p [34]. The latter was also confirmed with Western-blots. In yeast there are approximately 20 MDR genes underlying tolerance to toxic compounds that are all involved in membrane transport. The yeast Mdr proteins are generally referred to as Pleiotrophic drug resistance proteins or Pdr proteins. They are composed of 3 major classes; ATP-binding cassette (ABC) superfamily, major facilitator superfamily (MFS) and transcription factors, such as PDR1. Pdr12p belongs to the first class of proteins. Its membrane localisation was confirmed by immunofluorescence detection of a GFP fusion protein of Pdr12p. The importance of the action of Pdr12p in weak-organic acid resistance was underlined by studies in knock-out mutants by Piper et al. [34]. They showed that the (apparent) lag-time, growth-rate and, particularly in the case of benzoic acid, cell yield were decreased in Pdr12p deletion strains upon growth in the presence of increasing concentrations of these compounds. Meanwhile, Coote and co-workers showed that Pdr12p activity could be assessed using carboxyfluorescein loaded cells in a cellular fluorescence extrusion assay [18]. It was thus clearly demonstrated in a direct assay that the addition of sorbic acid at sub-lethal levels to a yeast culture led to a significant rise in Pdr12p activity. This fluorescence extrusion assay also showed that the activity of the pump was energy dependent as shown by its dependency on the presence of glucose [18]. In growing fully weak-acid adapted cells also Hsp30 is induced [32]. Hsp30p moderates in an as yet not fully elucidated manner the activity of the proton-pumping plasmamembrane ATPase, thus contributing to the establishment of a new level of cellular homeostasis in weak-acid adapted cells. Finally, a direct measurement of radioactively labelled benzoate efflux from the cells showed that in weak-acid adapted cells (i. e., with induced Pdr12p) such efflux was significantly increased [17]. In summary, the data showed that cells have to be adapted (pre-exposed) to sorbic acid before efflux is observed, efflux is energy-dependent, efflux does not occur in a strain where the PDR12 gene has been deleted and is competitively inhibited by sorbic/benzoic acid [18, 34]. Efflux is not due to a drop in intracellular pH (pHi), as little change in pHi is measured. Furthermore, the data suggests that transport proceeds directly from the cytosol to the external medium [17, 18].

All of this information, however, left various questions still unanswered. The most important one being how cells avoid futile cycling of the weak acid preserving compounds. Genome-wide transcript analyses and 2D gel-protein analyses of cellular proteins, have been performed on cells cultured in the absence and presence of sorbic acid [26]. Some novel interesting and thought provoking observations were made. As expected, genes encoding cellular proteins involved in energy metabolism were induced as were those involved in cellular stress responses known to play a role in responses to erroneous protein folding. The power of combining a full cellular protein analysis (superior physiological relevance) with a micro-array transcript analysis (superior comprehensiveness) was shown by the fact that the two sets of data provided clearly complementary and only to a limited extent overlapping data. Apart from Pdr12p the only protein identified both at the transcript and the protein level was the stress protein Hsp26p. Previously, only Cheng and Piper had alluded to a role for this general stress protein in their study of the effect of sorbic acid on cellular heat stress resistance [10]. DE NOBEL et al. [26] corroborated and extended these observations with their genomics analyses by showing that a knock-out of *HSP26* was hypersensitive to the action of sorbic acid. Thus this stress protein clearly has a central functional role to play in the cellular protection against the anti-microbial action of these compounds. Surprisingly, genes encoding proteins involved in transposition showed the highest induction level. The cells were cultured at a sub-lethal sorbic acid concentration ranging between 0.45 mM and 0.9 mM. One might speculate that depending on the state in their cell cycle at which the stress hits them or depending on their age, cells in the population try to counteract the imposed stress by altering their genome, i. e., by mutagenesis. How this should take place remains enigmatic and thus this observation calls for more detailed experimentation at various time-concentration ranges involving also other stress agents and conditions. Indeed genome instability in a population of cells responding to an external stress might be one of the reasons for the observed heterogeneity in stress response at the single cell level [reviewed in 43].

At the level of the plasmamembrane function, induction of the gene encoding Pdr12p was observed, but DE NOBEL et al. [26] also noted induction of genes involved in the cell wall and membrane stress response systems. It should be realized that the weak-organic acid food preservatives sorbic and benzoic acid are lipophilic and hence indeed have membrane perturbing effects. The nature and extent of the latter are currently unclear. Interestingly though, applying sorbic acid to cells at pH 5.0 prevents a proper activation of the heat-shock response, such as the synthesis of heat shock proteins upon a sub-lethal heat stress. At this pH, sorbic acid is expected to have both membrane perturbing and weak-acid antimicrobial actions [see 10]. In contrast, applying sorbic acid at pH 7.0, where the entire compound is expected to be in the dissociated form, led to an induction of heat stress protective mechanisms [10]. The hypothesis is that at the higher pH when the compound is present as sorbate it can only have a membrane perturbing effect but not a weak-acid anti-microbial effect. These observations are consistent with the notion that heat stress response is energetically costly [MENSONIDES et al., personal communication] and is often regulated at the level of the plasmamembrane [see e. g. 32]. Biochemical verification and validation of presumed changes at the plasmamembrane and cell-wall level of yeasts resistant against weak-organic acids needs to be performed. LOUREIRO-DIAS and co-workers already suggested several years ago that in benzoic acid adapted cells such a lowered rate of cellular uptake of labelled acid occurs [discussed in 17]. Our own unpublished work (DE KRUIJFF et al., pers. commun.) shows no major changes in phospholipid composition in weak-acid adapted cells. We nonetheless speculate that the net effect could be that cells effectively extrude the antimicrobial weak acid from the cytoplasm. The sequence of events might then be that upon imposing a sub-lethal stress with the acids on cells, these might respond first at the metabolic level with an induction of the plasma membrane proton pumping ATPase and Pdr12p, and a subsequent activation of the cell wall and membrane remodelling pathways. These events would be accompanied, depending on the level of stress imposed, by a redirection of the available cellular energy pools from cell-division to stress response. Indeed the stress response is often accompanied by an entry of the cells into lag (similar to stationary?)-phase. The molecular switches involved in switching cellular behaviour from cell division to damage-repair before further growth are subject to continued study [see e. g., 19]. At the end of lag-phase the cellular physiological situation should have been adapted such that cell division can be sustained without high risk of failure to complete it successfully. Energy can

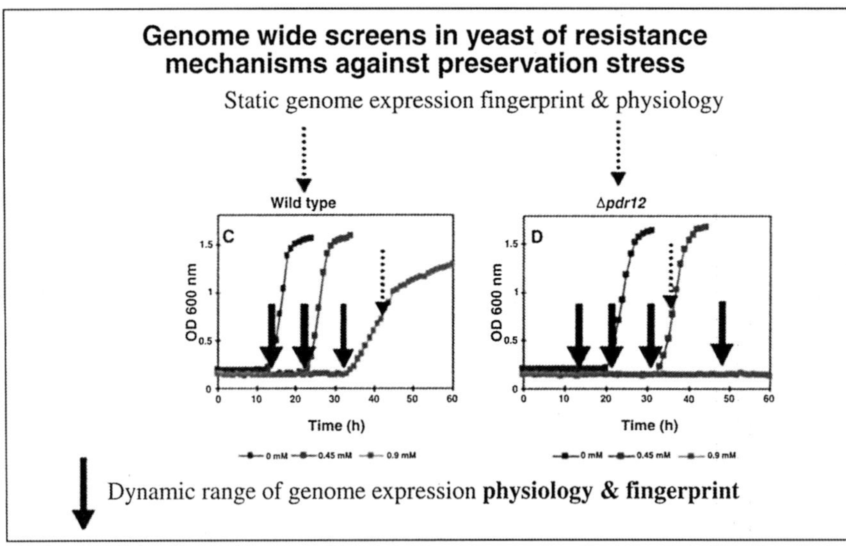

Fig. 7.2-1 Integrative genomics analysis of cellular stress response to weak-organic acids. On top it is indicated (dashed arrows) that sampling log phase cells, with or without the application of environmental stress, leads to a 'static expression finger print and physiological interpretation'. What is needed is a time-series of samples (bottom row of bold arrows) where both physiological parameters and the molecular expression profiles are measured such that a dynamic range of these cellular features is recorded leading to 'Integrative (functional) Genomics'. Shown are in the left hand panel wild-type and in the right hand panel *PDR12* Δ cells cultured in the absence and presence of sorbic acid (0.45 and 0.9 mM) in a rich medium (Yeast Pepton Dextrose: YPD) at pH 5.0.

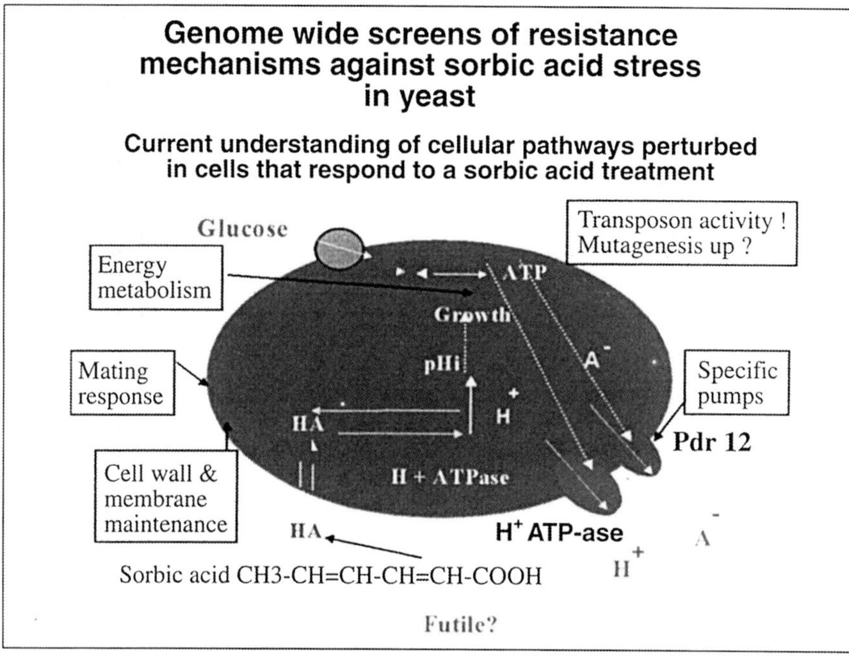

Fig. 7.2-2 Overview of the cellular processes modified by sorbic acid stress as monitored with genome-wide transcript analysis [taken from BRUL et al., 8].

than be set free from stress response reactions, i. e., the activity of the acid and anion pumps can be tempered, and may be used for cell division. The observed induction of Hsp30p, a stress protein tempering the activity of the plasma-membrane H^+-ATPase in fully adapted cells [32], points to the establishment of an intricate balance at the metabolic level between the cellular stress response and cell division machinery (see also above). In *Bacillus subtilis* it has been shown that a deletion of the general stress regulator σ^B enables cells cultured in the absence of stress under glucose limited conditions to sustain a higher growth rate [39].

Figure 7.2-1 describes our thinking behind the various events in cellular stress adaptation in a schematic fashion. Figure 7.2-2 summarizes experimental results pertaining to the activation of cellular pathways in response to sorbic acid stress. Many questions remain unanswered with respect to the reaction of yeasts against the currently used preservatives. The micro-array and other “omics” technologies allow us now to analyse on a global scale the cellular stress response rather than looking only at some, at first sight, more prominent aspects.

7.3 Novel food preservation systems

During the last years many studies have been devoted to the identification of novel food preservation strategies as an alternative to the classical systems. These include both physical means by which mainly closed shelf-life of foods can be guaranteed and (bio)chemical means, based on principles used by nature [see e. g., 4, 14, 30, 45]. A common denominator in this approach was the targeting of the treatments towards effects at the level of the plasmamembrane and cell wall, because these constitute both the Achilles heel of the cell (the membrane) and its first line of defence (the cell wall).

Figure 7.3-1 gives an artist impression of the cell envelope of fungi and lists the main novel, 'green' preservation routes. Many of the studies were accompanied by a search for the underlying mechanism of the cellular growth inhibitory effects such that a knowledge-based, rather than a trial and error new, anti-fungal system applicable in foods can emerge. It should be noted that many legislative hurdles now exist to the actual application of some of the systems discussed below. Indeed it is our believe that the application of the genome-wide transcript analysis, in combination with proteomics and metabolomics using various mass-spectrometry based detection systems, will help assessing the safety of novel food preservation systems and their equivalence to currently used food preservation systems [44].

Membrane-perturbing agents tested include antimicrobial peptides present in various natural sources, the most prominent of which being nisin in lactic acid bacteria, and naturally occurring antimicrobial flavours and fragrances [6, 11, 36]. Cell wall perturbing agents in-

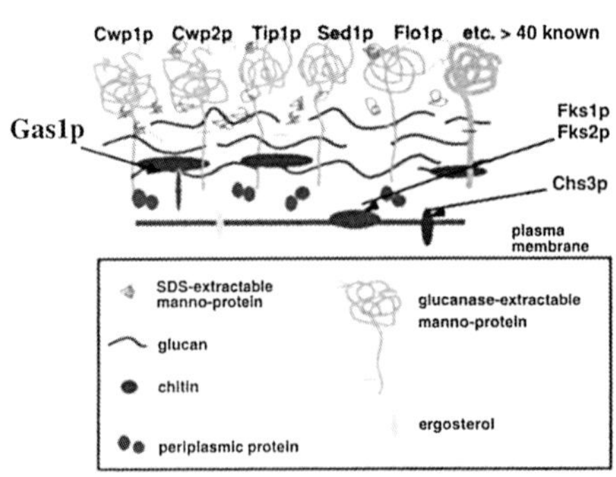

■ novel nature derived food preservatives

- alpha-helix peptides (nisin, MB-21)-membrane perturbers
- non-protein lipophilic organic biomolecules-membrane perturbers
- β1,3 & 1,6-glucanases, chitinases -cell wall degrading enzymes
- natural inhibitors of Fks (glucan synthase), Chs (chitine synthase), Gas (cross-linking enzyme)-cell wall perturbers
- combinations

Fig. 7.3-1 Schematic overview of the yeast cell wall with some targets and nature-derived ("green") fcod preservatives indicated.

cluded both enzymes acting on resting and actively dividing cells, and compounds (oligosaccharides) that interfere with the generation of a normal cell wall [23]. The latter included inhibitors of enzymes synthesizing building blocks of the cell wall and enzymes involved in wall cross-linking [3, 30]. The agents inhibiting cell wall biosynthesis have the disadvantage of only acting on actively dividing cells but they were generally significantly more heat stable. Finally, often combinations of agents have been tested since these are thought to allow for a lowering of the level of the individual compounds used according to the principle of 'hurdle technology' for food preservation [see e. g., 2, 4, 5].

In one attempt the action of extracts of plants on the outgrowth of fungi, including yeasts, was evaluated (Figure 7.3-2 gives an example of typical experimental results). It was found

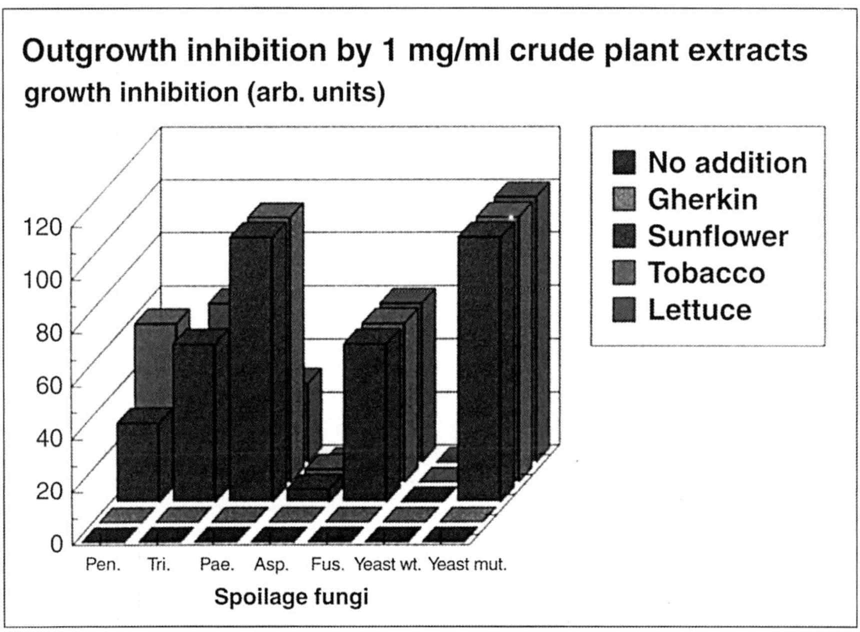

Fig.7.3-2 Outgrowth inhibition of fungi by extracts of the leaves of the plants indicated in the right-hand corner box. Inhibition was measured using a BioScreen reading system and is given as arbitrary units [28]. The horizontal axis shows the various fungi, *Penicillium roqueforti* (Pen.), *Trichoderma harzianum* (Tri.), *Paecilomyces variotii* (Pae), *Aspergillus niger* (Asp.), *Fusarium avenaceum* (Fus), *Saccharomyces cerevisiae* wild-type laboratory strain FY834 (Yeast wt), *Saccharomyces cerevisiae* mutant strain cwp1,2Δ with deletions of the genes encoding cell wall protein 1 and 2 (Yeast mut.). The strains have been described in Brul et al. [5, 12].

that specific compounds in the extracts including cell wall lytic β1,3- and β1,6-glucanases, chitinases and membrane perturbing proteins, including those from the pathogenesis-related protein-1 and pathogenesis-related protein-5 groups were responsible for the growth inhibitory effect [4]. It was also clear that the fungi had variable levels of sensitivity to the antifungal treatment, as may be expected since the level of specific cell wall proteins and the type of linkages to the cell wall glucan vary [e. g., 5]. It is presumably not realistic to expect to find a panacea for all problems without compromising the organoleptic quality of the foods in and on which unwanted growth of yeasts and moulds may occur.

In the same framework of studies, DIELBANDHOESING et al. [12] asked the question whether, yeasts could be made sensitive to the one currently available antimicrobial peptide which is food-grade, namely nisin. To that end they studied the sensitivity of yeast cells to the compound during various stages of the cell cycle. It was found that at the stage where the wall level of Cell Wall Protein 2 (Cwp2p) was relatively low, the cells were most sensitive to the membrane perturbing effect of nisin as assessed by the uptake of Propidium Iodide by the cells. Analyzing the effects of nisin in Cwp2p (and Cwp1p) deletion strains further corroborated this. Under normal culture conditions cells can apparently compensate for the absence of these proteins since a growth inhibitory phenotype is only observed upon challenging the cells with nisin or other antimicrobial peptides such as the synthetic MB-21. The latter peptide originates from a Unilever Home & Personal Care program (assessed for its mode-of-action and discussed in detail in [25]). A general observation made was that the cell wall harbours many covalently linked proteins with an important role in cellular defence. These include Cwp2p for antimicrobial peptide resistance, Fks2p involved in the stress response against cell wall damage, heat, and carbon source restriction, Tip1p involved in cold adaptation and adaptation to anaerobiosis, Cwp1p induced upon many forms of cell wall damage, and Sed1p involved prominently in β1,3-glucanase resistance during stationary phase [reviewed in 21]. The overall conclusion of these studies was that prevention of effective wall incorporation of these proteins may provide effective means of preventing cellular stress response against many classical (see the part on sorbic and benzoic acid in "Classical food preservatives") and novel antifungal stresses aimed at preserving foods. BOM et al. set out to device a system based on the competitive incorporation of an externally added β1,6-glucan sugar oligosaccharide as a means to prevent effective wall incorporation of stress proteins [2, 3]. They tested their system by assessing the secretion of a fusion protein between Cwp2p and β-galactosidase of which the wall localisation in yeasts had previously been shown. Upon addition of a β1,6-glucan dimer or oligomer, preferably trimer, that inhibited the biogenesis of a wild-type cell wall, presumably through competition for the β1,6-glucan cross-linking enzyme system, the fusion protein disappeared from the cell wall and appeared in the culture medium. Concomitantly, *Saccharomyces cerevisiae* cells became hypersensitive to the action of the membrane perturbing synthetic peptide MB-21. The system was also tested on the notoriously food spoiling yeast *Zygosaccharomyces bailii*. It was found that while quantitatively significant differences in sensitivity occurred, this yeast also turned out to be sensitive to the action of the cell wall perturbing oligosaccharide in combination with the membrane active peptide. Current ongoing studies at various institutes try to analyse the effects of other antifungal oligosaccharides including chitin and chitosan oligomers [see e. g., 15, 27, 29]. These compounds have a history in food consumption, and as a consequence are interesting for application as food preservatives or to impart other functionalities on the food such as proper structure and flavour retention or release characteristics. An issue of prime concern is the overall cellular reaction against these novel systems, as is the case of the currently used preservatives.

In order to analyse the cellular response towards treatment with a cell wall-lytic enzyme, DE NOBEL et al. have performed a genome-wide transcript analysis of cells growing in the presence of sub-lethal concentrations of the enzyme (DE NOBEL et al., 2002 submitted; pers. commun.). The results were analyzed with the algorithms described by BUSSEMAKER et al. [9] and VAN HELDEN et al. [16] (see Table 7.3-1). Both allow the researcher to correlate the expression of various genes upon culturing cells under specified environmental conditions to specific cellular signal transduction and stress response pathways on the basis of binding sites for transcription factors present in their promoters. The Bussemaker method does not use cut-off values and the information on the expression of all genes is used to come to the final clustering of the expression profile. It was shown by both methods that many genes with so-called canonical general stress responsive elements STRE (AGGGG) in their promoters, binding sites for the transcription factors Msn2 and Msn4, were activated upon adaptation to wall lytic stress conditions. Msn2 and Msn4 are known effectors of the High Osmolarity Glycerol (HOG) pathway [discussed in 21 and references therein]. The measurements were all done in actively growing and thus truly stress adapted cells. Subsequent verification with knock-out strains showed that *hog1* Δ strains were hypersensitive to enzymatic cell wall perturbation, as could be expected.

The HOG1 pathway signals a higher than normal extracellular osmolarity in conjunction with a lower value for the intracellular osmolarity. Its activity prepares the cell to counteract the imposed growth inhibition [discussed in 21, see also references therein]. The experiments by DE NOBEL et al. suggested that, in one way or another, this pathway also contributed to the cell's reaction to the cell wall weakening imposed upon cells by the wall lytic enzymes. It may be hypothesized that lytic enzyme stress initially induces swelling of the cells and consequently a drop in intracellular osmotic potential. This could be followed by an induction of trehalose accumulation in cells concomittantly with an activation of the protein kinase C1 (PKC1) pathway to repair the damage in the wall. In any case there are still a number of other promoter elements, some of which have not yet been described, that correlate with an induced expression of genes upon cell wall weakening and thus may play a role in the stress response against wall weakening (see Table 7.3-1).

Other well-known food spoiling organisms now become accessible to validation studies of the principles discussed. Through the application of comparative and functional genomics these often not yet fully sequenced food-borne microbes can be studied in their natural environment. This holds in particular for strains of the highly acid tolerant *Zygosaccharomyces bailii*, strains of the prominent filamentous spoilage fungus *Aspergillus niger* and strains of the medically important *Candida albicans*. The genomes of the latter two fungi are al-

Tab. 7.3-1 **The use of BioInformatics to reduce the complexity of data points in an analysis of the response of the yeast *Saccharomyces cerevisiae* to a stress with the cell wall intercalating agent Calcofluorwhite (CFW) and the cell wall lytic enzyme cocktail Zymolyase 100T. The results obtained with CFW and Zymolyase were essentially identical (taken with permission from DE NOBEL et al. 2002, submitted to Genome Biology, and slightly modified).**

Tab. 7.3-1a) **The promoter regions (–600 to –1) of the genes up-regulated in response to CFW- or Zymolyase-induced stress as analyzed according to the method described by BUSSEMAKER et al. [9]. Only the top-scoring motifs are shown and represent all other significant motifs (*P* value < 0.09). $\Delta\chi^2$ is the relative reduction in the error between the experimental data and a linear model based on a single motif. The fitting parameter, $F_{(single)}$, associated with each motif quantifies the change in expression level caused by the presence of the motif in the promoter of a gene. The sign indicates whether the motif acts as a positive or negative regulatory sequence. The count represents the number of times the motif occurs in the combined upstream regions of the yeast genome.**

Motif	$\Delta\chi^2$	*P*-value	$F_{(single)}$	Count
AAAATTT	0.020115	0.0000000000	–0.057854	1541
AGGGG	0.009927	0.0000000001	+0.046468	1057
CCCCT	0.005591	0.0000171804	+0.033079	1137
CGATGAG	0.009694	0.0000000024	–0.101204	251
CCGTACA	0.009387	0.0000000057	–0.146155	113
CCATACA	0.005599	0.0002689372	–0.089980	170
TGAAAAA	0.006811	0.0000084762	–0.032825	1601
GATC	0.004955	0.0000265131	+0.014502	5352
CAATTTT	0.005898	0.0001145217	–0.049800	614
ACGGCAC	0.005590	0.0002756341	+0.154442	62

most completely sequenced and will become available to the scientific community in a format currently under discussion (see http://www.tigr.org/tdb/mdb/mdb.html of the Institute for Genomic Research (TIGR) for a recent update of the sequencing effort).

A technique often used to identify transcripts involved in certain physiological actions of these species is the cDNA-AFLP technique which has currently continuously improving bioinformatics analysis methods linked to it [33]. As far as general aspects of comparative genomics of microorganisms are concerned, it is possible to fingerprint various strains against each other at the genome and genome-wide transcript levels under varying defined culturing conditions such that an unambiguous strain characterisation is possible. It is for instance well known that some "wild-type" laboratory strains are extremely sensitive to the

Tab. 7.3-1b) The promoter regions (–600 to –1) of all 51 genes up-regulated in response to CFW- or Zymolyase-induced stress were analyzed according to the method described by VAN HELDEN et al. (1998) for the presence of over-represented motifs of 4 up to 8 nucleotides in length. Only the motifs giving the highest significance for each oligo-length and representing either the canonical STRE (AGGGG) or the uncharacterized motif GGCCA are shown. Abbreviations: Ms, number of matching sequences (promoters) that contain at least one occurrence of the motif; Occ, number of occurrences of the motif among all 51 promoters; Exp, expected number of occurrences in non-coding regions; *sig*, significance index calculated as defined by VAN HELDEN et al. [16].

Length	Motif	Ms	Occ	Exp	Sig
4	**GGGG**	42	139	72.34	9.36
5	**AGGGG**	35	64	24.77	7.70
6	A**AGGGG**	24	31	10.23	3.57
7	A**AGGGG**C	10	12	2.27	1.40
8	A**AGGGG**CA	6	7	0.67	0.65
4	**GGCC**	39	174	74.40	9.41
5	**GGCCA**	29	61	24.21	6.89
6	**GGCCA**G	14	21	4.91	3.90
7	CT**GGCCA**	5	10	1.56	1.32
8	CCT**GGCCA**	4	5	0.34	0.00

action of cell wall lytic enzymes, while others are very resistant. The same holds for other imposed stresses (DE NOBEL H., MENSONIDES F. and RESENDE C., unpubl. observ.). Before the advent of the "omics" era, no comprehensive mechanistic insight in the background of these observations was available. Therefore, extensive screening of various "wild-type" strains had to be performed before the reference wild-type strain could be chosen for a given set of stresses and conditions. The power of comparative genomics may be illustrated by a non-yeast example. BRUL et al. [8] recently discussed the application of transcript analysis during sporulation of both laboratory *Bacillus subtilis* strains producing ordinary heat resistant spores, and strains producing highly heat resistant spores, isolated from processed (heat treated) foods. While most genes were similarly induced, a group of genes was significantly induced more strongly early during sporulation of the product isolate producing the heat resistant spores. A sub-set of these genes turned out to be essential for the development of the heat resistant spores and thus can be considered molecular markers for the originating strain.

7.4 Concluding remarks

The application of the modern molecular-biological tools allows us to study the behaviour of microorganisms at an unprecedented level of detail and sophistication. The microbe can be viewed as a system that aims at sustaining itself through energy generation and maintenance of essential structural parameters. The surplus of energy generated is then used for cell multiplication, cell division and responses to ever changing environments. This concept is not very different from any System description at the various levels of Biology. This immediately brings forward the fact that in order to have this integral molecular biological description of the cell, one needs to link the molecular to the physiological details (i. e., functional genomics). Such studies are currently undertaken for a continuous heat stress response in yeast. MENSONIDES et al. (pers. commun.) have grown yeast cells under controlled batch fermentation conditions both at regular culture temperatures (28 °C) and at elevated temperatures (37–43 °C). First results show that upon up-shift of the culture temperature, the cells tend to show a rapid rise in substrate consumption. This is followed upon restart of growth by a slight drop of substrate consumption, generally to just above pre-stress levels. Also a steady increase in both ATP and ADP absolute cellular levels was noted. These data point to an immediate metabolic response to heat shock by the cells. After a delay phase of about 0.5–1 hour the cell wall remodelling pKC1 pathway was activated as evidenced by the dual phosphorylation of its MAP-kinase output Slt2p. Interestingly, the end of the delay phase also coincides with the end of the actual growth lag-phase. Integration of such data will allow us to build up a 'continuum function' of cellular behaviour based on global cellular data for gene presence, transcription, protein presence, function and metabolic reactions. Such analyses will be based on many *in silico* calculations and will need, as such, large informatics input. Indeed bioinformatics is a rapidly evolving field, driven by the extremely fast rapid developments of the core genomics technologies. Various *in silico* cell initiatives are currently taken at various academic and institutional sites around the world. This is the case at TIGR (U.S.A.), at the Keio University (Japan) where Masaru Tomita initiated the e-cell project in 1999, at the Top Institute for Food Sciences WCFS in Wageningen (The Netherlands), and at the Advanced Genomics Initiative (AGI) in Amsterdam (The Netherlands). The AGI group aims towards an integration of both (molecular) biological and biophysical approaches and treats the cell as a micro-factory. This is an often used concept, which with the discussed 'omics' technologies is not just a nice concept anymore, but an actual experimental framework.

Major remaining issues are, which cellular factors constitute the switches between cell-division and stress response and what do they sense? Do they operate in a digital fashion or analogue mode? Now that we can dissect the cellular physiology at the molecular level, we should start to ask ourselves the question where in the cell all these reactions occur. A proper spatio-temporal understanding of stress response of microorganisms using genomics technologies is hereby a major challenge.

7.5 References

[1] ABEE, T.; WOUTERS, J.A.: Microbial stress response in minimal processing. Int. J. Fd Microbiol. **50** (1999) 65–91.

[2] BOM, I.J.; BRUL, S.: Composition for combating fungi. European-Patent 972040986.9 (1998).

[3] BOM, I.J.; KLIS, F.M.; DE NOBEL, H.; BRUL, S.: A new strategy for inhibition of the spoilage yeast *Saccharomyces cerevisiae* and *Zygosaccharomyces bailii* based on combination of a membrane-active peptide with an oligosaccharide that leads to an impaired glycosylphosphatidulinositol (GPI)-dependent yeast wall protein layer. FEMS Yeast Res. **1** (2001) 187–194.

[4] BRUL, S.; COOTE, P.J.; DIELBANDHOESING, S.K.; NAAKTGEBOREN, G.; STAM W.M.; STRATFORD, M.: Natural composition for combatting fungi. WO-Patent 97/16973 (1997).

[5] BRUL, S.; KING, A.; VAN DER VAART, J.M.; CHAPMAN, J.; KLIS, F.; VERRIPS, C.T.: The incorporation of mannoproteins in the cell wall of *S. cerevisiae* and filamentous ascomycetes. Antonie van Leeuwenhoek **72** (1997) 229–237.

[6] BRUL, S.; COOTE, P.J.: Preservative agents in foods. Mode of action and microbial resistance mechanisms. Int. J. Fd Microbiol. **50** (1999) 1–17.

[7] BRUL, S.; ROMMENS, A.; VERRIPS, C.T.: Mechanistic studies on the inactivation of *Saccharomyces cerevisiae* by high pressure. Innov. Fd Sci. Emerg. Technol. **1** (2000) 99–108.

[8] BRUL, S.; KLIS, F.M.; OOMES, S.J.C.M.; MONTIJN, R.C.; SCHUREN, F.; COOTE, P.; HELLINGWERF, K.J.: Genomics of survivors of food preservation processes for precision processing. Trends Fd Sci. Technol. (2002) (in press).

[9] BUSSEMAKER, H.J.; LI, H.; SIGGIA, E.D.: Regulatory element detection using correlation with expression. Nat. Genet. **27** (2001) 167–171.

[10] CHENG, L.; PIPER, P.W.: Weak acid preservatives block the heat shock response and heat-schock-element-directed lacZ expression of low pH *Saccharomyces cerevisiae* cultures, an inhibitory action partially relieved by respiratory deficiency. Microbiology **140** (1994) 1085–1096.

[11] CLEVELAND, J.; MONTVILLE, T.J.; NES, I.F.; CHIKINDAS, M.L.: Bacteriocins; safe natural antimicrobials for food preservation. Int. J. Fd Microbiol. **71** (2001) 1–20.

[12] DIELBANDHOESING S.K.; ZHANG, H.; CARO, L.H.P.; VAN DER VAART, J.M.; KLIS, F.M.; VERRIPS, C.T.; BRUL, S.: Specific cell wall proteins confer resistance to nisin upon yeast cells. Appl. Envir. Microbiol. **64** (1998) 4047–4052.

[13] GOULD, G.W.: Methods for preservation and extension of shelf-life. Int. J. Fd Microbiol. **33** (1996) 51–64.

[14] GOULD, G.W.: Preservation past, present and future. Br. Med. Bull. **56** (2000) 84–96.

[15] HELANDER, I.M.; NURMIAHO-LASSILA, E.L.; AHVENAINEN, R.; RHOADES, J.; ROLLER, S.: Chitosan disrupts the barrier properties of the outer membrane of gram-negative bacteria. Int. J. Fd Microbiol. **71** (2001) 235–244.

[16] HELDEN, J. VAN; ANDRE, B.; COLLADO-VIDES, J.: Extracting regulatory sites from the upstream region of yeast genes by computational analysis of oligonucleotide frequencies. J. Mol. Biol. **281** (1998) 827–842.

[17] HENRIQUES, M.; QUINTAS, C.; LOUREIRO-DIAS, M.C.: Extrusion of benzoic acid in *Saccharomyces cerevisiae* by an energy dependent mechanism. Microbiology **143** (1997) 1877–1883.

[18] HOLYOAK, C.D.; BRACEY, D.; PIPER, P.W.; KUCHLER, K.; COOTE, P.: The *Saccharomyces cerevisiae* weak-acid-inducible ABC transporter Pdr12 transports fluorescein and preservative anions from the cytosol by an energy-dependent mechanism. J. Bacteriol. **181** (1999) 4644–4652.

[19] HOLYOAK, C.D.; THOMPSON, S.; ORTIZ CALDERON C.; HATZIXANTHIS, K.; BAUER, B.; KUCHLER, K.; PIPER, P.W.; COOTE, P.J.: Loss of Cmk1 Ca(2+)-calmodulin-dependent protein kinase in yeast results in constitutive weak organic acid resistance, associated with a post-transcriptional activation of the Pdr12 ATP-binding cassette transporter. Mol. Microbiol. **37** (2000) 595–605.

[20] KAPTEYN, J.C.; TER RIET, B.; VINK, E.; BLAD, S.; DE NOBEL, H.; VAN DEN ENDE, H.; KLIS, F.M.: Low external pH induces HOG1-dependent changes in the organisation of the *Saccharomyces cerevisiae* cell wall. Mol. Microbiol. **39** (2001) 469–479.

[21] KLIS, F.M.; MOL. P.; HELLINGWERF, K.J.; BRUL, S.: Dynamics of cell wall structure in *Saccharomyces cerevisiae*. FEMS Microbiol. Rev. (2002) (in press).

[22] LEWIS, K.: In search of natural substrates and inhibitors of MDR pumps. J. Mol. Microbiol. Biotechnol. **3** (2001) 247–254.

[23] LORITO, M.; FARKAS, V.; REBUFFAT, S.; BODO, B.; KUBICEK, C.P.: Cell wall synthesis is a major target of mycoparasitic antagonism by *Trichoderma harzianum*. J. Bacteriol. **178** (1996) 6382–6385.

[24] MEMBRE, J.M.; KUBACZKA, M.; CHENE, C.: Combined effects of pH and sugar on growth rate of *Zygosaccharomyces rouxii*, a bakery product spoilage yeast. Appl. Envir. Microbiol. **65** (1999) 4921–4925.

[25] MOL, G.N.; BRUL, S.; KONINGS, W.N.; DRIESSEN, A.J.: Comparison of the membrane interaction and permeabilization by the designed peptide Ac-MB21-NH2 and truncated dermaseptin S3. Biochemistry **39** (2000) 11907–11912.

[26] NOBEL, H. DE; LAWRIE, L.; BRUL, S.; KLIS, F.M.; DAVIS, M.; ALLOUSH H.; COOTE, P.: Parallel and comparative analysis of the proteome and transcriptome of sorbic acid stressed *Saccharomyces cerevisiae*. Yeast **18** (2001) 1413–1428.

[27] OH, H.I.; KIM, Y.J.; CHANG, E.J.; KIM, J.Y.: Antimicrobial characteristics of chitosans against food spoilage microorganisms in liquid media and mayonnaise. Biosci. Biotechnol. Biochem. **65** (2001) 2378–2383.

[28] OOMES, S.; S. BRUL.: An improved method for the screening of fungal growth inhibition. Fd Technol. Biotechnol. **36** (1998) 79–84.

[29] OUATTAR, B.; SIMARD, R.E.; PIETT, G.; BEGIN, A.; HOLLEY, R.A.: Inhibition of surface spoilage bacteria in processed meats by application of antimicrobial films prepared with chitosan. Int. J. Fd Microbiol. **62** (2000) 139–148.

[30] PARKER, J.E.; AARTS, N.; AUSTIN, M.A.; FEYS, B.J.; MOISAN, L.J.; MUSKETT, P.; RUSTERUCCI, C.: Genetic analysis of plant disease resistance pathways. Novartis Fund Symp. **236** (2001) 153–161.

[31] PATTISON, T.L.; VON HOLY, A.: Effect of selected natural antimicrobials on Baker's yeast activity. Lett. Appl. Microbiol. **33** (2001) 211–215.

[32] PIPER, P.; ORTIZ-CALDERON, C.; HOLYOAK, C.; COOTE, P.; COLE, M.: Hsp30, the integral plasma membrane heat shock protein of *Saccharomyces cerevsiae*, is a stress-inducible regulator of plasma membrane H^+-ATPase. Cells Stress Chaperones **2** (1997) 12–24.

[33] QIN, L.; PRINS, P.; JONES, J.T.; POPEIJUS, H.; SMANT, G.; BAKKER, J.; HELDER, J.: GenEST, a powerful bidirectional link between cDNA sequence data and gene expression profiles generated by cDNA-AFLP. Nucl. Acid Res. **29** (2001) 1616–1622.

[34] PIPER, P.; MAHE, Y.; THOMPSON, S.; PANDJAITAN, R.; HOLYOAK, C.; EGNER, R.; MUHLBAUER, M.; COOTE, P.; KUCHLER, K.: The Pdr12 ABC transporter is required for the development of weak organic acid resistance in yeast. EMBO. J. **17** (1998) 4257–4265.

[35] ROLLER, S.; COVILL, N.: The antifungal properties of chitosan in laboratory media and apple juice. Int. J. Fd Microbiol. **47** (1999) 67–77.

[36] ROSAHL, S.: Lipoxygenases in plants – their role in development and stress response. Z. Naturforsch. **51** (1996) 123–138.

[37] RUSSEL, N.J.; COLLEY, M.; SIMPSON, R.K.; TRIVETT, A.J.; EVANS, R.I.: Mechanism of action of pulsed high electric field (PHEF) on the membrane of food-poisoning bacteria is an 'all-or-nothing' effect. Int. J. Fd Microbiol. **55** (2000) 133–136.

[38] SCHULLER, D.; CORTE-REAL, M.; LEAO, C.: A differential medium for the enumeration of the spoilage yeast *Zygosaccharomyces bailii* in wine. J. Fd Prot. **63** (2000) 1570–1575.

[39] SCHWEDER, T.; KOLYSCHKOW, A.; VÖLKER, U.; HECKER, M.: Analysis of the expression and function of the σ^B-dependent general stress regulon of *Bacillus subtilis* during slow growth. Arch. Microbiol. **171** (1999) 439–443.

[40] STEEG, P.F. TER; OTTEN, G.D.; ALDERLIESTEN, M.; DE WEIJER, R.; NAAKTGEBOREN, G.; BIJL, J.; VASBINDER, A.J.; KERSHOF, I.; VAN DUIJVENDIJK, A.M.: Modelling the effects of (green) antifungals, droplet size distribution and temperature on mould outgrowth in water-in-oil emulsions. Int. J. Fd. Microbiol. **67** (2001) 227–239.

[41] STEELS, H.; JAMES, S.A.; ROBERTS, I.N.; STRATFORD, M.: *Zygosaccharomyces lentus*: a significant new osmophilic, preservative resistant spoilage yeast, capable of growth at low temperature. J. Appl. Microbiol. **87** (1999) 520–527.

[42] STRATFORD, M.; ANSLOW, P.A.: Evidence that sorbic acid does not inhibit yeast as a classic 'weak acid preservative'. Lett. Appl. Microbiol. **27** (1998) 203–206.

[43] SUMNER, E.R.; AVERY, S.V.: Phenotypic heterogeneity: differential stress resistance among individual cells of the yeast *Saccharomyces cerevisiae*. Microbiology **148** (2002) 345–351.

[44] VOS, W.M. DE: Advances in genomics for microbial food fermentations and safety. Curr. Opin. Biotechnol. **12** (2001) 493–498.

[45] WOUTERS, P.C.: Pulsed Electric Field Inactivation of Microorganisms, Berlin, Germany: Ph.D thesis, Technical University of Berlin, Germany (2000).

[46] YALE, J.; BOHNERT, H.J.: Transcript expression in *Saccharomyces cerevisiae* at high salinity. J. Biol. Chem. **276** (2001) 15996–16007.

8 Yeasts in dairy products

MARIE-THERESE FRÖHLICH-WYDER

8.1 Introduction

In 1680 yeasts were discovered by the Dutch scientist Antonie van Leeuwenhoek. During the second half of the nineteenth century, the French biochemist Louis Pasteur showed that yeasts were responsible for the conversion of sugar to ethanol and carbon dioxide [111]. It was only with the development of a technique to isolate pure cultures on solid media by Robert Koch that it became possible to select yeast strains on the basis of their fermentative characteristics [111, 115]. Yeasts may be defined as unicellular fungi reproducing by budding or fission [80]. Some authors regard yeasts merely as fungi that produce unicellular growth, but that are otherwise not different from filamentous fungi. Consequently, yeasts are ascomycetous or basidiomycetous fungi that reproduce vegetatively by budding or fission and are capable of forming sexual states that are not enclosed in a fruiting body [9]. Fungi for which no sexual stage is known, are traditionally included into the deuteromycetes or fungi imperfecti [7]. However, due to recent molecular studies these asexual yeasts are currently placed in their proper phylogenetic position [40, 83–85]. At present, approximately 800 yeast species are recognized, but only a few are commonly used or isolated.

Yeasts are the most important microorganisms ever exploited by man, because they have been used during several thousand of years for the production of a wide range of foods such as bread, wine, beer and kefyr, and more recently for the production of ethanol for fuel, biochemicals for the pharmaceutical industry and many other substances. Yeasts present on fruits, vegetables, equipments, in homemade starters, and in all kinds of raw biological material, such as milk, are responsible for the occurrence of spontaneous fermentations. As an example, the raw milk and the environment of a cheese factory, such as the brine, are important sources for yeast contamination of the cheese surface. During ripening of smeared cheeses, these yeasts are indispensable.

8.2 Yeasts and dairy products

Dairy products provide a unique ecological niche for the selection and growth of specific yeast species [42]. Relatively few yeast species occur in dairy products such as milk, cream, yoghurt, butter, cheese and kefyr. Dairy yeasts share a number of physiological and biochemical characteristics [42], such as the fermentation or assimilation of lactose, a high proteolytic or lipolytic activity, utilization of lactic or citric acids, growth at low temperatures and tolerance to elevated salt concentrations. Among the most important dairy yeasts are *Kluyveromyces marxianus* (= *Candida kefyr*), *Kluyveromyces lactis*, *Debaryomyces hansenii* (= *C. famata*), *Yarrowia lipolytica* (= *C. lipolytica*), and *Saccharomyces cerevisiae*. Microbial interactions in which yeasts play a role are 1. the inhibition of disadvantageous microorganisms, such as pathogens (e. g. *Clostridium butyricum* and *C. tyrobutyricum*) [39], 2. the synergistic effects with lactic acid bacteria [21, 41], and 3. fermentations [65]. During the maturation of surface-ripened cheeses, *Debaryomyces hansenii* utilizes lactic acid resulting in an increase of the pH, which increases growth of *Brevibacterium linens* [65]. Yeasts are important in the production of kefyr and related fermentation products [148], but also in the making of cheeses, such as blue cheeses and white mould cheeses [65, 121 146, 149].

Yeasts play an important role as spoilage organisms in dairy products because of their ability to grow at low temperatures and low pH values, their resistance to physicochemical stresses and their metabolic activities [65]. For that reason, fermentative yeasts are often responsible for the spoilage of yoghurt and other sour milks. Blowing of packages is an undesirable consequence of yeast fermentation. Yeast spoilage is particularly important in fermented milk products and cheeses, and less in fresh or pasteurized milk, cream and butter. In fruit yoghurts, yeasts may be introduced by non-dairy ingredients such as fruits, sugar, honey and nuts [42, 65]. More generally, spoilage yeast can be introduced during the entire production chain, ranging from the farm, dairy plant, to the final product. Hygiene and sanitation measures are important to control contamination of dairy products with yeasts [56]. Dairy products may also be infected by human pathogenic yeasts, which are usually not transmitted through food [42].

8.3 Kefyr

8.3.1 The history of kefyr

Kefyr is an acidic, mildly alcoholic and very ancient fermented milk beverage originating from the northern slopes of Caucasus, more specifically, from the village of Karatschajeff (2500 m) at the foot of Elbrus (5600 m) [26, 75, 78, 126]. The root of the name "Kefyr" can be referred to the Turkish word "kef" meaning pleasant, delightful, well-being, making drunken, fermenting, or to the word "kiaf" meaning froth, or to the Caucasian word "kefy" meaning best quality [57, 75, 137]. All these different meanings reveal a distinctive feature of kefyr, i. e., it undergoes both a lactic acid and an alcoholic fermentation, and the latter is due to yeasts. The altitude of the region of origin and, therefore, the rather low temperatures, led to a selection of mesophilic microorganisms [78, 139].

For a long time, the manufacture of kefyr has been known only to members of the Ossete and Karabbiner tribes. They prepared kefyr from either cow, sheep or goat milk in bags of goat hides. In day time, due to the rather cold climate, the sacks were subjected to sunlight and at night, they were taken into the house and hung at the door. Every person who passed by, had to kick the sack in order to mix the content. Fresh milk was added when some of the fermented milk was removed, providing a continuous natural fermentation [26, 27, 78, 100]. Depending on the outside temperature, the product could be quite different. Low temperature led to a relatively high concentration of ethanol (up to 1 %) and carbon dioxide, whereas an elevated temperature to a more acidic product [139].

The actual starter culture of kefyr are the kefyr grains. But until today, nobody really knows where and how these grains appeared. Legends and presumptions are the only sources for an explanation of their formation. KUNTZE [82] and DUITSCHAEVER et al. [27] refer to the above-described manufacturing procedure of kefyr. During the ongoing spontaneous fermentation of kefyr, cauliflower-like aggregates are formed, consisting of a matrix of polysaccharides and coagulated proteins, in which a variety of microorganisms is embedded. BOUROUNOFF [129] has reported on a saga of the Caucasian people. The grains are said to originate from another fermented milk, called "Ayran" which is similar to kefyr. Ayran is made by natural souring of the milk in oak vats, or sacks of goat hides, with pieces of either calf's or camel's stomach. The grains have been collected from the walls of the vats and are added directly to fresh milk. The new sour milk, kefyr, is found to be much more pleasant than ayran. According to another story, the grains were found by shepherds in the bushes of the high mountains as a gift from heaven [82]. The best known and most legendary explanation for the origin of the grains, also called "grains of the prophet", is reported by PODWYSSOZKI [110] and KOROLEVA [78]. Allah himself gave the first grains to a chosen tribe as a symbol of immortality. According to another version, Mohammed was the bearer of the grains and he told the people how the grains have to be used. He strictly forbade the secret of kefyr preparation or the grains to be given away. Otherwise, if unbelievers got hold of the grains, these would lose their magic and healing power. This legend explains why the method of kefyr preparation has been kept secret for so long. Until now nobody has been able to disclose the secret of the formation of kefyr grains.

Dr. G. DZHOGAN in 1867 [137] has been the first one who reported on the beneficial effects of kefyr in the treatment of intestinal and stomach diseases. This was the end of the secrecy of kefyr and the start of its spreading through Europe. The owner of the Moscow Dairy got the idea to produce kefyr on an industrial scale. To obtain the grains, he sent a beautiful woman, one of his workers, to the Caucasian tribes. She was kindly received by their prince but did not get the grains. On her way back, she was kidnapped by the mountain people to become the prince's wife. The woman was then freed by the gendarmes and as a compensation, the prince had to give her 10 pounds of "Mohammed grains" [78]. This is the story of how the grains started to move westwards. In the former USSR and in Bavaria, kefyr started to be produced on an industrial scale in the 1930's [153].

8.3.2 The kefyr grain

Kefyr grains are characterized by an irregular form, a white or slightly yellow colour and by their elastic consistency. The various types of grains can range from flat sheets to scrolls and rolls, to the cauliflower floret forms and finally to millet-like grains [94]. They are the result of a strong and specific symbiotic relation of microorganisms and grow as biologically "independent units". In spite of much effort, all attempts to obtain new kefyr grains by various combinations of microorganisms isolated from them have failed so far [10, 77].

The following microorganisms may be part of the basic symbiotic microflora: lactic acid streptococci (*Streptococcus lactis, Str. cremoris, Str. diacetylactis*), lactobacilli (*Lactobacillus brevis, L. casei, L. delbrueckii, L. helveticus, L. acidophilus, L. kefir*), mesophilic heterofermentative lactic acid bacteria (LAB; *Leuconostoc mesenteroides*), yeasts (*Kluyveromyces marxianus, Torulaspora delbrueckii, Saccharomyces cerevisiae*) and acetic acid bacteria (*Acetobacter aceti, A. pasteurianus*) [77, 137]. Lactobacilli are found in the grain in a concentration of 10^9–10^{10} cfu/g, *Leuconostoc* in a concentration of 10^7 cfu/g. For yeasts the counts detected are 10^6–10^8 cfu/g and acetic acid bacteria range from 10^2 to 10^8 cfu/g [58, 72, 104, 124]. Only streptococci are not always detectable in the grain [29, 61, 72, 100].

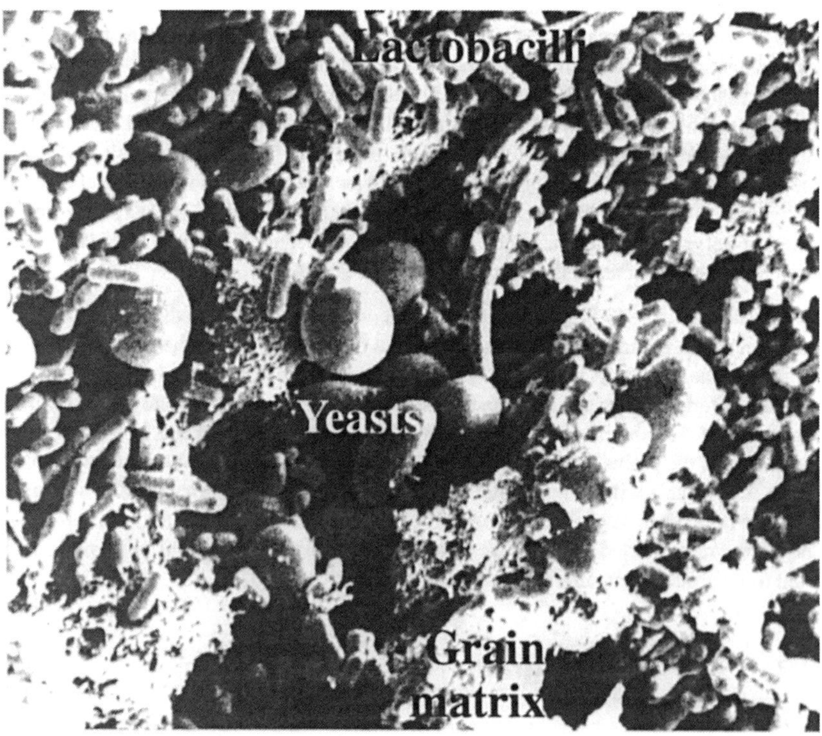

Fig. 8.3-1 Scanning electron micrograph of the intermediate zone of a kefyr grain [116] (x 3040).

The microorganisms are embedded in a fibrous grain matrix consisting of coagulated casein, polysaccharides, fat and lysed cells (Fig. 8.3-1). Investigations using scanning electron as well as optical microscopy revealed a specific distribution of the microorganisms in the grain that, however, can differ strongly. ROSI [124], BOTTAZI and BIANCHI [10] and BOTTAZI et al. [11] have observed the presence of yeasts particularly in the centre but also along the peripheral channels. This has not been confirmed by MOLSKA et al. [100] who found that yeasts are not as common inside as in the periphery. MANN [92] and KOROLEVA [77] have reported on the dominance of non-lactose fermenting yeasts in the centre of the grains, whereas the lactose positive species are located mostly on the surface, together with bacteria. Lactobacilli are often found to be associated with yeasts mostly on the surface, suggesting that they develop in micro-colonies building the grain mass [10, 87, 92, 100, 133]. However, TOBA et al. [133] did not find any particular arrangement of microorganisms in the grain.

The total dry matter of the grain is about 10 % with the following composition: protein 30–34 %, fat 3–4 %, ash 7–12 % and polysaccharides 45–60 % [11]. LA RIVIÈRE and KOOLMAN [87] have examined the composition, properties and the origin of the polysaccharides. Acid hydrolysis of the polysaccharides yields only D-glucose and D-galactose in approximately equal portions. The specific optical rotation is + 65 ° ± 4 °. No other polysaccharide is known to have the same characteristics, and the new polysaccharide was designated as "kefiran". LA RIVIÈRE and KOOLMAN [87] were able to isolate *L. brevis* as the responsible strain for the production of kefiran. *L. brevis* produces kefiran as capsular material only in the presence of the lactose negative yeast species *S. delbrueckii* (= *Torulaspora delbrueckii).* Kefiran is soluble in hot water but insoluble in cold water, has a constant viscosity over a wide pH range and cannot be hydrolyzed by enzymes [61]. All these properties are essential for kefyr grains to maintain their particular form through repetitive fermentation cycles. Further examinations of the grain by HIROTA [61] and KANDLER and KUNATH [72] revealed that the predominant lactobacillus is *L. kefir* (formerly *L. brevis*). These authors did not assume that *L. kefir* is responsible for kefiran production. In contrast, HOSONO et al. [62] and PINTADO et al. [109] claimed that *L. kefir* is the responsible strain for capsular kefiran production. PIDOUX et al. [108] re-identified *L. kefir* as *L. hilgardii* which produces kefiran, a gelling dextran. Finally, FUJISAWA et al. [45] and TOBA et al. [133] isolated an encapsulated *Lactobacillus* and proposed to name it *L. kefiranofaciens*. Thus, it still remains to be solved which microorganism is responsible for the production of kefiran.

8.3.3 The kefyr

In the preparation of kefyr, two phases, namely fermentation and ripening, can be distinguished. Fermentation is generally done at 18–22 °C for 18–20 hours. Lower temperatures favour the growth of yeasts and higher temperatures the LAB and, consequently, the acidification process [11, 74, 97, 152]. The quantity of inoculated kefyr grains (2–5 %) also affects the fermentation. Large inoculums shorten the fermentation process due to a rapid accumulation of lactic acid and result in a low content of streptococci and yeasts at the end of the fermentation. A low quantity of grains leads to an increasing number of the major groups of microorganisms. Stirring the inoculated milk during fermentation results in increased numbers of streptococci, yeasts and, if present, acetic acid bacteria. Finally, washing of the grains prior to inoculation results in a decrease of the main groups of microorganisms and, consequently, in a longer fermentation time [77]. Even though streptococci cannot be detected microscopically in the milk after inoculation with kefyr grains, they cause a rapid increase of acidity during the first hours of the fermentation and are found to dominate at the end of the process. After subsequent subculturing of kefyr starter without grains, lactobacilli and yeasts tend to disappear and streptococci become dominant [29, 58].

After separating the grains from the kefyr, ripening is performed at a temperature of 8–10 °C for 1–3 days. During this phase, the concentration of ethanol and other flavour components increase due to the fermenting activity of yeasts [33, 48, 74, 77].

According to KOROLEVA [77], a properly prepared kefyr should have the following composition of microorganisms:

- homofermentative mesophilic lactic acid streptococci 10^8–10^9 cfu/ml
- thermophilic lactobacilli 10^5 cfu/ml
- heterofermentative lactic acid streptococci 10^7–10^8 cfu/ml
- yeasts 10^5–10^6 cfu/ml
- acetic acid bacteria 10^5–10^6 cfu/ml

The pH of fermented milk prepared with kefyr grains varies around 4.3–4.6, the lactic acid content can vary between 8–11 g/L and the ethanol content between 0.1–5 g/L [11, 58]. The carbon dioxide content, which is formed during alcoholic fermentation and that is responsible for the prickly taste of kefyr, was reported to be around 1.33 g/L [17].

For the production of kefyr on an industrial scale, grains may be used to prepare the starter culture that is then used for the inoculation of milk for kefyr production. This method has a few disadvantages such as the large amount of grains needed, the fermentation procedure becomes time consuming, and the composition of the microflora in the product varies [26]. Therefore, kefyr grains are usually replaced by starters composed of pure microorganisms isolated from grains. Such cultures can be prepared as freeze-dried starters. The quality and the taste of the resulting kefyr product are found to be uniform [93, 106, 153]. Other procedures are based on two fermentation stages. The lactic acid fermentation with lactobacilli, leuconostoc and streptococci is performed at 25–32 °C until the pH is lowered to 4.4–4.7. The yeasts are incubated either separately in milk and then added to the sour milk, or they are added directly to the fermented milk and then incubated at 10 °C for 24 hours [11, 75]. This method results in a product of good quality and flavour [27, 28].

Investigations have shown that the composition of commercial kefyr can vary to a great extent. Lactobacilli are found either in counts of up to 10^5 cfu/g, or are absent in other cases. Yeasts are detected in a range of 0–10^8 cfu/g and ethanol is usually found in a concentration of 0–0.4 %. Thus, it is evident that the flavour strongly depends on the manufacturing procedure [33, 36, 50, 75, 140].

Many attempts have already been made to make kefyr on an industrial scale using pure cultures of microorganisms [11, 28, 77]. However, the use of starters other than the grain itself results in a different final product. Lactobacilli and yeasts tend to disappear, whereas streptococci become dominant [33].

8.3.4 The yeast flora of kefyr

The sharp acid and yeasty flavour together with the prickling sensation contributed by the carbon dioxide can be considered as the typical flavour of kefyr [27]. The yeasts (Fig. 8.3-2) play a leading role in the development of the characteristic taste and aroma because of their ability to ferment carbon sources releasing ethanol and carbon dioxide [77]. However, to obtain the best flavour, the count of yeasts should reach at least 10^3–10^5 cfu/ml kefyr [25, 49]. Also, the flavour characteristics are very much determined by the yeast species present in kefyr [33, 37, 93, 119].

Several working groups have reported on the yeast count in the grains and in the resulting kefyr, as well as in commercial kefyr products. The microbial counts in grain depend strongly on the method applied for their determination. By direct microscopic counting, 10^8 yeasts/g grain are detected, whereas by plate counting only 10^6 to 10^7 cfu/g are found [17,

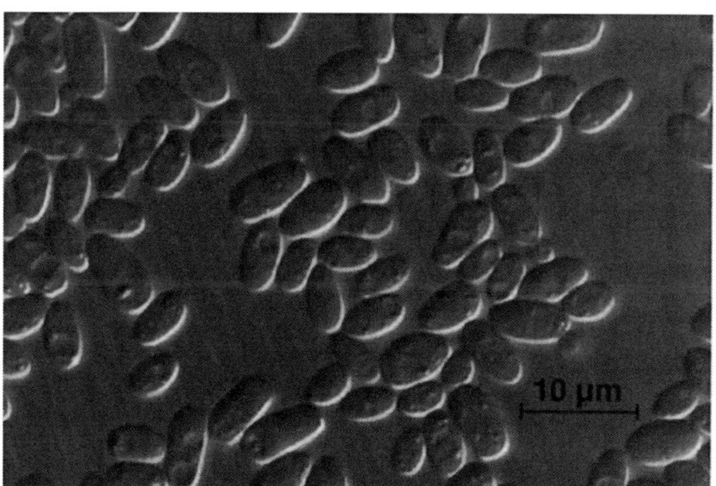

Fig. 8.3-2 Cells of *Saccharomyces turicensis*, a new yeast from kefyr [147].

72, 124]. After adding the grains to milk and stirring, a number of 10^5 cfu/ml milk is found [72]. In kefyr prepared with grains, the amount of yeasts (10^5–10^7 cfu/ml) is very similar to that in the grain itself [17, 25, 33, 72, 80, 124]. In the fermented milk made with kefyr (without using grains) the yeast count is 10^5 cfu/ml [81]. Commercial kefyr samples differ strongly from traditionally prepared kefyr. Many samples contain no yeasts at all, in others the count reaches 10^6 cfu/ml [25]. Manufacturers usually try to keep the yeast number as low as possible to avoid blowing of the packages [47]. In addition, there are no compelling regulations on the composition of the kefyr microflora except for the International Dairy Federation Standard that proposes a minimal yeast count of 10^4 cfu/g in kefyr [63].

A question often debated is whether all yeasts found belong to the specific kefyr yeast flora, or, if this is not the case, which yeasts must be considered as contaminants. Usually, yeasts found in kefyr are the same as those species causing spoilage in other milk products [48]. Some authors claim that only yeasts fermenting lactose should be considered as specific for the kefyr flora because of their leading role in the alcoholic fermentation [50, 140]. Nevertheless, a high percentage of the yeasts found in kefyr are lactose negative [25, 36, 63, 102, 113, 116].

The first who examined the microbial flora of kefyr grains was KERN [73]. He showed that a symbiosis between a yeast and a bacterium is involved. The yeast is *Saccharomyces cerevisiae*, a species that does not ferment lactose. Table 8.3-1 shows the yeasts that have been isolated from kefyr grains, and Table 8.3-2 the frequency of isolation of the yeast species from kefyr grains or kefyr products. The role of yeasts is not only limited to their contribution to kefyr flavour. For example, LA RIVIÈRE [86] reported that appreciable growth of *L. brevis* occurs only in the presence of a yeast. Therefore, yeasts also promote the symbiosis between microorganisms by providing lactic acid bacteria with growth stimulants. On the other hand, lactic acid bacteria produce β-galactosidase which splits lactose into glucose and galactose. Nearly all the yeasts are able to utilise either glucose or galactose or both [25, 33, 76, 77, 109, 124].

Table 8.3-1 Yeast species isolated from kefyr and kefyr grains

Yeast species	New nomenclature [according to CBS yeasts database]	Reference
K. bulgaricus	*K. marxianus* var. *marxianus*	[116]
K. fragilis	*K. marxianus* var. *marxianus*	[17, 25]
K. lactis	*K. lactis* var. *lactis*	[113, 1119
K. marxianus	*K. marxianus* var. *marxianus*	[25, 36, 63, 102, 104, 113, 116, 148]
S. carls-bergensis	S. cerevisiae	[61, 97, 113]
S. cerevisiae	S. cerevisiae	[25, 27, 36, 61, 63, 73, 76, 102, 116, 124]
S. delbrueckii	*T. delbrueckii*	[25, 109, 116]
S. exiguus	*S. exiguus*	[63, 94]

Table 8.3-1 Continued

Yeast species	New nomenclature [according to CBS yeasts database]	Reference
S. fragilis	*K. marxianus* var. *bulgaricus*	[61, 86, 116]
S. florentinus	*Zygosaccharomyces florentinus*	[98, 139]
S. italicus	*S. cerevisiae*	[61]
S. kefir	*K. marxianus* var. *marxianus*	[116]
S. lactis	*K. lactis* var. *lactis*	[61, 104]
S. unisporus	*Saccharomyces unisporus*	[25, 36, 63, 109, 113, 119, 148]
C. holmii	*S. exiguus*	[36]
C. kefyr	*K. marxianus* var. *marxianus*	[25, 36, 75, 94, 102, 113, 116, 133, 140, 152]
C. lambica	*P. fermentans* var. *fermentans*	[25]
C. lipolytica	*Y. lipolytica*	[25]
C. pseudotropicalis	*K. marxianus* var. *marxianus*	[25, 61, 86, 94, 104, 116, 139]
C. tenuis	*C. tenuis*	[37, 104, 113]
C. valida	*P. membranifaciens*	[113]
T. delbrueckii	*T. delbrueckii*	[25, 86, 87, 102, 109, 116, 124, 149]
Tor. holmii	*S. exiguus*	[86, 153]
Tor. kefir	*K. marxianus* var. *marxianus*	[97, 116]
B. anomalus	*D. anomala*	[113, 147]
G. candidum	*G. candidum*	[78, 87, 116, 139]
I. occidentalis	*I. occidentalis*	[36]
P. fermentans	*P. fermentans* var. *fermentans*	[65, 119]
Y. lipolytica	*Y. lipolytica*	[119]

B = Brettanomyces; C = Candida; D = Debaryomyces; G = Geotrichum; I = Issatchenkia; K = Kluyveromyces; P = Pichia; S = Saccharomyces; T = Torulaspora; Tor = Torulopsis; Y = Yarrowia; CBS = Centraalbureau voor Schimmelcultures (http://www.cbs.knaw.nl/)

Table 8.3-2 Utilisation of carbon compounds by yeast species isolated from kefyr and kefyr grains [4, 80]

*Yeast species**	Utilisation of carbon sources					Frequency of mentioning in literature **
	Lactose		Galactose		Lactic acid	
	A	F	A	F	A	
K. marxianus var. *marxianus*	+	+,–	+	+	+	58
S. cerevisiae	–	–	+,–	+,–	+,–	27
T. delbrueckii	–	–	+,–	+,–	+,–	15
S. unisporus	–	–	+	+	–	12
K. lactis var. *lactis*	+	+	+	+	+	9
S. exiguus	–	–	+	+	+,–	9
P. membranifaciens	–	–	–	–	+,–	8
G. candidum	–	–	+	–	+,–	6
P. fermentans var. *fermentans*	–	–	–	–	+	6
C. tenuis	+,–	–	+	+	+,–	4
Dek. anomala	+	+	+	+	+,–	4
Zygos. florentinus	–	–	+(–)	+,–	–	3
Y. lipolytica	–	–	–(+)	–	+,d	2
D. hansenii var. *hansenii*	+,–	–	+	–,d	+,–	1
D. polymorphus var. *polymorphus*	+,–	–	–	+	–	1
I. occidentalis	–	–	–	–	+	1
S. servazzii	–	–	+	+	–	1

A = Assimilation; F = Fermentation; + = reaction positive; – = reaction negative; +,– = reaction variable; +(–) = reaction positive, seldom negative and vice versa; d = reaction delayed positive; * nomenclature according to CBS yeast database [CBS = Centraalbureau voor Schimmelcultures (http://www.cbs.knaw.nl/)]; ** from references in Table 1 and others; for abbreviations of genera see Table 8.3-1; *Dek.* = *Dekkera*

8.4 Cheese

To make cheese, milk from domestic animals is transformed into a coagulum by the action of rennet and of LAB. Then, water is expelled by physical and microbial interactions in order to concentrate casein and fat selectively. During the ripening period, casein, fat and carbon sources are metabolized in a complex process by enzymes of the microorganisms from the starter culture. The end product is a cheese with characteristic flavour, taste, consistency and shape. According to their consistency, cheeses have been classified into extra-hard,

hard, semi-hard, semi-soft, soft and fresh cheeses [13]. Cheeses may also be grouped by the raw material, fat content, the ripening, etc.

8.4.1 Brief history

The rich and fertile agricultural area situated between the rivers Euphrates and Tigris in Iraq is known to be the cradle of civilisation. The staple foods were mainly bread and cheese. Remnants of material found during an archaeological survey, proved to have been cheese made from the milk of either cows or goats. From carvings and other findings it is also assumed that milk was stored in skin bags where a fermentation process took place. Most probably, either yoghurt, laban, koumiss or kefyr was produced, or the whey was drained off through a cloth or a perforated bowl and the solid curd then salted. The whey was usually used as a refreshing drink. The early coagulants for milk that were applied in addition to the fermentation process, were the juice of fig tree, vinegar and milk clotting enzymes from the stomach of hare or kid. The first written references to cheese can be found in the bible, and later Homer, Herodotus and others also referred to cheese [127]. The spread of cheese-making has probably followed the same paths as bread. This geographical migration has resulted in new varieties of cheese. At present, 2000 names applied to cheese can be found in the literature.

8.4.2 The yeast flora of cheese

A large number of varieties of cheese are characterized, as mentioned above, by the development of a specific surface microflora that is generally composed of moulds, yeasts, micrococci and coryneform bacteria. Yeasts, therefore, are frequently found within the microflora of many types of cheese. Their occurrence is not unexpected because of their tolerance to low pH and moisture, high salt concentration and low storage temperatures [42]. Also, they are widely dispersed in the dairy environment and appear as natural contaminants in the raw milk, the air, the dairy utensils, the brine, and in smear water [141–143]. The brine, being one of the most important sources of contamination, may be the vector of several yeast species such as *Debaryomyces hansenii, Candida versatilis, Kluyveromyces marxianus, Saccharomyces cerevisiae, Torulaspora delbrueckii, Trichosporon cutaneum* var. *cutaneum* and *Yarrowia lipolytica* [8, 128]. The following species are found in raw milk: *D. hansenii, Clavispora lusitaniae, Tr. cutaneum* var. *cutaneum, Rhodotorula mucilaginosa* and *K. marxianus* [59]. The utilization of lactic acid and the formation of alkaline metabolites by yeasts results in an increase of the pH value, which enables the growth of less acid tolerant microorganisms such as the micrococci and coryneform bacteria [32].

In the first few days of ripening, the yeast count on the surface of the cheese increases very rapidly until it reaches a maximum after 10 days [22]. The numbers can increase to 10^6–10^9 cfu/g [91, 121] or 10^7–10^8 cfu/cm^2 [32, 150]. There after, the population remains at a nearly

Table 8.4-1 Yeast species isolated from cheese

Yeast species	New nomenclature [according to CBS yeasts database]	Reference
K. marxianus	*K. marxianus* var. *marxianus*	[14, 16, 23, 31, 32, 42, 71, 89, 90, 118, 121, 128, 141–143, 150]
K. lactis	*K. lactis* var. *lactis*	[5, 14, 16, 42, 69, 71, 89–91, 101]
K. bulgaricus	*K. marxianus* var. *marxianus*	[16, 89]
K. fragilis	*K. marxianus* var. *marxianus*	[42, 91]
D. hansenii	*D. hansenii*	[5, 14, 16, 23, 32, 42, 44, 69, 71, 89–91, 101, 118, 121, 128, 141–143, 149]
G. candidum	*G. candidum*	[14, 23, 32, 71, 89, 90, 107, 118, 143, 150]
G. capitatum	*Dipodascus capitatus*	[16, 150]
S. cerevisiae	*S. cerevisiae*	[14, 16, 23, 31, 42, 69, 89, 90, 101, 118, 121, 141]
S. unisporus	*S. unisporus*	[14]
S. italicus	*S. cerevisiae*	[89]
S. exiguus	*S. exiguus*	[68]
S. fragilis	*K. marxianus* var. *marxianus*	[42]
S. lactis	*K. lactis* var. *lactis*	[42, 91]
C. catenulata	*C. catenulata*	[14, 31, 32, 121, 134, 141]
C. famata	*D. hansenii*	[5, 14, 42, 91, 134]
C. intermedia	*C. intermedia*	[14, 16, 31, 32, 71, 121, 128]
C. kefyr	*K. marxianus* var. *marxianus*	[14, 121]
C. krusei	*Issatchenkia occidentalis*	[14]
C. lipolytica	*Y. lipolytica*	[14, 42, 71, 121]
C. pseudotropicalis	*K. marxianus* var. *marxianus*	[42, 89]
C. robusta	*S. cerevisiae*	[14, 101]
C. rugosa	*C. rugosa*	[14, 128]
C. sake	*C. sake*	[16, 143]
C. sphaerica	*K. lactis var. lactis*	[5, 14, 101]
C. tenuis	*C. tenuis*	[128]
C. utilis	*P. jadinii*	[1, 16]
C. versatilis	*C. versatilis*	[23, 91, 128]
C. zeylanoides	*C. zeylanoides*	[112]
Cl. lusitaniae	*Cl. lusitaniae*	[14, 69, 150]

Table 8.4-1 Continued

Yeast species	New nomenclature [according to CBS yeasts database]	Reference
Cr. albidus	*Cr. albidus*	[121, 141, 142]
I. orientalis	*I. orientalis*	[31, 118, 128]
P. fermentans	*P. fermentans* var. *fermentans*	[14, 16, 42]
P. jadinii	*P. jadinii*	[149]
P. kluyveri	*Pichia kluyveri* var. *kluyveri*	[16]
P. membranaefaciens	*P. membranifaciens*	[14, 16, 42, 118]
P. pseudocactophila	*P. pseudocactophila*	[149]
Rh. glutinis	*Rh. glutinis* var. *glutinis*	[141, 142]
Rh. minuta	*Rh. minuta* var. *minuta*	[31, 32, 142]
Rh. rubra	*Rh. mucilaginosa* var. *mucilaginosa*	[69, 70, 150]
Tor. sphaerica	*K. lactis* var. *lactis*	[16, 42, 89]
Tor. mogii	*Zygos. rouxii*	[89]
Tor. versatilis	*C. versatilis*	[16, 89]
T. delbrueckii	*T. delbrueckii*	[14, 31, 32, 141–143]
Tr. cutaneum	*Tr. cutaneum* var. *cutaneum*	[69, 150]
Tr. beigelii	*Tr. cutaneum* var. *cutaneum*	[31, 32, 128]
Tr. pullulans	*Tr. pullulans*	[118]
W. californica	*W. californica*	[149]
Y. lipolytica	*Y. lipolytica*	[14, 16, 31, 32, 42, 44, 90, 91, 141–143, 150]
Zygos. rouxii	*Zygos. rouxii*	[16, 23, 89]

C = *Candida; Cl* = *Clavispora; Cr* = *Cryptococcus; D* = *Debaryomyces; G* = *Geotrichum; H* = *Hansenula; I* = *Issatchenkia; K* = *Kluyveromyces; P* = *Pichia; Rh* = *Rhodotorula; S* = *Saccharomyces; T* = *Torulaspora; Tor* = *Torulopsis; Tr* = *Trichosporon; W* = *Williopsis; Y* = *Yarrowia; Zygos* = *Zygosaccharomyces;* CBS = Centraalbureau voor Schimmelcultures (http://www.cbs.knaw.nl/)

constant level and decreases only slightly to a final number of about 10^7 cfu/g [90]. In the interior of soft cheeses, there is an almost parallel development of the yeast population but at a 100 or even 10,000-fold lower magnitude [23, 31, 89, 150]. In general, higher numbers are present in soft and blue-veined cheeses [14, 42].

Table 8.4-2 Utilization of carbon compounds by yeast species isolated from cheese [4, 80]

Yeast species *	Utilisation of carbon sources						Frequency of mentioning in literature **
	Lactose		Galactose		Glucose	Lactic acid	
	A	F	A	F	F	A	
K. marxianus var. *marxianus*	+	+,–	+	+	+	+	22
D. hansenii	–	–	+	–	+,–	+,–	19
K. lactis var. *lactis*	+	+	+	+	+	+	15
S. cerevisiae	–	–	+,–	+,–	+	+,–	13
Y. lipolytica	–	–	–,d	–	–	+,d	10
G. candidum	–	–	+	–	–,w	+,–	10
C. intermedia	–	–	+	+,d	+	–	6
C. versatilis	d	d	+	+,d	+	+,–	6
P. membranifaciens	–	–	–	–	–,d	+,–	5
Z. rouxii	–	–	+,–	–	+	–	4
C. catenulata	–	–	+	–,d	+,–	+	3
P. fermentans var. *fermentans*	–	–	–	–	+	+	3
Tr. cutaneum	+	–	+	–	–	+	3
I. orientalis	–	–	–	–	+	+	3
C. rugosa	–	–	+,d	–	–	+,d	2
P. jadinii	–	–	–	–	+	+	2
P. anomala	–	–	+,d	–,d	+	+	2
Rh. mucilaginosa	–	–	+,–	–	–	+,–	2
T. delbrueckii	–	–	+,–	+,–	+	+,–	2
K. marxianus var. *marxianus*	+	+,–	+	+	+	+	22

A = Assimilation; F = Fermentation; + = reaction positive; – = reaction negative; +,– = reaction variable; +(–) = reaction positive, seldom negative and vice versa; d = reaction delayed positive; w = reaction weak positive; * New nomenclature according to Yeasts CBS database [CBS = Centraalbureau voor Schimmelcultures (http://www.cbs.knaw.nl/)]; ** from references in Tab. 8.4-1; abbreviations of most genera are given in Table 8.4-1

Investigations of the composition of the yeast flora reveal a large diversity with more than 10 species among which *Kluyveromyces lactis, K. marxianus, Debaryomyces hansenii, Saccharomyces cerevisiae, Yarrowia lipolytica, Trichosporon cutaneum (=T. beigelii), Rhodotorula mucilaginosa* and *Torulaspora delbrueckii* are the most frequent [14, 16, 22, 69, 70, 71, 89, 90, 91, 101, 118]. In Table 8.4-1 and Table 8.4-2, the yeast species isolated from different types of cheese such as Cheddar, Gouda, Danablu, Roquefort, Tilsit, Tête de Moine, Gruyère, Münster, Brie, Camembert and many others are listed.

The composition of the yeast flora of young cheese seems to be rather heterogeneous and depends strongly on the cheese plant in which it has been produced [32]. In the cheese prior to brining, lactose positive species such as *K. lactis, K. marxianus* and *T. delbrueckii* are predominant. These species most probably also contribute to the characteristic open texture of blue-veined cheeses [42, 91]. The technology of cheese ripening also has an impact on the composition of the yeast flora. The typical yeast flora of mould-ripened cheeses seems to be mainly composed of *D. hansenii* and *G. candidum,* as well as of *K. marxianus* and *Y. lipolytica.* Smear-ripened cheeses contain mainly *D. hansenii,* but also *Y. lipolytica* and *G. candidum. D. hansenii* and *K. marxianus* occur mainly in blue-veined cheeses. *D. hansenii, K. marxianus* and *G. candidum* are mainly found in acid-curd cheeses and *K. marxianus* and *C. zeylanoides* are usually isolated from fresh cheeses [5, 31, 112, 134].

8.4.3 The role of yeasts during cheese ripening

The major recognized action of yeasts during cheese ripening is the metabolism of lactic acid with a consequent increase of pH values. This promotes the growth and action of cheese-ripening microorganisms sensitive to acid such as *Brevibacterium linens.* In addition, lactic acid bacteria show a higher survival rate which accelerates proteolysis, and consequently the ripening process [23, 42, 88].

As already mentioned above, the lactose fermenting species, for example *K. marxianus,* contribute to the open structure of mainly blue-veined cheeses. Their ability to ferment lactose results in the formation of carbon dioxide and also in flavouring compounds such as ethanol and acetaldehyde [23, 90]. However, it should be considered that there is a risk of a yeasty off-flavour and that the structure will be too open if the number of yeasts exceeds a certain level. The utilization of lactose also limits the acidification by lactic acid bacteria and, thus, affects the texture of the cheese [89].

Furthermore, yeasts contribute to the maturation of cheese by their lipolytic activity. Among the yeasts from cheese, *Y. lipolytica* is recognized as the species having the greatest lipolytic activity [16]. It is possible to accelerate ripening and to improve the quality of cheese by the addition of this yeast species [23, 90]. Esterase activity seems to be a common characteristic for many yeast isolates from cheese as well [105].

Many yeasts produce proteolytic enzymes. Species with a high proteolytic activity are *K. lactis, K. marxianus* and *D. hansenii* [30, 52]. *Y. lipolytica, G. candidum* and *C. catenulata* are species with a strong extracellular proteolytic and/or peptidolytic activity [3, 121, 122, 134]. Intracellular proteinases are detected in yeasts of the genera *Trichosporon* and *Debaryomyces.* The activity of these proteinases (pH_{opt} = 5.5–6; T_{opt} = 60 °C) is specific on caseins. Exopeptidases that are aminopeptidases and carboxypeptidases seem to play a major role in the proteolysis of milk proteins. The aminopeptidases with an optimum pH of

7.5–8 are present in nearly all the yeast species. The carboxypeptidases have an optimum pH of 4. Most of them are located inside the cell [16, 22, 89, 90]. All intracellular enzymes would be much more significant to the cheese ripening process if released by cell lysis [41, 121, 152].

The enzymatic activity of yeasts may also play an important role in the breakdown of bitter peptides, which are usually a result of an unbalanced activity of both proteinases and peptidases. By releasing smaller peptides and amino acids, the aminopeptidases and carboxypeptidases contribute essentially to the breakdown of bitter peptides [90]. Especially *G. candidum* is known to show such an activity [3, 22, 23]. Synergistic effects of yeasts with lactic acid bacteria result in a stronger proteolysis by *D. hansenii* [21, 41].

It can be concluded that yeasts are of importance during the maturation of cheese. However, only little is known about their specific proteolytic and lipolytic activity on milk proteins and fat. Further investigations on their physiological and biochemical characteristics are needed in order to select relevant strains for starter cultures [14, 42, 90, 118].

In the following, four important species used for cheese production are described. The names in brackets indicate the anamorphic state of the species.

8.4.3.1 Debaryomyces hansenii

D. hansenii (= *C. famata*) is one of the most prevalent yeast species in dairy products, especially on cheese surfaces. In Roquefort cheese, as an example, it is largely responsible for the formation of slime on the surface. At equivalent water activity, the species tolerates salt better than glucose [20]. Therefore, the high tolerance to salt is not surprising [123]. *D. hansenii* shows a maximal growth rate between 25 and 30 °C, and is able to grow at 5 and up to 32–37 °C [4, 42].

Intracellular proteinases (pH_{opt} 5.8), which preferably hydrolyse caseins, and extracellular proteinases, as well as leucinaminopeptidases and carboxypeptidases have been detected [16, 22]. It has been demonstrated that the proteolytic activity of *D. hansenii* cultured in skim milk together with LAB is greater than the sum of their activities when cultured separately. *D. hansenii* also prolongs the survival of LAB in cheese [151]. However, it does not hydrolyze casein at ripening temperatures of 10 °C [135].

This yeast utilizes lactic acid, preferably the L(+) isomer, aerobically and anaerobically, as well as acetic acid. Therefore, its role in de-acidifying the surface of cheese is apparent. The anaerobic pathway is much slower, but there is still evidence that the reduction in concentration of the lactic and acetic acids in cheese may inhibit the growth of *Clostridium tyrobutyricum* [39]. Furthermore, good growth on citrate, even in the presence of salt, is observed [44, 123, 135].

Also, a higher amount of free fatty acids can be detected in cheese inoculated with *D. hansenii* [21], even though only little release of free fatty acids from butterfat at 10 °C has been observed in another work [135]. In general, it is found that this yeast contributes to a more rapid proteolysis, as well as overall ripening [21]. In cheese curd slurries, *D. hansenii* increases significantly the pH and shows proteolytic activity. It generates an alcoholic, acidic, fruity and also cheesy aroma [95, 149]. The species is very heterogeneous and consists of several phenons [128, 150].

8.4.3.2 *Yarrowia lipolytica*

Interest in *Y. lipolytica* (= *C. lipolytica*) arose from its rather uncommon physiological characteristics. Strains are more frequently isolated from lipid- and protein-containing substrates than from sugar-containing substrates, because it has a strong extracellular lipolytic and proteolytic activity [135]. Thus, the species often occurs in dairy products such as cheese, yoghurts, or salads containing meat or shrimps [145], as well as in spoiled food [20].

Y. lipolytica is strictly aerobic, and utilizes lactic and citric acids [44, 122]. A concentration of 1 % citric or lactic acid (pH 4.5) does not inhibit growth, whereas 1 % acetic acid is lethal [20]. It is able to release high amounts of formic acid [44]. This may be the reason why, in cheese curd slurries, the species does not affect pH at all [149]. Growth is observed at 5–10 °C, but the optimal temperature lies between 25–30 °C and the maximal temperature ranges from 33 to 37 °C [121].

Due to its strong extracellular enzymatic activity, *Y. lipolytica* is a good candidate for the production of a cheese-flavoured base [12]. In cheese model systems, it exhibits a putrid, cabbage and a strong cheesy (Parmesan, Sbrinz, Munster), but not a fruity aroma [95, 149]. The consequence of using *Y. lipolytica* as an adjunct culture, is that the flavour of cheese is favourably affected [56, 146]. It is thought to produce volatile flavour compounds such as methanthiol, dimethylsulfide, and to enhance the synthesis of aroma compounds by bacteria [95].

8.4.3.3 *Pichia jadinii*

P. jadinii (= *C. utilis*) is known for its strong fermentative ability (facultative fermentation). Thus, the response of a glucose grown culture to oxygen limitation is alcoholic fermentation after a lag phase of about 1 h, during which glycerol, pyruvate and D-lactate are formed as the main fermentation products [68]. The species utilizes lactate but not lactose. Contrary to the three other yeast species described in more details, growth is possible up to 44 °C [20]. Concerning its role in cheese production, it is added to mesophilic starters to enhance flavour development and to improve the texture [130]. Nevertheless, *P. jadinii* is also found to cause blowing in young cheese [1].

8.4.3.4 *Geotrichum candidum*

G. candidum is considered either as a yeast or as a yeast-like fungus depending on the morphology of the colonies [23]. Two main biotypes may, therefore, be distinguished. One is characterized by clearly white strains, more or less felt, a rapid growth, an optimal temperature of growth at 25–30 °C, a strong proteolytic activity, the formation of a true mycelium and an alkalizing action. The other type forms cream-coloured colonies, has a yeast-like appearance, a weak growth and weak proteolytic activity, an optimal temperature of growth between 22–25 °C and an acidifying action [53, 54, 89].

Growth can be observed in the range of 5–38 °C, with an optimum at 25 °C, and at pH 5–5.5. The species is sensitive to salt and growth is limited at concentrations above 1 % NaCl [107]. The yeast produces extracellular lipases and proteinases, and two endopeptidases with an optimum pH of 5.5 to 6 [3, 90]. In the production of Camembert cheese, it has been shown that *G. candidum* considerably decreases bitterness by the breakdown of bitter peptides through aminopeptidase activity [99], as well as by the inhibition of *Penicillium* growth [136].

Strains of *G. candidum* strains are also able to deaminate glutamic and aspartic acids [51], as well as tryptophan, leucine, methionine and phenylalanine [55]. The catabolism of amino acids by *G. candidum* strains can produce alcohols and volatile sulphur compounds such as dimethyldisulfide, methanethiol and various S-methyl thioesters that are important for flavour development [67]. In cheese model systems, *G. candidum* yields cheesy, sulphur-like and alcoholic odours [95, 149].

8.4.4 Industrial use of whey

Whey is a waste product of cheese making that is produced in large amounts. For instance, a medium sized dairy plant produces c. 1.4×10^7 kg of whey per year [66]. In whey, *K. marxianus, D. hansenii* and *Sporobolomyces roseus* are dominant yeast species, but *Rhodotorula mucilaginosa* and *S. cerevisiae* occur in high numbers as well [60]. Fermentative production of ethanol from whey by *C. tropicalis* or genetically modified *S. cerevisiae* resulted in the production of 12.4 % (v/v) and 5 % ethanol (w/v), respectively [24, 64]. The commercial production of glycerol from whey using *K. marxianus* (= *K. fragilis*) seems promising as well [66, 114]. Other approaches are the production of food and feed as biomass and lipids as single cell oil by *Trichosporon beigelii* (= *Trichosporon cutaneum*) [132], and the production of citric acid by *Y. lipolytica* (= *C. lipolytica*) [2].

8.5 Yeasts as spoilage organisms in dairy products

The ability to grow at low pH, low temperatures, low water activities, high salt concentrations and certain enzymatic activities make yeasts not only desirable for dairy products as we have seen in the cases of cheese and kefyr. Yeasts can also cause spoilage in dairy products because of these very characteristics. The most common defects are gas production causing blowing of packages, yeasty and other off-flavours, discolorations and changes of texture [65].

The following examples illustrate the potential role of yeasts as spoilage organisms in fermented dairy products due to some growth properties. Species like *D. hansenii, G. candidum, K. marxianus, P. membranifaciens, Y. lipolytica* grow well at 6–10 °C [37]. Several important spoilage yeasts, such as *S. exiguus, I. orientalis, C. guilliermondii (= P. guilliermondii), C. parapsilosis, Debaryomyces* spp., *P. anomala,* and *S. cerevisiae*, have been tested for their ability to grow at different temperatures (8 and 22 °C), pH (2.5–6.0) and NaCl concentrations (0.4–8 % w/v). Most of the tested species are able to grow at the low temperatures (optimal pH and NaCl-concentration) and pH (optimal temperature and NaCl-concentration) and at high salt concentrations (optimal temperature and pH). However, a synergistic effect occurs between NaCl and pH. This means that if NaCl-concentration increases and pH decreases, growth of the yeasts is slowed down [6]. *Debaryomyces hansenii* seems to be the most tolerant species with respect to the combined effects of pH, salt and temperature. Other physiological properties of the species to understand their role in the spoilage of dairy products are the fermentation and assimilation of lactose (*K. lactis* and *K. marxianus*), the assimilation of lactic acid (a.o. *D. hansenii, K. lactis, K. marxianus, S. cerevisiae*), and the hydrolysis of casein (*D. hansenii, K. marxianus*) [131].

In raw and pasteurized milk yeasts usually occur in low numbers (i. e., 10^3 cells per ml), but the numbers may increase when bacterial growth is inhibited due to the presence of residual antibiotics [36]. *Cr. albidus, Cr. flavus, Cr. curvatus, Cr. diffluens, Cr. humicola, D. hansenii, K. marxianus, Rh. glutinis, Rh. mucilaginosa, Trichosporon* species, and *Y. lipolytica* are frequently isolated species [18, 34, 43, 138]. Inoculation experiments with a number of dairy yeasts isolated from cheese, such as *Y. lipolytica, K. marxianus, D. hansenii* and *S. cerevisiae*, showed that these species are able to grow in high numbers in milk (10^8–10^9 cells per ml) [122]. This growth is explained by intricsic properties of these species, such as assimilation of lactose and lactic acid, β-galactosidase activity (EC 3.2.1.23), utilization of organic acids such as citric, lactic, and succinic acids, and lipolytic and proteolytic activities. The inoculation of milk with *Y. lipolytica* and *Candida catenulata* resulted in solidification and glycerol production [122].

Sweetened and soured creams may be prone to spoilage by yeast species, resulting in a foamy appearance and yeasty odours. These products ideally contain less than 10 yeast cells per gram product [42]. Lipolytic yeast species, such as *D. hansenii, Cr. diffluens, Cr. laurentii* (including one of its synonym *Torula aurea*), *Rh. glutinis* and *Rh. rubra*, are most frequently isolated [43], but also *C. parapsilosis, I. orientalis* (= *C. krusei*) and *Trichosporon* species may occur [18]. Sweetened and condensed milk may be spoiled by fermentative yeasts [103]. Spoilage of butter by yeasts seems rather rare. Fleet and Mian [43] could not detect any yeast in 9 out of 16 butter samples, and in the remaining 7 samples only low yeast counts (less than 10^3 cells per gram) were observed. Lipolytic species, belonging to *Rhodotorula*, and *Y. lipolytica*, are most important in this respect [18, 42].

Yeasts and moulds are considered to be the most common spoilage organisms in fermented milk products (yoghurt, quark) causing blowing of the packages and off-flavours [117]. Not only the lactose fermenting species *K. marxianus* and *K. lactis* are responsible for spoilage of yoghurts, but also *H.* spp, *S. cerevisiae*, *P. membranifaciens*, *P. guilliermondii* and *G. candidum* [35]. Basidiomycetous yeasts of the genera *Sporobolomyces* and *Rhodotorula* have been reported to occur in yoghurt as well [120, 131]. Some of the spoilage yeasts, *Hansenula* spp. and *S. cerevisiae*, are able to ferment galactose, which is produced through the hydrolysis of lactose by lactic acid bacteria [46]. The fermentation of carbohydrates results in the production of carbon dioxide, causing blowing of the packages. Off-flavours, such as fruity, bitter and yeasty notes, are also generated [42].

Threshold values for the sensorial detection of a spoilage by yeasts in quark may vary between ~10^4 and 10^6 cfu/g, depending on the yeast species [37]. Under Good Manufacturing Practice (GMP) the number of yeast cells at the time of production should be less than 10 cells per gram product, and preferably less than 1 cell per gram product [19, 42], and at the time of retailing the yeast count must be les than 100 cells per gram product [19]. However, the number of yeast cells frequently varies around 10^3 cells per gram product in shelved products [42], and has been found to be as high as 10^5 cells per gram product [131]. In particular fruit yoghurts are prone to spoilage by yeasts [19, 42]. Major sources of contamination are the environment of the production and packaging sites, but also the fruit bases used for yoghurt production. In the latter case, the following yeast species may be isolated: *C. magnoliae, C. parapsilosis, P. holstii, P. membranifaciens, Z. bailii, M. pulcherrima* and *I. orientalis* [96]. Recently, molecular identification techniques of yoghurt spoiling yeasts have been developed [79]. Digoxygenin labelled probes of the 18S rDNA have been used in a dot blot hybridization and allow the identification of most yoghurt spoiling yeast species. Also, *in situ* hybridization assays have been developed using the 3' ends of the 18S rDNA to detect *S. cerevisiae, D. hansenii, P. anomala* and *Dekkera bruxellensis*. The practical detection limit of this latter approach is 10^3 cfu per gram product, and therefore it is recommended to start the *in situ* detection with an enrichment step [79].

Yeasts are also encountered in the spoilage of soft and fresh cheese causing gassy and flavour defects. *Torulaspora delbrueckii, C. parapsilosis, C. sake, Cr.* spp., *D. hansenii, K. marxianus, P. fermentans, P. guilliermondii, P. membranifaciens, P. norvegensis, Rh.* spp. and *Y. lipolytica* are most commonly isolated [143, 144]. The swelling of cheeses due to yeasts are mostly related to the fermentative action of *K. marxianus* and *Dek. anomala* [38, 42]. On the other hand, *K. marxianus* is desired, for example, in the production of blue-veined cheeses in order to facilitate the development of an open texture. Gas production is the main reason for changes of texture, which is a defect caused frequently by yeasts.

Another common defect caused by yeasts is the discoloration of cheeses, in particular the oxidative browning of the cheese surface after a comparatively long ripening period. The browning defect is due to the oxidation of tyrosine by tyrosinase (EC 1.14.18.1), an enzyme mainly produced by *Yarrowia lipolytica*. It appears usually at the end of the ripening when the pH is becoming more alkaline [15, 125]. Even though the overall flavour and texture qualities of the cheeses are not affected, consumers do not appreciate the appearance. Yeast occurrence is usually due to recontamination from the production and the packaging area.

8.6 Conclusion

By means of their metabolic pathways, the very same yeast species can play either an important beneficial role, e. g., as ripening agents, or a rather unfavourable role in the spoilage of dairy products. Even though in recent years the interest in dairy yeasts has increased considerably, only little knowledge is available on their beneficial contribution to the quality of dairy products such as cheese.

8.7 References

[1] ALEKSIEVA, V.; DUPARINOVA, M.; KOSTOVA, N.; MLADENOVA, I.: Blowing of processed cheese due to *Candida utilis*. Vet. Sbirka **82** (1984) 31–33.

[2] ABOU-ZEIS, A.A.; BAGHLAF, A.O.; KHAN, J.A.; MAKHASHIN, S.S.: Utilization of date seeds and cheese whey in production of citric acid by *Candida lipolytica*. Agric. Wastes **8** (1983) 131–142.

[3] AUBERGER, B.; LENOIR, J.; BERGERE, J.L.: Partial characterization of exopeptidases produced by a strain of *Geotrichum candidum*. Sci. Aliments **17** (1997) 655–670.

[4] BARNETT, J.A.; PAYNE, R.W.; YARROW, D.: YEASTS: characteristics and identification, 3rd ed. Cambridge, U.K.: Cambridge University Press (2000).

[5] BESANCON, X.; SMET, C.; CHABALIER, C.; RIVEMALE, M.; REVERBEL, J.P.; RATOMAHENINA, R.; GALZY, P.: Study of surface yeast flora of Roquefort cheese. Int. J. Fd Microbiol. **17** (1992) 9–18.

[6] BETTS G.D.; LINTON P.; BETTERIDGE R.J.: Food spoilage yeasts: effects of pH, NaCl and temperature on growth. Fd Contr. **10** (1999) 27–33.

[7] BEUCHAT, L.R.: Food and beverage mycology, 2nd ed. New York: Van Nostran Reinhold (1987).

[8] BINTSIS, T.; LITOPOULOU TZANETAKI, E.; DAVIES, R.; ROBINSON, R.K.: Microbiology of brines used to mature feta cheese. Int. J. Dairy Technol. **53** (2000) 106–114.

[9] BOEKHOUT, T.; KURTZMAN, C.P.: Principles and methods used in yeast classification, and an overview of currently accepted yeast genera. In: Non-conventional yeasts in biotechnology: a handbook (edited by Wolf, K.). Berlin: Springer Verlag (1996) 1–81.

[10] BOTTAZZI, V.; BIANCHI, F.: A note on scanning electron microscopy of microorganisms associated with the kefir granule. J. Appl. Bacteriol. **48** (1980) 265–268.

[11] BOTTAZZI, V.; ZACCONI, C.; SARRA, P.G.; DALLAVALLE, P.; PARISI, M.G.: Kefir. Microbiology, chemistry and technology. Ind. Latte **30** (1994) 41–62.

[12] BOUDREAUX, D.P.: Cheese-flavoured substance. Eur. Patent Appl. EP 0 137 536 A1 (1985).

[13] BURKHALTER, G.: IDF-Catalogue of cheeses. Int. Dairy Fed. IDF document **141** (1981).

[14] CALLON, C.; CHATAUD, J.; VANDERBECKEN, F.; LARPENT, J.P.: Isolation and identification of yeasts from the internal microflora of various cheese types. Microbiol. Alimen. Nutr. **12** (1994) 23–29.

[15] CARREIRA, A.; PALOMA, L.; LOUREIRO, V.: Pigment producing yeasts involved in the brown surface discoloration of ewes' cheese. Int. J. Fd Microbiol. **41** (1998) 223–230.

[16] CHOISY, C.; GUEGUEN, M.; LENOIR, J.; SCHMIDT, J.L.; TOURNEUR, C.: Microbiological aspects. In: Cheesemaking. Science and technology (edited by ECK, A.). Paris, France: Technique et Documentation Lavoisier (1987) 259–292.

[17] CLEMENTI, F.; GOBETTI, M.; ROSSI, J.: Carbon dioxide synthesis by immobilized yeast cells in kefir production. Milk Sci. Int. **44** (1989) 70–74.

[18] COOKE, W.B.; BRAZIS, A.R.: Occurrence of molds and yeasts in dairy products. Mycopath. Mycol. Appl. **35** (1968) 281–289.

[19] DAVIS, J.G.: Laboratoy control of yogurt. Dairy Ind. **35** (1970) 139–144.

[20] DEAK, T.; BEUCHAT, L.R.: Handbook of food spoilage yeasts. Boca Raton, U.S.A.: CRC Press (1996).

[21] DEIANA, P.; FATICHENTI, F.; FARRIS, G. A.; MOCQUOT, G.; LODI, R.; TODESCO, R.; CECCHI, L.: Metabolization of lactic and acetic acids in Pecorino Romano cheese made with a combined starter of lactic acid bacteria and yeast. Lait **64** (1984) 380–394.

[22] DESMAZEAUD, M.J.: The role of cultures of microorganisms in the flavour and texture of cultured milk products. In: Proc. XXIII Int. Dairy Congr., October 8–12, Montreal, Canada (1990) 1555–1577.

[23] DEVOYOD, J.J.: Yeasts in cheese-making. In: Yeast technology (edited by Spencer, J.F.T.; Spencer, D.M.). Berlin, Germany: Springer Verlag (1990) 228–240.

[24] DOMINGUES, L.; LIMA, N.; TEIXEIRA, J.A.: Alcohol production from cheese whey permeate using genetically modified flocculent yeast cells. Biotechnol. Bioeng. **72** (2001) 507–514.

[25] DRATHEN, M.: Assessment of the microbiological, nutritional and sensory quality of kefir. Ph. D. Thesis, Justus-Liebig-Universitat Giessen, Germany (1987).

[26] DUITSCHAEVER, C.L.: What is kefir and how can it be made? Modern Dairy **68** (1989) 18–19.

[27] DUITSCHAEVER, C.L.; KEMP, N.; EMMONS, D.: Pure culture formulation and procedure for the production of kefir. Milk Sci. Int. **42** (1987) 80–82.

[28] DUITSCHAEVER, C.L.; KEMP, N.; EMMONS, D.: Comparative evaluation of five procedures for making kefir. Milk Sci. Int. **43** (1988) 343–345.

[29] DUITSCHAEVER, C.L.; KEMP, N.; SMITH, A.: Microscopic studies of the microflora of kefir grains and of kefir made by different methods. Milk Sci. Int. **43** (1988) 479–481.

[30] EL SODA, M.: Acceleration of cheese ripening: Recent advances. J. Fd Prot. **49** (1986) 395–399.

[31] ELISKASES-LECHNER, F.: Yeasts in selected cheese varieties – occurrence, identity and biochemical characteristics. Int. Dairy Fed. Spec. Iss. **9801** (1998) 88–96.

[32] ELISKASES-LECHNER, F.; GINZINGER, W.: The yeast flora of surface-ripened cheeses. Milk Sci. Int. **50** (1995) 458–462.

[33] EMBACHER, C.: Die Problematik des industriell hergestellten Kefirs. Deutsche Molkereiztg **101** (1980) 1696–1701.

[34] ENGEL, G.: Yeast in silage and raw milk. Milchwissenschaft **41** (1986) 633–637.

[35] ENGEL, G.: Schnellnachweis von Hefen in Quark. Milchwissenschaft **47** (1992) 435–437.

[36] ENGEL, G.; KRUSCH, U.; TEUBER, M.: Mikrobiologische Zusammensetzung von Kefir. I. Hefen. Milk Sci. Int. **41** (1986) 418–421.

[37] ENGEL, G.; EINHOFF, K.; PROKOPEK, D.; TEUBER, M.: Beziehungen zwischen Hefenart, Hefenzahl und hefenbedingten sensorischen Fehlern in Magerquark nach Zugabe definierter Hefenarten. Milk Sci. Int. **42** (1987) 428–430.

[38] FADDA M.E.; COSENTINO S.; DEPLANO M.; PALMAS F.: Yeast populations in Sardinian feta cheese. Int. J. Fd Microbiol. **69** (2001) 153–156.

[39] FATICHENTI, F.; BERGERE, J.L.; DEIANA, P.; FARRIS, G.A.: Antagonistic activity of *Debaryomyces hansenii* towards *Clostridium tyrobutyricum* and *Clostridium butyricum*. J. Dairy Res. **50** (1983) 449–457.

[40] FELL, J.W.; BOEKHOUT, T.; FONSECA, A.; SCORZETTI, G.; STATZELL-TALLMAN, A.: Biodiversity and systematics of basidiomycetous yeasts as determined by large-subunit rDNA D1/D2 domain sequence analysis. Int. J. Syst. Evol. Microbiol. **50** (2000) 1351–1371.

[41] FERNANDEZ DEL POZO, B.; GAYA, P.; MEDINA, M.; RODRIGUEZ MARIN, M.A.; NUNEZ; M.: Changes in chemical and rheological characteristics of La Serena ewes' milk cheese during ripening. J. Dairy Res. **55** (1988) 457–464.

[42] FLEET, G. H.: Yeasts in dairy products. J. Appl. Bacteriol. **68** (1990) 199–211.

[43] FLEET, G.H.; MIAN, M.A.: The occurrence and growth of yeasts in dairy products. Int. J. Fd Microbiol. **4** (1987) 145–155.

[44] FREITAS, A.C.; PINTADO, A.E.; PINTADO, M.E.; MALCATA, F.X.: Organic acids produced by lactobacilli, enterococci and yeasts isolated from Picante cheese. Eur. Food Res.Technol. **209** (1999) 434–438. 72409.

[45] FUJISAWA, T.; ADACHI, S.; TOBA, T.; ARIHARA, K.; MITSUOKA, T.: *Lactobacillus kefiranofaciens* sp. nov. isolated from kefir grains. Int. J. Syst. Bacteriol. **38** (1988) 12–14.

[46] GIUDICI P.; MASINI G.; CAGGIA C.: The role of galactose fermenting yeast in plain yogurt spoilage. Ann. Microbiol. Enzimol. **46** (1996) 11–19.

[47] GLAESER, H.: Kefir: Kulturen, Herstellung, chemische Zusammensetzung, ernährungsphysiologischer Wert. Ernährungs-Umschau **28** (1981) 156–158.

[48] GLAESER, H.; BEITER, M.: Enzymatische Bestimmung des Aethanolgehaltes in handelsüblichem und selbst hergestelltem Kefir. Deutsche Milchwirtschaft **50** (1979) 1865–1868.

[49] GLAESER, H.; HANGST, E.: Mikrobiologisch-hygienische Aspekte der Herstellung von Sauermilch im Haushalt. Ernährungs-Umschau **32** (1985) 399–402.

[50] GLAESER, H.; HANGST, E.; ZIEGLER, K.: Biochemische Charakterisierung der in Molkereikefir und Kefirkulturen vorkommenden Hefen. Deutsche Molkereiztg **107** (1986) 483–490.

[51] GREENBERG, R.S.; LEDFORD, R.A.: The deamination of glutamic and aspartic acids by *Geotrichum candidum* and its possible role in Limburger cheese ripening. J. Dairy Sci. **61** (1978) 106.

[52] GRIEVE, P.A.; KITCHEN, B.J.; DULLEY, J.R.; BARTLEY, J.: Partial characterization of cheese-ripening proteinases produced by the yeast *Kluyveromyces lactis*. J. Dairy Res. **50** (1983) 469–480.

[53] GUEGUEN, M.: Aptitude of the species *Geotrichum candidum* to the production of proteolytic enzymes. Lait **543/544** (1975) 145–159.

[54] GUEGUEN, M.: Etudes sur les caractères culturaux et la morphologie de *Geotrichum candidum* Link. Lait **62** (1982) 625–644.

[55] GUÉHO, E.; BUISSIÈ J.: Méthode d'identification biochimique de champignons filamenteux arthrosporés appartement au genre *Geotrichum* Link ex Pers. Ann. Microbiol. **126A** (1975) 483–500.

[56] GUERZONI, M.E.; GOBBETTI, M.; LANCIOTTI, R.; VANNINI, L.; CHAVES LOPEZ, C.: *Yarrowia lipolytica* as potential ripening agent in milk products. In: Yeasts in the dairy industry: Positive and negative aspects. Int. Dairy Fed. Spec. Iss. **9801** (1998) 23–33.

[57] HACCIUS, C.: Kefir. Schweiz. Milchztg **14** (1885) 19–21.

[58] HÄFLIGER, M.J.: Die Bedeutung der Essigsäurebakterien im Kefir, Ph.D. thesis ETH No. 9270. Zürich, Switzerland: Swiss Federal Institute of Technology (1990).

[59] HARIDY, M.S.A.: Yeast flora of raw milk in El-Minia City, Egypt. Cryptog. Mycol. **13** (1992) 321–326.

[60] HARIDY, M.S.A.: Occurrence of yeasts in yoghurt, cheese and whey. Cryptog. Mycol. **14** (1993) 255–262.

[61] HIROTA, T.: Microbiological studies on kefir grains. Technical Research Institute. Snow Brand Milk Prod. Co. Ltd., Saitam, Japan **84** (1987) 67–69.

[62] HOSONO, A.; TANABE, T.; OTANI, H.: Binding properties of lactic acid bacteria isolated from kefir milk with mutagenic amino acid pyrolyzates. Milk Sci. Int. **45** (1990) 647–651.

[63] International Dairy Federation Standard: Norme générale de composition pour les laits fermentés, Vol. **163** (1992).

[64] IZAGUIRRE, M.E.; CASTILLO, F.J.: Selection of lactose-fermenting yeasts for ethanol production from whey. Biotechnol. Lett. **4** (1982) 257–262.

[65] JAKOBSEN, M.; NARVHUS, J.: Yeasts and their possible beneficial and negative effects on the quality of dairy products. Int. Dairy J. **6** (1996) 755–768.

[66] JENQ, W.; SPECKMAN, R.A.; CRANG, R.E.; STEINBERG, M.P.: Enhanced conversion of lactose to glycerol by *Kluyveromyces fragilis* utilizing whey permeate as a substrate. Appl. Environ. Microbiol. **55** (1989) 573–578.

[67] JOLLIVET, N.; CHATAUD, J.; VAYSSIER, Y.; BENSOUSSON, M.; BELIN, J.-M.: Production of volatile compounds in model milk and cheese media by eight strains of *Geotrichum candidum* Link. J. Dairy Res. **61** (1994) 241–248.

[68] KALITERNA, J.; WEUSTHUIS, R.A.; CASTRILLO, J.I.; VAN DIJKEN, J.P.; PRONK, J.T.: Transient responses of *Candida utilis* to oxygen limitation: regulation of the Kluyver effect for maltose. Yeast **11** (1995) 317–325.

[69] KAMINARIDES, S.E.; ANIFANTAKIS, E.M.: Evolution of the microflora of Kopanisti cheese during ripening. Study of the yeast flora. Lait **69** (1989) 537–546.

[70] KAMINARIDES, S.E.; ANIFANTAKIS, E.M.; BALIS, C.: Changes in Kopanisti cheese during ripening using selected pure microbial cultures. J. Dairy Technol. **45** (1992) 56–60.

[71] KAMMERLEHNER, J.: Cheese with smear rind. Deutsche Milchwirtschaft **46** (1995) 1084–1086.

[72] KANDLER, O.; KUNATH, P.: *Lactobacillus kefir* sp.nov., a component of the microflora of kefir. Syst. Appl. Microbiol. **4** (1983) 286–294.

[73] KERN, E.: *Dispora caucasica* nov. g. et nov. sp., eine neue Bakterienform. Biol. Zbl. **2** (1882) 135–136

[74] KLEEBERG, J.: Studien über Yoghurt und Kefir. I. Untersuchungen an Handelspräparaten. Centralbl. Bakt. Parasitenk. Infektionkr. II. Abt. **68** (1926) 321–326.

[75] KLUPSCH, H.: Produktverbesserung am Beispiel Kefir. Deutsche Molkereiztg **105** (1984) 466–468.

[76] KOROLEVA, N. S.: Special products (Kefir, Koumyss, etc.). XXI Int. Dairy Congr. **2** (1984) 146–152.

[77] KOROLEVA, N.S.: Starters for fermented milks. Section 4: Kefir and kumys starters. Int. Dairy Fed. IDF Bulletin **227** (1991) 35–40.

[78] KOROLEVA, N.S.: Technology of kefir and kumys. Int. Dairy Fed. IDF Bulletin **227** (1991) 96–99.

[79] KOSSE, D.; SEILER, H.; AMANN, R.; LUDWIG, W.; SCHERER, S.: Identification of yoghurt-spoiling yeasts with 18S rRNA-targeted oligonucleotide probes. Syst. Appl. Microbiol. **20** (1997) 468–480.

[80] KREGER-VAN RIJ, N.J.W.: The yeasts – a taxonomic study, 3rd ed. Amsterdam, The Netherlands: Elsevier (1984).

[81] KRUSCH, U.: Mikrobiologische Charakterisierung von Kefir. Deutsche Molkereiztg **105** (1984) 1332–1336.

[82] KUNTZE, W.: Studien über fermentierte Milch. Centralblatt für Bakt. Parasitenk Infektionkr. II. Abt. **24** (1909) 101–122.

[83] KURTZMAN, C.P.; ROBNETT, C.J.: Molecular relationships among hyphal ascomycetous yeasts and yeastlike taxa. Can. J. Bot. **73 Suppl. 1** (1995) S824–S830.

[84] KURTZMAN, C.P.; ROBNETT, C.J.: Identification of clinically important ascomycetous yeasts based on nucleotide divergence in the 5' end of the large-subunit (26S) ribosomal DNA gene. J. Clin. Microbiol. **35** (1997) 1216–1223.

[85] KURTZMAN, C.P.; ROBNETT, C.J.: Identification and phylogeny of ascomycetous yeasts from analysis of nuclear large subunit (26S) ribosomal DNA partial sequences. Antonie van Leeuwenhoek **73** (1998) 331–371.

[86] LA RIVIÈRE, J.W.M.: Ecology of yeasts in the kefir grain. Antonie van Leeuwenhoek **35 Suppl.:** Yeast Symposium (1969) D15–D16.

[87] LA RIVIÈRE, J.W.M.; KOOLMAN, P.: Kefiran, a novel polysaccharide produced in the kefir grain by *Lactobacillus brevis*. Arch. Mikrobiol. **59** (1967) 269–278.

[88] LEE, J.T.; LIM, J.W.: Studies on the ripening of cheese made with lactic acid bacteria and yeasts. I. Changes of physico-chemical composition. Kor. J. Anim. Sci. **30** (1988) 170–177.

[89] LENOIR, J.: The surface flora and its role in the ripening of cheese. Int. Dairy Fed. IDF Bulletin No. **171** (1984) 3–20.

[90] LENOIR, J.; LAMBERET, G.; SCHMIDT, J.L.; TOURNEUR, C.: Microbial activity dominates cheese ripening. Rev. Lait. Franç. **444** (1985) 54–62.

[91] LOPEZ DIAZ, T.M.; SANTOS, J.A.; PRIETO, M.; GARCIA LOPEZ, M.L.; OTERO, A.; DIAZ, T.M.L.; LOPEZ, M.L.G.: Mycoflora of a traditional Spanish blue cheese. Neth. Milk Dairy J. **49** (1995) 191–199.

[92] MANN, E.J.: Kefir und Koumiss – Eine Literaturübersicht. Die Molkerei-Ztg, Welt der Milch **43** (1989) 917–919.

[93] MANN, E.J.; BOTTAZZI, V.; ZACCONI, C.; SARRA, P.G.; DALLAVALLE, P.; PARISI, M.G.: Kefir and Koumiss. Microbiology, chemistry and technology. Dairy Ind. Int. **61** (1994) 9–10.

[94] MARSHALL, V.M.; COLE, W.M.; BROOKER, B.E.: Observations on the structure of kefir grains and the distribution of the microflora. J. Appl. Bacteriol. **57** (1984) 491–497.

[95] MARTIN, N.; SAVONITTO, S.; MOLIMARD, P.; BERGER, C.; BROUSSE, M.; SPINNLER, H.E.: Flavour generation in cheese curd by coculturing with selected yeast, mold, and bacteria. J. Dairy Sci. **82** (1999) 1072–1080.

[96] MEIER, C.; WYDER, M.T.: Klassifizierung von Hefeisolaten aus Joghurt mittels Sequenzierung der D2 Domäne auf der 28S rDNA. In: Miescher, S.: Antimicrobial and autolytic systems of dairy propionibacteria. Ph.D. DISS. ETH No. 13486. Zürich, Switzerland: Swiss Federal Institute of Technology (1999) 1–138.

[97] MERILAINEN, V.T.: Microorganisms in fermented milks: other microorganisms. Int. Dairy Fed. IDF Bulletin No. **179** (1984) 89–93.

[98] MITEVA, E.; KIROVA, E.; GADJEVA, E.; RADEVA, M.: Sensory aroma and taste profiles of raw-dried sausages manufactured with a lipolytically active yeast culture. Nahrung **30** (1986) 829–832.

[99] MOLIMARD, P.; LESSCHAEVE, I.; BOUVIER, I.; VASSAL, L.; SCHLICH, P.; ISSANCHOU, S.; SPINNLER, H.E.: Bitterness and nitrogen fractions of soft cheeses of the Camembert type: Role of the combination of *Penicillium camembertii* and *Geotrichum candidum*. Lait **74** (1994) 361–374.

[100] MOLSKA, I.; KOCON, J.; ZMARLICKI, S.: Electron microscopic studies on structure and microflora of kefir grains. Acta Alim. Pol. **6** (1980) 145–154.

[101] NAHABIEH, F.; SCHMIDT, J. L.: Contribution to the study of the yeast flora in several major types of goat cheese. Lait **70** (1990) 325–343.

[102] NEVE, H.: Analysis of kefir grain starter cultures by scanning electron microscopy. Milk Sci. Int. **47** (1992) 275–278.

[103] OLSON, H.C.; HAMMER, B.W.: Observations on yeasts causing gas in sweetened condensed milk. Iowa State Coll. J. Sci. **10** (1935) 37–43.

[104] OTTOGALLI, G.; GALLI, A.; RESMINI, P.; VOLONTERIO, G.: Composizione microbiologica, chimica ed ultrastruttura dei granuli di Kefir. An. Microbiol. Enzim. **23** (1973) 109–121.

[105] PEREIRA DIAS, S.; POTES, M.E.; MARINHO, A.; MALFEITO FERREIRA, M.; LOUREIRO, V.: Characterisation of yeast flora isolated from an artisanal Portuguese ewes' cheese. Int. J. Fd Microbiol. **60** (2000) 55–63.

[106] PETTERSSON, H.E.; CHRISTIANSSON, A.; EKELUND, K.: Making kefir without grains using freeze-dried cultures. Nordisk Mejeriindustri **12** (1985) 58–60.

[107] PHILIPP, S.: 'Special cultures' – significance and application in cheese factories. Deutsche Molkereiztg **106** (1985) 1706–1710.

[108] PIDOUX, M.; DE RUITER, G.A.; BROOKER, B.E.; COLQUHOUN, I.J.; MORRIS, V.J.: Microscopic and chemical studies of a gelling polysaccharide from *Lactobacillus hilgardii*. Carbohydr. Polymers **13** (1990) 351–362.

[109] PINTADO, M.E.; DA SILVA, J.A.L.; FERNANDEZ DEL POZO, B.; MALCATA, F.X.; HOGG, T.A.: Microbiological and rheological studies on Portuguese kefir grains. Int. J. Fd Sci. Technol. **31** (1996) 15–26.

[110] PODWYSSOZKI, W.: Kefyr. Kaukasisches Gärungsferment und Getränk aus Kuhmilch. St. Petersburg: Verlag Carl Richner (1884).

[111] PRETORIUS, J.S.; VAN DER WESTHUIZEN, T.J.: The impact of yeast genetics and recombinant DNA technology on the wine industry – A review. S. Afr. J. Enol. Vitic. **12** (1991) 3–31.

[112] PRILLINGER, H.; MOLNAR, O.; ELISKASES-LECHNER, F.; LOPANDIC, K.: Phenotypic and genotypic identification of yeasts from cheese. Antonie van Leeuwenhoek **75** (1999) 267–283.

[113] PUHAN, Z.; VOGT, O.: Hefehaltige Sauermilchprodukte – Technologie und Stoffwechsel. Milchw. Ber. **78** (1984) 7–14.

[114] RAPIN, J.-D.; MARISON, I.W.; VON STOCKAR, U.; REILLY, P.J.: Glycerol production by yeast fermentation of whey permate. Enz. Microb. Technol. **16** (1994) 143–150.

[115] ROBINSON, R.K.; TAMINE, A.Y.: Microbiology of fermented milks. In: Dairy Microbiology (edited by Robinson, R.K.). London, U.K. and New Jersey, U.S.A.: Applied Science Publishers (1981) 245–278.

[116] ROGINSKI, H.: Fermented milks. Austr. J. Dairy Technol. **43** (1988) 37–46.

[117] ROHM, H.: Zur Bedeutung von Hefen und Schimmelpilzen für die Milchwirtschaft, Teil I und II. Deutsche Milchwirtschaft **14/16** (1991) 404–406/476–480.

[118] ROHM, H.: Importance of yeasts and moulds in the dairy industry. Part II. Deutsche Milchwirtschaft Hildesheim **42** (1991) 478–480.

[119] ROHM, H.; LEHNER, M.: Charakteristische Eigenschaften von österreichischem Kefir. Ernährung **14** (1990) 571–574.

[120] ROHM, H.; ELISKASES-LECHNER, F.; BRÄUER, M.: Diversity of yeasts in selected dairy products. J. Appl. Bacter. **72** (1992) 370–376.

[121] ROOSTITA, R.; FLEET, G.H.: The occurrence and growth of yeasts in Camembert and blue-veined cheeses. Int. J. Fd Microbiol. **28** (1996) 393–404.

[122] ROOSTITA, R.; FLEET, G.H.: Growth of yeasts in milk and associated changes to milk composition. Int. J. Fd Microbiol. **31** (1996) 205–219.

[123] ROOSTITA, R.; FLEET, G.H.: Growth of yeasts isolated from cheeses on organic acids in the presence of sodium chloride. Fd Tech. Biotechnol. **37** (1999) 73–79.

[124] ROSI, J.: The Kefir grain beverage microorganisms: The yeasts *Saccharomyces delbrueckii* and *Saccharomyces cerevisiae*. Sci. Tecn. Latt. Casearia **29** (1978) 59–67.

[125] ROSS, H.M.; HARDEN, T.J.; NICHOL, A.W.; DEETH, H.C.: Isolation and investigation of microorganisms causing brown defects in mould-ripened cheeses. Austr. J. Dairy Technol. **55** (2000) 5–8.

[126] SCHULZ, M.E.: 100 Jahre Kefir in Nordeuropa und die heutige Bedeutung der Kefir-Sauermilch. Deutsche Molkereiztg **13** (1968) 489–492.

[127] SCOTT, R.: Cheesemaking practice. Barking, U.K.: Applied Science Publishers Ltd (1981).

[128] SEILER, H.; BUSSE, M.: The yeasts of cheese brines. Int. J. Fd Microbiol. **11** (1990) 289–303.

[129] BOUROUNOFF, B.: Recherches sur le kéfir. Le Lait, **14** (1934) 819–827.

[130] STANLEY, G.: New developments in functional starters. The dairy industry, Symp. Proc. 24–25 April **53** (1991) 162–191.

[131] SURIYARACHCHI, V.R.; FLEET, G.H.: Occurrence and growth of yeasts in yogurts. Appl. Environ. Microbiol. **42** (1981) 574–579.

[132] TAHOUN, M.K.; EL-MERHEB, Z.; SALAM, A.; YOUSSEF, A.: Biomass and lipids from lactose or whey by *Trichosporon beigelii*. Biotechnol. Bioeng. **29** (1987) 358–360.

[133] TOBA, T.; ARIHARA, K.; ADACHI, S.: Distribution of microorganisms with particular reference to encapsulated bacteria in kefir grains. Int. J. Fd Microbiol. **10** (1990) 219–224.

[134] VAN DEN TEMPEL, T.; JAKOBSEN, M.: Yeasts associated with Danablu. Int. Dairy J. **8** (1998) 25–31.

[135] VAN DEN TEMPEL, T.; JAKOBSEN, M.: The technological characteristics of *Debaryomyces hansenii* and *Yarrowia lipolytica* and their potential as starter cultures for production of Danablu. Int. Dairy J. **10** (2000) 263–270.

[136] VASSAL, L.; GRIPON, J.C.: Bitterness in Camembert cheese: Role of rennet and *Penicillium caseicolum* and methods of control. Lait **64** (1984) 397–417.

[137] VON WIESE, W.: Kefir – ein Sauermilcherzeugnis im Widerstreit zwischen Hersteller, Lebensmittelüberwachung und Verbraucher. Deutsche Milchwirtschaft **6** (1986) 227–229.

[138] WALKER, H.W.; AYRES, J.C.: Yeasts as spoilage organisms. In: The yeasts, vol. 3, Yeast technology (edited by Rose, A.H.; Harrison, J.S.). London, U.K.: Academic Press (1970) 464–527.

[139] WEGNER, K.: Mikrobiologie tierischer Lebensmittel, Kefir. Leipzig, Germany: VEB Fachbuchverlag (1981) 211–217.

[140] WEIS, W.; BURGBACHER, G.: 100 Jahre Kefir in Deutschland – Nach wie vor ein aktuelles Thema. Untersuchung von "Kefir" aus Molkereien und Handel sowie dessen Problematik. Deutsche Milchwirtschaft **37** (1986) 81–90.

[141] WELTHAGEN, J.J.; VILJOEN, B.C.: Yeast profile in Gouda cheese during processing and ripening. Int. J. Fd Microbiol. **41** (1998)185–194.

[142] WELTHAGEN, J.J.; VILJOEN, B.C.: The isolation and identification of yeasts obtained during the manufacture and ripening of Cheddar cheese. Fd Microbiol. **16** (1999) 63–73.

[143] WESTALL, S.; FILTENBORG, O.: Yeast occurrence in Danish feta cheese. Fd Microbiol. **15** (1998) 215–222.

[144] WESTALL, S.; FILTENBORG, O.: Spoilage yeasts of decorated soft cheese packed in modified atmosphere. Fd Microbiol. **15** (1998) 243–249.

[145] WOLF, K.: Nonconventional yeasts in biotechnology: A handbook. Berlin, Germany: Springer Verlag (1996).

[146] WYDER, M.T.; BACHMANN, H.P.; PUHAN, Z.: Role of selected yeasts in cheese ripening: An evaluation in foil wrapped Raclette cheese. Fd Sci.Tech.-Lebensm.-Wiss. Technol. **32** (1999) 333–343.

[147] Wyder, M. T.; Meile, L.; Teuber, M.: Description of *Sacharomyces turicensis* sp nov., a new species from Kefyr. Syst. Appl. Microbiol. **22** (199) 420–425.

[148] WYDER, M.T.; PUHAN, Z.: Investigation of the yeast flora in dairy products: A case study of Kefyr. Fd Technol. Biotechnol. **35** (1997) 299–304.

[149] WYDER, M.T.; PUHAN, Z.: Role of selected yeasts in cheese ripening: an evaluation in aseptic cheese curd slurries. Int. Dairy J. **9** (1999) 117–124.

[150] WYDER, M.T.; PUHAN, Z.: Investigation of the yeast flora in smear ripened cheeses. Milk Sci. Int. **54** (1999) 330–333.

[151] YAMAUCHI, K.; KANG, K.H.; KAMINOGAWA, S.; KOMAGATA, K.; FATICHENTI, F.; BERGERE, J.L.; DEIANA, P.; FARRIS, G.A.: Effects of yeasts isolated from cheese on the growth of lactic acid bacteria in skim-milk. Jpn J. Zootechn. Sci. **46** (1975) 73–80.

[152] ZAMBONELLI, C.; RAINIERI, S.; CHIAVARI, C.; MONTANARI, G.; BENEVELLI, M.; GRAZIA, L.: Autolysis of yeasts and bacteria in fermented foods. Ital. J. Fd Sci. **12** (2000) 9–21.

[153] ZOURARI, A.; ANIFANTAKIS, E.M.: Le kefir. Caractères physico-chimiques, microbiologiques et nutritionnels. Technologie de production. Une revue. Lait **68** (1988) 373–392.

9 Yeasts in meat and meat products

JOHN SAMELIS and JOHN N. SOFOS

9.1 Introduction

Meat and poultry products are prone to microbial contamination and support growth of spoilage or pathogenic microorganisms [96, 109, 141]. Among microorganisms contaminating meat products [108, 141], yeasts normally constitute a minor part of the flora [47, 74, 108, 155]. While this mixed microbial association is changing in response to intrinsic, extrinsic and processing factors [60, 103, 154], yeasts generally have limited opportunities to grow and spoil the meat. This is because yeasts proliferate at slow rates and compete poorly with bacteria [155]. Most yeasts, however, are more resistant than bacteria to several food-related stresses such as low water activity (a_w), low pH, high salinity, and chemical preservatives, while certain species are extremely psychrotrophic [18, 23, 53, 62]. Thus, yeasts may opportunistically become important spoilage agents in meat products, especially when bacterial growth is retarded due to the inhibitory effects of the above stress factors [47, 74, 155]. Also yeasts are aerobic or microaerophilic organisms, and, therefore their competitiveness and spoilage potential increase under aerobic conditions.

Compared to bacteria, published research on the ecology, taxonomy, physiology and beneficial or detrimental effects of yeasts in meat products is limited. Most studies are aimed at enumerating yeast populations on fresh or processed meats, without any further testing due to their low incidence and potentially low practical significance compared to bacteria. Studies on the ecological diversity and succession of yeasts from the field to meat products are even less common. In addition, there are difficulties in comparing studies due to differences in methodology used for enumeration, isolation and identification. Numerous yeast species have been reported as isolated from meat and meat products [47, 74]. However, this complexity has been lessened following major reclassifications in yeast taxonomy that have led to combining previously separated, closely-related species or genera. For example, *Debaryomyces hansenii*, the most prevalent and important yeast in processed meats, currently includes this species in addition to the former species *Debaryomyces kloeckeri, D. nicotianae* and *D. subglobsus* [81]. Also, *Torulopsis* have been included in *Candida* [81]. Advances in yeast taxonomy have mainly been based on molecular identification methods [38, 82], which are discussed in Chapters 3 and 4.

The identification and distribution of yeasts isolated from meat and meat products have been discussed in previous reviews [45, 47, 53, 55, 73, 74, 155]. In addition, excellent reviews have been published on the physiological and biochemical tests and differentiation criteria for foodborne yeasts, including those found in meat [32, 40]. The scope of this chapter is to present a brief overview of this knowledge and update it by discussing more recent studies, without attempting to provide a detailed presentation of the extreme diversity of meat yeasts. Instead, the aim is to emphasize the most important yeast genera or species in the context of their natural succession as affected by meat processing, and to discuss their beneficial versus detrimental effects and associated physiological properties, and the methods to control yeast spoilage in meat products.

9.2 Yeast biodiversity in meat products

Like any food, meat is colonized by a product-specific microbial flora originally based on natural selection and further subjected to dynamic changes as a response to meat processing [60, 103, 141]. As it happens in most food commodities, depending on processing methods, microbial populations may fluctuate considerably, while some of the types of microorganisms initially present usually decrease in proportion to a dominating microflora consisting of a few well-adapted and environmentally favoured species [103, 129, 130]. Similarly to bacteria, the biodiversity of yeasts on meat is affected by environmental and processing factors, which jointly influence the occurrence and succession of yeast species in meat products [47, 53].

9.2.1 Fresh meats

Yeasts are found in soil, water, air, plant surfaces and living animals, from where they may be transmitted into the slaughterhouse and onto meat processing equipment to eventually contaminate carcasses and fresh meat cuts. DILLON and BOARD [45] reviewed studies on the incidence of yeasts in the field (e. g., soil samples from sheep and cattle stockyards and paddocks or under pasture, leaves and roots of pasture plants, air samples) and noted that anamorphic yeasts dominate the yeast flora from such environments. Mainly, species belonging to the genera *Cryptococcus, Rhodotorula* and *Candida* have been found. Overall, yeasts represent a minor part (< 3 %) of the total microbial flora in the field [45, 46]. The distribution of yeast species in the field may be subjected to seasonal changes as, for example, a high (> 80 %) as compared to a low (< 8 %) incidence of carotenoid-pigmented yeasts occured in hay and soil samples from the United Kingdom between March and December [46]. *Candida, Cryptococcus*, and *Rhodotorula* were recovered from pastures, fleece, carcass surfaces, minced lamb and lamb products, indicating a direct transfer of contamination from the field to the lamb processing environment and products [45, 46]. On lamb carcasses in particular, *Candida famata* (anamorph of *D. hansenii*), *C. glabrata, C. mesenterica, C. curvata* (reclassified as *Cryptococcus curvatus* [12]) *Cr. albidus, Cr. laurentii, Rh. minuta* and *Rh. rubra* (reclassified as *Rh. mucilaginosa* [12]) were present [46]. Interestingly though, despite their high incidence in the field, *Rhodotorula* species were consistently low on lamb carcasses and products [46], indicating selection by the low temperature of meat plants. In accordance, *R. glutinis* was common on lamb carcasses in the summer, but not in the winter when *Candida* and *Cryptococcus* spp. were predominant [88]. In fact, the yeast flora of lamb loins stored at –5 °C did not include any *Rhodotorula*, but it was composed of *Cr. laurentii* var. *laurentii*, *Cystofilobasidium infirmo-miniatum*, *Trichosporon pullulans* and *C. zeylanoides*, with the former species accounting for more than 90 % of the population [89]. This yeast population was favoured because bacterial growth ceased at –5 °C, while the yeast population increased from 10 to 10^6 cfu/cm^2 within 20 weeks [89]. This yeast predominance, however, is unlikely to occur at temperatures above 0 °C as under these conditions bacteria dominate and yeasts comprise less than 5 % of the microbial population on fresh meats [73, 74]. Indeed, the yeast population on fresh meat cuts is normally in the range of 10^2 to 10^3 cfu/g, while mincing results in higher numbers (10^3–10^5 cfu/g) [75, 84]. Minced meat stored aerobically (0–5 °C) may develop even higher (10^6–10^7 cfu/g) yeast numbers, suggesting a potential role of yeasts in its spoilage [31, 74].

The yeast flora of fresh beef and pork appears to be similar with that of lamb. AYRES [5] isolated *Candida, Torulopsis* (reclassified as *Candida*) and *Rhodotorula* species, and HSIEH and JAY [69] recovered these genera in addition to strains of *Cryptococcus* and *Trichosporon* from refrigerated fresh beef. Notably, a succession of yeasts of the above genera as well as of different species within each genus occurred while fresh beef was being processed (e. g., minced) or spoiled [69]. Indeed, *Candida* spp. comprised 82 % of the yeast flora on fresh beef with *C. lipolytica* and *C. lambica* (= *Pichia fermentans* var. *fermentans*) being predominant, while, except for one sample, *Candida* spp. were the only yeasts recovered from spoiled beef, with *C. lipolytica* (= *Yarrowia lipolytica*) and *C. zeylanoides* being predominant [69]. *Candida sake*, a species infrequently reported to dominate the yeast flora of fresh meats, was found to be predominant in South African minced beef before radurization [76]. After 8 days of storage of irradiated samples at 4 °C, however, *C. zeylanoides* succeeded *C. sake* as the predominant yeast in antibiotic-containing media, but not on acidified media, while both yeasts lost their dominance to *C. famata* after 14 days of storage [76]. The yeast flora of minced beef in the U.K. was also dominated (60 %) by *Candida* spp., while *Cryptococcus* (10 %) and *Rhodotorula* spp. (3 %) in association with few strains of *Trichosporon, Pichia* and *Debaryomyces* were also present [112]. Likewise, minced lamb contained 73 % *Candida*, 21 % *Cryptococcus* and only 6 % *Rhodotorula,* with the *Candida* flora being more diverse than that on lamb carcasses [46]. A similar succession of yeasts was noted in fresh pork [30, 31]. While the isolated species from the slaughter area and the lairage in a pig processing plant were *Cryptococcus, Rhodotorula, Torulopsis and Trichosporon* [30], pig (24-h post-mortem) carcasses were mainly colonized by *C. zeylanoides* and *D. hansenii* [31]. Thus, *Candida* spp. may have better chances to survive during further processing of slaughtered animals and the "ageing" conditions of minced fresh meats under refrigeration than other anamorphic "wild" yeasts found in the field.

Yeasts are minor components of the natural flora of fresh poultry [96]. BARNES et al. [11] reported a large (e. g., 10^7 cfu/cm^2) increase in the numbers and proportion of yeasts present on spoiled, polyethylene-wrapped, air chilled turkey carcasses stored at –2 °C for 35 days. Predominant yeasts under these conditions were *Cr. laurentii* var. *laurentii* and *C. zeylanoides,* while prior to storage yeasts found on carcasses were *P. membranifaciens*, *Tr. pullulans, Candida* (formerly *Torulopsis*) and *Rh. glutinis* [11]. This yeast activity, similar with that observed in lamb loins stored at –5 °C [89], was attributed to restriction of growth of gram-negative bacteria by low a_w as a result of partial freezing of the poultry skin [11]. Other studies have shown increases (ca. 2 logs) of yeasts on skin from fresh and spoiled poultry carcasses stored under refrigeration [56, 59] with final populations exceeding 5 log cfu/g [59]. When 159 yeast isolates from fresh and spoiled poultry carcasses [59] were identified, an interesting succession of genera and species was observed [152]. *Candida* (57.5 % and 51.8 %), *Cryptococcus* (7.5 % and 2.6 %), *Debaryomyces* (20.0 % and 29.1 %) and *Yarrowia lipolytica* (2.5 % and 11.4 %) are isolated from both fresh and spoiled poultry, respectively, with the former two genera tending to decrease and the latter two genera tending to increase at spoilage [152]. *Candida zeylanoides* is prevalent throughout storage, and, unlike other *Candida* spp., increases from fresh (35 %) to spoiled poultry (45.5 %). Likewise, 58.3 % of yeast isolates from frozen chicken carcasses are *C. zeylanoides*, 16.7 % are *Y. lipolytica*, and 16.7 % are *Rh. rubra* (≡ *Rhodotorula mucilaginosa*) [48]. Among 152 strains isolated from raw poultry products (e. g., whole chicken, ground chicken, liver, heart, gizzard and ground turkey) stored at 5 °C, *Y. lipolytica* and *C. zeylanoides* account for 39.5 % and 26.3 % of total isolates assigned to 12 different species. These include 24 % of basidiomycetous yeasts, mainly *Cryptococcus* spp. [70]. It is concluded that yeasts, particularly *Y. lipolytica*, may play an important role in fresh poultry [70].

9.2.2 Cured fresh and cooked meats

When meat is cured or brined, the a_w is lowered from 0.98–0.99 to ca. 0.96, the salt/brine concentration increases while nitrites/nitrates and spices are introduced as additional hurdles [90, 103]. These conditions induce natural selection resulting in a gradual, partial to complete, replacement of the gram-negative by a gram-positive flora, consisting mainly of lactic acid bacteria (LAB) [90, 103, 129–131]. If fresh cured meats are stored in air or in air permeable packages under refrigeration, abundant growth of yeasts may occur, and the product may be spoiled by yeast activity. Typical examples are the fresh British sausage [50, 51] and the traditional Greek country-style sausage [128].

Growth of yeasts in fresh British sausage is favoured because bacteria are inhibited by sulphite [9, 44]. Yeast populations range from 10 to 10^5 cfu/g during a study that indicated their potential role in spoilage [50]. Other studies underscore the importance of yeasts in British sausage and clarify their taxonomy and physiological attributes associated with spoilage [30, 31, 43, 47, 51]. Most common species in unsulphited and sulphited sausages are *D. hansenii* followed by *C. zeylanoides* and *P. membranifaciens* (teleomorph of *C. valida*) [31]. *Cryptococcus* and *Rhodotorula* spp. are also present, but their proportion is reduced in sulphited sausages due to their lower tolerance to the preservative [31].

The yeast flora isolated from "serwolatka", a Polish fresh pork sausage, consists mainly of *Candida* and *Debaryomyces* spp., while members of *Brettanomyces, Cryptococcus, Rhodotorula, Sporobolomyces,* and *Trichosporon* are also present [145]. Populations of unidentified yeasts of 10^6–10^7 cfu/g are also found in "loukaniko choriatiko", a country-style Greek fresh pork sausage, stored aerobically at 3 and 12 °C, with the higher storage temperature accelerating growth and the development of "malty" off-odours [128].

Yeasts have been commonly isolated from cooked cured meat products, but normally they are not important in spoilage. This is because heat processing followed by refrigerated storage in vacuum packages inactivates initial yeast contamination and prevents growth of cross-contaminating yeasts due to low oxygen availability [107, 129, 131, 132]. Likewise, packaging under nitrogen prevents yeast growth on frankfurters [137] and pastrami [83]. Overall, when contamination is high before vacuum packaging of cooked meats or the final product is in air permeable films or following opening of vacuum packages and storage in domestic refrigerators, yeasts may become important members of the spoilage association [131, 132].

JAY [73, 74] reviewed studies on yeasts isolated from frankfurters, hams and sausages, and highlighted the high incidence of *Debaryomyces*, *Candida* and *Torulopsis* (i. e., *Candida*). Indeed, only these genera, represented by *D. hansenii*, *C. catenulata*, *C. lipolytica*, *C. zeylanoides* and *C. saitoana* (formerly *Torulopsis candida*) along with *Tr. pullulans* have been isolated from the surface of packaged frankfurters [49]. Likewise, *C. saitoana* and *D. hansenii* were the only yeasts recovered from various cured meats [139]. Also, *C. famata* (= *Debaryomyces hansenii*) predominated on bacon with a salt content higher than 4 %, while bacon with low salt content was colonized by *C. tropicalis* and *C. krusei* (= *Issatchenkia orientalis*) [57]. Notably, pathogenic *Candida* spp., such as *C. parapsilosis* and *C. tropicalis*, were isolated from Bologna-type sausages and smoked ham [144]. *Candida* and *Debaryomyces* spp. were also the most prevalent yeasts among 123 isolates from a Vienna sausage processing plant [151]. Fewer isolates were assigned to *Rhodotorula*, *Yarrowia*, *Pichia*, *Galactomyces*, *Cryptococcus*, *Trichosporon* and *Torulaspora* [151].

Studies have examined yeast types and behaviour in processed poultry products, such as marinated and roasted chicken and turkey and sausages made from poultry meat [25, 70]. CARLOS and HARRISON [25] reported that marination of chicken with 0.2 % pimento leaf oil significantly reduced the total yeast counts (0.5 to > 2.0 logs) during storage at 4 °C for one week. This effect, however, was not evident at 12 °C. ISMAIL et al. [70] noted that marination of chicken breasts (e. g., teriyaki, Italian style blue cheese, barbeque, hickory, and roasted chicken sauces) had little influence on numbers of yeasts detected, while heating tended to eliminate yeasts from marinated products. *Yarrowia lipolytica* and *C. zeylanoides* were the most prevalent species in marinated (45.5 % and 24.2 % of isolates) and roasted (54.5 % and 21.2 %) poultry meats, respectively, indicating a more prominent role of these species in spoilage of processed poultry products than previously recognized [70].

9.2.3 Dried and fermented meats

Meat curing and drying, with or without an intermediate fermentation step, represents a traditional processing sequence, which may favour yeast growth [6, 90, 92]. This is because it creates an environment of reduced moisture (e. g., < 35 %), low a_w (e. g., < 0.92), increased salt concentration (e. g., > 4 %) and, when carbohydrates are added to ferment, increased acid concentration (pH < 5.3). This broad category of processed meat products includes dry fermented sausages and salamis, dry cured hams and other specific types of traditional intermediate moisture meats, such as basturma and jerky. During meat fermentation, in particular, yeasts may increase from initial numbers of 10^2–10^3 cfu/g in the raw batter to 10^4–10^7 cfu/g in the fermenting sausage, depending on the factory flora, processing conditions and use of starter cultures [29, 92, 118, 126, 127, 130, 135]. In general, naturally fermented sausages at low temperatures (e. g., < 25 °C) normally yield higher numbers of yeasts compared to starter-mediated products [87, 117, 118, 126, 127]. This is due to reduced fermentation rates that increase the potential for survival and growth of yeasts in naturally fermented sausages [127, 130]. Otherwise, yeasts may increase slightly, but eventually they usually decline to levels similar or lower than initial contamination, thus, becoming insignificant as part of the beneficial flora (see section 9.3) or as potential spoilage agents (see section 9.4) with prolonged ripening of fermented sausage [87, 98, 117].

Fermented sausage manufacture has been shown to select for only few yeasts from the diverse pre-fermentation flora (e. g., acid- and nitrate-tolerant yeasts that can also tolerate the low a_w, such as *Debaryomyces* and *Candida*) [14]. Indeed, among 33 yeast species isolated from German dry fermented sausages and other cured meats, *D. hansenii* is the most prevalent, while several *Candida* spp., including *C. lipolytica, C. parapsilosis, C. rugosa* and *C. famata,* are also numerous [86]. Other less frequently isolated genera are *Bullera, Cryptococcus, Pichia, Rhodotorula, Saccharomyces, Sporobolomyces* and *Trichosporon* [86].

The yeast flora of Italian dry sausages includes mostly *Debaryomyces* spp. (mainly *D. hansenii*), while *Candida, Geotrichum, Rhodotorula, Torulaspora,* and *Trichosporon* spp. are the other yeasts present [28]. Likewise, Grazia et al. [61] identified 84 % of yeasts from Italian salami as *Debaryomyces* (e. g., 82 % *D. hansenii*), while 8 % were *Candida*, 5 % were *Kloeckera apiculata* (= *Hanseniaspora uvarum*), 2 % were *Metschnikowia pulcherrima* and 1 % others. Buzzini and Haznedari [24] also observed a high incidence of *Debaryomyces* (62.1 %), particularly *D. hansenii* (50 %), in Italian fermented sausages, while *Candida* (18.1 %; mainly *C. zeylanoides*, 11.2 %), *Rhodotorula* (12.9 %), *Zygosaccharomyces rouxii* (4.3 %) and *Trichosporon* (2.6 %) comprised the remaining isolates. More recently, the yeast flora of Naples-type salami was found to match previous studies from Italy as far as an overall predominance of *D. hansenii* (39.3 %) was concerned [29]. However, the distribution of other yeasts was quite different in this Southern Italian salami, with a high incidence of *Trichosporon* spp. (31.6 %; *Tr. terrestre* (= *Arxula terrestris*), 17.7 % and *Tr. pullulans*, 13.9 %) and *Cr. albidus* (21.5 %) and a low incidence (7.6 %) of *C. incommunis* [29].

The yeast flora isolated (100 strains) during the natural fermentation and ripening of Greek dry salami consists mainly of both anamorphic and teleomorphic states of ascomycetous yeasts [99]. Similarly to German and Italian dry sausages, *D. hansenii* (48 %) predominates, while *D. maramus* (16 %) and *D. polymorphus* (2 %) are also isolated. *Candida spp.* are the second highest category of yeasts occurring in Greek sausage, and is represented by a broader spectrum of species, namely *C. famata* (= *Debaryomyces hansenii*) (7 %), *C. zeylanoides* (6 %), *C. guilliermondii* (= *Pichia guilliermondii*) (6 %), *C. parapsilosis* (5 %) and *Candida* (formerly *Torulopsis*) *kruisii* (2 %). Basidiomycetous yeasts are represented by *Cr. humicolus* (formerly *C. humicola*) (2 %), *Cr. albidus* var. *albidus* (3 %), *Cr. skinneri* (1 %) and *Tr. pullulans* (2 %) [99].

Candida iberica (reclassified as *C. zeylanoides*) [12] has been isolated from Spanish fermented sausages [121], but it is not predominant [52]. Instead, dominance of *D. hansenii* is reported at all stages of sausage processing; *Tr. ovoides* (formerly *Tr. beigelii*), *Y. lipolytica* (teleomorph of *C. lipolytica*), *C. intermedia/curvata*, *C. parapsilosis* and *Citeromyces matritensis* (teleomorph of *C. globosa*) are other yeast species identified [52].

Yeasts are commonly found as part of the natural flora of dry-cured hams and other intermediate moisture meats [67, 101, 110, 125, 147, 150, 158]. For example, yeasts are found at $> 10^6$ cfu/g in 26 % of samples of commercial South African non-fermented dried sausage with an average a_w-value of 0.692 [67]. Populations of 10^4–10^7 log cfu/g of yeasts are present on the surface of dry-cured lacón, a traditional Spanish meat product [150]. Yeast contamination levels of Portuguese country-cured hams and bacon ranged from 10^3 to 10^9 cfu/g, and the predominant species are *D. hansenii, Cr. laurentii, Cr. humicolus, D. polymorphus,* and *P. guilliermondii* [125]. In contrast, in Spanish dry cured hams *D. hansenii* accounts for only 5 % of the isolated yeasts while *Pichia* spp. (*P. ciferii, P. holstii* and *P. sydowiorum*) are predominant (67 %). *Rh. glutinis* (19 %) and *Cr. albidus* (9 %) are also detected [101]. Yeasts associated with "biltong", a traditional African dried meat, are *C. zeylanoides, D. hansenii* and *Tr. beigelii* [147]. Recently, Wolter et al. [158] have isolated *D. hansenii* as the predominant yeast, together with *Cr. laurentii, Cr. hungaricus, T. delbrueckii, Rh. mucilaginosa, Sporobdomyces roseus, D. vanrijiae, Tr. beigelii* (= *Tr. cutaneum*), *Y. lipolytica, S. cerevisiae* and *C. zeylanoides* from "biltong", "cabanossi" and other dried meats [158]. Salt-tolerating yeasts tend to multiply during processing of basturma, but their growth is hampered by garlic [77].

9.3 Beneficial aspects of yeasts in meat products

Compared to other foods such as bread and bakery products (Chapter 11), beer (Chapter 13) and wine (Chapter 14), the beneficial effects of meat yeasts are less prominent but still present in products such as dry fermented sausages and raw hams [6, 90, 92]. In fact, among the microbial groups involved in the fermentation and ripening of sausages, yeasts seem to play a secondary role behind that of LAB, non-pathogenic staphylococci/micrococci and fungi. These desirable groups of microorganisms activate distinct metabolic activities, which contribute to product quality and safety [6, 90, 92, 122, 130]. Yeasts, in particular, require oxygen for growth and, thus, proliferate close to the edge or on the surface of products [63, 64, 91, 150]. They may breakdown lipids and potentially proteins, but may also exert antioxidative effects by destroying peroxides and depleting oxygen from the product surface [91, 92, 123]. Consequently, yeasts are considered to positively affect sausage colour and flavour due to their oxygen-scavenging and lipolytic activities. They may also delay rancidity and further catabolize products of fermentation, such as lactate produced by meat lactobacilli, to other by-products, thereby increasing the pH and contributing to the development of less tangy and more aromatic sausages [6, 29, 64, 91, 126, 127]. It needs to be stressed though that similar activities are strongly exerted by surface-growing fungi in mould-ripened sausages, such as most traditional types of Italian, Hungarian and French salamis [6, 28, 92, 122]. This active mycoflora, which mainly consists of naturally occurring or inoculated *Penicillium* spp. [4, 91, 117, 122, 135], may mask yeast activity and its beneficial effects. Therefore, yeast activity may be more accurately specified in fermented sausages without mould coverings, such as most types of German, Spanish, Greek and other Balkan salamis, where the use of selected yeasts as starter cultures may have more prominent effects on product quality.

ROSSMANITH et al. [124] were the first to use *D. hansenii* as a yeast starter culture after this species was described as the predominant yeast in German fermented sausages [86]. Positive effects on the development of a characteristic yeast flavour and stabilization of red colour in those sausages has been observed [124]. Also, a study by LANGNER [85] has highlighted the correlation between aromatization of yeast-ripened sausages and the presence of carbonyls, especially long-chain aldehydes. MITEVA et al. [100] demonstrated that Bulgarian dry sausages inoculated with a lipolytic *Candida* strain undergoes greater lipolysis compared to non-inoculated controls, as indicated by sensory aroma and taste profiles. Notably, a *D. hansenii* starter strain in German sausages is inhibitory towards *Staphylococcus aureus,* an effect attributed to the oxygen depletion by the yeast further enhanced by lactic and micrococcal starter cultures [97]. Moreover, *D. hansenii* causes an increased ammonia concentration and pH, and decreases lactate and acetate content of the sausages [58]. Accordingly, *D. hansenii,* and its anamorph *C. famata,* are promoted as commercial starter cultures in Northern Europe [64]. Based on the hypothesis that 10^6 yeasts/cm^2 is equivalent to the biomass of 10^8 bacteria/cm^2 [45] and that yeasts are mainly distributed close to the sausage surface, a 6-log inoculation level with *D. hansenii* may be adequate to enhance sausage aromatization. However, care is needed to maintain the antimicrobial effect of organic acids and to avoid nitrate accumulation in fermented sausages by the addition of yeast culture [45, 97].

The use of *D. hansenii* and other yeast species as starter cultures may also be introduced in other than German-type fermented sausages. This is particularly true for the Balkan and Iberian countries where mould-ripened salamis are not traditionally produced, and yeasts, mainly *D. hansenii,* may be found in levels higher than 10^5 cfu/g in lightly smoked, naturally fermented or dry-cured products [52, 99, 125, 126, 127]. In fact, this natural selection of salt-tolerant, non-nitrate reducing, lipolytic or proteolytic *D. hansenii* strains in these salamis indicates that this species merits attention as far as its effect on salami aromatization [52, 99, 134]. Furthermore, direct addition of lipases produced by *D. hansenii* [142, 143] or other lipolytic species such as *C. cylindracea* [159] in sausage mixtures may be another approach to enhance lipolysis and, thus, aroma formation.

Recent studies [114, 142, 143], however, have questioned whether *D. hansenii* and other lipolytic yeasts, such as *C. utilis* (= *P. jadinii*), may indeed promote lipolysis *in situ* in fermented sausage when used as starter cultures. For example, when the ability of lipases from starter strains of *D. hansenii* and *Staphylococcus* to hydrolyze pork fat is tested, results show that lipolysis is strongly inhibited at sausage fermentation (10–30 °C, pH 4.7–6.0, NaCl 2.5–7.0 %) compared to optimal (37 °C, pH 6.5–7.0, no NaCl) conditions [142, 143]. More recently, it has been demonstrated that *D. hansenii* and *C. utilis* die out before the end of ripening following inoculation in fermented sausages, while sensory analysis show a slight difference between control and inoculated sausages [114]. Moreover, the garlic powder added to the sausage mix inhibits the growth of yeast cultures [114]. These findings are supportive of an increasing scientific opinion based on recent research indicating that aroma formation due to lipolysis and proteolysis in fermented meats may be primarily due to endogenous enzymes [115]. Additional studies, with yeast starter cultures are needed in this field.

9.4 Detrimental aspects of yeast in meat products

Detrimental effects of yeasts in meat products can be categorised as spoilage or pathogenic. To our knowledge, meat yeasts have not been reported to have a direct negative impact on human health (e. g., formation of toxic or allergic catabolic compounds), although pathogenic species such as *C. tropicalis* and *C. parapsilosis* have been isolated from meat [144]. Yeasts, however, may compromise safety of cured meats by depleting nitrite [156] and organic acids and increasing pH to enhance survival and growth of pathogenic bacteria, or leading to accumulation of nitrates [2, 45, 53, 64, 97]. Conversely, the negative impact of yeasts through meat spoilage is established [47, 53].

Yeast spoilage of fresh meats is limited and associated mostly with uncommonly applied conditions in meat handling or storage that are inhibitory to bacterial growth. For example, despite the high populations of yeasts, mainly *Cr. laurentii* var *laurentii*, on lamb stored at –5 °C [89], spoilage flavours were not detected [157]. Johannsen et al. [76] have reported enhanced growth of yeasts on irradiated meat as due to bacterial inhibition by irradiation, and considered yeasts as possible spoilage flora. However, no sensory evaluation data have been presented to support this possibility. Yeast numbers of 10^5–10^6cfu/g and yeast flora shifts resulting in prevalence of certain species, such as *C. zeylanoides* or *Y. lipolytica*, in spoiled beef [69] and poultry [70, 152] have been suggested to play a role in spoilage. Bacterial growth in fresh meats, however, is normally far more pronounced to mask any sensory defects caused by yeasts. Bacterial spoilage in chilled aerobically stored meat is initiated at the expense of the limiting amounts of endogenous glucose, which is rapidly converted to gluconate offering a competitive advantage to psychrotrophic, gram-negative bacteria [60, 84, 113]. After the depletion of glucose, this predominant gram-negative flora, mainly consisting of *Pseudomonas* spp., attacks amino acids resulting in the formation of malodorus sulphides, esters and amines that eventually cause putrefactive spoilage [60, 113, 141]. DILLON and BOARD [45] postulated that a similar situation may be the case for yeasts in meat, supporting this by a study [27] on the spoilage pattern of frozen peas by *Rh. glutinis* (e. g., at expense of carbohydrates the yeast catabolized leucine to form 2-methyl-furan to cause an offensive off-odour). However, in practice, it is not obvious how, under normal circumstances, the competitive bacterial flora of fresh meats allows faster uptake of glucose by yeasts. NWACHUKWU and AKPATA [111] have reported spoilage of snail meat by *C. famata* (= *D. hansenii*). The growth of the yeast is associated with a decrease of carbohydrate and protein content from 16 to 7 % and from 2.5 to 0.4 %, respectively, after incubation at 28–30 °C for 4 days. The meat pH decreases from 9.5 to 7.4 and slime and off-odours develop at spoilage [111]. However, conditions of high carbohydrate content and alkaline pH characterizing snail meat are absent in meat from animals or birds. Nevertheless, CHABELA et al. [26] observed that yeasts play an important role in spoilage of retail meats in Mexico City by altering flavour characteristics.

Yeast spoilage phenomena have been more prominent, but are still occurring rarely in processed meats [47, 53]. In these products, yeast spoilage is normally delayed but it may be unmasked because putrefactive *Pseudomonas* spp. are absent or inhibited while the predominant LAB do not cause offensive primary spoilage [131, 132]. Reviews [45, 47] have discussed data from pioneer studies (1940 or earlier) describing surface slime caused by yeasts on spoiled sausages. This situation, however, may be indicative of aerobic storage or insufficient vacuuming, conditions that presently are unlikely to occur or are associated with faulty packaging of meat. Indeed, all later studies that have reported a major involvement of yeasts in spoilage of cured meats have been dealing with specifically treated [49] or aerobically stored products [50, 51, 128, 131, 145]. DOWDELL and BOARD [51] associated spoilage of fresh British sausage with yeasts, after skins of stale sausages were covered in a thick yellow-green film of yeasts. As mentioned, yeasts gain a competitive advantage in British sausage due to their natural resistance to sulphite used as a preservative to restrict bacterial (e. g., *Pseudomonas*) growth [9, 45]. Yeast resistance to sulphite may be due to the formation of acetaldehyde by some yeasts, which binds the sulphite [43, 44]. Thus, yeasts outcompete bacteria by assimilating carbohydrates readily available in British sausage due to addition of biscuits and bread-crubs, while acetaldehyde and secondary lipolytic activities of yeasts appear to contribute to typical off-odour formation at spoilage [31, 45]. Furthermore, because only free sulphite acts as an antimicrobial [80], the potential of yeasts to enhance survival and growth of bacterial pathogens in British sausage deserves investigation.

A similar type of yeast spoilage occurs in fresh Greek country-style sausage stored at 3 and 12 °C in air [128]. This product is made from cured coarse-cut pork meat and lard with added spices and sugars and stuffed in natural casings; it is not typically fermented or ripened but subjected to a 24-h drying and cold-smoking treatment [128]. Thus, sausages undergo a "hidden" fermentation resulting in decreased pH and a_w to eventually favour growth of psychrotrophic yeasts (at 3 °C, dryer surface due to cold air circulation) accompanied by growth of surface moulds (at 12 °C, moister product surface) and an increase in pH [128].

The greater the increase of yeasts, the greater the raise of pH and the faster the development of "malty" off-odours, especially at 12 °C [128]. Also, when "taverna" sausage, a traditional cooked Greek cured meat product, is stored in air permeable packages at 10 °C, more than 10^6 cfu/g of yeasts develop after 18 to 30 days [131]. The sausages develop unpleasant vinegar like smell and a sticky surface slime due to surface-growing yeasts [131]. Similar phenomena are also observed in various types of Greek cooked cured meats stored at 4 °C, but no growth of yeasts and associated spoilage are observed when such products are vacuum packaged and stored at 4, 10 or 12 °C [131, 132].

9.5 Physiological characteristics of yeasts in meat

Survival and growth of yeasts in meat depends on their physiological capabilities in terms of: (i) coping successfully with environmental or processing stresses; and, (ii) utilizing efficiently carbon and nitrogen sources available in the substrate. Stresses that yeasts may encounter during processing and storage include, cold, heat, acid, salt, low a_w or dry surfaces, starvation, low redox potential, carbon dioxide and other gases, chemical preservatives, natural antimicrobials, sanitizers and physical processes such as irradiation or high pressure. Nutrients that may be utilized include carbohydrates, lipids and proteins as primary substrates, and secondary by-products of bacterial catabolism, such as organic acids, glycerides, peptides, amino acids, amines, etc. Differences in inherent properties among meat yeasts affect their stress responses and efficacy of nutrient utilization; therefore, such differences are important in indicating which genera or species eventually predominate in meat products.

Based on yeast biodiversity in meat discussed in section 9.2, it becomes evident that there is a shift from the basidiomycetous type occurring in the field and in refrigerated or frozen meats to the ascomycetous type occurring in spoiled, minced, cured, fermented or otherwise processed meat products. In addition, as indicated, certain yeast species specifically predominate in distinct types of meat products, such as *Cr. laurentii* var. *laurentii* in fresh meats stored at temperatures below 0 °C, *Y. lipolytica* on poultry, and *D. hansenii* in salted, cured, dried or fermented meats. Thus, these dominant species possess a spoilage potential probably linked to different inherent physiological and biochemical properties, such as growth temperature, proteolytic and lipolytic activities, hydrophobicity, a_w and preservative tolerance, and production or assimilation of organic acids [53, 62].

The high occurrence of teleomorphic basidiomycetous yeasts in pastures, on fleece/carcasses [45] and on fresh refrigerated meats [31, 76, 89] may be due to their non-fastidious nature and lower minimum growth temperature, compared to the ascomycetous group. Storage of carcass meats at low refrigeration temperatures reduces yeast growth rates and slows down sexual replication and, thus, it selects for yeasts with an asexual life cycle [7]. In addition, the intracellular lipid accumulation offers protection against membrane damage of such yeasts due to cold [7]. Conversely, the poor competitiveness of basidiomycetous yeasts in fermented sausages at temperatures above 15 °C compares favourably with their inability to ferment sugars and to grow in the presence of 10 % salt in broth [99]. As indicated, conditions of low pH, high salt content and low a_w prevailing in cured dried and fermented meat products are favourable to ascomycetous yeasts, mainly *Debaryomyces* and particularly *D. hansenii*. Consistent with this ecological trend, several studies have shown that the lag phase of broth cultures of *D. hansenii* is shorter and their growth potential is greater at a_w values < 0.90 compared to other meat yeasts [14, 68], including *Y. lipolytica* and *P. membranifaciens* [62]. Indeed, *C. zeylanoides, Rh. mucilaginosa* and *D. hansenii* isolated from cured meats show an increased minimum inhibitory concentration (MIC) of NaCl, in the above order, from 1571 to 1873 mM (e. g., permitting growth at a_w < 0.94), while yeasts isolated from beer, soft drinks and mayonnaise-based salads have lower MICs [68]. These data underscore the high osmotic tolerance of *D. hansenii*, which is due to induction of intracellular protective mechanisms to salt stress [3, 106, 120]. *Debaryomyces hansenii* also has a short lag phase at temperatures 3 to 10 °C, which further enhances its increased spoilage potential in refrigerated cured meats [62]. Overall, the lag phase has been reported as the most important factor affecting the spoilage potential of low pH and high salt foods by yeasts, which is affected mainly by temperature on which major synergistic effects of NaCl and pH are evident [16, 120].

GUERZONI et al. [62] have studied yeast hydrophobicity, which is defined as the ability of cells to migrate from a polar to a non-polar phase in a two-phase system, namely a yeast suspension and a heptane aqueous layer. Based on the ratio of absorbances between the two phases associated with levels of adherent cells, it is shown that 94 % of *Y. lipolytica* strains are hydrophobic in contrast to *D. hansenii* and *P. membranifaciens* which have low frequency of hydrophobicity [62]. The ability of *Y. lipolytica* to migrate in a non-polar phase suggests that it may prefer the lipid phase of foods [62]. In fact, this property may explain the predominance and high spoilage potential of *Y. lipolytica* on the skin of fresh poultry [70].

As indicated, the lipolytic and proteolytic activities of yeasts are important in spoilage and during ripening of dry fermented sausages. DALTON et al. [31] have reported that 30 % of yeasts isolated from fresh or spoiled British sausage are lipolytic. *Y. lipolytica* possess stronger lipase (94 % of strains at 5 °C) and protease (100 % at 25 °C, 86 % at 5 °C) activities than *D. hansenii* (80 % of stains lipolytic at 5 °C, no strain proteolytic) or *P. membranifaciens* (20 % lipolytic at 5 °C, none proteolytic) [62]. More recently, however, a pronounced proteolytic activity against pork muscle sarcoplasmic protein extracts is attributed to *D. hansenii* isolates from Spanish dry fermented sausages. The latter also demonstrated *in vitro* proteinase and aminopeptidase activities [134]. On the other hand, the lipase activity of *D. hansenii* on pork fat is established, although, as indicated, it is more pronounced at neutral pH and 37 °C rather than at pH below 6.0 and temperatures above 25 °C (conditions prevailing in meat fermentations) [142, 143]. Also, *D. hansenii* as well as *Y. lipolytica* possess esterase activities [15, 62] that, in fact, may be greater at 5 °C if compared to 20 °C [62].

Most meat yeasts have restricted fermentative capabilities, but they may produce organic acids and alcohols. For example, *D. hansenii, Y. lipolytica* and *P. membranifaciens* produce mainly citrate, but also oxalate and succinate, at different concentrations dependent on the strain [62]. Citrate, in particular, may be important due to its acidifying capacity and as a precursor of oxidation by-products affecting flavour. Indeed, yeasts are known for their ability to oxidize sugars, alcohols and organic acids to produce aldehydes, ketones and acids [53]. Lactate, in particular, accumulated in cured/fermented meats due to sugar breakdown by LAB, may be oxidized by spoilage yeasts, which may proliferate at the expense of this acid [45, 53]. We have recently modeled [2] this yeast activity in fresh Greek country-style sausage stored aerobically at 3 and 12 °C as a function of pH, water content (W), concentration of sodium chloride [NaCl] and the amount of organic acids (A*c*) from sugar breakdown accumulated in the product to be used as carbon source by yeasts:

$$r_E = r_{Es} \times [NaCl]/pH \times W - r_{Eg} \times Ac/Aci,$$

where r_E is the yeast growth rate, r_{Es} is the yeast population sensitivity due to the destructive effect of the meat ecosystem, r_{Eg} is the initial specific growth of yeasts and A*ci* is the initial organic acid concentration in the sausage mix. Following certain mathematical transformations [2], the above equation was successfully used to predict yeast growth in the sausage, which resulted in increase of product pH and acceleration of spoilage. Lactate oxidation may also compromise safety of cured meat products, an important issue that needs to be evaluated and, if possible, predicted with the use of models [2].

Yeasts are also resistant to more than 10 % (w/v) of sodium lactate [68], indicating that this flavouring and preservative agent may be inactive against yeasts at the concentrations added to meat products. Yeast resistance to weak organic acids and their salts (e. g., acetates, sorbates and benzoates) has also attracted attention in terms of their ability to overcome inhibition and cause spoilage [17, 23, 93, 105, 140]. Most work on yeast resistance to these weak-acid preservatives has been done with *S. cerevisiae* and related species [23, 93] because sorbates and benzoates are certainly more important to the dairy, beverage or salad industries [21, 53]. Dipping in potassium sorbate (up to 5 %) solutions, however, may be used to inhibit surface growth of yeasts and moulds during sausage fermentation and ripening [126, 127, 140]. Among yeasts commonly found in meat, *Y. lipolytica* is reported to be the most resistant to benzoic and sorbic acids as compared to *D. hansenii* and *P. membranifaciens* [62]. Conversely, the latter species has an increased resistance to acetic acid up to a concentration of 1.2 % (w/w) [62]. Overall, the resistance of spoilage yeasts to weak organic acids depends on the H+-pumping P-type membrane ATPase [66]. However, expelling protons (e. g., H+ pumping) as a response to maintain yeast cell homeostasis interrupted by acid dissociation inside the cell is an energetically expensive protective mechanism [66]. As a result, at increased levels of weak organic acids, yeast cells become exhausted due to reduction of their energy pools for growth, and this may irreversibly affect essential metabolic functions [23, 66].

Natural resistance of yeasts to organic acids, mainly lactate and acetate, may have significant practical implications in meat plants. The increasing use of acid decontamination technologies to reduce surface microbial contamination and pathogenic bacteria on animal and poultry carcasses and fresh meat cuts in the United States [141] may selectively favor growth of acid-tolerant yeasts. Recent research in our laboratory has shown that meat decontamination waste fluids that contained different proportions of 2 % lactic or 2 % acetic acid solutions used to spray fresh beef top round cuts are selective for yeasts, while they inhibit the normal *Pseudomonas*-like meat spoilage flora [133]. More specifically, lactate-containing (0.02 to 0.1 %) washings are highly selective for carotenoid *Rhodotorula*-like yeasts, while acetate-containing washings are mainly selective for *Candida*- and *Debaryomyces*-like yeasts [133]. Previous studies have reported a similar selection of yeasts during refrigerated storage of fresh pork previously treated with lactate [149], beef treated with citrate or citrate plus lactate [148] or veal tongues treated with lactic acid [153]. Primary effects of various decontamination interventions have recently been evaluated against *Y. lipolytica* due to its high spoilage potential in fresh poultry [71]. However, long-term effects of use of organic acids as decontamination agents on the attributes of the meat spoilage or pathogenic flora on products or in the meat plant environment have been largely overlooked [133].

The ability of certain yeasts, such as *D. hansenii, C. zeylanoides, C. saitoana* and *P. membranifaciens*, to resist sulphite (450 µg SO_2/g) added as a preservative in British fresh sausage is associated with their ability to form acetaldehyde that binds available sulphite [9, 47]. In contrast, the low incidence of *Cr. albidus* and *Rh. mucilaginosa* can be due to their inability to form acetaldehyde [47]. Acetaldehyde production by *C. norvegica* is sulphite-induced and occurs during the exponential phase in sulphited (500 µg SO_2/ml) lab lemco glucose broth buffered at pH 5, 6 or 7. Growth at pH 4, however, is inhibited by sulphite, indicating that sulphite tolerance of yeasts is pH-dependent [44]. Moreover, acetaldehyde production occurs in glucose-, fructose- or ethanol-containing broth, but not in the presence of lactate or other assimilable compounds, indicating a substrate-dependent sulphite resistance of *C. norvegica* [44]. On the other hand, the non-acetaldehyde-forming *C. vini* (≡ *Pichia fluxuum*) grows with 500 µg SO_2/ml in broth at pH 6 and 7, but not at pH 4 and 5, suggesting that binding may not be the only mechanism for sulphite resistance in yeasts [44].

Another concern associated with yeasts is the enhancement of their spoilage potential following treatment of meat with emerging preservation methods such as irradiation and high hydrostatic pressure (HP), which inactivate bacteria. Indeed, yeasts are significant in the spoilage of irradiated (2-5 kGy) frankfurters [49] and, unlike bacteria, they are also unaffected by irradiation (2.5 kGy) of minced beef [76]. Similarly, while yeasts are not detected in unirradiated fresh poultry meat, *Y. lipolytica, C. zeylanoides* and *Tr. beigelii* are isolated from the corresponding irradiated samples, with the former species being present after irradiation in high numbers due to its apparent higher resistance [138]. *Candida zeylanoides* and *Tr. beigelii* (= *Tr. cutaneum*) are also found to be the most resistant yeasts in irradiated (3 kGy) British fresh sausage, whereas *D. hansenii* is reduced by 1.5 kGy [94]. The greater sensitivity of *D. hansenii* to irradiation, as compared to *Candida* spp., is also observed in frankfurters [49]. Notably, *C. zeylanoides* is the yeast that could better sustain the combined effects of irradiation (3 kGy) and sulphite in British fresh sausages stored at 4 °C for 14 days [94]. In a later study, McCarthy and Damoglou [95] reported that *D* values (e. g., irradiation dose required to reduce yeast population by one log cycle) of the above yeasts at higher (> 2 kGy) irradiation doses increased in sausage as compared to phosphate buffered saline. This indicates a protective effect of meat proteins and polysaccharides to irradiated yeast cells [95]. Recently, an irradiation dose of 5.7 kGy was reported to inhibit yeasts in addition to the complete inhibition of *Listeria monocytogenes* in a pre-prepared meat meal [54]. Unlike irradiation, the resistance of yeast species commonly associated with meat to high pressure (HP) has yet to be addressed. For example, HP of 300 MPa at 25 °C or 250 MPa at 45 °C, inactivate *S. cerevisiae* and *Z. bailii* in meat spaghetti sauce, with inactivation being enhanced by mild heat treatment and increased acidity [116].

9.6 Specific methods for analysis of yeasts in meats

The detailed consideration of methods for yeast detection, isolation, enumeration, identification, and their current systematic classification is beyond the scope of this chapter. The reader may be referred to excellent reviews by DEAK and BEUCHAT [32, 40] dealing with simplified keys for identification of foodborne yeasts, previous [12, 81] and current [13] yeast taxonomic studies including description of testing methods [146], and Chapters 1–5 of this book. Below is a brief overview of methods specifically used in studies dealing with meat yeasts which, overall, do not deviate significantly from the respective general methods [146].

Classical analysis of meat products for yeasts involves suspension of a pre-weighed sample (e. g., 25 g) in sterile diluents (e. g., 0.1 % peptone water) followed by plating of serial dilutions on agar media selective for yeasts. Initially, all-purpose media acidified to an approximate pH of 3.5 were used to selectively isolate and enumerate yeasts. Examples are malt extract agar (adjusted to pH 3.5 with 10 % tartaric acid) [72], plate count agar (pH 3.5 with 10 % citric acid) [50] or potato dextrose agar (pH 3.7 with 10 % tartaric acid) [67]. Subsequent studies showed that antibiotic-containing media at neutral pH, such as oxytetracycline glucose yeast extract [102] or Rose-Bengal-chloretetracycline [72] agars, were superior to acidified media in recovering more meat yeasts, and at higher numbers due to the absence of acid stress [10, 53]. Among the dyes used in yeast-selective media to restrict growth of moulds [47], Rose Bengal has been the most effective, although it may cause cytotoxic and photodynamic inactivation of microorganisms, including yeasts [8, 10]. Nevertheless, Rose Bengal combined with one or two of the above antibiotics or additional ones, such as chloramphenicol, gentamycin and streptomycin, at 25–100 ppm have been preferred in yeast enumeration studies [10, 53, 79]. Initially, in our studies, we have used potato dextrose agar with 100 mg/l chloramphenicol [126–128]. Later, this has been replaced by Rose Bengal chloramphenicol agar [129–132]. The latter medium is commonly used in recent years [117, 118, 135], although oxytetracycline glucose agar [150] and malt extract agar [29] are still preferred by some workers. Overall, the stress history (acid, desiccation, etc.) of yeasts in foods, including meat, has to be accounted for when choosing a medium, because certain selective media such as dichloran 18 % glycerol agar may be inhibitory to stressed cells [20, 42]. Also, the incubation temperatures for isolation and enumeration of yeasts have been variable (e. g., 5 °C to 37 °C) [10, 45, 104], but 22 to 28 °C for 5–7 days seem to be the most appropriate [10, 104].

Alternative rapid or easy-to-perform methods for yeast detection and enumeration [34, 65] have been developed in recent years, such as the Petrifilm™ (3M Company, Saint Paul, Minnesota, USA) method [19], the indirect conductance method [36, 37], the MicroScan enzyme-based system [39], and the direct epifluorescent filter technique [53, 136]. These methods have shown promise along with certain limitations. Their application, however, has mainly been associated with laboratory media or liquid foods and beverages [36, 37, 53], and rarely with meat products.

DEAK and BEUCHAT [32] have constructed a simplified identification key for foodborne yeasts based on 4 to 7 tests only to routinely characterize an isolate at the genus level and 10–15 tests at the species level. It has to be emphasized that yeast identification is a tedious and time-consuming task requiring skills based on experience for recognising phenotypic criteria. Furthermore, discrepancies from identification keys may occur due to abnormal yeast behaviour or variations within media and methods. For example, we have experienced difficulties in identifying yeast isolates from Greek dry salami, particularly those assigned to the genus *Debaryomyces*, because sporulation of the majority of these isolates was difficult to induce [99]. Also, despite *D. hansenii* is reported to assimilate D-xylose and raffinose [81], most *D. hansenii* isolates from Greek salami [99] or starter culture preparations in Germany [63] were negative. In addition, strain differences in assimilation reactions are observed between testing with the classical plate or the API 20C methods, frequently leading to erroneous identifications [99]. DEAK and BEUCHAT [33, 35] have shown that D-xylose and raffinose, included in the API 20C kit, are not assimilated by certain yeast species listed as positive. They [33, 35] concluded that the API 20C system is suitable for identifying foodborne yeasts, provided that additional tests such as nitrate assimilation, glucose fermentation and urease reaction are performed. See Chapter 5 for more detailed developments on these matters.

As indicated, recent advances in yeast taxonomy have been based on molecular and biochemical methods using sophisticated techniques [38, 41, 82]. Such methods include the determination of fatty acid profiles [22, 119] and several molecular techniques such as DNA fingerprinting by polymerase chain reaction (PCR), pulsed-field gel electrophoresis (PFGE), random amplified polymorphic DNA (RAPD) analysis, and restriction fragment length polymorphism (RFLP) analysis [38, 41]. Detailed descriptions of these techniques are provided in Chapters 3 and 4.

9.7 Quality control

Controlling contamination and growth of yeasts in meat products is easy and feasible provided that preventive measures are applied from the abattoir to the final product. As environmental contaminants, yeasts may be controlled by proper application of preventive measures in the context of hazard analysis critical control point (HACCP), good manufacturing practices (GMP) and good sanitary practices (GSP). For example, animal washing before slaughter, use of clean chlorinated water, cleaning and sanitation of processing equipment and utensils, personnel hygiene, and monitoring of air movement to restrict airborne contamination can contribute to further reductions of the expectedly low initial yeast numbers in meat processing plants. Given such preventive measures are in-place, maintaining a low redox potential through proper packaging, whenever possible, is the most effective measure for achieving yeast control.

ADAMS et al. [1] reported that vacuum packaging of British fresh sausage extended shelf life at 6 °C to more than 20 days compared with 9 to 14 days in conventional packs. This extension was primarily attributed to the slower growth rate and the reduction in numbers of yeasts by approximately 2 logs after 10 days of storage, which resulted in higher amounts of free sulphite in vacuum compared to air packages [1]. Likewise, we have repeatedly shown that several different types of Greek cooked cured meat products contained less than 2.0 log cfu/g of yeasts during storage at refrigerated temperatures for up to 30 days [129, 131, 132]. In contrast, the same products packed in air developed yeast populations ranging from 3.5 to 6.7 log cfu/g at 30 days, depending on the product type [132] and storage temperature [131]. SAMELIS and METAXOPOULOS [128] also recommended packaging of Greek country-style sausage under vacuum to extend shelf life. Overall, packaging in vacuum, CO_2 or nitrogen is of primary importance in the control of yeasts.

Additional methods for yeast control in cured, dried and fermented meats may include smoking of products, immersion or spraying with potassium sorbate solutions, use of starter or protective cultures, and cleaning (e.g., brushing) of the product surface prior to sale. Smoking is effective in retarding growth of surface-growing yeasts, as it may favor core-growing LAB during sausage fermentation [98], or during the cold-smoking and drying process of fresh country-style sausages [128]. In addition, as indicated, starter-mediated sausage fermentations restrict the numbers of yeasts during product ripening due to greater bacterial competition at the first stages of processing [87, 117, 118]. The potential use of protective cultures in cured fresh or cooked meat primarily targeted against bacterial pathogens may also have inhibitory effects on yeasts.

Yeast control on animal carcasses, primal cuts or minces of fresh meat can be achieved provided that storage temperatures are above freezing to allow for competitive bacterial growth. Further control can be provided by storage under vacuum or modified atmosphere packaging, when feasible. The increasing use of decontamination interventions and technologies (e.g., irradiation, HP, ozone) on fresh or processed meat products with the main aim to inactivate bacterial pathogens [54, 78, 141] requires careful consideration because the altered microbial ecology may allow growth of resistant yeasts surviving or introduced after decontamination [133, 149].

9.8 Future prospects and conclusions

Fresh meat acquires a variety of contaminating microorganisms during processing and handling, but only a fraction of these develop and dominate the "spoilage association". Gram negative bacteria, mainly pseudomonads, dominate in aerobically stored, refrigerated meats, while gram-positive bacteria, mainly LAB, dominate in fresh or cured meats packaged under vacuum or modified atmospheres and in fermented or dry cured products. Yeasts cannot compete against bacteria under favorable conditions since their growth rate is lower. However, yeasts gain advantage for growth when the intrinsic (pH, acidity, a_w, salt concentration, preservatives, etc.) or extrinsic (low temperatures, high redox potential, physical hurdle treatments) factors and their interactions in meat change in a way that bacteria cannot grow any further. In most such cases, spoilage yeasts proliferate at the expense of certain bacterial metabolites, like organic acids produced by LAB.

The most important yeasts in meat products belong to the teleomorphic ascomycetous genera *Debaryomyces, Pichia* and *Yarrowia*, the ascomycetous anamorphic genus *Candida*, and the basidiomycetous anamorphic genera *Rhodotorula*, *Cryptococcus*, and *Trichosporon*. Fresh meat processing and storage causes a progressive replacement of basidiomycetous yeasts by ascomycetous yeasts, and an apparent predominance of *Candida* spp. at spoilage, while meat salting, curing and fermentation are selective for *Debaryomyces* (mainly) and *Candida* spp. Overall, some important physiological and biochemical characteristics of meat yeasts and their interactions result in the selection of certain yeast species in specific meat products. The best examples are *Cr. laurentii* var. *laurentii* on carcass meats stored below 0 °C, *Y. lipolytica* in fresh and spoiled poultry, and *D. hansenii* in cured dried and fermented meats.

Spoilage caused by yeasts is mainly due to their lipolytic and proteolytic activities, although their action on carbohydrates and associated by-products of bacterial metabolism may also lead to the formation of compounds reducing the sensory quality of meat products, such as organic acids, alcohols, esters and others. Spoilage is manifested by the development of off-odors, slime formation, discoloration and surface-product colonization. When yeast growth and metabolic activity are monitored and controlled, yeasts may exert significant beneficial effects on the sensory quality of certain meat products, such as fermented sausages and dried cured meats.

Future research on yeasts associated with meat products should be focused on: (i) improving yeast detection and enumeration by rapid methods; (ii) increasing efficient use of molecular and biochemical techniques on yeast identification and classification; (iii) better understanding of yeast responses to meat-related stresses as affected by species, competition of each species with other yeasts or bacteria and environmental conditions; (iv) elucidation of instant and long-term effects of meat decontamination, irradiation, HP and other emerging technologies or preservation methods on yeast survival, growth and spoilage potential; (v) selection and genetic improvement of yeasts as starter cultures for fermented meat products; (vi) optimization of production and utilization of yeast enzymes (e. g., lipases, proteases) to monitor aromatization of fermented meat products; (v) use of yeast autolysate extracts as "meat flavouring" ingredients [160]; and, (vi) investigating potential pathogenicity of yeasts in meat products.

9.9 References

[1] ADAMS, M.R.; BAKER, T.; FORREST, C.L.: A note on shelf-life extension of British fresh sausage by vacuum packaging. J. Appl. Bacteriol. **63** (1987) 227–232.

[2] AGGELIS, G.; SAMELIS, J.; METAXOPOULOS, J.: A novel modelling approach for predicting microbial growth in a raw cured meat product stored at 3 °C and at 12 °C in air. Int. J. Fd Microbiol. **43** (1998) 39–52.

[3] ALMAGRO, A.; PRISTA, C.; CASTRO, S.; QUINTAS, C.; MADEIRA-LOPES, A.; RAMOS, J.; LOUREIRO-DIAS, M.C.: Effects of salts on *Debaryomyces hansenii* and *Saccharomyces cerevisiae* under stress conditions. Int. J. Fd Microbiol. **56** (2000) 191–197.

[4] ANDERSEN, S.S.: Compositional changes in mycoflora during ripening on naturally fermented sausages. J. Fd Protect. **58** (1995) 426–429.

[5] AYRES, J.C.: Temperature relationships and some other characteristics of the microflora developing on refrigerated beef. Fd Res. **25** (1960) 1–18.

[6] BACUS, J.N.: Fermented meat and poultry products. In: Advances in meat research: meat and poultry microbiology, Vol. 2 (edited by Pearson, A.M.; Dutson, T.R.) London, U.K.: Macmillan Publishers (1986) 123–164.

[7] BANKS, J.G.: Yeast spoilage of some chilled foods, meat and dairy products. In: Proceedings of Campden Food Preservation Research Association, 7–9th May 1985, Stratford upon Avon, U.K. (1985).

[8] BANKS, J.G.; BOARD, R.G.; CARTER, J.; DODGE, A.D.: The cytotoxic and photodynamic inactivation of micro-organisms by Rose Bengal. J. Appl. Bacteriol. **58** (1985) 391–400.

[9] BANKS, J.G.; DALTON, H.K.; NYCHAS, G.J.; BOARD, R.G.: Review – Sulphite, an elective agent in the microbiological and chemical changes occurring in uncooked comminuted meat products. J. Appl. Biochem. **7** (1985) 161–170.

[10] BANKS, J.G.; BOARD, R.G.: Some factors influencing the recovery of yeasts and moulds from chilled foods. Int. J. Fd Microbiol. **4** (1987) 197–206.

[11] BARNES, E.M.; IMPEY, C.S.; GEESON, G.D.; BUHAGIAR, R.W.M.: The effect of storage temperature on the shelf-life of eviscerated air-chilled turkeys. Brit. Poultry Sc. **19** (1978) 77–84.

[12] BARNETT, J.A.; PAYNE, R.W.; YARROW, D.: Yeasts: characteristics and identification, 2nd ed. Cambridge, U.K.: Cambridge University Press (1990).

[13] BARNETT, J.A.; PAYNE, R.W.; YARROW, D.: Yeasts: characteristics and identification, 3rd ed. Cambridge, U.K.: Cambridge University Press (2000).

[14] BEM, Z.; LEISTNER, L.: Die wasseractivitatstoleranz der bei Pokelfleischwarren vorkommenden hefen. Fleischwirtschaft **50** (1970) 492–493.

[15] BESANCON, X.; RATOMAHENINA, R.; GALZY, P.: Isolation and partial characterization of an esterase (EC 3.1.1.1) from *Debaryomyces hansenii* strain. Neth. Milk Dairy J. **49** (1995) 97–110.

[16] BETTS, G.D.; LINTON, P.; BETTERIDGE, R.J.: Synergistic effect of sodium chloride, temperature and pH on growth of a cocktail of spoilage yeasts: a research note. Fd Microbiol. **17** (2000) 47–52.

[17] BEUCHAT, L.R.: Effects of potassium sorbate and potassium benzoate on inactivating yeasts heated in broths containing sodium chloride and sucrose. J. Fd Protect. **44** (1981) 765–769.

[18] BEUCHAT, L.R.: Influence of water activity on growth, metabolic activities and survival of yeasts and molds. J. Fd Protect. **45** (1983) 135–141.

[19] BEUCHAT, L.R.; NAIL, B.V.; BRACKETT, R.E.; FOX, T.L.: Comparison of the petrifilm yeast and mold culture film method to conventional methods for enumerating yeasts and molds in foods. J. Fd Protect. **54** (1991) 443–447.

[20] BEUCHAT, L.R.; FRANDBERG, E.; DEAK, T.; ALZAMORA, S.M.; CHEN, J.; GUERRERO, S.; LOPEZ-MALO, A.; OHLSSON, I.; OLSEN, M.; PEINADO, J.M.; SCHNURER, J.; DE SILONIZ, M.I.; TORNAI-LEHOCZKI, J.: Performance of mycological media in enumerating desiccated food spoilage yeasts: a interlaboratory study. Int. J. Fd Microbiol. **70** (2001) 89–96.

[21] BONESTROO, M.H.; DE WIT, J.C.; KUSTERS, B.J.M.; ROMBOUTS, F.M.: Inhibition of the growth of yeasts in fermented salads. Int. J. Fd Microbiol. **17** (1993) 311–320.

[22] BOTHA, A.; KOCK, J.L.F.: Application of fatty acid profiles in the identification of yeasts. Int. J. Fd Microbiol. **19** (1993) 39–51.

[23] BRUL, S.; COOTE, P.: Preservative agents in foods. Mode of action and microbial resistance mechanisms. Int. J. Fd Microbiol. **50** (1999) 1–17.

[24] BUZZINI, P.; HAZNEDARI, S.: Caratterizzazione de lieviti isolati da insaccati fermentati prodotti in Umbria. Valutazione preliminare della loro attivita proteolitica e lipolitica. Ind. Aliment. **34** (1995) 620–625.

[25] CARLOS, A.M.A.; HARRISON, M.A.: Inhibition of selected microorganisms in marinated chicken by pimento leaf oil and clove oleoresin. J. Appl. Poultry Res. **8** (1999) 100–109.

[26] CHABELA, M.L.P.; SERRANO, G.M.R.; CALDERON, P.L.; GUERRERO, I.: Microbial spoilage of meats offered for retail sale in Mexico City. Meat Sc. **51** (1999) 279–282.

[27] COLLINS, M.A.; BUICK, R.K.: Effect of temperature on the spoilage of stored peas by *Rhodotorula glutinis*. Fd Microbiol. **6** (1989) 135–141.

[28] COMI, G.; CANTONI, C.: I lieviti in insaccati crudi stagionati. Ind. Aliment. **19** (1980) 857–860.

[29] COPPOLA, S.; MAURIELLO, G.; APONTE, M.; MOSCHETTI, G.; VILLANI, F.: Microbial succession during ripening of Naples-type salami, a southern Italian fermented sausage. Meat Sc. **56** (2000) 321–329.

[30] DALTON, H.K.: The yeasts and their chemical changes in the British fresh sausage. Ph.D. Thesis, University of Bath, Bath, U.K. (1984).

[31] DALTON, H.K.; BOARD, R.G.; DAVENPORT, R.R.: The yeasts of British fresh sausage and minced beef. Antonie van Leeuwenhœk **50** (1984) 227–248.

[32] DEAK, T.; BEUCHAT, L.R.: Identification of foodborne yeasts. J. Fd Protect. **50** (1987) 243–264.

[33] DEAK, T.; BEUCHAT, L.R.: Evaluation of simplified and commercial systems for identification of foodborne yeasts. Int. J. Fd Microbiol. **7** (1988) 135–145.

[34] DEAK, T.: Simplified techniques for identifying foodborne yeasts. Int. J. Fd Microbiol. **19** (1993) 15–26.

[35] DEAK, T.; BEUCHAT, L.R.: Comparison of the SIM, API 20C, and ID 32C systems for identification of yeasts isolated from fruit juice concentrates and beverages. J. Fd Protect. **56** (1993) 585–592.

[36] DEAK, T.; BEUCHAT, L.R.: Comparison of conductimetric and traditional plating techniques for detecting yeasts in fruit juices. J. Appl. Bacteriol. **75** (1993) 546–550.

[37] DEAK, T.; BEUCHAT, L.R.: Evaluation of the indirect conductance method for the detection of yeasts in laboratory media and apple juice. Fd Microbiol. **10** (1993) 255–262.

[38] DEAK, T.: Methods for the rapid detection and identification of yeasts in foods. Trends Fd Sc. & Technol. **6** (1995) 287–292.

[39] DEAK, T.; BEUCHAT, L.R.: Evaluation of the MicroScan enzyme-based system for the identification of foodborne yeasts. J. Appl. Bacteriol. **79** (1995) 439–446.

[40] DEAK, T.; BEUCHAT, L.R.: Handbook of food spoilage yeasts. Boca Raton, FL, U.S.A.: CRC Press (1996).

[41] DEAK, T.; CHEN, J.; BEUCHAT, L.R.: Molecular characterization of *Yarrowia lipolytica* and *Candida zeylanoides* isolated from poultry. Appl. Environ. Microbiol. **66** (2000) 4340–4344.

[42] DEAK, T.; CHEN, J.; GOLDEN, D.A.; TAPIA, M.S.; TORNAI-LEHOCZKI, J.; VILJOEN, B.C.; WYDER, M.T.; BEUCHAT, L.R.: Comparison of dichloran 18 % glycerol (DG18) agar with general purpose mycological media for enumerating food spoilage yeasts. Int. J. Fd Microbiol. **67** (2001) 49–53.

[43] DILLON, V.M.: Sulphite tolerance of yeasts from comminuted lamb products. Ph.D. Thesis University of Bath, Bath, U.K. (1988).

[44] DILLON, V.M.; BOARD, R.G.: A study of sulphite-tolerant yeasts from comminuted lamb products. Biotechnol. Appl. Biochem. **12** (1990) 99–115.

[45] DILLON, V.M.; BOARD, R.G.: Yeasts associated with red meats. J. Appl. Bacteriol. **71** (1991) 93–108.

[46] DILLON, V.M.; DAVENPORT, R.R.; BOARD, R.G.: Yeasts associated with lamb. Mycol. Res. **95** (1991) 57–63.

[47] DILLON, V.M.: Yeasts and moulds associated with meat and meat products. In: The Microbiology of Meat and Poultry (edited by Davies, A.; Board, R.). London, U.K.: Blackie Academic and Professional (1998) 85–117.

[48] DIRIYE, F.U.; SCORZETTI, G.; MARTINI, A.: Methods for the separation of yeast cells from surfaces of processed, frozen foods. Int. J. Fd Microbiol. **19** (1993) 27–37.

[49] DRAKE, S.D.; EVANS, J.B.; NIVEN, C.F.: The identity of yeasts in the surface flora of packaged frankfurters. Fd Res. **24** (1959) 243–246.

[50] DOWDELL, M.J.; BOARD, R.G.: A microbiological survey of fresh British sausage. J. Appl. Bacteriol. **31** (1968) 378–396.

[51] DOWDELL, M.J.; BOARD, R.G.: The microbial associations in British fresh sausages. J. Appl. Bacteriol. **34** (1971) 317–337.

[52] ENCINAS, J-P.; LOPEZ-DIAZ, T-M.; GARCIA-LOPEZ, M-L.; OTERO, A.; MORENO, B.: Yeast populations on Spanish fermented sausages. Meat Sc. **54** (2000) 203–208.

[53] FLEET, G.H.: Food spoilage yeasts. In: Yeast Technology. (edited by Spencer, J.F.T.; Spencer, D.M.). Berlin, Heidelberg, Germany: Spinger-Verlag (1990), 124–166.

[54] FOLEY, D.M.; REHER, E.; CAPORASO, F.; TRIMBOLI, S.; MUSHERRAF, Z.; PRAKASH, A.: Elimination of *Listeria monocytogenes* and changes in physical and sensory qualities of a prepared meal following gamma irradiation. Fd Microbiol. **18** (2001) 193–204.

[55] FUNG, D.Y.C.; LIANG, C.: Critical review of isolation, detection, and identification of yeasts from meat products. Crit. Rev. Fd Sc. Nutrit. **29** (1990) 341–379.

[56] GALLO, L.; SCHMITT, R.E.; SCHMIDT-LORENZ, W.: Microbial spoilage of refrigerated fresh broilers. I. Bacterial flora and growth during storage. Lebensmittel Wissensch. Technol. **21** (1988) 216–233.

[57] GARDNER, G.A.: Microbiological and chemical changes in lean Wiltshire bacon during aerobic storage. J. Appl. Bacteriol. **34** (1971) 645–654.

[58] GEHLEN, K.H.; MEISEL, C.; FISCHER, A.; HAMMES, W.P.: Influence of the yeast *Debaryomyces hansenii* on dry sausage fermentation. In: Proc. 37th Int. Congr. Meat Sci. Technol., **2** (1991) 871–876.

[59] GEORNARAS, I.; DYKES, G.A.; VON HOLY, A.: Microbial populations associated with refrigerated poultry. S. Afric. J. Sc. **90** (1994) 570–582.

[60] GILL, C.O.: Microbial interaction with meats. In: Meat Microbiology (edited by Brown, M.H.). London, U.K.: Applied Science Publishers (1982), 225–264.

[61] GRAZIA, L.; SUZZI, G.; ROMANO, P.; GIUDICI, P.: The yeasts of meat products. Yeast **5** (1989) S495–S499.

[62] GUERZONI, M.E.; LANCIOTTI, R.; MARCHETTI, R.: Survey of the physiological properties of the most frequent yeasts associated with commercial chilled foods. Int. J. Fd Microbiol. **17** (1993) 329–341.

[63] HAMMES, W.P; ROLZ, I.; BANTLEON, A.: Microbiologische Untersuchung der auf dem deutschen Markt vorhandenen Starterkulturpraparate fur die Rohwurstbereitung. Fleischwirtschaft **65** (1985) 729–734.

[64] HAMMES, W.P.; KNAUF, H.J.: Starters in the processing of meat products. Meat Sc. **36** (1994) 155–168.

[65] HEARD, G.M.; FLEET, G.H.: A convenient microtitre tray procedure for yeast identification. J. Appl. Bacteriol. **68** (1990) 447–451.

[66] HOLYOAK, C.D.; STRATFORD, M.; MCMULLIN, Z.; COLE, M.B.; CRIMMINS, K.; BROWN, A.J.P.; COOTE, P.: Activity of the plasma-membrane H+ -ATPase and optimal glycolytic flux are required for rapid adaptation and growth in the presence of the weak acid preservative sorbic acid. Appl. Environ. Microbiol. **62** (1996) 3158–3164.

[67] HOLZAPFEL, W.H.; HALL, A.N.: The microbiology of South African dried sausage. S. Afric. J. Animal Sc. **6** (1976) 199–206.

[68] HOUTSMA, P.C.; DE WIT, L.C.; ROMBOUTS, F.M.: Minimum inhibitory concentration (MIC) of sodium lactate for pathogens and spoilage organisms occurring in meat products. Int. J. Fd Microbiol. **20** (1993) 247–257.

[69] HSIEH, D.Y.; JAY, J.M.: Characterization and identification of yeasts from fresh and spoiled ground beef. Int. J. Fd Microbiol. **1** (1984) 141–147.

[70] ISMAIL, S.A.S.; DEAK, T.; ADD EL-RAHMAN, H.A.; YASSIEN, M.A.M.; BEUCHAT, L.R.: Presence and changes in populations of yeasts on raw and processed poultry products stored at refrigeration temperature. Int. J. Fd Microbiol. **62** (2000) 113–121.

[71] ISMAIL, S.A.S.; DEAK, T.; ADD EL-RAHMAN, H.A.; YASSIEN, M.A.M.; BEUCHAT, L.R.: Effectiveness of immersion treatments with acids, trisodium phosphate, and herb decoctions in reducing populations of *Yarrowia lipolytica* and naturally occurring aerobic micropropagation on raw chicken. Int. J. Fd Microbiol. **64** (2001) 13–19.

[72] JARVIS, B.: Comparison of an improved Rose-Bengal-chloretetracycline agar with other media for the selective isolation and enumeration of moulds and yeasts in foods. J. Appl. Bacteriol. **36** (1973) 723–727.

[73] JAY, J.M.: Meats, poultry and seafoods. In: Food and beverage mycology (edited by Beuchat, L.R.). New York, U.S.A.: AVI Publishing (1978) 129–144.

[74] JAY, J.M.: Occurrence and significance of specific yeasts and molds in meats, poultry and seafoods. In: Food mycology. (edited by Rhodes M.E.). Boston, U.S.A.: Hall G.K. & Company (1979) 44–55.

[75] JAY, J.M.; MARGITIC, S.: Incidence of yeasts in fresh ground beef and their ratios to bacteria. J. Fd Sc. **46** (1981) 648–649.

[76] JOHANNSEN, E.; NIEMAND, J.G.; EAGLE, L.; BREDENHANN, G.: Yeast flora of non-radurised and radurised minced beef – a taxonomic study. Int. J. Fd Microbiol. **1** (1984) 217–227.

[77] KATSARAS, K.; LAUTENSCHLAGER, R.; BOSCHKOVA, K.: The performance of microflora and starter cultures during curing, drying, and storing of pasterma. Fleischwirtschaft **76** (1996) 308–310.

[78] KIM, J.G.; YOUSEF, A.E.; DAVE, S.: Application of ozone for enhancing the microbiological safety and quality of foods: a review. J. Fd Protect. **62** (1999) 1071–1087.

[79] KING, J.R.; HOCKING, A.D.; PITT, J.I.: Dichloran – Rose Bengal medium for enumeration and isolation of molds from foods. Appl. Environ. Microbiol. **37** (1979) 959–964.

[80] KING, J.R.; PONTING, J.D.; SANSHUCK, D.W.; JACKSON, R.; MIHARA, K.: Factors affecting death of yeast by sulphur dioxide. J. Fd Protect. **44** (1981) 92–97.

[81] KREGER-VAN RIJ, N.J.W.: The Yeasts. A Taxonomic Study, 3rd ed. Amsterdam, The Netherlands: Elsevier Science Publishers (1984).

[82] KURTZMAN, C.P.: Molecular taxonomy of the yeasts. Yeast **10** (1994) 1727–1740.

[83] LALEYE, L.C.; LEE, B.H.; SIMARD, R.E.; CARMICHAEL, L.; HOLLEY, R.A.: Shelf life of vacuum- or nitrogen-packed pastrami: effects of packaging atmospheres, temperature and duration of storage on microflora changes. J. Fd Sc. **49** (1984) 827–831.

[84] LAMBROPOULOU, K.A.; DROSINOS, E.H.; NYCHAS, G.J.E.: The effect of glucose supplementation on the spoilage microflora and chemical composition of minced beef stored aerobically or under a modified atmosphere at 4 °C. Int. J. Fd Microbiol. **30** (1996) 281–291.

[85] LANGNER, H.J.: Aromastoffe in der Rohwurst. Fleischwirtschaft **52** (1972) 1299–1306.

[86] LEISTNER, L.; BEM, Z.: Vorkommen und Bedeutung von Hefen bei Pökelfleischwaren. Fleischwirtschaft **50** (1970) 350–351.

[87] LIZASO, G.; CHASCO, J.; BERIAIN, M.J.: Microbiological and biochemical changes during ripening of salchichon, a Spanish dry cured sausage. Fd Microbiol. **16** (1999) 219–228.

[88] LOWRY, P.D.: Limiting conditions of yeast growth on frozen meat. In: Proc. 30th Eur. Meet. Meat Res. Workers. Bristol, U.K.: University of Bristol (1984) 255–256.

[89] LOWRY, P.D.; GILL, C.O.: Development of a yeast microflora on frozen lamb stored at –5 °C. J. Fd Protect. **47** (1984) 309–311.

[90] LUCKE, F.K.: Microbiological processes in the manufacture of dry sausage and raw ham. Fleischwirtschaft International **2** (1988) 23–28.

[91] LUCKE, F.K.; HECHELMANN, H.: Starter cultures for dry sausages and raw ham. Composition and effect. Fleischwirtschaft **67** (1987) 307–314.

[92] LUCKE, F-K.: Fermented Meats. In: The microbiological safety and quality of food, Vol. I (edited by Lund, B.M.; Baird-Parker, T.C.; Gould, G.W.). Maryland, U.S.A.: Aspen Publications (2000) 420–444.

[93] LUND, B.M.; EKLUND, T.: Control of pH and use of organic acids. In: The Microbiological Safety and Quality of Food, Vol. I (edited by Lund, B.M.; Baird-Parker, T.C.; Gould, G.W.). Maryland, U.S.A.: Aspen Publications (2000) 175–199.

[94] MCCARTHY, J.A.; DAMOGLOU, A.P.: The effect of low-dose gamma irradiation on the yeasts of British fresh sausage. Fd Microbiol. **10** (1993) 439–446.

[95] MCCARTHY, J.A.; DAMOGLOU, A.P.: The effect of substrate on the radiation resistance of yeasts isolated from sausage meat. Lett. Appl. Microbiol. **22** (1996) 80–84.

[96] MEAD, G.C.: Microbiology of poultry and game birds. In: Meat Microbiology (edited by Brown, M.H.). London, U.K.: Applied Science Publishers (1982) 67–101.

[97] MEISEL, C.; GEHLEN, K.H.; FISCHER, A.; HAMMES, W.P.: Inhibition of the growth of *Staphylococcus aureus* in dry sausages by *Lactobacillus curvatus, Micrococcus varians* and *Debaryomyces hansenii.* Food Biotechnol. **3** (1989) 145–168.

[98] METAXOPOULOS, J.; SAMELIS, J.; PAPADELLI, M.: Technological and microbiological evaluation of traditionai processes as modified for the industrial manufacturing of dry fermented sausage in Greece. Ital. J. Fd Sc. **13** (2001) 3–18.

[99] METAXOPOULOS, J.; STAVROPOULOS, S.; KAKOURI, A.; SAMELIS, J.: Yeasts isolated from traditional Greek dry salami. Ital. J. Fd Sc. **8** (1996) 25–32.

[100] MITEVA, E.; KIROVA, E.; GADJEVA, D.; RADEVA, M.: Sensory aroma and taste profiles of raw dried sausages manufactured with a lipolytically active yeast culture. Nahrung **30** (1986) 829–832.

[101] MOLINA, I.; SILLA, H.; FLORES, J.: Study of the microbial flora in dry-cured ham. 4. Yeasts. Fleishwirtschaft **70** (1990) 74–76.

[102] MOSSEL, D.A.A.; KLEYNEN-SEMMELING, A.M.C.; VINCENTIE, H.H.; BEERENS, H.; CATSARAS, M.: Oxytetracycline-glucose-yeast extract agar for selective enumeration of moulds and yeasts in foods and clinical material. J. Appl. Bacteriol. **33** (1970) 454–457.

[103] MOSSEL, D.A.A.; CORRY, J.E.L.; STRUIJKAND, C.B.; BAIRD, R.M.: Essentials of the Microbiology of Foods. Chichester, U.S.A.: John Wiley & Sons (1995).

[104] MURRAY, K.A.; GILMOUR, A.; MADDEN, R.H.: Microbiological quality of chilled beef carcasses in northern Ireland: A baseline study. J. Fd Protect. **64** (2001) 498–502.

[105] NEVES, L.; PAMPULHA, M.E.; LOUREIRO-DIAS, M.C.: Resistance of food spoilage yeasts to sorbic acid. Lett. Appl. Microbiol. **19** (1994) 8–11.

[106] NEVES, M.L.; OLIVEIRA, R.P.; LUCAS, C.M.: Metabolic flux response to salt-induced stress in the halotolerant yeast *Debaryomyces hansenii.* Microbiology **143** (1997) 1133–1139.

[107] NIELSEN, H.J.S.: Influence of temperature and gas permeability of packaging film on development and composition of microbial flora in vacuum packaged bologna-type sausage. J. Fd Protect. **46** (1983) 693–698.

[108] NORTJE, G.L.; NEL, L.; JORDAAN, E.; BADENHORST, K.; GOEDHART, E.; HOLZAPFEL, W.H.: The aerobic psychrotrophic populations on meat and meat contact surfaces in a meat production system and on meat stored at chill temperatures. J. Appl. Bacteriol. **68** (1990) 335–344.

[109] NOTTINGHAM, P.M.: Microbiology of carcass meats. In: Meat Microbiology (edited by Brown, M.H.). London, U.K.: Applied Science Publishers (1982) 13–65.

[110] NUNEZ, F.; RODRIGUEZ, M.M.; CORDOBA, J.J.; BERMUDEZ, M.E.; ASENSIO, M.A.: Yeast population during ripening of dry-cured Iberian ham. Int. J. Fd Microbiol. **29** (1996) 271–280.

[111] NWACHUKWU, S.U.; AKPATA, T.V.I.: Utilization of carbohydrate and protein by *Candida famata* during spoilage of snail meat (*Archachatina marginata* Swainson). J. Fd Agric. **1** (1987) 27–30.

[112] NYCHAS, G.J.: Microbial growth in minced meat. Ph.D. Thesis, University of Bath, Bath, U.K. (1984).

[113] NYCHAS, G.-J.E.; DROSINOS, E.H., BOARD, R.G.: Chemical changes in stored meat. In: The Microbiology of Meat and Poultry (edited by Davies, A.; Board, R.). London, U.K.: Blackie Academic and Professional (1998) 288–326.

[114] OLESEN, P.T.; STAHNKE, H.: The influence of *Debaryomyces hansenii* and *Candida utilis* on the aroma formation in garlic spiced fermented sausages and model minces. Meat Sc. **56** (2000) 357–368.

[115] ORDONEZ, J.A.; HIERRO, E.M.; BRUNA, J.M.; DE LA HOZ, L.: Changes in the components of dry-fermented sausages during ripening. Crit. Rev. Fd Sc. Nutrit. **39** (1999) 329–367.

[116] PANDYA, Y.; JEWETT, F.F.; HOOVER, D.G.: Concurrent effects of high hydrostatic-pressure, acidity and heat on the production and injury of yeasts. J. Fd Protect. **58** (1995) 301–304.

[117] PAPA, F.; ZAMBONELLI, C.; GRAZIA, L.: Production of Milano style salami of good quality and safety. Fd Microbiol. **12** (1995) 9–12.

[118] PARENTE, E.; MARTUSCELLI, M.; GARDINI, F.; GRIECO, S.; GRUDELE, M.A.; SUZZI, G.: Evolution of microbial populations and biogenic amine production in dry sausages produced in Southern Italy. J. Appl. Microbiol. **90** (2001) 882–891.

[119] PERRIER, V.; DUBREUCQ, E.; GALZY, P.: Fatty acid and carotenoid composition of *Rhodotorula* strains. Arch. Microbiol. **164** (1995) 173–179.

[120] PRISTA, C.; MADEIRA-LOPES, A.: Thermokinetic and energetic profiles of the yeast *Debaryomyces hansenii* in the presence of sodium chloride. Biotechnol. Lett. **17** (1995) 1233–1236.

[121] RAMIREZ, C.; GONZALEZ, C.: *Candida iberica* sp. n. A new species isolated from Spanish sausages. Can. J. Microbiol. **18** (1972) 1778–1780.

[122] RODEL, W.; STIEBING, A.; KROCKEL, L.: Ripening parameters for traditional dry sausages with a mould covering. Fleischwirtschaft **73** (1993) 848–853.

[123] RODRIGUEZ, M.; NUNEZ, F.; CORDOBA, J.J.; BERMUDEZ, M.E.; ASENSIO, M.A.: Evaluation of proteolytic activity of micro-organisms isolated from dry cured ham. J. Appl. Microbiol. **85** (1998) 905–912.

[124] ROSSMANITH, E.; MINTZLAFF, H.-J.; STRENG, B.; CHRIST, W.; LEISTNER, L.: Hefen als Starterkulturen für Rohwurste Jahresber. Bundesanst. Fleischforsch. (1972) I47–I48.

[125] SALDANHA DA GAMA, A.; MALFEITO FERREIRA, M.; LOUREIRO, V.: Characterization of yeasts associated with Portuguese pork-based products. Int. J. Fd Microbiol. **37** (1997) 201–207.

[126] SAMELIS, J.; AGGELIS, G.; METAXOPOULOS, J.: Lipolytic and microbial changes during the natural fermentation and ripening of Greek dry sausages. Meat Sc. **35** (1993) 371–385.

[127] SAMELIS, J.; STAVROPOULOS, S.; KAKOURI, A.; METAXOPOULOS, J.: Quantification and characterization of microbial populations associated with naturally fermented Greek dry salami. Fd Microbiol. **11** (1994) 447–460.

[128] SAMELIS, J.; METAXOPOULOS, J.: The microbiology of traditional Greek country-style sausage during manufacture followed by storage at 3 ° and 12 °C in air. Italian J. Fd Sc. **10** (1998) 155–163.

[129] SAMELIS, J.; KAKOURI, A.; GEORGIADOU, K.G.; METAXOPOULOS, J.: Evaluation of the extent and type of bacterial contamination at different stages of processing of cooked ham. J. Appl. Microbiol. **84** (1998) 649–660.

[130] SAMELIS, J.; METAXOPOULOS, J.; VLASSI, M.; PAPPA, A.: Stability and safety of traditional Greek salami – a microbiological ecology study. Int. J. Fd Microbiol. **44** (1998) 69–82.

[131] SAMELIS, J.; GEORGIADOU, K.G.: The microbial association of Greek taverna sausage stored at 4 and 10 °C in air, vacuum or 100 % carbon dioxide, and its spoilage potential. J. Appl. Microbiol. **88** (2000) 58–68.

[132] SAMELIS, J.; KAKOURI, A.; REMENTZIS, J.: Selective effect of the product type and the packaging conditions on the species of lactic acid bacteria dominating the spoilage microbial association of cooked meats at 4 °C. Fd Microbiol. **17** (2000) 329–340.

[133] SAMELIS, J.; SOFOS, J.N.; KENDALL, P.A.; SMITH, G.C.: Effect of acid adaptation on survival of *Escherichia coli* O157:H7 in meat decontamination washing fluids and potential effects of organic acid interventions on the microbial ecology of the meat plant environment. J. Fd Protect. **65** (2002) 33–40.

[134] SANTOS, N.N.; SANTOS-MENDONCA, R.C.; SANZ, Y.; BOLUMAR, T.; ARISTOY, M.C.; TOLDRA, F.: Hydrolysis of pork muscle sarcoplasmic proteins by *Debaryomyces hansenii*. Int. J. Fd Microbiol. **68** (2001) 199–206.

[135] SANZ, Y.; TOLDRA, V.F.; FLORES, J.: Effect of nitrate and nitrite curing salts on microbial changes and sensory quality of non-fermented sausages. Int. J. Fd Microbiol. **42** (1998) 213–217.

[136] SEDDON, B.; MCKAY, T.; MACKENZIE, L.; KEMP, J.S.; FENLON, D.R.; FARR, L.: The use of viablue and deft for the detection of yeasts, molds and bacteria. Int. J. Biodeter. Biodegrad. **32** (1993) 3–18.

[137] SIMARD, R.E.; LEE, B.H.; LALEYE, C.L.; HOLLEY, R.A.: Effects of temperature, light and storage time on the microflora of vacuum- or nitrogen-packed frankfurters. J. Fd Protect. **46** (1983) 199–205.

[138] SINIGAGLIA M.; LANCIOTTI, R.; GUERZONI M.E.: Biochemical and physiological characteristics of *Yarrowia lipolytica* strains in relation to isolation source. Can. J. Microbiol. **40** (1994) 54–59.

[139] SMITH, M.T.; HADLOK, R.: Hefen und Fleisch: Vorkommer, Systematic und Differenzierung. Fleischwirtschaft **56** (1976) 379–384.

[140] SOFOS, J.N.: Sorbate Food Preservatives. Boca Raton, FL, U.S.A.: CRC Press (1989).

[141] SOFOS, J.N.: Microbial growth and its control in meat, poultry and fish. In: Advances in meat research, Vol. 9, Quality Attributes and their Measurements in Meat, Poultry and Fish Products (edited by Pearson, A.M.; Dutson, T.R.). Glasgow, U.K.: Chapman Hall (1994) 359–403.

[142] SORENSEN, B.B.; SAMUELSEN, H.: The combined effects of environmental conditions on lipolysis of pork fat by lipases of the meat starter culture organisms *Staphylococcus xylosus* and *Debaryomyces hansenii*. Int. J. Fd Microbiol. **32** (1996) 59–71.

[143] SORENSEN, B.B.: Lipolysis of pork fat by the meat starter culture *Debaryomyces hansenii* at various environmental conditions. Int. J. Fd Microbiol. **34** (1997) 187–193.

[144] STAIB, F.; MISHRA, S.K.; TOMPAH, B.; GROSSE, G.; ABEL, T.; BLISSE, A.; FOLKEN FROHLICH, B.: Pathogenic yeast-like fungi in meat products. Zentralbl. Bakt. Mikrobiol. Hyg. Orig. A **248** (1980) 422–429.

[145] SZCZEPANIAK, B.; RZEZNICZAK, J.; PEZACKI, W.: Higiena I technologia zywnosci zwirzecego pochodzenia. Zmiennosc drozdzy W wedlinach surowych II characterystyka jakosciowa. Med. Weteryn. **31** (1975) 623–626.

[146] VAN DER WALT, J.P.; YARROW, D.: Methods for isolation, maintenance, classification and identification of yeasts. In: The yeasts. A taxonomic study, 3rd ed. (edited by Kreger-van Rij, N.J.W.). Amsterdam, The Netherlands: Elsevier Science Publishers (1984) 45–104.

[147] VAN DER RIET, W.B.: Studies on the mycoflora of biltong. S. Afric. Fd Rev. **3** (1976) 105–111.

[148] VANDERZANT, C.; CHESSER, L.K.; SAVELL, J.W.; GARDNER, F.A.; SMITH, G.C.: Effect of addition of glucose, citrate and citrate-lactic acid on microbiology and sensory characteristics of steaks from normal and dark, firm and dry, beef carcasses displayed in polyvinyl chloride film and in vacuum packages. J. Fd Protect. **46** (1983) 775–780.

[149] VAN NETTEN, P.; HUIS IN'T VELD, J.H.; MOSSEL, D.A.A.: The effect of lactic acid decontamination on the microflora on meat. J. Fd Saf. **14** (1994) 243–257.

[150] VILAR, I.; GARCIA FONTAN, M.C.; PRIETO, B.; TORNADIJO, M.E.; CARBALLO, J.: A survey on the microbiological changes during the manufacture of dry-cured lacón, a Spanish traditional meat product. J. Appl. Microbiol. **89** (2000) 1018–1026.

[151] VILJOEN, B.C.; DYKES, G.A.; GALLIS, M.; VON HOLLY, A.: Yeasts associated with Vienna sausage packaging. Int. J. Fd Microbiol. **18** (1993) 53–62.

[152] VILJOEN, B.C.; GEORNARAS, I.; LAMPRECHT, A.; VON HOLY, A.: Yeast populations associated with processed poultry. Fd Microbiol. **15** (1998) 113–17.

[153] VISSER, I.J.R.; KOOLMEES, P.A.; BIJKER, P.G.H.: Microbiological conditions and keeping quality of veal tongues as affected by lactic acid decontamination and vacuum packaging. J. Fd Protect. **51** (1988) 208–213.

[154] VON HOLY, A.; HOLZAPFEL, W.H.: The influence of extrinsic factors on the microbiological spoilage pattern of ground beef. Int. J. Fd Microbiol. **6** (1988) 269–280.

[155] WALKER, H.W.; AYRES, J.C.: Yeasts as spoilage organisms. In: The yeasts, yeast technology, Vol.3 (edited by Rose, A.H.; Harrison, J.S.). London, U.K.: Academic Press (1970) 463–527.

[156] WICKERHAM, L.J.: Presence of nitrite-assimilating species of *Debaryomyces* in lunch meats. J. Bacteriol. **74** (1957) 823–833.

[157] WINGER, R.J.; LOWRY, P.D.: Sensory evaluation of lamb after growth of yeasts at –5 °C. J. Fd Sc. **48** (1983) 1883–1885.

[158] WOLTER, H.; LAING, E.; VILJOEN, B.C.: Isolation and identification of yeasts associated with intermediate moisture meats. Fd Technol. Biotechnol. **38** (2000) 69–75.

[159] ZALACAIN, I.; ZAPELENA, M.J.; ASTIASARAN, I.; BELLO, J.: Addition of lipase from *Candida cylindracea* to a traditional formulation of a dry fermented sausage. Meat Sc. **42** (1996) 155–163.

[160] ZAMPONELLI, C.; RAINIERI, S.; CHIAVARI, C.; MONTANARI, G; BENEVELLI, M.; GRAZIA, L.: Autolysis of yeasts and bacteria in fermented foods. Ital. J. Fd Sc. **12** (2000) 9–21.

10 Yeasts in fruit and fruit products

GRAHAM H. FLEET

10.1 Introduction

The "spontaneous" fermentation of grape juice into wine, as observed by Louis Pasteur about 150 years ago, established the broadly accepted view that yeasts have a natural association with fruits and fruit products. Many ecological and biochemical studies have now described the seminal role of yeasts in the production of wine from grapes and other fruits, and in the spoilage of a vast range of fruits and fruit products. More recently, interest has focussed on yeasts associated with the surface of fruits, because some of these species have excellent properties for the biocontrol of filamentous fungi that invade and spoil the product at both the pre-harvest and post-harvest stages.

Yeasts are not the only organisms of importance in the microbiology of fruits and fruit products. The spoilage of fruits by moulds and the potential of moulds to produce mycotoxins in these products are well recognized. Bacteria, especially lactic acid bacteria and acetic acid bacteria, are prominent in the spoilage of some fruits and fruit products, and certain species of lactic acid bacteria can have a positive contribution in the production of wines. In recent years, fresh and processed fruits have been increasingly recognized as vectors in outbreaks of food-borne disease that are attributable to a range of bacterial, viral and protozoan pathogens [13]. Consequently, the significance of yeasts in fruits and fruit products should be considered in relation to their total microbiology, good discussions of which are provided by the International Commission on Microbiological Specifications for Foods (ICMSF) [67] and LUND and SNOWDON [82].

10.2 Fruits as a habitat for yeast diversity

The diversity of yeast species associated with various fruits and fruit products will be discussed throughout this Chapter, with further information being presented in Chapters 12, 14 and 16. However, it is appropriate to consider, here, some of the broader issues that lead to fruits and fruit products becoming ecosystems for the growth and survival of yeasts, generally, and more specifically. In this context, several questions emerge. Where do the yeasts come from or how do they contaminate the product, what chemical and physical properties of the product select for yeast colonization, how do environmental factors affect colonization, and what is the spatial distribution of yeasts throughout the product? Answers to these questions will vary according to whether the fruit is fresh (intact) or processed. In the case of fresh produce, the outer surface of the fruit will be the primary habitat and the ecological principles associated with phyllosphere microbiology will largely apply [39]. For processed produce, where the physical integrity of the fruit surface has been destroyed, the ecological principles associated with food microbiology will be most relevant [54].

10.2.1 Yeasts associated with fresh fruits

The surface of fresh, undamaged fruits have yeast populations that range from 10^2–10^6 cfu/ cm^2. For the one type of fruit, many factors affect the populations and species that are present. These factors include fruit cultivars, geographic location, fruit development and maturity, seasonal and climatic conditions and application of agrochemicals. Controlled, quantitative studies of these variables have not been done, and current understanding is largely based on qualitative, descriptive data. Most studies have been done with wine grapes and apples used in cider production because of the significance of indigenous yeasts in these alcoholic fermentations.

At the stage of pre-harvest in the field, fruits become exposed to yeasts that occur in the air, soil, wind-blown dust, rain, irrigation water, fertilizers and on the surfaces of leaves [40, 75, 98]. Insects are significant in the dispersal of yeasts to fruits [7, 9, 28, 98]. Bees that pollinate flowers evolving into fruits, *Drosophila* fruit flies, fruit wasps, moths, spiders, and beetles have all been implicated. It is well established that fruit flies feed on the yeasts associated with flowers and fruits, and that similar yeast species can be isolated from the fruit and the crop of the insect [30, 74, 91, 115, 130].

Many chemical and physical factors affect the survival and growth of yeasts on arrival at the fruit surface (Table 10.2-1) [39, 76]. Some mechanisms of entrapment, adsorption or attachment of the yeast cells to the outer surface are necessary to avoid them being washed away by rainfall, dew or irrigation waters. Generally, most fruits have an epidermal layer of skin which is covered by a cuticle and epicuticular layer of wax. The chemical composition of these materials are broadly known [61]. This cuticular material represents the first point of contact of yeasts with the fruit surface. The affinity of yeasts towards this surface material may determine the specificity of the association. Such associations need to be established, along with the potential of yeasts to cause degradation of the cuticular layer by the production of extracellular lipases, cutinases and other enzymes [39, 112]. The attachment of mi-

Tab. 10.2-1 Factors affecting the survival and growth of yeasts on the surfaces of fruits

- Attachment to fruit surface
- Availability of nutrients
- Water availability and desiccation
- Tolerance of the stresses of temperature, sunlight and irradiation
- Tolerance of naturally occurring inhibitors and agrochemicals
- Competition with other microbial species

crobial cells to phyllospheric locations has been correlated with their production of extracellular gums and mucilages [39]. Species of *Rhodotorula*, *Cryptococcus* and *Sporobolomyces* are well known for these properties and this might be one factor that explains their frequent association with the surfaces of many fruits. The ability to produce pseudohyphae and initiate invasive growth may be another mechanism by which yeasts attach to the surface of fruits [68].

The fruit surface presents an environment of limited nutrient availability, depending on the concentration of sugars, organic acids and amino acids, which leach from the underlying tissue [69]. Any damage to the skin will increase nutrient availability. Wind-borne dust and vegetation that becomes attached to the fruit surface could be another source of nutrients. Water availability will be a determining factor, but is likely to be a cyclical event depending on rainfall, irrigation, dew formation, humidity and degree of exposure to the sun. Prolonged exposure to the sun might lead to desiccation of yeast cells, and capsular gum formation could be a protective factor in this context [39]. The surface temperature of the fruit will vary on a daily and seasonal basis and, in warmer climates, it could exceed 40 °C. Tolerance of sunlight and irradiation are other survival factors that could be enhanced by the production of carotenoid pigments. Species of *Rhodotorula*, *Sporobolomyces* and *Cryptococcus* that occur on fruits, generally produce red, pink or yellow pigments. Additional, selective or survival properties include the ability to compete with other organisms for nutrients and resistance to naturally occurring antagonistic substances (e. g., essential oils, phytoalexins), fungicides and pesticides, and antagonists produced by other microbial species including killer toxins of different yeasts species or strains [117]. DI MENNA [38] has noted the sensitivity of phyllosphere yeasts to antagonistic substances produced by phyllosphere bacteria and fungi, while MCCORMACK et al. [88] reported the production of antibacterial compounds by phyllosphere strains of *S. roseus*, *Rh. glutinis* and *Cr. laurentii*.

The physical location of microorganisms on or in fresh and intact fruits is attracting renewed attention, mainly because of food safety issues and the potential for these species to escape destruction by procedures that rely on surface decontamination [82]. While foodborne yeasts do not present a significant public health issue, they do pose a spoilage threat. The majority of the microflora occur on the fruit surface but there is increasing realization that a small proportion is located internally. This concept is not new, since the early literature (reviewed in [66, 75, 83, 94]) makes significant reference to the isolation of yeasts from the inner tissues of fresh apples, grapes and other fruits, especially from the regions of the core and seeds. These findings have been substantiated in later studies [109, 113]. It is likely that these organisms have their origins as contaminants from the budding and flowering processes [9, 28, 78]. Sequestering and internalization of surface contaminants through the stem end and other surface openings are also possibilities. More rigorous study of these internalization concepts is needed to better define the yeast population and species that occur at internal locations and to understand those factors that determine their origin and survival. Electron microscope studies clearly showed that yeast colonization of the surfaces of grapes [10], apples [89] and oranges [4] is more dense at the sites of stem removal, stomata openings and wounds, where there would be increased nutrient availability.

10.2.1.1 Grapes

Many researchers have isolated yeasts from the surface of wine grapes (reviewed in [2, 71, 86, 126]). Total yeast populations vary from as few as 10 to 10^6 cfu/cm^2 or g. Some authors have observed the population and species of these yeasts to vary with cultivars of grape [28, 31, 56, 58, 86, 120, 137] but no trends can be drawn from these data. Presently, it is not possible to explain these cultivar influences in relation to the physical and chemical properties of the grape surface [61]. However, yeast populations on the grape surface increase as the grape matures on the vine. Generally, very few yeast cells (10–10^3 cfu/cm^2) are detected on immature grape berries, but they increase to 10^4–10^6 cfu/cm^2 during ripening [28, 29, 32, 56, 97, 111, 120, 133]. It is assumed that more sugars leach from the underlying tissues to the surface during ripening to account for this increase. Nevertheless, climatic conditions at the time of maturity and harvest can have significant influences on the yeast population. Cold, rainy conditions tend to favour higher populations, while dry, hotter conditions could prevent the growth of some yeasts or even result in a decrease in population if the temperature exceeds 38–40 °C [32, 56, 79, 137]. Pre-harvest applications of fungicides may also influence the yeast populations of grapes, but systematic, quantitative studies are required [18, 19, 58, 97, 133].

Data are not consistent but, generally, unripe grape berries harbour a predominance of *Rhodotorula*, *Cryptococcus* and *Candida* species (e. g., *Rh. glutinis*, *Rh. rubra* (= *Rh. mucilaginosa*), *Rh. mucilaginosa*, *Cr. albidus*, *Cr. laurentii*, *Cr. diffluens*, *C. famata* (= *Debaryomyces hansenii*), *Candida pulcherrima* (= *Metschnikowia pulcherrima*). In some cases, species of *Sporobolomyces* (*S. roseus*), *Sporidiobolus* and *Rhodosporidium* are found [29, 31, 97, 111, 120, 133]. Most of these species can be isolated from mature, ripe grape berries but, at this stage, *Kloeckera apiculata* (= *Hanseniaspora uvarum*) and *Metschnikowia pulcherrima* are often the predominant species [5, 20, 46, 56, 77, 86, 90, 111, 120, 133, 137]. Nevertheless, there are reports where no *Hanseniaspora uvarum* have been found on mature grapes [31, 96, 137].

Damage to the skin of grape berries increases the availability of nutrients for microbial growth and encourages a greater (>10^6 cfu/cm^2) population of yeasts that need to coexist with the development of increased populations of filamentous fungi, acetic acid bacteria and lactic acid bacteria. Sour rots may develop and these are accompanied by increased populations of species such as *C. stellata*, *C. krusei* (= *Issatchenkia orientalis*), *H. uvarum* and *M. pulcherrima* and decreased incidence of *Rhodotorula* and *Cryptococcus* spp. Fermentative species of *Saccharomyces* and *Zygosaccharomyces* are also prominent in such grapes [59]. Grapes infected with *Botrytis cinerea* have a similar yeast flora. It has been suggested that this fungus produces botrycin that is inhibitory to some yeasts [77].

There is much debate in the literature about the occurrence of *Saccharomyces cerevisiae* on grapes because of its principal role in the alcoholic fermentation of grape juice into wine. While some authors believe that the grape berry is the natural origin of this species, others do not agree with this view [86, 92, 126]. Many authors have not been able to isolate

S. cerevisiae from either immature or mature grapes using direct agar plating methods, or they have found it very infrequently compared with other species. The probability of its isolation by these methods increases as the grapes mature and progress to overripe and damaged condition. However, *S. cerevisiae* is readily isolated from mature grapes by enrichment culture [93, 126, 131, 132]. Overall, the majority of studies suggest that *S. cerevisiae* does occur on grapes in the vineyard, but its populations are too low (<10–100 cells/cm^2) to be detected by direct plating procedures. Nevertheless, there are reports where it has not been found after enrichment, and these data suggest that climatic factors, such as rainfall and temperature, and viticultural factors, such as agrochemical applications, may be significant in affecting its survival and occurrence [86, 105, 131, 132]. Electrophoretic karyotyping and other molecular methods have been used to demonstrate that a diversity of different strains of *S. cerevisiae* occurs on grapes, with some suggestions of regional or geographical specificity [68, 105, 131, 132, 134]. It is likely that similar strain diversity will apply to the other yeast species associated with grapes.

10.2.1.2 Apples

The yeasts associated with apples have been thoroughly reviewed [7, 9, 83] although it is worthwhile to read some of the earlier papers [8, 15, 16, 25, 85, 136]. The surfaces of sound, intact apples have yeast populations ranging from 10–10^5 cfu/cm^2 or g [7, 109]. The population generally increases as the fruits approach maturity, but variations in yeast counts occur during fruit growth on the tree and probably reflect the influences of climate, farming practices and fungicide applications [9, 85, 122, 136]. Apples that have dropped to the ground or those that are infected with filamentous fungi have higher yeast populations [16, 85]. There is evidence that the yeast ecology of apples may vary with geographical location, but the data are limited [7, 15, 16]. *Metschnikowia pulcherrima* was the most prevalent species on apples harvested in the United Kingdom, and was followed by isolations of *Rh. glutinis*, *S. roseus*, *D. hansenii* and, to a lesser extent, *H. uvarum*, which becomes more dominant during the post-harvest storage of apples. Apples harvested from orchards in Canada showed a predominance of *C. malicola* (= *Aureobasidium pullulans* var. *pullulans*) followed by *Rh. glutinis*, *Cr. laurentii*, *Cr. albidus* and *D. hansenii* [25, 136]. *Metschnikowia pulcherrima* was notably absent in these studies. Similar yeasts have been found on apples harvested from Spanish and French orchards [122]. Extensive surveys of the yeasts associated with English apple orchards provide an insight of the yeast ecology of soils, air, insects, apple trees (bark, leaves, buds, flowers) as well as the apple fruit [9, 28]. Recent studies suggest that some yeasts such as *Rh. glutinis* may be associated with development of the russet disorder in apples [65]. Apples, artificially damaged by small puncture wounds, were rapidly colonized by *S. roseus*, *Bullera alba* (= *Bulleromyces albus*) and *Cryptococcus* spp., as well as various bacteria and filamentous moulds [89].

10.2.1.3 Citrus fruit

The surfaces of fresh, undamaged citrus fruits (oranges, grapefruit, lemons) have yeast populations of 10^2–10^4 cfu/cm^2. Species identified include *C. guilliermondii* (= *Pichia guilliermondii*), *C. oleophila*, *C. sake*, *Cr. albidus*, *Cr. laurentii*, *D. hansenii*, as well as *Pichia* and *Rhodotorula* species. In some cases, *C. guilliermondii* has been the most frequently isolated species [22, 41, 43, 107]. Damaged, necrotic oranges and grapefruit had a predominance of *H. uvarum*, *P. fermentans* and *P. kluyveri* that could also be isolated from *Drosophila* flies feeding on the fruit [130].

10.2.1.4 Strawberries

Freshly harvested strawberries have yeast populations of 10^4–10^6 cfu/g. Most prevalent species are *Cr. albidus*, *Cr. laurentii*, *H. uvarum*, *C. reukaufii* (= *Metschnikowia gruessii*) and *M. pulcherrima*, depending on the study [17, 37, 60, 87]. *Hanseniaspora uvarum* has been associated with the spoilage of damaged strawberries [81].

10.2.1.5 Other fruits

Qualitative observations on the isolation of yeasts from other fruits are scattered throughout the literature. These include bananas [101], avocados [118], pears [23], guavas [1], amazon fruit [91] and tomatoes [1, 30]. Generally, populations vary between 10^2–10^5 cfu/g, with a prevalence of *Cryptococcus*, *Rhodotorula* and *Candida* species that give way to higher populations and more fermentative species if the fruit tissue is damaged.

10.2.2 Yeasts associated with processed fruits

The processing of fruits generally causes loss of natural integrity and biological structure, thereby making sugary substrates readily available for yeast growth. In fact, it is well recognized that the high sugar content and low pH of many fruit products creates a most favourable and selective environment for yeasts. Left unchecked, these products will quickly undergo a natural yeast fermentation, the outcome of which could be desirable or undesirable, depending on the product. The yeasts responsible for the fermentation originate from the surface of the fruit, contact with processing equipment and other environmental sources, and grow to populations as high as 10^7–10^8 cfu/ml or g. The main products of interest include wines, fruit juices, fruit juice and fruit pulp concentrates, canned fruits, dried fruits, glaced fruits and ready-to-eat fruit slices and fruit salads. Carbonated soft drinks could also be included because many of these use fruit extracts as a base ingredient [51, 82, 119] (see also Chapter 12).

The diversity of yeast species associated with processed fruits has been extensively reviewed, and catalogued, and will be further discussed in other Chapters of this book. In particular, the reader is referred to articles by WALKER and AYRES [135], DEAK [33], FLEET [51], TUDOR and BOARD [128], STRATFORD et al. [119] and DEAK and BEUCHAT [36]. As mentioned already, these products will select for a predominance of fermentative yeast species in genera such as *Saccharomyces, Zygosaccharomyces, Hanseniaspora/Kloeckera, Candida* and *Pichia*, in contrast to the *Cryptococcus* and *Rhodotorula* species that dominate on undamaged fresh fruits. Specific associations will evolve, depending on how the fruit is processed. Fruits fermented into wines will harbour a dominance of ethanol tolerant strains of *S. cerevisiae* [53]. Fruit juice concentrates, dried fruits and glaced fruits that are characterized by very high sugar concentrations are likely to develop high populations of osmotolerant/xerotolerant species such as *Zygosaccharomyces rouxii, Z. bailii, Schizosaccharomyces pombe* and *H. valbyensis*. *Zygosaccharomyces bailii* is well known for its tolerance of preservatives (e. g., benzoates and sorbates), and low temperatures, and is frequently isolated from fruit products processed under these conditions [44, 100].

10.3 Beneficial aspects of fruit yeasts

There are three broad areas where yeasts associated with fruits lead to economic and social benefit. The first area is widely known and extensively developed, and involves the production of fruit-based alcoholic beverages (see Chapter 14). The second area concerns the contribution of yeasts to the processing of some fruits into other commodities such as chocolate and coffee (see Chapter 16). The third area is one of emerging interest and recognizes the potential of fruit-associated yeasts to biologically control the spoilage of fruits by filamentous fungi (moulds).

10.3.1 Alcoholic beverages

Almost any sugary fruit can be processed into an alcoholic beverage. The fermentation of grape juice into table wine is the best known example, and has given rise to an extensive, international industry. Other well known products derived from these wines include sparkling wines (Champagne), fortified wines (Sherry, Porto) and brandy (Cognac, Armagnac) which is distilled from grape wine. Cider and perry are other wines produced from the fermentation of apple and pear juices respectively. The underlying role of yeasts in producing all these products has been extensively described [7, 50, 53, 70, 71]. Essentially, the harvested fruit is crushed and the juice (or juice plus skins and pulp) is allowed to undergo alcoholic fermentation. The fermentation commences, naturally, being conducted by yeast species that originate from the surface of the fruit and the surfaces of juice processing equipment. An ecological succession of yeast species generally occurs throughout fermentation. The initial stages are characterized by the growth and dominance of various species of *Hanseniaspora/Kloeckera, Candida, Pichia, Kluyveromyces*, and *Metschnikowia*. As the ethanol concentration in the juice increases, these species progressively die off, allowing more ethanol tolerant strains of *Saccharomyces cerevisiae/bayanus* to predominate and complete the fermentation. However, many processing variables, such as temperature, can influence this succession. The eventual dominance of *Saccharomyces cerevisiae* in virtually all wine fermentations has lead to its commercialisation as active dried preparations, which wine makers can purchase to inoculate into the juice to induce or selectively guide the fermentation. The aim, here, is to give a more predictable, rapid fermentation. The merits of this technological approach in comparison to those fermentations allowed to proceed naturally, without intervention, are widely debated, especially with respect to yeast impact on the complexity and individuality of wine flavour (see Chapter 14) [53, 64, 104].

10.3.2 Processing

Yeasts are involved in the processing of two major commodities, namely, cocoa beans and coffee beans (Chapter 16). Chocolate is derived from cocoa beans which are the fruit of the tree *Theobroma cacao* grown in plantations of the tropical areas of Africa, South America and South East Asia. The first stage in chocolate processing involves fermentation of freshly harvested beans. The fermentation functions to solubilize the mucilaginous pulp surrounding the beans and to produce ethanol and acetic acid which kill the bean and initiate endogenous biochemical processes that produce precursors of chocolate flavour and colour [116, 124]. The fermentation is a natural, traditional process that involves the successive growth of a diversity of fungal, bacterial and yeast species. Yeast species from the genera *Kloeckera, Saccharomyces, Candida* and *Kluyveromyces* have been implicated in the fermentation. Coffee beans originate as fruit of the tree, *Coffea* spp., also grown in plantations in tropical regions of the world. These beans are also surrounded by a mucilaginous pulp that needs to be removed as part of the process to produce coffee. In some cases, fermentation is used to achieve this goal. Again, fungi, bacteria and yeasts are responsible for the fermentation, but details of the ecology and biochemistry of the process are not well defined [124]. Olives are another fruit that is processed by fermentation. While lactic acid bacteria are the principal agents of fermentation, there are suggestions that yeasts also make a positive contribution [62].

10.3.3 Yeasts as biocontrol agents

The concept of using yeasts as natural agents for the biocontrol of fruit spoilage fungi emerged in the mid-1980s, in response to two commercial pressures, namely, concerns about the safety of chemical fungicides, and the increasing resistance of some fungi to these fungicides. Initially, it was observed that some yeast species, isolated from the surfaces of fruits, had strong antagonistic activity towards major fruit spoilage fungi. Further research subsequently revealed the diversity of yeast species with anti-fungal activity (Table 10.3-1), and has demonstrated that these species can significantly decrease the incidence of mould

Tab. 10.3-1 **Yeast species with biocontrol activity against fruit spoilage fungi**

Yeast Species	Spoilage fungi	Fruit	Reference
Cryptococcus infirmo-miniatus (= *Cystofilobasidium infirmo-miniatum*)	*Penicillium expansum* *Botrytis cinerea*	Pears, apples, cherries	11, 23
Cryptococcus laurentii	*Penicillium expansum* *Botrytis cinerea*	Pears, apples, cherries	11, 23, 110
Cryptococcus albidus	*Penicillium expansum* *Botrytis cinerea*	Apples	47, 49
Cryptococcus humicolus	*Botrytis cinerea*	Apples	49
Filobasidium floriforme	*Botrytis cinerea*	Apples	49
Rhodotorula glutinis	*Penicillium expansum* *Botrytis cinerea*	Pears, apples, cherries	11, 23
Rhodosporidium toruloides	*Botrytis cinerea*	Apples	49
Sporobolomyces roseus	*Botrytis cinerea*	Apples	49
Candida sake	*Penicillium expansum* *Penicillium digitatum* *Botrytis cinerea*	Apples, citrus, kiwifruit	26, 43, 129
Candida oleophila	*Penicillium expansum* *Penicillium digitatum* *Penicillium italicum* *Botrytis cinerea*	Apples, pears, citrus	11, 22, 42, 43, 89
Candida pulcherrima (= *Metschnikowia pulcherrima*)	*Botrytis cinerea*	Kiwifruit, strawberries, apples	26, 60, 99
Candida reukaufii (= *Metschnikowia gruessii*)	*Botrytis cinerea*	Strawberries	60
Candida saitoana	*Penicillium expansum* *Penicillium digitatum* *Botrytis cinerea*	Apples, citrus	44
Candida guilliermondii (= *Pichia guilliermondii*)	*Penicillium digitatum*	Grapefruit	43
Candida famata (= *Debaryomyces hansenii*)	*Penicillium digitatum*	Citrus	4, 41

spoilage when applied to fruits at both the pre-harvest and post-harvest stages. One of these species, *Candida oleophila*, has been commercialised and is known as Aspire [42]. These yeasts inhibit mould growth and activity on fruits by a variety of mechanisms which include: competition for nutrients and space; production of anti-fungal substances; production of enzymes such as glucanases that destroy fungal cell walls; and induction of host (plant) defences activities that have anti-fungal properties (e. g., chitinase production) [21]. Many factors affect the efficacy of yeasts used as fungal biocontrol agents and these include: ability to colonize and survive on fruits, especially at the sites of tissue damage; time and concentration of application; and application in combination with other anti-fungal strategies. In addition to anti-fungal efficacy, their commercial acceptance will require economic methods for large-scale culture, packaging and distribution, and government approval that they are safe with respect to public health and the environment.

10.4 Detrimental aspects of fruit yeasts

The most significant, adverse impact of yeasts in or on fruits or fruit products is their potential to cause spoilage and consequent economic loss. Yeast spoilage of fruits has been thoroughly described and reviewed [33, 51, 119, 128]. In most instances, this spoilage is characterized by fermentation and the production of off-flavours and gassiness. With packaged products, gas formation leads to swelling and explosion of containers that could cause physical injury. Turbidity and sediment due to the presence of yeast biomass become evident in clarified, liquid products such as fruit juices and wines [123]. Fruits are used as ingredients in many commodities such as yoghurts, fruit pies, other bakery products and jams, and they can be a major source of yeast contamination and potential product spoilage.

In the context of foodborne disease, yeasts are not usually considered as a public health risk. Nevertheless, it is necessary to view this perception with increasing caution. An increasing number of yeast species, once thought as harmless, are now being considered in the category of emerging pathogens [63]. Such species include many that are commonly found in foods, including fruit products. The incidence of yeast infections of the blood and various body organs is increasing, particularly in patients who have immunodeficiency or are being treated with various immunosuppressive agents. The significance of foods as a source of pathogenic yeasts will come under increasing question.

10.5 Physiological and biochemical background

The physiological and biochemical properties of yeasts associated with foods and beverages, generally, have been discussed in previous reviews [51, 119]. This section will focus on those properties more relevant to fruit ecosystems.

The unique properties of phyllospheric yeasts associated with the surfaces of fruits have been mentioned in an earlier section (see also Table 10.2-1). For yeasts responsible for desirable fermentations and processes, the focus is on those properties that facilitate rapid growth, complete fermentations and biochemical activities that generate appropriate flavour and textural changes. Such properties have been extensively discussed for wine yeasts [14, 52, 53, 104]. Some key properties include ethanol tolerance, and metabolism of fruit juice sugars, amino acids, proteins and organic acids to produce a profile of flavour volatiles (esters, aldehydes, higher alcohols, fatty acids, thiols) that enhance wine character. More specific properties include the production of extracellular enzymes such as glycosidases that might contribute to wine flavour, and proteases that might impact on haze stability [24]. In the case of cocoa and coffee bean processing, production of extracellular pectic enzymes that degrade the bean pulp becomes a key yeast property [124].

With respect to yeasts that cause spoilage of fruit products, the focus is on conditions that limit or inactivate yeast growth. Key properties to be considered include: growth at low temperatures; inactivation by heat and other processes; response to water activity; and tolerance to preservatives. It is now widely recognized that most yeasts are psychrotrophic and are able to grow at refrigeration temperatures (5–10 °C). Consequently, refrigerated fruit products (especially, juices, pulps and concentrates) will generally encounter yeast spoilage if stored for extended time. Many species, including those often associated with fruit products (e. g., *S. cerevisiae, H. uvarum, Z. rouxii, Z. bailii*) have minimal growth temperatures in the range 0–5 °C (51) and some *Cryptococcus, Candida* and *Rhodotorula* species grow at less than 0 °C [27].

Outbreaks of yeast spoilage in heat-processed soft drinks, fruit juices and fruit concentrates have stimulated much interest in the susceptibility/resistance of yeasts to heat destruction. Thermal death curves and decimal reduction (D) values have been published for a range of yeasts such as *S. cerevisiae, H. uvarum, Z. bailii* and *Z. rouxii*, and various *Candida* species, that are associated with fruit products [51, 106, 121]. Under normal, non-stressful environmental conditions, the vegetative cells of these yeasts are quickly inactivated by heating at 60–65 °C, with D values generally less than 1 minute. However, these D values are significantly influenced by various environmental factors. They are decreased (i. e., the yeasts are more sensitive to destruction) when the cells are heated in the presence of weak acid preservatives, ethanol, anti-oxidants and low pH. In contrast, the cells are more resistant to inactivation when heated in environments containing high concentrations of sugars (as found in fruit juice concentrates) or salt. Thus, D values can increase to 5–10 minutes or more in the presence of 60 % glucose, depending on the synergistic influence of other factors such as pH. Yeast ascospores are about 50–100 times more heat-resistant than vegetative cells, depending on species and conditions [51, 106]. These heat-resistant data have important practical implications in designing processing conditions to prevent the spoilage of fruit juices and fruit concentrates or pulps. To decrease the intensity of such treatments, high hydrostatic pressure and pulsed electric field processes, combined with temperature, are being investigated for their ability to inactivate yeasts in various fruit products [51, 106, 108, 138]. However, the response of yeasts to these new processing technologies requires more investigation.

Fruit juice or fruit pulp concentrates generally have sugar concentrations of 50–70 % and water activity values of 0.85–0.90 or less, and are selectively spoiled by the growth of osmotolerant or xerotolerant yeasts (e. g., *Z. rouxii; Z. bailii; Z. bisporus; Sch. pombe*) [125]. The tolerance of these species to high sugar concentrations is not diminished by the low pH of the products [103]. These same species are also renowned for their resistance to permitted levels of weak acid preservatives (sorbate, benzoate, sulphur dioxide) [45, 51, 84, 103]. The biochemical and physiological mechanisms of tolerance to these extreme environments have been investigated [119, 125].

10.6 Specific methods of analysis for fruit-associated yeasts

The isolation, enumeration and identification of yeasts from fruits and fruit products follow the same principles and protocols for yeasts generally (see Chapters 2–4). These involve the sequential operations of: rinsing or maceration of the sample; dilution of the suspension; enumeration of the yeast cells in suspension by agar plating, most probable number, membrane filtration or microscopic methods; purification of isolates; and, identification of isolates to either genus, species or strain level. Details of these procedures can be found elsewhere [12, 51, see Chapters 2–4]. Nevertheless, particular considerations can apply to yeasts associated with fruit ecosystems.

MARTINI et al. [86] have emphasized the importance of thorough rinsing, maceration or sonication to dislodge and isolate all the yeast species associated with the surfaces of fresh fruits. Various authors have commented on the need to incorporate propionate, benomyl or biphenyl into agar plating media to suppress or restrict the growth of filamentous fungi (moulds) which could overgrow yeast colonies [32, 56, 58, 133].

Dichloran Rose Bengal Chloramphenicol Agar (DRBC) was developed to restrict the lateral growth of mould colonies and allow the simultaneous culture of yeasts and moulds from foods [100], but there is increasing evidence that this medium does not give maximum enumeration of yeasts, including those associated with fruits [3]. Special consideration is needed to maximize the recovery of osmotolerant or xerophilic yeasts from fruit concentrates. Additional sugar or glycerol should be incorporated into the diluents and agar media used to enumerate yeasts in these products, because they may be sensitive to low osmotic environments [3]. Yeasts in heat processed and frozen fruit products may be sublethally injured or stressed and require resuscitation for their recovery and enumeration [55]. Some authors have observed an increased diversity of yeasts isolated from the surfaces of fruits when isolation plates were incubated at lower temperatures (5–15 °C) compared to higher temperature [17, 29].

The identification of fruit associated yeasts follows the traditional morphological, biochemical and physiological tests outlined in KURTZMAN and FELL [72] and BARNETT et al. [6], but now supplemented with sequence data for ribosomal DNA subunits [48, 73, see Chapter 3]. Rapid kit identification systems (SIM, API 20C, ATB32, Biolog, see Chapter 5) have been successfully applied to species of fruit yeasts [34, 35, 102, 127]. Analysis of fatty acid composition and various PCR-based methods are now available for the identification of yeasts obtained from fruit ecosystems [80, 114]. Impedimetric and ATP-bioluminescence methods have been evaluated for the more rapid enumeration of yeasts in fruit products but these methods have not received wide acceptance [34, 51].

10.7 Quality control

The main concern, here, is prevention of spoilage. Since yeasts have a natural association with fruits, attention needs to be focussed on the application of Hazard Analysis of Critical Control Points (HACCP) and other quality management programs from raw material to final product. Pre-harvest management practices to prevent damage to fruit in the field, harvesting before over-maturity and separation of damaged fruit from non-damaged fruit are obvious strategies to minimize the load of yeasts on raw material prior to further processing. For heat-processed fruit products, it is essential that the time-temperature combinations be effective at yeast inactivation [119]. Where refrigeration is used, it should be recalled that many yeasts are psychrotrophic and can grow well at low temperatures. Refrigeration, therefore, will not prevent yeast spoilage, but only slows down the process. The lower the temperature, the longer the shelf-life, so good temperature management becomes very important. Weak acid preservatives are used to control yeasts in many heat-processed and refrigerated fruit products. For maximum efficiency of these agents, good management of their concentration and product pH is essential. Nevertheless, it needs to be recognized that preservative-resistant species may compromise this initiative. Good hygiene practices throughout all stages of processing are needed to prevent their access to the product. This will include effective cleaning and sanitation of process equipment, and preventing access of insects such as bees and fruit flies that are major vectors of yeasts. Poor cleaning and sanitation is one of the major causes of yeast spoilage of fruit products. Specifications (e. g., yeasts not detected in 1–10 g or 100–200 ml of product) may be adopted to manage product quality in some cases. Routine monitoring of raw materials or end products to ensure that they meet appropriate specifications is usually part of an overall strategy to manage product quality [80].

10.8 Future prospects and conclusions

Further research is needed to better understand the biodiversity of yeasts associated with the surfaces of fresh fruits and the mechanisms by which these yeasts survive and grow at this phyllospheric habitat throughout the pre-harvest stage. Present knowledge about this biodiversity comes from the use of standard cultural methodology to determine the yeast microflora. This information needs to be supported by studies using newer, molecular ecological approaches that examine the total microbial DNA with techniques such as denaturing or temperature gradient gel electrophoresis. By analogy with other ecosystems, such strategies are likely to reveal a wider biodiversity than currently known [95]. Greater understanding of this phyllospheric ecology should advance the prospect of using yeasts for the biocontrol of fruit spoilage fungi, and enable more creative exploitation of indigenous yeasts in wine fermentation and fruit processing. With respect to yeasts used in wine fermentation and fruit processing, further study is needed to link the biochemical and physiological activities of individual species and strains with particular sensory attributes in the final product. Such data will enable the selection of a broader diversity of strains in order to develop commercially available starter cultures. Strategies for controlling fruit spoilage yeasts will be advanced by developing a better understanding of their resistance or susceptibility to processing operations, especially some of the newer processes [57]. More rapid, automated systems, probably based on nucleic acid probes, for the detection of individual spoilage species are required.

10.9 References

[1] ABRANCHES, J.; STARMER, W.T.; HAGLER, A.N.: Yeast – yeast interactions in guava and tomato fruits. Microb. Ecol. **42** (2001) 186–192.

[2] AMERINE, M.; KUNKEE, R.E.: Microbiology of winemaking. Ann. Rev. Microbiol. **22** (1968) 323–358.

[3] ANDREWS, S.; DE GRAAF, H.; STAMOTION, H.: Optimisation of methodology for enumeration of xerophilic yeasts from foods. Int. J. Fd Microbiol. **35** (1997) 109–116.

[4] ARRAS, G.: Mode of action of an isolate of *Candida famata* in biological control of *Penicillium digitatum* in orange fruits. Postharvest Biol. Technol. **8** (1996) 191–198.

[5] BARNETT, J.A.; DELANEY, M.A.; JONES, E.; MAGSON, A.B.; WINCH, B.: The numbers of yeasts associated with wine grapes of Bordeaux. Arch. Mikrobiol. **83** (1972) 52–55.

[6] BARNETT, J.A.; PAYNE, R.W.; YARROW, D.: Yeasts: characteristics and identification, 3rd ed. Cambridge, U.K.: Cambridge University Press (2000).

[7] BEECH F.W.: Yeasts in cider-making. In: The Yeasts, Vol. 5, 2nd ed.. (edited by Rose, A.H.; Harrison, J.S.). London, U.K.: Academic Press (1993) 169–214.

[8] BEECH, F.W.: The yeast flora of apple juices and ciders. J. Appl. Bacter. **21** (1958) 257–266.

[9] BEECH, F.W.; DAVENPORT, R.R.: The role of yeasts in cider-making. In: The yeasts, Vol. 3, 1st ed. (edited by Rose, A.H.; Harrison, J.S.). London, U.K.: Academic Press (1970) 73–146.

[10] BELIN, J.M.: Recherches sur la répartition des levures à la surface de la grappe de raisin. Vitis **11** (1972) 135–145.

[11] BENBOW, J.; SUGAR, D.: Fruit surface colonization and biological control of post-harvest diseases of pear by pre-harvest yeast applications. Pl. Dis. **83** (1999) 839–844.

[12] BEUCHAT, L.R.: Selective media for detecting and enumerating foodborne yeasts. Int. J. Fd Microbiol. **19** (1993) 1–14.

[13] BEUCHAT, L.R.: Pathogenic microorganisms associated with fresh produce. J. Fd Protect. **59** (1996) 204–216.

[14] BISSON, L.F.: Stuck and sluggish fermentations. Am. J. Enol. Viticult. **50** (1999) 107–119.

[15] BOWEN, J.F.; BEECH, F.W.: The distribution of yeasts on cider apples. J. Appl. Bacteriol. **27** (1964) 333–341.

[16] BOWEN, J.F.; BEECH, F.W.: Yeast flora of cider factories. J. Appl. Bacteriol. **30** (1967) 475–483.

[17] BUHAGIAR, R.W.M.; BARNETT, J.A.: The yeasts of strawberries. J. Appl. Bacteriol. **34** (1971) 727–739.

[18] CABRAS, P.; ANGIONI, A.: Pesticide residues in grapes, wine and their processing products. J. Agricult. Fd Chem. **48** (2000) 967–973.

[19] CABRAS, P.; ANGIONI, A.; GARAU, V.L.; PIRISI, F.M.; FARRIS, G.A.; MADOU, G.; EMONTI, G.: Pesticides in fermentative processes of wine. J. Agricult. Fd Chem. **47** (1999) 3854–3857.

[20] CASTELLI, T.: Les agents de la fermentation vineire. Arch. Mikrobiol. **20** (1956) 323–342.

[21] CHALUTZ, E.; DROBY, S.: Biological control of post-harvest disease. In: Plant-microbe interactions and biological control (edited by Boland, G.J.; Kuykendall, L.D.). New York, U.S.A.: Marcel Dekker (1998) 157–170.

[22] CHAND-GOYAL, T.; ECKERT, J.W.; DROBY, S.; ATKINSON, K.: A method for studying the population dynamics of *Candida oleophila* on oranges in the grove, using a selective isolation medium and PCR technique. Microbiol. Res. **153** (1998) 265–270.

[23] CHAND-GOYAL, T.; SPOTTS, R.A.: Enumeration of bacterial and yeast colonists of apple fruits and identification of epiphytic yeasts on pear fruits in the Pacific Northwest United States. Microbiol. Res. **151** (1996) 427–432.

[24] CHAROENCHAI, C.; FLEET, G.H.; HENSCHKE, P.; TODD, B.: Screening of non-*Saccharomyces* wine yeasts for the presence of extracellular hydrolytic enzymes. Austr. J. Grape Wine Res. **3** (1997) 2–8.

[25] CLARK, D.S.; WALLACE, R.H.; DAVID, T.J.: Yeasts occurring on apples and in apple cider. Can. J. Microbiol. **1** (1954) 145–149.

[26] COOK, D.W.M.; LONG, P.G.; GANESH, S.: The combined effect of delayed application of yeast biocontrol agent and fruit curing for the inhibition of the post-harvest pathogen *Botrytis cinerea* in kiwi fruit. Post harvest Biol. Technol. **16** (1999) 233–243.

[27] DAVENPORT, R.R.: Cold-tolerant yeasts and yeast like organisms. In: Biology and activities of yeasts (edited by Skinner, R.A.; Passmore, S.M.; Davenport, R.R.). London, U.K.: Academic Press (1980) 215–230.

[28] DAVENPORT, R.R.: Ecological concepts in studies of microorganisms on aerial plant surfaces. In: Microbiology of aerial plant surfaces (edited by Dickinson, C.H.; Preece, T.F.). London, U.K.: Academic Press (1976) 200–215.

[29] DAVENPORT, R.R.: Microecology of yeasts and yeast-like organisms associated with an English vineyard. Vitis **13** (1974) 123–130.

[30] DE CAMARGO, R.; PHAFF, H.J.: Yeasts occurring in *Drosophila* flies and in fermenting tomato fruits in Northern California. Fd Res. **22** (1957) 367–372.

[31] DE LA TORRE, M.J.; MILLAN, M.C.; PEREZ-JUAN, P.; MORALES, J.; ORTEGA, J.M.: Indigenous yeasts associated with *Vitis vinifera* grape varieties cultured in southern Spain. Microbios **100** (1999) 27–40.

[32] DE LA TORRE, M.J; MILLAN, M.C.; PEREZ-JUAN, P.M.; MORALES, J.; ORTEGA, J.M.: Changes in the microbiota during ripening of two *Vitis vinifera* grape varieties growing in Southern Spain. Microbios **96** (1998) 165–176.

[33] DEAK, T.: Foodborne yeasts. Adv. Appl. Microbiol. **36** (1991) 179–278.

[34] DEAK, T.: Methods for the rapid detection and identification of yeasts. Trends Fd Sc. Technol. **6** (1995) 287–292.

[35] DEAK, T.; BEUCHAT, L.R.: Comparison of the SIM, API 20C and ID32C systems for identification of yeasts from fruit juice concentrates and beverages. J. Fd Protect. **56** (1993) 585–592.

[36] DEAK, T.; BEUCHAT, L.R.: Yeasts associated with fruit juice concentrates. J. Fd Protect. **56** (1993) 777–782.

[37] DENNIS, L.; DAVIS, R.P.: Susceptibility of strawberry varieties to post-harvest fungal spoilage. J. Appl. Bacteriol. **42** (1977) 197–206.

[38] DI MENNA, M.E.: The antibiotic relationships of some yeasts from soil and leaves. J. Gen. Microbiol. **27** (1962) 249–257.

[39] DICKINSON, C.H.: Adaptations of microorganisms to climatic conditions affecting aerial plant surfaces. In: Microbiology of the phyllosphere (edited by Fokkema, N.J.; van den Heuvel, J.). Cambridge, U.K.: Cambridge University Press (1986) 77–100.

[40] DO CARMO-SOUSA, L.: Distribution of yeasts in nature. In: The yeasts, Vol. 1, 1st ed. (edited by Rose, A.H.; Harrison, J.S.). London, U.K.: Academic Press (1969) 79–106.

[41] DROBY, S.; CHALUTZ, E.; WILSON, C.L.; WISNIEWSKI, M.: Characterization of the biocontrol activity of *Debaryomyces hansenii* in the control of *Penicillium digitatum* on grapefruit. Can. J. Microbiol. **35** (1989) 794–800.

[42] DROBY, S.; COHEN, L.; DAVIS, A.; WEISS, B.; HOREV, B.; CHALUTZ, E.; KATZ, H.; KEREN-TZUR, M.; SHACHNAI, A.: Commercial testing of Aspire: a yeast preparation for the biological control of postharvest decay of citrus. Biol. Control **12** (1998) 97–101.

[43] DROBY, S.; LISCHINSKI, S.; COHEN, L.; WEISS, B.; DAVIS, A.; CHAND-GOYAL T.; ECKERT, J.W.; MANULIS, S.: Characterization of an epiphytic yeast population of grapefruit capable of suppression of green mold decay caused by *Penicillium digitatum.* Biol. Control **16** (1999) 27–34.

[44] EL-GHAOUTH, A.; SMILANICK, J.L.; BROWN, C.E.; IPPOLITO, A.; WILSON, C.L.: Control of decay of apple and citrus fruits in semi-commercial tests with *Candida saitoana* and 2-deoxy-D-glucose. Biol. Control **20** (2001) 96–101.

[45] EL-HALOUAT, A.; GOURAMA, H.; UYTTENDAELE, M.; DEBEVERE, J.M.: Effects of modified atmosphere packaging and preservatives on the shelf-life of high moisture prunes and raisins. Int. J. Fd Microbiol. **41** (1998) 177–184.

[46] ETHIRAJ, S.; SURESH, E.R.; ONKARAYYA, H.: Nature of yeasts present on grapes grown in south India and their wines. Vitis **13** (1979) 161–164.

[47] FAN, Q.; TIAN, S.: Post-harvest biological control of grey mold and blue mold on apple by *Cryptococcus albidus* (Saito) Skinner. Post-harvest Biol. Technol. **21** (2001) 341–350.

[48] FELL, J.W.; BOEKHOUT, T.; FONSECA, A.; SCORZETTI, G.; STATZELL-TALLMAN, A.: Biodiversity and systematics of basidiomycetous yeasts as determined by large subunit rDNA D1/D2 domain sequence analysis. Int. J. Syst. Evol. Microbiol. **50** (2000)1351–1371.

[49] FILONOW, A.B.; VISHNIAC, H.S.; ANDERSON, J.A.; JANISIEWICZ, W.J.: Biological control of *Botrytis cinerea* in apple by yeasts from various habitats and their putative mechanisms of antagonism. Biol. Control **7** (1996) 212–220.

[50] FLEET, G.H.: Microbiology of alcoholic beverages. In: Microbiology of fermented foods, Vol. 1, 2nd ed. (edited by B.J. Wood). London, U.K.: Blackie Academic & Professional (1998) 217–262.

[51] FLEET, G.H.: Spoilage yeasts. Crit. Rev. Biotechnol. **12** (1992) 1–44.

[52] FLEET, G.H.: Wine microbiology and biotechnology. Chur, Switzerland: Harwood Academic Publishers (1993).

[53] FLEET, G.H.: Wine. In: Food Microbiology Fundamentals and Frontiers, 2nd ed. (edited by Doyle, M.P.; Beuchat, L.R.; Montville, T.J.). Washington, D.C., U.S.A.: ASM Press (2001) 747–772.

[54] FLEET, G.H.: Microorganisms in food ecosystems. Int. J. Fd Microbiol. **50** (1999) 101–117.

[55] FLEET, G.H.; MIAN, M.: Induction and repair of sublethal injury in food spoilage yeasts. J. Fd Mycol. **2** (1998) 85–94.

[56] GOTO, S.; YOKOTSUKA, I.: Wild yeast populations in fresh grape musts of different harvest times. J. Ferment. Technol. **55** (1977) 417–422.

[57] GOULD, G.W.: New and emerging physical methods of preservation. In: The microbiological safety and quality of food, Vol. 1 (edited by Lund, B.M.; Baird-Parker, T.C.; Gould, G.W.). Gaithersburg, Maryland, U.S.A.: Aspen Publisher (2000) 277–293.

[58] GUERRA, E.; SORDI, G.; MANNAZZO, I.; CLEMENTI, F.; FATICHENTI, F.: Occurrence of wine yeasts on grapes subjected to different pesticide treatments. Ital. J. Fd Sc. **11** (1999) 221–230.

[59] GUERZONI, E.; MARCHETTI, R.: Analysis of yeast flora associated with grape sour rot and of chemical disease markers. Appl. Envir. Microbiol. **53** (1987) 571–576.

[60] GUINEBRETRIERE, M.H.; NGUYEN-THE, C.; MORRISON, N.; REICH, M.; NICOT, P.: Isolation and characterization of antagonists for the biocontrol of post-harvest wound pathogen *Botrytis cinerea* on strawberry fruits. J. Fd Protect. **63** (2000) 386–394.

[61] HARDIE, W.J.; O'BRIEN, T.P.; JAUDZEMS, V.G.: Morphology, anatomy and development of the pericarp after anthesis in grape *Vitis vinifera*. Austr. J. Grape Wine Res. **2** (1996) 97–142.

[62] HARRIS, L.J.: The microbiology of vegetable fermentations. In: Microbiology of fermented foods, 2nd ed. (edited by Wood, B.J.). London, U.K.: Blackie Academic & Professional (1998) 45–72.

[63] HAZEN, K.C.: New and emerging yeast pathogens. Clin. Microbiol. Rev. **8** (1995) 462–478.

[64] HEARD, G.M.: Novel yeasts in winemaking – looking to the future. Fd Austr. **51** (1999) 347–352.

[65] HEIDENREICH, M.C.M.; CORRAL-GARCIA, M.R.; MOMOL, E.A.; BURR, T.J.: Russet of apple fruit caused by *Aureobasidium pullulans* and *Rhodotorula glutinis*. Plant Dis. **81** (1997) 337–342.

[66] INGRAM, M.: Yeasts in food spoilage. In: The chemistry and biology of yeasts (edited by Cook, A.H.). New York, U.S.A.: Academic Press (1958) 603–635.

[67] International Commission on Microbiological Specifications for Foods. In: Microorganisms in food, Vol. 6, Microbial ecology of food commodities. London, U.K.: Blackie Academic & Professional (1998) 252–273.

[68] KHAN, W.; AUGUSTYN, O.P.H.; VAN DER WESTHUIZEN, T.J.; LAMBRECHTS, M.G.; PRETORIUS, I.S.: Geographic distribution and evaluation of *Saccharomyces cerevisiae* strains isolated from vineyards in the warmer, in-land regions of the Western Cape in South Africa. S. Afr. J. Enol. Viticult. **21** (2000) 17–31.

[69] KOSUGE, T.; HEWITT, W.B.: Exudates of grape berries and their effect on germination of conidia of *Botrytis cinerea*. Phytopathology **54** (1964) 167–172.

[70] KUNKEE, R.E.; AMERINE, M.: Yeasts in winemaking. In: The yeasts, Vol. 3, 1st ed. (edited by Rose, A.H.; Harrison, J.S.). London, U.K.: Academic Press (1970) 5–71.

[71] KUNKEE, R.E.; BISSON, L.F.: Wine-making yeasts. In: The yeasts, Vol. 5, 2nd ed. (edited by Rose, A.H.; Harrison J.S.). London, U.K.: Academic Press (1993) 69–128.

[72] KURTZMAN, C.P.; FELL, J.W.: The yeasts: a taxonomic study, 4th ed. Amsterdam, The Netherlands: Elsevier (1998).

[73] KURTZMAN, C.P.; ROBNETT, C.J.: Identification and phylogeny of ascomycetous yeasts from analysis of large subunit (26S) ribosomal DNA partial sequences. Antonie van Leeuwenhoek **73** (1998) 331–378.

[74] LACHANCE, M.A.; STARMER, W.J.; ROSA, C.A.; BOWLES, J.M.; BARKER, J.S.F.; JANZEN, D.H.: Biogeography of the yeasts of ephemeral flowers and their insects. FEMS Yeast Research **1** (2001) 1–8.

[75] LAST, F.T., PRICE, D.: Yeasts associated with living plants and their environs. In: The yeasts, Vol. 1, 1st ed. (edited by Rose, A.H. and Harrison, J.S.) London, U.K.: Academic Press (1969) 183–218.

[76] LAST, F.T.; WARREN, R.C.: Non-parasitic microbes colonising green leaves: their form and functions. Endeavour **31** (1972) 143–150.

[77] LE ROUX, G.; ESCHENBRUCH, R.; DE BRUIN, S.I.: The microbiology of South African winemaking. Part VIII. The microflora of healthy and *Botrytis cinerea* infected grapes. Phytophylactica **5** (1973) 51–54.

[78] LEBEN, C.: Microorganisms associated with plant buds. J. Gen. Microbiol. **71** (1972) 327–331.

[79] LONGO, J.; CANGADO, J.; AGRELO, D.; VILLA, T.G.: Effect of climatic conditions on yeast diversity in grape musts from north west Spain. Am. J. Enol. Viticult. **42** (1991) 141–144.

[80] LOUREIRO, V.; QUEROL, A.: The prevalence and control of spoilage yeasts in foods and beverages. Trends Fd Sc. Technol. **10** (1999) 356–365.

[81] LOWINGS, P.H.: The fungal contamination of Kentish strawberry fruits in 1955. Appl. Microbiol. **4** (1956) 84–88.

[82] LUND, B.M.; SNOWDON, A.L.: Fresh and processed fruit. In: The microbiological safety and quality of food (edited by Lund, B.M.; Baird-Parker, T.C.; Gould, G.W.). Gaithersbrug, Maryland, U.S.A.: Aspen Publishers (2000) 738–758.

[83] LÜTHI, H.: Microorganisms in non-citrus juices. Adv. Fd Res. **9** (1959) 221–284.

[84] MAIMER, E.; BUSSE, M.: Growth properties and gas formation by yeasts isolated from processed fruits in media with various Brix values and sorbic acid conents. J. Fd Protect. **55** (1992) 192–197.

[85] MARSHALL, C.R.; WALKLEY, V.J.: Some aspects of microbiology applied to commercial apple juice production. I. Distribution of microorganisms on the fruit. Fd Res. **16** (1951) 448–456.

[86] MARTINI, A.; CIANI, M.; SCORZETTI, G.: Direct enumeration and isolation of wine yeasts from grape surfaces. Am. J. Enol. Viticult. **47** (1996) 435–440.

[87] MARTINI, A.; FEDERICI, F.: A new approach to the study of the yeast flora associated with strawberry surfaces. Giorn. Bot. Ital. **110** (1976) 297–301.

[88] MCCORMACK, P.J.; WILDMAN, H.G.; JEFFRIES, P.: Production of antibacterial compounds by phylloplane-inhibiting yeasts and yeast-like fungi. Appl. Environ. Microbiol. **60** (1994) 927–931.

[89] MERCIER, J.; WILSON, C.L.: Colonisation of apple wounds by naturally occurring microflora and introduced *Candida oleophila* and their effect on infection by *Botrytis cinerea* during storage. Biol. Control **4** (1994) 138–144.

[90] MOORE, K.J.; JOHNSON, M.G.; MORRIS, J.R.: Indigenous yeast microflora on Arkansas white Riesling (*Vitis vinifera*) grapes and in model must systems. J. Fd Sc. **53** (1988) 1725–1728.

[91] MORAIS, P.B.; MARTINS, M.B.; KLACZKO, L.B.; MENDONÇA-HAGLER, L.C.; HAGLER, A.N.: Yeast succession in the Amazon fruit *Parahancornia amapa* as resource partitioning among *Drosophila* spp. Appl. Environ. Microbiol. **61** (1995) 4251–4257.

[92] MORTIMER, R.K.; POLSINELLI, M.: On the origins of wine yeast. Res. Microbiol. **150** (1999) 199–204.

[93] MRAK, E.M.; MCCLUNG, L.S.: Yeasts occurring on grapes and in grape products in California. J. Bacteriol. **40** (1940) 395–400.

[94] MRAK, E.M.; PHAFF, H.J.: Yeasts. Ann. Rev. Microbiol. **2** (1948) 1–46.

[95] MUYZER, G.; SMALLA, K.: Applications of denaturing gradient gel electrophoresis (DGGE) and temperature gradient gel electrophoresis (TGGE) on microbial ecology. Antonie van Leeuwenhoek **73** (1998) 127–141.

[96] PARISH, M.E.; CARROLL, D.E.: Indigenous yeasts associated with Muscedina (*Vitis rotundifolica*) grapes and musts. Am. J. Enol. Viticult. **36** (1985) 165–167.

[97] PARLE, J.N.; DI MENNA, M.E.: The source of yeasts in New Zealand wines. N. Z. J. Agricult. Res. 9 (1966) 98–107.

[98] PHAFF, H.J; STARMER, W.T.: Yeasts associated with plants, insects and soil. In: The yeasts, Vol. 1, 2nd ed. (edited by Rose, A.H.; Harrison, J.S.) London, U.K.: Academic Press (1987) 123–180.

[99] PIANO, S.; NEYROTTI, V.; MIGHELI, Q.; GULLINO, M.L.: Biocontrol capability of *Metschnikowia pulcherrima* against *Botrytis* post-harvest rot of apple. Post-harvest Biol. Technol. **11** (1997) 131–140.

[100] PITT, J.I.; HOCKING, A.: Fungi and food spoilage, 2nd ed. London, U.K.: Blackie Academic & Professional (1997).

[101] POSTMASTER, A.; SIVASITHAMPARUM, K.; TURNER, D.W.: Enumeration and identity of microorganisms isolated from the surface of banana fruits at three developmental stages. Sc. Horticult. **69** (1997) 189–197.

[102] PRAPHAILONG, W.; VAN GESTEL, M.; FLEET, G.H.; HEARD, G.M.: Evaluation of the Biolog system for identification of food and beverage yeasts. Lett. Appl. Microbiol. **24** (1997) 455–459.

[103] PRAPHAILONG, W.; FLEET, G.H.: The effect of pH, sodium chloride, sucrose, sorbate and benzoate on the growth of spoilage yeasts. Fd Microbiol. **14** (1997) 459–468.

[104] PRETORIUS, I.S.: Tailoring wine yeasts for the new millennium: novel approaches to the ancient art of winemaking. Yeast **16** (2000) 675–729.

[105] PRETORIUS, I.S.; VAN DER WESTHUIZEN, T.J.; AUGUSTYN, O.P.H.: Yeast biodiversity in vineyards and wineries and its importance to the South African wine industry: a review. S. Afr. J. Enol. Viticult. **20** (1999) 61–74.

[106] RASO, J.; CALDERON, M.C.; GONGORA, M.; BARBOSA-CANOVAS, G.V.; SWANSON, B.G.: Inactivation of *Zygosaccharomyces bailii* in fruit juices by heat, high hydrostatic pressure and pulsed electric fields. J. Fd Sc. **63** (1998) 1042–1044.

[107] RECCA, J.; MRAK, E.M.: Yeasts occurring in citrus products. Fd Technol. December (1952) 450–454.

[108] REYNS, K.M.F.A.; SOONTJENS, C.C.F.; CORNELIS, K.; WEEMOES, C.A.; HENDRICKX, E.M.; MICHIELS, C.W.: Kinetic analysis and modelling of combined high-pressure-temperature inactivation of the yeast *Zygosaccharomyces bailii*. Int. J. Fd Microbiol. **56** (2000) 198–210.

[109] Riordan, D.C.R.; Sapero, G.M.; Hankinson, T.R.; Magee, M.; Mattrazzo, A.M.; Annous, B.A.: A study of U.S. orchard to identify potential sources of *Escherichia coli* 0157: H7. J. Fd Protect. **64** (2001) 1320–1327.

[110] Roberts, R.G.: Post-harvest biological control of grey mold of apple by *Cryptococcus laurentii.* Phytopathol. **80** (1990) 526–530.

[111] Rosini, G.; Federici, F.; Martini, A.: Yeast flora of grape berries during ripening. Microb. Ecol. **8** (1982) 83–89.

[112] Sagt, C.M.; Müller, W.H.; Boonstra, J.; Verkleij, A.J.; Verrips, C.T.: Impaired secretion of a hydrophobic cutinase by *Saccharomyces cerevisiae* correlated with an increase associated with immunoglobulin heavy chain binding protein. Appl. Environ. Microbiol. **64** (1998) 316–324.

[113] Samish, Z.; Etinger-Tulczynska, R.; Bick, M.: The microflora within the tissue of fruits and vegetables. J. Fd Sc. **28** (1963) 259–266.

[114] Sancho, T.; Gimenez-Jurado, G.; Malfeito-Ferreira, M.; Loureiro, V.: Zymological indicators: a new concept applied to detection of potential spoilage yeast species associated with fruit pulps and concentrates. Fd Microbiol. **17** (2000) 613–624.

[115] Sandhu, D.K.; Waraich, M.K.: Yeasts associated with pollinating bees and flower nectar. Microb. Ecol. **11** (1985) 51–58.

[116] Schwan, R.F.; Wheals, A.E.: Mixed microbial fermentations of chocolate and coffee. In: Yeasts in food (edited by Boekhout, T.; Robert, V.). Hamburg, Germany: Behr's Verlag (2003) 429–449.

[117] Starmer, W.T.; Ganter, P.F.; Aberdeen, V.; Lachance, M.A.; Phaff, H.J.: The ecological role of killer yeasts in natural communities of yeasts. Can. J. Microbiol. **33** (1987) 783–796.

[118] Stirling, A.M.; Pegg, K.G.; Hayward, A.C.; Stirling, G.R.: Effect of copper fungicide on *Colletotrichum gloesporioides* and other microorganisms on avocado leaves and fruit. Austr. J. Agric. Res. **50** (1999) 1459–1468.

[119] Stratford, M.; Hofman, P.O.; Cole, M.B.: Fruit juice fruit drinks and soft drinks. In: The Microbiological safety and quality of food (edited by Lund, B.M.; Baird-Parker, T.C.; Gould, G.W.). Gaithersburg, Maryland, U.S.A: Aspen Publishers (2000) 836–869.

[120] Suarez, J.A.; Gonzalez, M.C.; Callejo, M.J.; Colomo, B.; Gonzalez, A.: Contribution to the study of varietal wines from Rioja and Navara. (1) Microbial growth trends during grape maturation. Bull. L'OIV **759–760** (1994) 389–407.

[121] Tchango Tchango, J.; Tailliez, R.; Eb, P.; Njine, T.; Hornez, J.P.: Heat resistance of the spoilage yeasts *Candida pelliculosa* and *Kloeckera apis* and pasteurisation for some tropical fruit juices and nectars. Fd Microbiol. **14** (1997) 93–99.

[122] Teixido, N.; Usall, J.; Magan, N.; Vinaqs, I.: Microbial population dynamics on Golden Delicious apples from bud to harvest and effect of fungicide applications. Annls Appl. Biol. **134** (1999) 109–116.

[123] Thomas, S.: Yeasts as spoilage organisms in beverages. In: The yeasts, Vol. 5, 2nd ed. (edited by Rose A.H.; Harrison, J.S.) London, U.K.: Academic Press (1993) 517–562.

[124] Thompson, S.S.; Miller, K.; Lopez, A.: Cocoa and coffee. In: Food microbiology fundamentals and frontiers, 2nd edition (edited by Doyle, M.P.; Beuchat, L.R.; Montville, T.J.). Washington, D.C., U.S.A.: ASM Press (2001) 721–734.

[125] Tokuoka, K.: A review: sugar and salt tolerant yeasts. J. Appl. Bacteriol. **74** (1993) 101–110.

[126] TOROK, T.; MORTIMER, R.K.; ROMANO, P.; SUZZI, G.; POLSINELLI, M.: Quest for wine yeasts – an old story revisited. J. Ind. Microbiol. **17** (1996) 303–313.

[127] TOROK, T.; KING, A.D.: Comparative study on the identification of foodborne yeasts. Appl. Envir. Microbiol. **57** (1991) 1207–1212.

[128] TUDOR, E.A.; BOARD, R.G.: Food spoilage yeasts. In: The yeasts, Vol. 5, 2nd ed. (edited by Rose, A.H.; Harrison, J.S.). London, U.K.: Academic Press (1993) 435–516.

[129] USALL, J.; TEXIDO, N.; TORRES, R.; DE ERIBE, X.O.; VINAS, I.: Pilot tests of *Candida sake* (CPA-1) applications to control post-harvest blue mold on apple fruit. Post-harvest Biol. Technol. **21** (2001) 147–156.

[130] VACEK, D.C.; STARMER, W.T.; HEED, W.B.: Relevance of the ecology of *Citrus* yeasts to the diet of *Drosophila*. Microbiol. Ecol. **5** (1979) 43–49.

[131] VAN DER WESTHUIZEN, T.J.; AUGUSTYN, O.P.H.; KHAN, W.; PRETORIUS, I.S.: Seasonal variation of indigenous *Saccharomyces cerevisiae* strains isolated from vineyards in the coastal regions of the Western Cape in South Africa. S. Afr. J. Enol. Viticult. **21** (2000) 10–16.

[132] VAN DER WESTHUIZEN, T.J.; AUGUSTYN, O.P.H.; PRETORIUS, I.S.: Geographical distribution of indigenous *Saccharomyces cerevisiae* strains isolated from vineyards in the coastal regions of the Western Cape in South Africa. S. Afr. J. Enol. Viticult. **21** (2000) 3–9.

[133] VAN ZYL, J.A.; DU PLESSIS, L.D.W.: The microbiology of South African winemaking. Part 1. The yeasts occurring in vineyards musts and wines. S. Afr. J. Agric. Sc. **4** (1961) 393–402.

[134] VERSAVAUD, A.; COURCOUX, P.; ROULLAND, C.; DULAU, L.; HALLET, J.N.: Genetic diversity and geographical distribution of wild *Saccharomyces cerevisiae* strains from the wine producing area of Charentes, France. Appl. Envir. Microbiol. **61** (1995) 3521–3529.

[135] WALKER, H.W.; AYRES, J.C.: Yeasts as spoilage organisms. In: The yeasts, Vol. 3, 1st ed. (edited by Rose, A.H.; Harrison, J.S.). London, U.K.: Academic Press (1970) 463–527.

[136] WILLIAMS, A.J.; WALLACE, R.H.; CLARK, D.S.: Changes in the yeast population on Quebec apples during ripening. Can. J. Microbiol. **2** (1956) 645–648.

[137] YANAGIDA, F.; ICHINOSE, F.; SHINOHARA, T.; GOTO, S.: Distribution of wild yeasts in the white grape varieties of central Japan. J. Gen. Appl. Microbiol. **38** (1992) 501–504.

[138] ZOOK, C.D.; PARISH, M.E.; BRADDOCK, R.J.; BALABON, M.D.: High pressure inactivation kinetics of *Saccharomyces cerevisiae* ascospores in orange and apple juices. J. Fd Sc. **64** (1999) 533–535.

11 Yeasts in bread and baking products

BERNARD BONJEAN and LUC-DOMINIQUE GUILLAUME

11.1 Introduction

A recent archaeological discovery showed that the ancient inhabitants of Asia Minor already utilized yeast as a fermentation agent. *Saccharomyces cerevisiae* is one of the best studied eukaryotes, both from academic and industrial points of view, and is the yeast species used in the bakery. The publication of the complete genome of *S. cerevisiae* [22] strongly boosted the study and understanding of the cellular functions of this unicellular organism. In this chapter we will review the use of yeast in the bakery, and describe the industrial production, the different types of yeasts used by small and large-scale bakeries, the genetic improvements obtained recently, and spoilage of baking products.

11.2 Properties of baking yeast

Leavening systems are used to raise bakery and pastry products. This can be achieved in several ways:

a) By fermentation
Sugar + yeasts $\rightarrow$ ethanol + CO_2 + ATP (energy)

b) By decomposition of ammonium bicarbonate
$NH_4HCO_3 \rightarrow NH_3 + H_2O + CO_2$

c) By a chemical reaction between a base and an acid
$HX + NaHCO_3 \rightarrow NaX + H_2O + CO_2$

d) By a change of the condition of water, for example in puff pastry (Fig. 11.2-1).

e) By incorporation of air in the batter, for example in Biscuit-genoises.

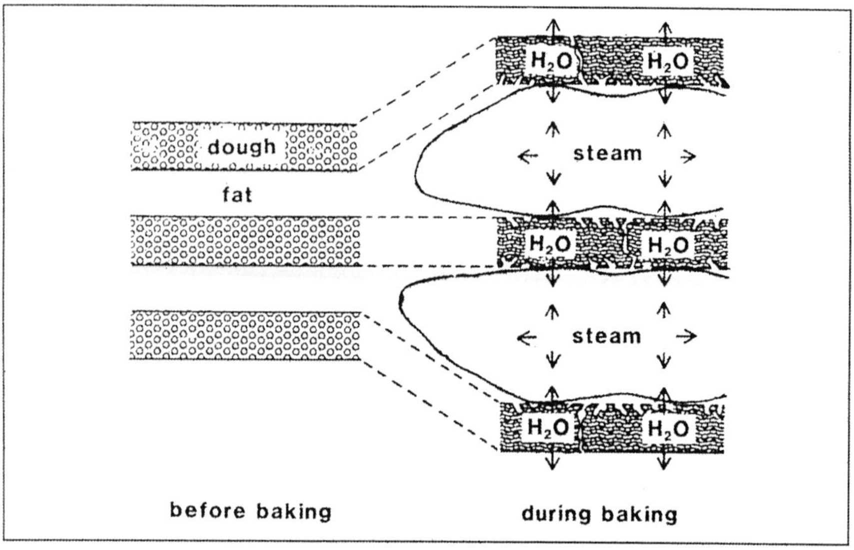

Fig. 11.2-1 **The principle of puff pastry aeration. In the aeration mechanism, the steam is moving from the dough layer into the liquid fat layers, forcing the dough layers to separate and lift in the pastry.**

11.2.1 Yeast in bread making process

11.2.1.1 Yeast as a fermentation agent

In 1941, BAKER [5] performed comparative tests between stiff doughs packed in vacuum versus those kept in the open air. His results indicated that stiff doughs packed in vacuum resulted in a smaller volume of the baked product, a coarser structure of the soft part and a more intense coloration of the crust. This phenomenon was explained by the lower number of alveoli with gas and by the smaller diameter of the alveoli resulting in a higher internal pressure. Also, the concentration of dissolved CO_2 in water is important for the development of the dough. Baker's conclusion was that yeast is not able to generate any additional gas bubbles during fermentation, but that only pre-existing gas bubbles will further develop.

In 1979, CHAMBERLAIN and COLLINS [10] stiffed dough in oxygen. During the first 10 minutes of the stiffening the volume of the doughs diminished. Their explanation was that yeast transformed the oxygen present in the gaseous alveoli into dissolved CO_2, thus causing the disappearance of gaseous nuclei.

During the fermentation, yeast releases CO_2 gas. This gas dissolves into the water and saturates the aqueous phase. After this saturation has been reached, all the CO_2 produced will diffuse into the unsaturated gas phase and allow the increase of the volume of the bread. This phenomenon of solubilization of CO_2 into water causes a lowering of the pH and a subsequent increase of the acidity of the dough.

Yeast contributes to the flavour of the baking products as well. The alcohol produced, the lower pH and the synthesis of different metabolites during fermentation, contribute to the development of flavour and aroma of the bread. However, the flavour depends on the recipe used (e. g., the proportion of yeast to flour) and how the product is being proofed and baked (e. g., the temperature and time of fermentation and baking).

The baking yeast is usually used in doughs to produce CO_2, but ethanol and aromas are formed as well. The living yeast cells are a rich source of glutathione, tripeptides, glutamic acid, casein and glycine. The glutathione occurs inside the living yeast cell, and, therefore, has no direct contact with the dough under normal processing conditions. If one wants to release the glutathione, it is necessary to break the cell membrane and the cell wall. This can be achieved by heating the yeast above 80 °C. At this temperature the yeast is inactivated and no longer produces any CO_2. The glutathione is released in its reduced form GSH, and acts as a dough relaxant like L-cystein HCl. However, GSH is less aggressive than this latter compound.

During mixing, the gluten network develops by oxidation reactions between flour proteins. In the case of high quality flours and good gluten development during mixing, the dough may become excessively strong due to excess amounts of disulfide bounds, resulting from the oxidation of individual cysteine residues (R-SH) into disulfide (R-S-S-R) bridges. In this case, the glutathione from the yeast may relax the dough, by breakage of disulfide bonds, resulting in better dough handling characteristics.

During fermentation, the dough becomes stronger as well. This can be explained by the formation of CO_2 that will strengthen the gluten system, thus making the doughs more tenacious. The lowering of the pH will also contribute to modifications in the physico-chemical bonds.

11.2.1.2 Factors affecting the fermentation activity

Two main factors are important during the bread making process, namely the formation of CO_2 gas and the gluten network. Yeast as a living organism, is strongly influenced by its environment. The volume of the bread is directly linked to the fermentation activity of yeast (= formation of CO_2), and also to the capacity of the dough to retain this gas (retention of gas). The latter is achieved by the gluten network. Therefore, these two elements have to be in balance to obtain a good final product.

The osmotic pressure, which is defined as the difference in pressure between one environment and another, strongly influences the fermentation activity of yeast. This metabolic activity is affected by the migration of water from the yeast cells to the external environment. The osmotic pressure in dough depends mainly on the percentages of salt and sugar, and these ingredients can, if their level is too high, reduce the fermentation activity of the yeast. Generally, sugar concentrations below 5 % have a positive effect on the fermentation rate.

Sugar concentrations above 5 % reduce the fermentation rates, and the same occurs with salt concentration above 1 %. Different possibilities exist to overcome these problems, namely:

a) the use of an osmotolerant yeast strain, which is resistant and active in sugar-rich environments
b) increase the quantity of yeast (Table 11.2-1)
c) increase the temperature of the fermentation (Table 11.2-2)
d) work with a sponge and dough
e) change the proportion between salt and sugar

Table 11.2-1 shows the effects of the percentage of yeast used on the fermentation time and the bread volume obtained. It can be clearly seen that increasing the amount of yeasts results in shorter fermentation times to obtain a certain volume of bread. Consequently, the fermentation time required can be changed by the quantity of yeast used. However, one has to keep in mind that yeast also contributes to the flavour of the bread.

Temperature has a profound effect on the fermentation rate as well (Table 11.2-2). We compared the effect of two temperatures, namely 24 and 30 °C on the fermentation time. Our results clearly indicate that the fermentation time required is a function of the temperature. This is often neglected by bakers. At higher temperatures, doughs have the tendency to develop more quickly. This fact explains why in certain doughs, such as those used to produce baguettes, it is important to use lower dough temperatures (23–24 °C). This temperature will limit the rising of the dough and shape the baguette without tearing the bread apart. For oth-

Tab. 11.2-1 Impact of yeast quantity on fermentation time

Yeast percentage (%)	Fermentation time (minutes)	Bread volume (litre/kg)
2	155	5.6
4	100	5.5
8	60	5.7

Tab. 11.2-2 Impact of temperature on fermentation time

Dough temperature (°C)	Fermentation time (minutes)	Bread volume (litre/kg)
24	200	5.6
30	141	5.5

er applications, such as those for brioches, a higher temperature is required. Raw and frozen doughs need to be handled at temperatures of about 18 °C.

The pH has an effect on the fermentation as well. The intracellular pH of yeast cells varies between 5.6 and 5.8. Therefore, a pH between 4 and 6 is optimal for the fermentation activity of yeast. This occurs usually during a large part of the baking process. However, towards the end of the baking process the pH decreases till about 5.

11.3 Physiological aspects of baking yeast

Baker's yeast contains 46 % carbon, 32 % oxygen, 8,5 % nitrogen, 6 % hydrogen and 7.5 % ash (dry weight) [23, 24, 11]. A stoechiometric equation can be made when taking into account that 200 g of sucrose are required for the production of 100 g of solid yeast.

200 g sucrose + 10.32 g NH_3 + 100.44 g oxygen + 7.5 g " ash "
$\rightarrow$ 100 g solid yeast + 140.14g CO_2 + 78.12 g H_2O

The growth environment of the yeast must contain the different nutrients required, which need to be transported into the cell. Therefore, the yeast uses a variety of transporter systems. At least 271 membrane transporter genes occur in *S. cerevisiae* [70], which belong to several different gene families (see http://alize.ulb.ac.be/YTPdb).

11.3.1 Assimilation of carbon

The Carbon sources traditionally used for the production of baker's yeast are beet and/or cane molasses. These by-products of the sugar industry contain on average 50–55 % carbohydrates, which can be assimilated, and sucrose is the most abundant sugar (Table 11.3-1). *S. cerevisiae* has a variety of genes encoding for the proteins used in hexose transport. They are part of the Hxt protein family, with the genes *HXT1* to *HXT17* [9, 32, 2]. Two genes, *RGT2* and *SNF3*, encode protein sensors of extra cellular glucose [34, 51]. When glucose is present in the environment of the cell, the sensors generate an intracellular signal that acts on the transcription of the *HXT* gene family [9, 52, 13, 33]. The Hxt proteins belong to the major facilitator superfamily (MFS), which transport substrates by passive energy – independent facilitated diffusion.

The members of the Hxt family have different physiological functions [52]. The transcription of *HXT3* is induced by glucose and is independent of the sugar concentration. Transcription of the *HXT2* and *HXT4* genes is induced by low levels of glucose, that of *HXT6* and *HXT7* is repressed by high levels of glucose, whereas *HXT1* is induced by a high glucose concentration. The other members of the *HXT* family have no known physiological functions, except *HXT14*, which allows a slight growth of yeast on galactose [70].

The proteins are essential for the cell to grow in the presence of sugars. A strain with a mutation in the genes *HXT1* to *HXT7* can not grow in an environment containing glucose, fruc-

Tab. 11.3-1 Nutrient composition of molasses as percentage of total solids [56]

Composition	Cane molasses	Beet molasses
Sugars	73.1	66.5
sucrose	45.5	63.5
raffinose	0	1.5
invert sugar	22.1	0
other	5.5	1.5
Organic	15.5	23.0
other N	3.1	0
other amino acids	0	3.0
betaine	0	5.5
organic acids	7.0	5.5
pectin	2.7	5.0
glutamic acid + pyrrolidin carboxylic acid	2.4	4.0
Inorganic	11.7	10.5
K_2O	5.3	6.0
Na_2O	0.1	1.0
CaO	0.2	0.2
MgO	1.0	0.2
$Al_3O_3.Fe_2O_3$	0	0.1
SiO_2	0	0.1
Cl	1.1	1.7
$SO_2 + SO_3$	2.3	0.5
P_2O_5	0.8	0.1
N_2O_5	0	0.4
other	0.9	10.2

tose or mannose, and no glycolic flux is observed in these mutants [52]. A protein encoded by the *GAL2* gene transports galactose, whereas maltose is transported by the product of the *MAL61* gene. These two genes are induced by their respective substrates and are repressed when another preferential source, such as glucose, is available [9].

11.3.2 Assimilation of nitrogen

Different sources of nitrogen, such as ammonia, urea, and ammonia salts, can be used for the production of baking yeast. *S. cerevisiae* contains a number of transporters for ammonium ions [39]. Two of them are of high affinity and a third one seems to be of low affinity [18]. The *MEP1* gene encodes transporters of high capacity [40], *MEP2* encodes transporters of high affinity homologous to Mep1p, whereas *MEP3* encodes low affinity transporters. These three genes are subject to control by nitrogen, and have different levels of expression. Urea can be assimilated by yeast and crosses the cell membrane via a permease [19].

Amino acids are taken in charge by general transporters [31, 29]. All known amino-acid permeases belong to a single family of homologous proteins [57]. Agp1p is a general permease for most uncharged amino acids. Gnp1p, which is closely related to Agp1p, is only able to transport leucine (Leu), serine (Ser), threonine (Thr), cysteine (Cys), methionine (Met), glutamine (Gln) and asparagine (Asn), while Bap2p and Bap3p (also similar to Agp1p) are able to transport isoleucine (Ile), Leu, valine (Val), Cys, Met, phenylalanine (Phe), tyrosine (Tyr) and tryptophan (Trp) [57].

Others permeases have a higher specificity [57]: Put4p can transport alanine (Ala), glycine (Gly) and proline (Pro), Tat2p can take Phe, Trp and Tyr, while Can1p, Lyp1p and Alp1p are specific for the cationic amino acids. Dip5p shows a transport capacity and a high affinity for glutamine acid (Glu) and aspartic acid (Asp) [58].

11.3.3 Assimilation of inorganic elements

Certain elements are required by the yeast cells to assure optimal growth. Some of these elements are present in the molasses, but phosphor, magnesium and trace elements must be added separately. Phosphate is essential for growth and survival of the yeast cells, because it is involved in all phases of the metabolism. Phosphor is added as phosphoric acid or ammonium phosphate. Magnesium is essential because it acts as a cofactor in many enzymatic reactions, and is added as magnesium sulphate.

11.3.4 Assimilation of vitamins

Biotin is essential for the growth of yeast [41, 73]. This vitamin is present in cane molasses, but not in beet molasses, and consequently these two molasses may be mixed. Thiamine is not essential for growth, but is an activator of gene expression [30]. The importance of panthotenate has been demonstrated by comparing cultures that contained the vitamin and those that were panthotenate deficient [66]. The study showed that, in panthotenate poor media (less than 30 μg panthotenate per litre), the glucose consumption rate was 50% lower than the consumption in rich panthotenate media (more than 60 μg panthotenate per litre). All these vitamins are transported into the cells by permeases.

11.4 Production of baking yeast

11.4.1 Preservation of strains, preparation of the inoculum and raw materials used

The yeast strains utilized in the production process are preserved under strict conditions in the laboratory or cryopreserved in a culture collection, in order to guarantee stability of the required strain characteristics. The cultures are regularly subcultured with the greatest care possible, which may be carried out by a few well-trained people to ensure optimal renewal.

The production process starts with the inoculation of the production strain in small volumes using Erlenmeyer, Carlberg or Pasteur flasks. Secondly, one or more, easy to sterilize fermenters are prepared to grow sufficient yeast cells to inoculate fermenters of greater capacity. The small volume fermentations are performed in batch cultures (maximum 30 m^3), in which all the ingredients are added at the beginning of the fermentation. These batch fermentations may take up to 24 hours. Thereafter, these cultures are transferred to the factory.

Raw materials used in the fermentations do not require a special treatment, except for the molasses. These undergo a strict treatment to standardize their quality. The molasses are clarified in order to remove inorganic and non-sugar organic compounds. This process can also be coupled with an acidification of the molasses. Nitrogen may be added either as an ammonium solution, ammonium salts or urea in order to obtain 6.5 to 10 % of the yeast solids. Besides, one has to add a phosphor source, such as phosphoric acid. The P_2O_5 content in baker's yeast ranges from 1.5 to 3.5 % of the yeast solids. Some small quantities of minerals (K, Mg, Zn) are also supplied. As shown by Oura *et al.* [50], biotin (100 µg biotin / 100 g sugar) is required for aerobic yeast growth. Other vitamins, like thiamine and pantothenic acid are added as well.

11.4.2 Fed-batch fermentations

The main fermentations are of the fed-batch type. This means that only a small part of the substrate is added at the beginning of the fermentation (i. e., before the inoculation), and the majority of substrates during the process of fermentation according to predefined and strictly controlled profiles. The number of fed-batch cultures varies between producers. One may inoculate a single fed-batch fermentation using a big quantity of the pure yeast culture, or proceed in two or more fed-batch fermentations. Another method used, in order to limit the number of pure cultures required, is to use a parent-yeast culture that serves as an inoculum in the final fermentations. This approach has the advantage that the number of aseptic preparations is limited, which reduces the risk of contamination. In all procedures, correct storage of the parent-yeast culture is very important.

After the last phase of multiplication, the yeast is concentrated and washed. This results in a yeast cream that has to be stored between 2 and 4 °C in tanks under agitation. The washing step eliminates traces of molasses, and results in a light cream coloured yeast product. After this step, several treatments can be applied to the yeast resulting in different final products.

11.4.3 Bakery yeast products

Four different main products are being produced for the bakery market, namely liquid yeast, compressed yeast, active dry yeast and instant active dry yeast. Some main characteristics

Tab. 11.4-1 Characteristics of different types of bakery yeast products

Yeast type	Dry matter (%)	Shelf life (month)
Liquid yeast / yeast cream	19 – 21	1
Fresh / compressed yeast	29 – 34	1
Active dry yeast (ADY)	92 – 94	12
Instant active dry yeast (IADY)	95 – 97	24

of these products are presented in Table 11.4-1. Two types of yeasts can be distinguished, based on their performance under different sugar contents of the doughs. Both belong to *S. cerevisiae*, but their performance is different in terms of osmotolerance. The first one is used for the lean dough and performs well in doughs with 0 to 10 % sugar. The second type is used for the sugared dough with 5 % or higher sugar contents.

11.4.3.1 Liquid yeast

The yeast cream produced can be directly used as a final product. Liquid yeast is used mainly by industrial bakers. The main advantage of this product is that it can be pumped and dosed automatically. Liquid yeast should be stored under cooled conditions, and the dry matter content varies about 20 %.

11.4.3.2 Compressed yeast

Compressed yeast is produced from the yeast cream. Filtration using a filter press or a rotary vacuum filter is necessary in order to eliminate part of the water contained in the yeast cream. The dry matter of this product varies between 29 and 34 %, and depends on the country and the habits of the bakers. Compressed yeast with a lower dry matter content is darker in colour and has a kneadable, rather plastic consistency. At higher dry matter contents the compressed yeast is more whitish and becomes crumbly. Fresh yeast must be preserved under cooled conditions, at a temperature of 4 °C. If the cold chain is disturbed and the temperature is raised, the fresh yeast looses its activity rapidly. This product can be produced in different formats, such as small cubes for the home bakery, blocks of 500 grams or 1 kg, and bags of 25 kg. The shell life of the product is around 5 weeks.

11.4.3.3 Active dry yeast

Active dry yeast (ADY) is produced as round granules. The drying process for ADY is relatively long. The yeast suspension is dried at an inlet temperature of 100–120 °C and an outlet temperature of 65–67 °C, thus resulting in hard particles. This explains why it is needed to dissolve ADY in lukewarm water (10 % ADY in H_2O) during at least 15 minutes before adding it to the flour. Adding some sugar (10 %) to this solute helps to activate the yeast, since it already can start fermenting. ADY has a good stability at room temperature and is not negatively influenced by the presence of oxygen and humidity. The shelf life of the product is around 1 year.

11.4.3.4 Instant active dry yeast

Instant active dry yeast (IADY) does not need to be rehydrated prior to adding it to the other ingredients of the dough. The product consists of dried, fine granules, which sediment rapidly when they come in contact with water. After filtration on a rotary drum filter or filter-press, the yeast is extruded in fine vermicelli and dried in a fluid-bed. Heated air is blown from the bottom trough the yeast vermicelli at velocities capable of suspending the yeast particles. The incoming air temperature (100–150 °C) and the yeast temperature (24–40 °C) are well controlled during the process. A rehydratation agent (i. e., an emulsifier) is added to the yeast in order to facilitate rehydratation when the dough is being prepared. The instant yeast is vacuum packed to protect it from oxygen during its conservation, and the shelf life of the product is around 2 years.

11.5 Genetic improvement of baking yeast

The reviews of RANDEZ-GIL et al. [55] and DEQUIN [16] present good summaries on the possibilities to improve yeast strains by genetic engineering. OSTERGAARD [49] and NIELSEN [46] give an overview of the metabolic engineering of *Saccharomyces cerevisiae* as well. Two approaches are followed by different laboratories and industrials, to improve baking qualities and to produce a low-cost baker's yeast without affecting the quality of the end product.

11.5.1 Efficiency of biomass production

As described above, beet molasses are used as a main substrate for the production of yeast biomass. Beet molasses are largely composed of sucrose, but contains other sugars as well (see Table 11.3-1). Baking yeast is unable to use melibiose, a sugar resulting from the hydrolysis of raffinose by invertase. Only one-third of a raffinose molecule can be metabolized by the yeast, because raffinose is cleaved in one molecule of fructose and one molecule of melibiose. The fructose is assimilated by the cell, but the melibiose cannot be used due to the lack of the enzyme α-galactosidase (EC 3.2.1.22). The cloning of the gene encoding for this α-galactosidase in the baker's yeast [36] resulted in a construct able to metabolize melibiose, thus leading to an increase of the biomass yield by 8 % [37, 21].

Other authors showed that alternative carbon sources might be used for the production of baker's yeast. Lactose, a by-product of the cheese industry, is a disaccharide like maltose whose uptake is under the control of a polymeric gene system. This sugar is hydrolyzed intracellularly after transport into the cell. A minimum of two genes is required for this sugar utilization: a permease to transport the lactose across the membrane and β-galactosidase (EC 3.2.1.23) to hydrolyse the lactose into two assimilable sugars: glucose and galactose. The cloning of genes of β-galactosidase (EC 3.2.1.23) [60] and lactose permease (EC 3.6.3.18) allows the use of lactose as a suitable substrate. This modified strain proved to be genetically stable and has the desired characteristics of baking strains [1, 59].

The assimilation of starch is possible after the cloning of genes coding for the *a*-amylase (EC 3.2.1.1) and the glucoamylase (EC 3.2.1.3). Constructions based on the *a*-amylase genes originating from *Aspergillus shirousamii* [62], *Aspergillus oryzae* [54], *Bacillus subtilis* [15], or *Lipomyces kononenkoae* [64], and constructions involving the glucoamylase gene from *Aspergillus awamori* [15] or *Schwanniomyces occidentalis* (= *Debaryomyces occidentalis*) [17] are described in the literature.

11.5.2 Improvement of fermentation characteristics

The performances and activities of baker's yeast in doughs are intimately related to the glycolytic activity, the capacity to adapt to substrate changes, and to the ability to synthesize enzymes and coenzymes under anaerobic conditions. The key to the success of baker's yeast resides in an optimal combination of these different properties. Different strategies have been used to improve these characteristics.

Derepression of genes coding for two key enzymes involved in the gluconeogenesis, the fructose-1,6- diphosphatase (EC 3.1.3.11) and the phosphoenolpyruvate carboxykinase (EC 4.1.1.38), resulted in an increase of the consumption of the glucose, thus causing an increase of the CO_2 produced [45]. Unfortunately, the generation time is increased by about 20 % and the biomass yield is around 30 % lower than that of the parent strain. Maltose is the main source of fermentable sugar. This sugar is produced by the action of amylolytic enzymes from damaged starch granules. Generally, the maltose utilization by *S. cerevisiae* requires the induction of two proteins, namely α-D-glucosidase (maltase) (EC 3.2.1.20) and maltose-permease (EC 3.6.3.19).

When glucose is absent or consumed, few molecules of maltose are transported into the cell and induce the synthesis of these two enzymes. Such a phenomenon permits the maltose to enter the cell rapidly by the action of the MalT gene (maltose transporter). Maltase (MalS) cleaves the maltose intracellularly into two glucose molecules, which are metabolized by the cell [49].

The development of cells utilizing maltose is crucial and can be achieved by either constitutive expression of maltase and maltose-permease in the cells or by multiplication of the *MAL* gene copy number. The *MAL* genes comprise also a gene (*MALR)* encoding a protein responsible for the regulation of the induction of the maltose transporter gene (*MALT*). Examples are given by Olsson and Nielsen [47], who discussed the application of metabolic engineering for the construction of glucose-derepressed strains.

OSINGA et al. [48] transformed baker's yeast with a construction that alter gene expression. They exchanged the original promoters and a part of the untranslated leader sequences for those of alcohol dehydrogenase I (*ADHI*) and translation elongation factor *EF1αA* derived from the host strain. As a consequence, the expression of *MALR* (MAL regulatory protein) became insensitive to glucose repression and independent of maltose for induction, and the CO_2 production increased by 18 % in the presence of glucose.

The difference between lagging and non-lagging strains is caused by the level of expression of *MAL* genes and the levels of maltose permease and maltase in cells at the moment of inoculation into a mixed sugar environment. A yeast pulse with maltose prior to inoculation into the sugar medium enhances sugar fermentation [25]. Improvements may be obtained by modulating the expression of the *MALx3* gene, which is the activator of the other *MAL* genes. Increasing the expression of this *MALx3* gene in the strain presenting this lag-phase may improve the metabolism of maltose in lean dough [27].

11.5.3 Resistance to stress

Baker's yeast cells are submitted to hyper-osmolarity during the different stages of production and utilization, such as the production of the yeast biomass, the subsequent treatments of this biomass, the drying process, the baking process, and during the preparation of the frozen bakery products [3, 12, 56, 8].

Yeast cells respond and adapt to environmental changes. In fact, the stress response processes involve normal physiological transitions. The responses of the yeast cell to a stress factor will be repeated at the level of control of growth, the signal transduction, the transcriptional control, etc. [61, 38, 68].

Many studies demonstrated a strong correlation between the intracellular concentration of trehalose and stress resistance (i. e., heat and freeze resistance) of yeasts in the absence of fermentation [71]. This resistance to stress is largely diminished when glucose is added. It has also been shown that trehalase mutants of baker's yeast have a higher resistance to freezing than the parent strains used to obtain the mutants [63].

Trehalose plays an important role in the freezing resistance of the yeast cells [71]. The disruption of genes encoding acid trehalase (*Δathl*) and/or the neutral trehalase (*Δnthl*) (EC 3.2.1.28) results in an intracellular accumulation of trehalose [63, 67]. The fermentation capacity of frozen doughs prepared with these mutant yeast strains is 9 to 14 % higher than those prepared with the wild type strains and the double mutant *Δnthl Δathl*, respectively. The single mutants showed an increased resistance to drying, which was not observed in the double mutant.

However, the yeast cells loose quickly their resistance to stress when they are fermenting, which is a disadvantage when one wishes to freeze partially fermented doughs. Isolated *fil* mutants (*F*ermentation *I*nduced *L*oss of stress resistance) maintained their fermentation performance during freezing [72, 69].

Glycerol seems to have an important function in stress resistance as well. Yeast cells able to accumulate glycerol have a higher fermentation activity in doughs rich in sugar. This was not observed in cells lacking this capacity [4, 28].

11.5.4 Enzymatic synthesis

The expression of endogenously produced α-amylase (EC 3.2.1.1) permits to augment the volume of the bread produced and gives a softer crumb [53]. Baking using a strain that expressed an endoxylanase (EC 3.2.1.8) of *Aspergillus nidulans* showed a 5 % augmentation of the volume of the bread [43]. The genes encoding for the endoxylanase X24 of *A. nidulans* and the α-amylase of *A. oryzae*, placed under the control of the actin gene promoter from *S. cerevisiae*, were cloned in baking yeast. The bread produced with this transformed strain showed an increase of the volume of 30 % and a decreased firmness if compared with commercial baker's yeast strains [42]. Recently, the expression of lipase in baking yeast strains resulted in bread with a higher loaf volume and a more uniform structure of the soft part in comparison to breads produced with commercial strains or control strains transformed with an empty plasmid [44].

11.6 Typing of baking yeast

Discrimination between strains of baking yeast is very important, because one wants to be sure that the correct strain is being used in the production process. The traditional techniques, based on fermentation and assimilation characteristics, do not allow the discrimination between commercial strains. All baker's yeast strains have a wide capacity to use sugars, are ethanol tolerant, have a highly efficient fermentation capacity, are genetically stable, and resist freezing and drying.

Only molecular techniques permit the efficient discrimination between strains. The molecular techniques of choice to discriminate between baking yeast strains are the study of chromosomal polymorphisms, the analysis of the mobile Ty and Y' elements, and the copy number of SUC genes encoding invertase [7].

The utilization of techniques related to the analyses of mitochondrial DNA (mtDNA) and karyotypes are not fully evaluated. This is due to the small number of strains studied, and to the absence of studies on the stability and reproducibility of these techniques [65]. Moreover, the discriminating power of certain methods turned out to be insufficient when they are tested alone. The study of BALEIRAS COUTO et al. [6] showed that the Random Amplified Polymorphic DNA (RAPD) technique permits discrimination at intra-species levels. However, this discrimination by RAPD depends strongly on the primers used during the PCR.

PCR fingerprinting allows the separation of strains within a species as well. The enzymatic restriction of the Internal Transcribed Spacer (ITS) domain of the ribosomal DNA (rDNA) showed discrimination at the sub-species level, but the resolution of the method depends also on the number and the types of restriction enzymes used. The enzymatic restriction of the Non Transcribed Spacer (NTS = InterGenic spacer (IGS)) domain of the rDNA did not allow the discrimination of the strains studied [6].

An interesting alternative to electrophoretic karyotyping is the Restriction Endonuclease Analysis of Genomic DNA (REAG) technique, combining endonuclease restriction (*Not*I in the study) of genomic DNA with pulsed field gel electrophoresis [74]. This technique allowed the separation of different isolates of *S. cerevisiae* isolated from patients. However, the approach should be tested on industrial strains of *S. cerevisiae* in order to evaluate its discriminatory power.

The use of Restriction Fragment Length Polymorphism (RFLP), as described by CLEMONS et al. [14], differentiated 41 out of 60 isolates of *S. cerevisiae* from different origin. Moreover, this study classified the strains into two groups that confirmed a previous classification based on the virulence of the strains.

An analysis of microsatellites in industrial and pathogenic strains showed that this technique is highly discriminating and distinguished between the 4 studied bakery strains (1 from Europe and 3 from Asia) [26]. This technique is based on the amplification by PCR of repeated sequences of 5 trinucleotides and 2 dinucleotides followed by electrophoresis in a polyacrylamide gel. The authors have also checked the performance of the method by doing the same analysis on a sample collected from a cultured fermentor (690 generations in 35 weeks). The reliability of the method was confirmed by the comparison of PCR experiments, using the same DNA and the same primers, performed three years before the actual study. PCR analysis using the same DNA and the same primers gave the same pattern. At the strain level, this technique showed a discriminatory power that has never been obtained by any other typing technique before.

11.7 Spoilage yeast of baking products

Two types of yeast contamination of baking products occur, namely growth of yeasts on the surface of the product and fermentation within the product. Growth of yeasts on the surface of the product is revealed by the appearance of small white, cream or rose spots [35]. The contaminating yeasts were identified as *Zygosaccharomyces bailii, Saccharomycopsis fibuligera, Pichia burtonii,* and *P. anomala.* These yeasts usually occur on the outer surface of products with a high water activity (a_w), but they may show a wide osmotolerance as well, in particular strains of the osmophilic species *Z. bailii.*

Fermentation within baking products occurs frequently in products with a high sugar content and low water activity [20]. The responsible yeast produces off-flavours [20]. For example, *Hyphopichia burtonii* (= *Pichia burtonii*) can produce styrene [35], and *Hansenula anomala* (= *Pichia anomala*) produces ethyl acetate in the presence of high concentrations of glucose or ethanol [35].

11.8 References

[1] ADAM, A.C.; PRIETO, J.A.; RUBIO-TEXEIRA, M.; POLAINA, J.: Construction of a lactose-assimilating strain of baker's yeast. Yeast **15** (1999) 1299–1305.

[2] ANDRÉ, B.: An overview of membrane transport proteins in *Saccharomyces cerevisiae*. Yeast **11** (1995) 1575–1611.

[3] ATTFIELD, P.V.: Stress tolerance: The key to effective strains of industrial baker's yeast. Nature Biotechnol. **15** (1997) 1351–1357.

[4] ATTFIELD, P.V.; KLETSAS, S.: Hyperosmotic stress response by strains of baker's yeasts in high sugar concentration medium. Lett. Appl. Microbiol. **31** (2000) 323–327.

[5] BAKER, J.C.; MIZE, M.D.: The origin of the gas cell in bread dough. Cereal Chem. **18** (1941) 19–34

[6] BALEIRAS COUTO, M.M.; EIJSMA, B.; HOFSTRA, H.; HUIS IN'T VELD, J.H.J.; VAN DER VOSSEN, J.M.B.M.: Evaluation of molecular typing techniques to assign genetic diversity among *Saccharomyces cerevisiae*. Appl. Environ. Microbiol. **62** (1996) 41–46.

[7] BENÍTEZ, T.; MARTÍNEZ, P.; CODÒN, A.C.: Genetic constitution of industrial yeast. Microbilogía SEM **12** (1996) 371–384.

[8] BEUDEKER, R.F.; VAN DAM, H.W. ; VAN DER PLAAT, J.B.; VELLENGA, K.: Developments in baker's yeast production. In: Yeast biotechnology and biocatalysis (edited by Verachtert, H.; De Mot, R.). New York, U.S.A.: Marcel Dekker (1990) 103–146.

[9] BISSON, L.F.; COONS, D.M.; KRUCKEBERG, A.L.; LEWIS, D.A.: Yeast sugar transporters. Crit. Rev. Biochem. Mol. Biol. **28** (1993) 259–308.

[10] CHAMBERLAIN, N.; COLLINS, T.H.: The Chorleywood bread process: The roles of oxygen and nitrogen. Baker's Digest. **53** (1979) 18– 24.

[11] CHEN, S.L.; CHIGER, M.: Production of baker's yeast. In : Comprehensive biotechnology, Vol. 3 (edited by Moo-Young, M.). Oxford, U.K.: Pergamon Press (1985) 429–461.

[12] CHEN, S.L.; GUTMANIS, F.: Process for production of osmotolerant yeast. United States patent (1983) US 4 220 563.

[13] CIRIACY, M.; REIFENBERGER, E.: Hexose transport. In: Yeast sugar metabolism: biochemistry, genetics, biotechnology, and applications (edited by Zimmermann F.K.; Entian K.-D.). Lancaster, U.S.A: Technomic Publishing Co. (1997) 45–65.

[14] CLEMONS, K.V.; PARK, P.; MCCUSKER J.H.; MCCULLOUGH M.J.; DAVIS R.W.; STEVENS D.A.: Application of DNA typing methods and genetic analysis to epidemiology and taxonomy of *Saccharomyces* isolates. J. Clin. Microbiol. **35** (1997) 1822–1828.

[15] DE MORAES, LM.; ASTOLFI-FILHO, S.; OLIVER, S.G.: Development of yeast strains for the efficient utilisation of starch: evaluation of constructs that express alpha-amylase and glucoamylase separately or as bifunctional fusion proteins. Appl. Microbiol. Biotechnol. **43** (1995) 1067–1076.

[16] DEQUIN, S.: The potential of genetic engineering for improving brewing, wine-making and baking yeasts. Appl. Microbiol. Biotechnol. **56** (2001) 577–588.

[17] DOHMEN, R.J.; STRASSER, A.W.; DAHLEMS, U.M.; HOLLENBERG, C.P.: Cloning of the *Schwanniomyces occidentalis* glucoamylase gene (*GAM1*) and its expression in *Saccharomyces cerevisiae*. Gene **30** (1990) 111–121.

[18] DUBOIS, E.; GRENSON M.: Methylamine/ammonia uptake systems in *Saccharomyces cerevisiae*. Multiplicity and regulation. Mol. Gen. Genet. **175** (1979) 67–76.

[19] ELBERRY, H.M.; MAJUMDAR, M.L.; CUNNINGHAM, T.S.; SUMRADA, R.A.; COOPER, T.G.: Regulation of the urea active transporter gene (*DUR3*) in *Saccharomyces cerevisiae*. J. Bacteriol. **15** (1993) 4688–4698.

[20] FLEET, G.: Spoilage yeasts. Crit. Rev. Biotechnol. **12** (1992) 1–44.

[21] GASENT–RAMIREZ, J.M.; CODON, A.C.; BENITEZ, T.: Characterisation of genetically transformed *Saccharomyces cerevisiae* baker's yeast able to metabolize melibiose. Appl. Environ. Microbiol. **61** (1995) 2113–2121.

[22] GOFFEAU, A. et al.: The yeast genome directory. Nature **387** (suppl.) (1997) 1–105.

[23] HARRISON, J.S.: Aspects of commercial yeast production. Process Biochem. **2** (1967) 41–45.

[24] HARRISON, J.S.: Yeast production. Progr. Industr. Microbiol. **10** (1971) 129–177.

[25] HAZELL, B.W.; ATTFIELD, P.V.: Enhancement of maltose utilisation by *Saccharomyces cerevisiae* in medium containing fermentable hexoses. J. Ind. Microbiol. Biotechnol. **22** (1999) 627–632.

[26] HENNEQUIN, C.; THIERRY, A.; RICHARD, G.F.; LECOINTRE, G.; NGUYEN, H.V.; GAILLARDIN, C.; DUJON, B.: Microsatellite typing as a new tool for identification of *Saccharomyces cerevisiae* strains. J. Clin. Microbiol. **39** (2001) 551–559.

[27] HIGGINS, V.J.; BRAIDWOOD, M.; BELL, P.; BISSINGER, P.; DAWES, I.W.; ATTFIELD, P.V.: Genetic evidence that high noninduced maltase and maltose permease activities, governed by *Malx3*-encoded transcriptional regulators, determine efficiency of gas production by baker's yeast in unsugared dough. Appl. Environ. Microbiol. **65** (1999) 680–685.

[28] HIRASAWA, R.; YOKOIGAWA, K.: Leavening ability of baker's yeast exposed to hyperosmotic media. FEMS Microbiol. Lett. **194** (2001) 159–162.

[29] HORÁK, J.: Yeast nutrient transporters. Biochem. Biophys. Acta **1331** (1997) 41–79.

[30] ICHIKAWA, K.; SHIBA, Y.; YAMAZAKI, M.; SERIZAWA, N.: Thiamine increases expression of yeast gene. Biosci. Biotechnol. Biochem. **61** (1997) 1221–1224.

[31] JAUNIAUX, J.C.; GRENSON, M.: *GAP1*, the general amino acid permease gene of *Saccharomyces cerevisiae*. Nucleotide sequence, protein similarity with the other bakers yeast amino acid permeases, and nitrogen catabolite repression. Eur. J. Biochem. **190** (1990) 39–44.

[32] KO, C.H.; GABER, R.F.: Roles of multiple glucose transporters in *Saccharomyces cerevisiae*. Mol. Cell. Biol. **13** (1993) 638–648.

[33] KRUCKEBERG, A.L.: The hexose transporter family of *Saccharomyces cerevisiae*. Arch. Microbiol. **166** (1996) 283–292.

[34] KRUCKEBERG, A.L.; WALSH, M.C.; VAN DAM, K.: How do yeast cells sense glucose? Bioessays **20** (1998) 972–976.

[35] LEGAN, J.D.; VOYSEY, P.A.: Yeast spoilage of bakery products and ingredients. J. Appl. Bacteriol. **70** (1991) 361–371.

[36] LILJESTRÖM-SUOMINEN, P.L.; JOUTSJOKI, V.; KORHOLA, M.P.: Construction of a stable α-galactosidase producing baker's yeast. Appl. Environ. Microbiol. **54** (1988) 245–249.

[37] LILJESTRÖM-SUOMINEN, P.L.; TUBB, R.S.; KORHOLA, M.P.: Construction of new α-galactosidase producing yeast strains and the industrial application of these strains. United States patent (1991) US 5 055 041.

[38] MAGER, W.H.; DE KRUIJFF, A.J.J.: Stress-induced transcriptionnal activation. Microbiol. Rev. **59** (1995) 506–531.

[39] MARINI, A.-M.; SOUSSI-BOUDEKOU, S.; VISSERS, S.; ANDRÉ, B.: A family of ammonium transporters in *Saccharomyces cerevisiae*. Mol. Cell. Biol. **17** (1997) 4282–4293.

[40] MARINI, A.-M.; VISSERS, S.; URRESTARAZU, A.; ANDRÉ, B.: Cloning and expression of the *MEP1* gene encoding a transporter of ammonium in *Saccharomyces cerevisiae*. EMBO J. **13** (1994) 3456–3463.

[41] MARTINEZ-FORCE, E.; BENITEZ, T.: Effects of varying media, temperature, and growth rates on the intracllular concentrations of yeast amino acids. Biotechnol. Prog. **11** (1995) 386–392.

[42] MONFORT, A.; BLASCO, A.; PRIETO, J.A.; SANZ, P.: Combined expression of *Aspergillus nidulans* endoxylanase X24 and *Aspergillus oryzae* α-amylase in industrial baker's yeasts and their use in bread making. Appl. Environ. Microbiol. **62** (1996) 3712–3715.

[43] MONFORT, A.; BLASCO, A.; PRIETO, J.A.; SANZ, P.: Construction of baker's yeast strains that secrete different xylanolytic enzymes and their use in bread making. J Cer. Sci **26** (1997) 195–199.

[44] MONFORT, A.; BLASCO, A.; SANZ, P.; PRIETO, J.A.: Expression of *LIP1* and *LIP2* genes from *Geotrichum* species in baker's yeast strains and their application to the bread-making process. J. Agric. Fd Chem. **47** (1999) 803–808.

[45] NAVAS, M.A.; CERDAN, S.; GANCEDO, J.M.: Futile cycles in *Saccharomyces cerevisiae* strains expressing the gluconeogenic enzymes during growth on glucose. Proc. Natl Acad. Sci. USA **90** (1993) 120–124.

[46] NIELSEN, J.: Metabolic engineering. Appl Microbiol Biotechnol **55** (2001) 263–283.

[47] OLSSON, L.; NIELSEN, J.: The role of metabolic engineering in the improvement of *Saccharomyces cerevisiae*: utilization of industrial strain. Enz. Microb. Technol. **26** (2000) 785–792.

[48] OSINGA, K.A.; BEUDEKER, R.F.; VAN DER PLAAT, J.B.; DE HOLLANDER, J.A.: New yeast strains providing for an enhanced rate of the fermentation sugars, a process to obtain such strains and the use of these strains. European patent (1988) EP 0306107A2.

[49] OSTERGAARD, S.; OLSSON, L.; NIELSEN, J.: Metabolic engineering of *Saccharomyces cerevisiae*. Microbiol. Mol. Biol. Rev. **64** (2000) 34–50.

[50] OURA, E.; SUOMALAINEN, H.: Biotin and the metabolism of baker's yeast. J. Inst. Brew. **84** (1978) 283–287.

[51] ÖZCAN, S.; DOVER, J.; JOHNSTON, M.: Glucose sensing and signaling by two glucose receptors in the yeast *Saccharomyces cerevisiae*. EMBO J. **17** (1998) 2566–2573.

[52] ÖZCAN, S.; JOHNSTON, M.: Function and regulation of yeast hexose transporters. Microbiol. Mol. Biol. Rev. **63** (1999) 554–569.

[53] RANDEZ-GIL, F.; PRITO, J.A.; MURCIA, A.; SANZ, P.: Construction of baker's yeast strains that secrete *Aspergillus oryzae* alpha-amylase and their use in breadmaking . J. Cer. Sci. **21** (1995) 185–193.

[54] RANDEZ-GIL, F.; SANZ, P.: Expression of *Aspergillus oryzae* alpha-amylase gene in *Saccharomyces cerevisiae*. FEMS Microbiol. Lett. **112** (1993) 119–123.

[55] RANDEZ-GIL, F.; SANZ, P.; RIETO, J.A.: Engineering baker's yeast: room for improvement. Trends Biotechnol. **17** (1999) 237–244.

[56] REED, G.; NAGODAWITHANA, T.W.: Yeast technology, 2nd ed. New York, U.S.A.: Van Nostrand Reinhold (1991).

[57] REGENBERG, B.; DURING-OLSEN, L.; KIELLAND-BRANDT, M.C.; HOLMBERG, S.: Substrate specificity and gene expression of the amino-acid permeases in *Saccharomyces cerevisiae*. Curr. Genet. **36** (1999) 317–328.

[58] REGENBERG, B.; HOLMBERG, S.; OLSEN L.D.; KIELLAND-BRANDT, M.C.: Dip5p mediates high-affinity and high-capacity transport of L-glutamate and L-aspartate in *Saccharomyces cerevisiae*. Curr. Genet. **33** (1998) 171–177.

[59] RUBIO-TEXEIRA, M.; AREVALO-RODRIGUEZ, M.; LEQUERICA, J.L.; POLAINA, J.: Lactose utilization by *Saccharomyces cerevisiae* strains expressing *Kluyveromyces lactis LAC* genes. J. Biotechnol. **84** (2001) 97–106.

[60] RUBIO-TEXEIRA, M.; CASTRILLO, J.I.; ADAM, A.C.; UGALDE, U.O.; POLAINA, J.: Highly efficient assimilation of lactose by a metabolically engineered strain of *Saccharomyces cerevisiae*. Yeast **14** (1998) 827–837.

[61] RUIS, H.; SCHÜLLER, C.: Stress signalling in yeast. Bioessays **17** (1995) 959–965.

[62] SHIBUYA, I.; TAMURA, G.; ISHIKAWA, T.; HARA, S.: Construction of an alpha-amylase/glucoamylase fusion gene and its expression in *Saccharomyces cerevisiae*. Biosci. Biotechnol. Biochem. **56** (1992) 884–889.

[63] SHIMA, J.; HINO, A.; YAMADA-IYO, C.; SUZUKI, Y.; NAKAJIMA, R.; WATANABE, H.; MORI, K.; TAKANO, H.: Stress tolerance in doughs of *Saccharomyces cerevisiae* trehalase mutant derived from commercial baker's yeast. Appl. Environ. Microbiol. **65** (1999) 2841–2846.

[64] STEYN, A.J.; PRETORIUS, I.S.: Characterization of a novel alpha-amylase from *Lipomyces kononenkoae* and expression of its gene (LKA1) in *Saccharomyces cerevisiae*. Curr. Genet. **28** (1995) 526–533.

[65] STRUELENS, M.: the Members of the European Study Group on Epidemiological Markers: Consensus guidelines for appropriate use and evaluation of microbial epidemiologic typing systems. Clin. Microbiol. Infect. **2** (1996) 2–11.

[66] TAHERZADEH, M.J.; LIDEN, G.; GUSTAFSSON, L.; NIKLASSON, C.: The effects of pantothenate deficiency and acetate addition on anaerobic batch fermentation of glucose by *Saccharomyces cerevisiae.* Appl. Microbiol. Biotechnol. **46** (1996) 176–182.

[67] TAKANO, H.; HINO, A.; YAMADA-IYO, C.; SUZUKI, Y.; NAKAJIMA, R.: Frozen dough-resistant, practical baker's yeast. European patent (1998) EP 0838520A2.

[68] THEVELEIN, J.M.: Signal transduction in yeast. Yeast **10** (1994) 1753–1790.

[69] THEVELEIN, J.M.; GORWA, M-F.; VAN DIJCK, P.; DUMORTIER, F.; TEUNISSEN, A.; LEMAIRE, K.; COLAVIZZA, D.; VERSELE, M.; LOIEZ, A.: Nouvelles souches « fil » résistantes au stress en condition de fermentation et/ou de croissance. European patent (1999) EP 0967280A1.

[70] VAN BELLE, D.; ANDRÉ, B.: A genomic view of yeast membrane transporters. Curr. Opin. Cell Biol. **13** (2001) 389–398.

[71] VAN DIJCK, P.; COLAVIZZA, D.; SMET, P.; THEVELEIN, J.M.: Differential importance of trehalose in stress resistance in fermenting and nonfermenting *Saccharomyces cerevisiae* cells. Appl. Environ. Microbiol. **61** (1995) 109–115.

[72] VAN DIJCK, P.; GORWA, M.F.; LEMAIRE, K.; TEUNISSEN, A.; VERSELE, M.; COLOMBO, S.; DUMORTIER, F.; MA, P.; TANGHE, A.; LOIEZ, A.; THEVELEIN, J.M.: Characterization of a new set of mutants deficient in fermentation-induced loss of stress resistance for use in frozen dough applications. Int. J. Fd Microbiol. **55** (2000) 187–192.

[73] WOEHRER, W.; ROEHR, M.: Regulatory aspects of baker's yeast in aerobic fed-batch cultures. Biotechnol. Bioeng. **23** (1981) 567–581.

[74] ZERVA, L.; HOLLIS, R.J.; PFALLER, M.A.: In vitro susceptibility testing and DNA typing of *Saccharomyces cerevisiae* clinical isolates. J. Clin. Microbiol. **34** (1996) 3031–3034.

12 Non-alcoholic beverages and yeasts

MALCOLM STRATFORD and STEPHEN A. JAMES

The authors wish to acknowledge a dept of gratitude to the late Professor BOB DAVENPORT for sharing his insight and experience of the ecology of yeasts in soft drinks environments, and to Hazel Steels for access to her unpublished data concerning the physiology of spoilage yeasts.

12.1 Introduction

Soft drinks in general are sugar-containing, low-pH beverages that constitute a hostile environment in which the vast majority of microbes are either killed or unable to grow. Pathogenic bacteria lose viability in soft drinks and fruit juices, although *E. coli* O157 and *Salmonellae* spp. can persist for weeks in chilled un-pasteurized or un-preserved juices [65, 238]. The ability to grow in soft drinks [198] is limited to a number of non-pathogenic yeasts, moulds, and a few acid-tolerant bacterial species (e. g., *Lactobacillus* and *Acetobacter* spp.).

Significant growth of yeast in soft drinks inevitably results in spoilage of the beverage either as a result of yeast metabolism or simply as a direct consequence of the visible biomass of yeast cells. In clear beverages, yeast growth to greater than 10^5 cells/ml becomes visible to the consumer; growth to a higher degree results in the formation of clouds of cells, and on occasion surface-films, clumps, flocs or particulates where yeasts flocculate or form pseudohyphae. Major spoilage yeasts such as *Saccharomyces cerevisiae* and *Zygosaccharomyces bailii*, while normally found as single cells, have been reported to flocculate and form pseudohyphae [17, 61, 194, 204]. Other yeasts may form surface films, notably *Issatchenkia orientalis* (= *Candida krusei*) and *Pichia membranifaciens* [140]. Extracellular polysaccharide can result in rope-formation and hazes may be formed by proteolytic or pectolytic degradation of fruit components [57]. Taints and off-flavours may be formed, typically resulting from ethanol formation, and excessive gas is often formed as a consequence of fermentation. Spoilage of soft drinks has been known to cause cans to split or burst, packaging to distort and metal kegs and glass bottles to explode, inflicting physical injury, particularly to the eyes [68].

However of the *circa* 800 yeast species known [17], relatively few yeasts are able to grow in non-alcoholic beverages, and these are responsible for most incidents of spoilage. PITT and HOCKING [140] list only 13 species, of which 10 are described as being responsible for spoilage of foods processed according to normal standards of good manufacturing practice (GMP). This limitation to the spoilage yeast flora is imposed, partly by the constraints of hygienic manufacture, but largely as a result of chemical and physical factors that form the yeast environment within each container of soft drink.

12.1.1 Definitions

Non-alcoholic beverages (better known as soft drinks) is a term encompassing a wide variety of products. It is generally accepted that soft drinks are sweetened, water-based beverages, usually with a balancing acidity [11]. For the purpose of this review, soft drinks are defined as beverages that are generally acidic, either still or carbonated, containing sugar, and flavoured with fruit or vegetable extracts. This includes colas, lemonades, fruit-flavoured drinks, tonics and ginger-ale mixers, fruit-flavoured sweetened mineral waters, isotonic sports drinks, fibre drinks, niche drinks, flavoured ice teas and functional "nutraceutical" drinks [131]. "Soft drinks" is often used to include fruit juices, fruit nectars, fruit cordials and concentrates. However in the current review, fruit juices and similar beverages are largely excluded as these are covered in Chapter 10, "Yeasts in fruit and fruit products". Also excluded for obvious reasons are alcoholic beverages (covered in Chapter 13 and Chapter 14), alcohol-fortified soft drinks and mixers (alcopops), milk-based beverages and hot beverages, such as tea and coffee.

12.1.2 Composition of soft drinks – yeast nutrients and inhibitors

Soft drinks in themselves form an environment suitable for the proliferation of some yeast species, but inimical or hostile to many others. Different types of soft drinks may support the growth of different yeast species. For example, a simple soft drink such as cheap "lemonade", containing sugar and citric acid only in carbonated water, requires little in terms of preservation, being only susceptible to spoilage by carbonation-resistant yeast species able to grow in poor nutrient conditions (e. g., *Dekkera anomala* [82]). Addition of real fruit juices to such a beverage would not only organoleptically improve its quality, but will also greatly increase the level of yeast nutrients present and consequently, the susceptibility to spoilage [140].

Soft drinks as an environment can be described simply from the chemical standpoint; what chemicals are present and at what concentration, for both nutrients and inhibitors. Due consideration must also be given to the physical attributes of soft drinks, such as storage temperature, pressure in carbonated beverages, and to the gas-permeability or otherwise of the container. A medium designed to be suitable for yeast growth usually contains hexose sugars as carbon sources, ammonium salts or amino acids as nitrogen sources, phosphate, sulphate, magnesium and potassium ions, trace elements and vitamins [8, 85], with oxygen available during incubation at 25–30 °C.

12.1.2.1 Sugars

Soft drinks usually contain sugars at concentrations similar to those found in fruit juices, typically 7 to 12 % w/w [209]. Diet formulations of soft drink may contain circa 1 % w/w sugar, while zero calorie formulations contain none. Concentration of sugar in soft drinks is customarily measured in degrees Brix, the w/w percentage sugar concentration. Sugars used are typically sucrose, glucose or fructose, added as sucrose granules, or glucose-type syrup produced from maize, or high-fructose syrup from wheat [19]. Addition of fruit juices may also contribute sugars, fructose, xylose and sorbitol from apple and pear juice [102, 207, 234], glucose and pentose sugars from grapes [4, 233], and sucrose from apricot, peach and pineapple juice [182].

12.1.2.2 Nitrogen- and phosphorus-containing compounds

Soft drinks are often very low in the nitrogen sources needed for growth of most yeast species. When available, ammonium salts and amino acids in soft drinks are derived from fruit components or other organic flavourings. Fruit juices do contain amino acids, typically 3–4 g/l in orange juice and 1 g/l in apple juice [13, 28, 58], although the great majority of this is as proline, a poor nitrogen source for yeast proliferation [58, 78]. The artificial sweetener aspartame, used in diet and zero calorie formulations can breakdown to yield phenylalanine, but in such low calorie drinks, sugars are absent thereby limiting yeast growth. In the absence of better nitrogen sources, certain yeasts possess the ability to utilize nitrate, commonly present in water (e. g., *Dekkera bruxellensis* [= *Dekkera intermedia* and *Brettanomyces lambica*], *Rhodotorula glutinis* and *Candida utilis* (= *Pichia jadinii*) [77, 85]).

Phosphates are present at low concentrations in most soft drinks, unless made with water containing high phosphate levels. The notable exceptions to this are cola-type beverages, which are acidified using phosphoric acid, giving them their distinctive flat sour flavour [48, 209]. Phosphates can also be derived from fruit components, which may contain 100 to 500 ppm phosphorus [3, 48, 92].

12.1.2.3 Metal salts, trace elements and vitamins

Water, particularly "hard water", contains substantial quantities of metal salts, particularly those of calcium and magnesium. Soft drinks are often prepared using softened water, but this still contains up to 50 ppm hardness as magnesium and calcium [209]. Similarly, trace elements such as copper, zinc and iron are likely to be present in small quantities. Such metal ions may also be readily derived from fruit or vegetable extracts, being present in the following ranges: potassium, 1000 ppm; sodium, calcium and magnesium 1.5–4.7 ppm; manganese, copper and zinc at 0.22–0.41 ppm [28]. Vitamins, such as inositol and B-group vitamins that are essential for growth of many yeasts, are unlikely to be present in soft drinks, unless fruit juices or organic flavourings are added.

12.1.2.4 Acids and acidulants

The refreshing nature of many soft drinks is largely due to the presence of various fruit acids, which both lower the pH and raise the titratable acidity. Such acids may comprise part of the fruit flavouring, or may be added as acidulants or acidity regulators. These include malic acid, phosphoric acid in colas, fumaric acid or adipic acid notably in the US market, and most commonly citric acid [181, 208]. The final concentrations are typically between 1 to 4 g/l, expressed in citric acid equivalents of titratable acidity. At such low concentrations as these, acidulant acids have little impact in themselves on the growth of yeasts [195], either as inhibitors or potential carbon sources. Similarly, the pH of soft drinks, typically 2.5–3.8, is also not sufficiently low to inhibit yeast growth, [139, 198]. However, multivalent acids, such as citric, malic, fumaric or adipic acids, readily form stable complexes with metal ions [110, 111], as does the better-known chelator EDTA. The highest affinity and most stable complexes are formed with highly charged transition metal ions, such as Fe^{3+}. In soft drinks prepared using softened water and containing only low levels of trace metal salts, it is possible that these acids contribute to the microbial stability of the beverage by removing key metal ion nutrients, such as iron, zinc, copper, or manganese [56, 66, 195].

12.1.2.5 Oxygen and carbon dioxide

In soft drinks, oxygen concentrations are usually very low, and designed to be so. Soft drinks deteriorate over time by chemical oxidation of the flavours, leading to a flat, dull, pruney taste in older beverages. Oxygen is largely eliminated from soft drinks at high temperature during pasteurization, and oxidation is further delayed by use of non-oxygen permeable packaging such as glass and metal cans, filling to minimize headspace, nitrogen flushing and by addition of antioxidants [76], notably ascorbic acid (vitamin C). As a result, growth of oxygen-requiring yeast species is likely to be suppressed in these micro-aerobic or anaerobic conditions [205]. Any yeast able to proliferate under these conditions is likely to be a facultative anaerobe. *Saccharomyces cerevisiae* is known to grow in anaerobic conditions when supplemented with trace amounts of unsaturated fatty acids and sterols [6, 7], whereas the factors required for anaerobic growth of *Z. bailii* are different and have not, as yet, been defined [149]. However, plastic packaging of soft drinks is gaining in popularity [170], and certain plastic bottles, such as PET, polyethylene tetraphthalate, allow slow oxygen permeation [76, 150]. Consequently, use of such packaging necessitates the addition of antioxidants to prevent oxidation and loss of freshness in soft drinks, particularly where fruit juices are added. Oxygen-permeable packaging has also been reported to permit spoilage by more aerobic micro-organisms [150, 156, 159, 160].

Carbonation involves dissolving carbon dioxide in the beverage under slight pressure, to form sparkling or fizzy soft drinks. Carbonation is measured in Volumes Bunsen, CO_2 volumes at 0 °C and 1 bar pressure dissolved per volume liquid, such that 1 volume = 1.96 g CO_2/l [119]. Carbonation varies from 1.5 volumes in lightly sparkling fruit juices to 5 volumes in soda water. Three volumes carbonation exerts a pressure of 2.6 bar. Carbonation has a major impact on the growth of microbes, with bacteria and moulds being particularly sensitive [32, 47, 49, 86], and yeasts affected at higher pressure [3, 120, 171, 177]. While yeast growth is inhibited by pressurized CO_2, cells are not killed. Nitrogen at similar pressure neither inhibits growth nor kills cells [106], demonstrating action by CO_2 and not by the pressure per se. Carbonation does not remove oxygen from solution as is popularly supposed, but inhibits directly as the partial pressure of CO_2 increases. Yeast cell division is inhibited [106] at pressures greater than 2 bar, particularly at low temperature [49, 54] possibly by interference with amino acid uptake [90].

12.2 Yeast biodiversity in non-alcoholic beverages

Non-alcoholic beverages represent a set of environmental conditions in which certain yeast species are commonly found but in which conditions for many species are entirely alien. The flora of soft drinks, the biodiversity, is controlled to a large extent by the ability of yeasts to gain access to this environment, acting as sources of infection of products, and is also controlled by the properties of individual species enabling certain yeasts to survive and proliferate. While a wide variety of yeasts may initially, and occasionally, gain assess to soft drinks, most will die within a few days. Indeed after prolonged incubation, relatively few species can be recovered as viable cells from non-alcoholic beverages.

Yeasts from many environments, from dust and water, from fruit and vegetables, syrups and sugars, from insects and even animals, all have the potential to gain access to soft drinks. It is hardly surprising then that large numbers of yeast species from a variety of different genera, have been isolated from soft drinks [15, 33, 45, 52, 143, 147, 155, 161, 162, 163, 166, 179, 219, 220, 237]. Many of these yeast isolates may have little significance in soft drinks, being accidental drop-ins, unable to grow, or aliens, shortly to die in this environment. It is important to recognize the yeast species of significance in soft drinks, amidst the background yeast flora. It must also be considered that a proportion of the yeast species reported in soft drinks may have been misidentified. Before modern DNA sequencing methods, yeast identification depended largely on assimilation tests and morphology; the accuracy of such tests depending in turn on the experience of the individual. Given that a proportion of yeasts even in recognized culture collections have been misidentified, it is quite possible that the degree of wrongly identified species isolated from soft drinks may be as high as 10–30%. However, the presence of large numbers of yeast cells of any species in non-alcoholic beverages or soft drinks is rarely beneficial, is almost always undesirable (even if it passes undetected) and occasionally will cause spoilage.

12.2.1 Soft drinks manufacture and sources of yeast infection

Soft drinks manufacture is essentially a simple process consisting of dissolving flavours, juices, acids, antioxidants and sugars into water, to form a beverage that is then dispensed into appropriate packaging. Sugar enters the factory either as granules or as 67 Brix syrup, and is dissolved in hot water to form the "simple syrup". Separately, other components of the soft drink, such as acids, flavours, fruit juices, vegetable extracts, antioxidants etc, are weighed, dissolved in water, and added to the simple sugar syrup, forming the final syrup at 40–50 Brix. This is directed to the proportioning pump where it is mixed with water to form the beverage, typically 7–10 Brix. The soft drink is then filled into bottles, tetrapaks or cans, which are capped or sealed (Fig. 12.2-1).

Microbes may enter soft drinks and their production facilities from a number of sources (Fig. 12.2-1). Entry of these organisms into the factory is most common on raw materials,

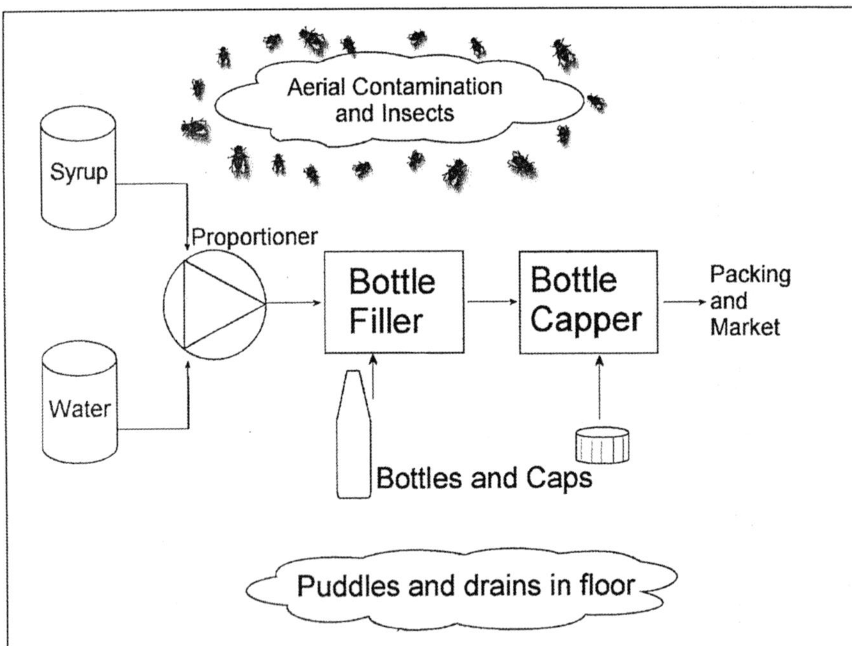

Fig. 12.2-1 Schematic diagram of filling operations in a soft drinks factory. Syrup and water are mixed at the proportioning pump, filled into bottles, which are then capped. Possible routes of yeast infection include aerial and insect vectors, raw materials such as sugar, and yeast growth in the proportioner, on filler heads and capper, in splashes of soft drink and within drains.

returned bottles, or via the air or aerial vectors [29, 157, 213, 214, 215]. Aerial contamination depends on the microbial loading and composition in the air, with ascomycetous yeasts such as *S. cerevisiae* found mostly associated with fruit harvesting [1]. Nearby sources of microbes greatly increases the microbial load, e. g., breweries, dust sources (harvesting or building work), fruit and vegetable storage/processing. Fresh fruit are routinely heavily surface-contaminated with yeasts and moulds, 10^3–10^6/gm. Insects are increasingly recognized as a vector for yeasts [95]. Many insects carry yeasts, not only consuming them as part of their diet [34, 60], but also benefiting from their presence (i. e., through colonization) in the digestive system. Insect frass, notably from fruit flies (*Drosophila* spp.), is particularly rich in soft drinks spoilage yeasts [17, 96].

Build up or colonization of spoilage flora within the factory greatly increases the likelihood of infection and spoilage [164, 165]. It has been estimated that 95 % of yeast infections of soft drinks are caused by poor plant hygiene [52]. Soft drinks manufacture is a very high volume operation, with up to 36,000 litres per hour passing through each filling line. Little time is allowed for microbial growth in the bulk liquid beverage, before pasteurization or addition of preservatives. Microbial build-up is therefore most significant in dead spaces or on surfaces proximate to the filling lines. Areas most at risk include the proportioning pump, filler and filler heads and capper [38, 141, 157]. Other areas of concern include old, dried splashes of soft drink, spilled damp ingredients, cracked flooring and drains (Fig. 12.2-1).

12.2.2 The significance of yeasts in the soft drinks environment

The microbial ecology of soft drinks and associated production facilities is a complex phenomenon. A review of the *circa* 800 yeast species discovered hitherto, shows that many have been reported in soft drinks, fruits and fruit juices [17]. Surveys of the ecology of soft drinks factories carried out over many years have revealed large numbers of yeast isolates and extensive lists of the species found therein (e. g., 155, 158, 161, 162]. In a recent "Forensic Approach" to soft drinks microbiology, Professor DAVENPORT has proposed a simplified taxonomy (Table 12.2-1) based on the behaviour of microbes, rather than on their specific names [10, 37, 38, 39, 40]. Group 1 were defined as spoilage microorganisms, well-suited to proliferate in soft drinks with the potential to cause spoilage from as few as 1 cell per container [38, 52]. Group 1 spoilage yeasts (e. g., *Z. bailii*) are characteristically osmotolerant, highly fermentative, vitamin-requiring and resistant to preservatives, such as acetic, sorbic and benzoic acids. These yeasts (Table 12.2-1) correspond largely with those listed by PITT and HOCKING [140] as being responsible for the vast majority of spoilage of foods prepared according to normal standards of GMP. Group 2 microbes were described by DAVENPORT [38] as Spoilage/Hygiene, capable of causing spoilage of soft drinks but only as a result of mistakes occurring during manufacture. These can be regarded as opportunistic spoilage organisms, ready to exploit errors such as under-dosing of preser-vatives, failure of pasteurization, hygiene failures, or ingress of oxygen. Such organisms are often present in soft

Tab. 12.2-1 Examples of yeasts found in soft drinks and production environments. Grouping of yeast species is as defined by the late Professor DAVENPORT [38] and the assignation of species to groups is based on discussions with Professor DAVENPORT [198], together with the experience of the authors. Group 1 are significant spoilage yeasts, usually absent in soft drinks manufacturing facilities. Group 2 are spoilage/hygiene species. These yeasts are often present in manufacturing facilities and opportunistically cause spoilage following manufacturing errors. Group 3 yeast species are hygiene indicators.

Group 1	Group 2	Group 3
Dekkera anomala	*Candida davenportii*	*Aureobasidium pullulans*
Dekkera bruxellensis	*Candida parapsilosis*	*Candida sake*
Dekkera naardenensis	*Debaryomyces hansenii*	*Candida solani*
Saccharomyces cerevisiae (abnormal)	*Hanseniaspora uvarum*	*Candida tropicalis*
Saccharomyces exiguus	*Issatchenkia orientalis*	*Clavispora lusitania*
Schizosaccharomyces pombe	*Lodderomyces elongisporus*	*Cryptococcus albidus*
Zygosaccharomyces bailii	*Pichia anomala*	*Cryptococcus laurentii*
Zygosaccharomyces bisporus	*Pichia membranifaciens*	*Debaryomyces etchellsii*
Zygosaccharomyces lentus	*Saccharomyces bayanus*	*Rhodotorula glutinis*
Zygosaccharomyces rouxii	*Saccharomyces cerevisiae*	*Rhodotorula mucilaginosa*

drinks factories, usually in low numbers and restricted by good hygiene procedures. Group 3 microbes are hygiene indicators and will not cause spoilage even if inoculated into soft drinks in large numbers. Most yeasts found in soft drinks production and products are Group 3 organisms. Their numbers can be used as a measure of the general hygiene of the facility, and certain species may indicate the origin of contamination. For example, the yeast-like mould *Aureobasidium pullulans* can be used to indicate air-borne dust contamination. Group 4 organisms were described as aliens, microbes completely out of their normal environment, but whose presence can often be used to forensically detect yeast origins and past events. As an example, the presence of *Kluyveromyces lactis*, an alien in soft drinks environments, might indicate past contact with a dairy or dairy products [38]. Examples of yeast species reported in soft drinks are listed in Table 12.2-1 together with their likely spoilage significance.

The individual characteristics of *Zygosaccharomyces* spoilage yeasts are discussed in detail in Chapter 6. In this review, attention is focussed on the detailed characteristics of a number of the more significant yeasts other than *Zygosaccharomyces* spp., isolated from, and causing spoilage of, soft drinks.

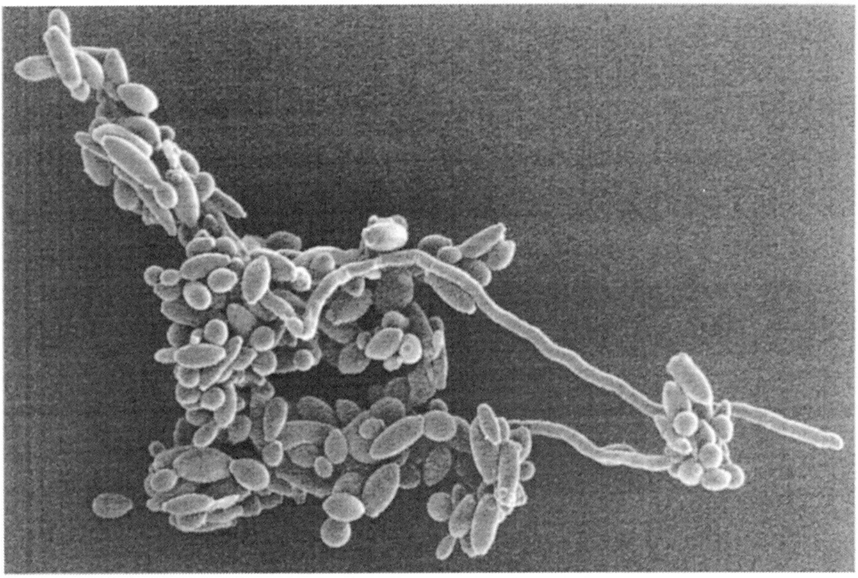

Fig. 12.2-2 Scanning electron micrograph of *Dekkera bruxellensis* strain 148, isolated from a spoiled diet-formulation soft drink. Cell morphology is highly variable showing hyphal, elongated and round forms (Micrograph taken by Mark Kirkland, Unilever R & D, Colworth House).

12.2.2.1 *Dekkera* (*Brettanomyces*) species

Dekkera is an ascospore-forming genus, and is the teleomorph of the genus *Brettanomyces*. Barnett et al. [17] lists 5 species, of which 3 are commonly associated with soft drinks spoilage, namely *D. anomala* (= *Br. anomalus*, syn. *Br. claussenii*) [15, 115]; *D./Br. bruxellensis*; *D. /Br. naardenensis*. Under the microscope these yeasts frequently appear to be a mixed culture, with a variety of cell morphologies in evidence (e. g., hyphal, pseudohyphal, elongated or round – Fig. 12.2-2). *D. naardenensis* is generally restricted to spoilage of soft drinks and carbonated beverages [42]. Whereas *D. anomala* and *D. bruxellensis* can also cause spoilage of wines, cider, and beer, as well as making a positive contribution to the acetic flavour of lambic beers [14, 18, 63, 89, 93, 101, 143, 225]. *D. anomala* has been described as most notorious for spoilage in non-alcoholic beverages, along with *S. cerevisiae* and *Z. bailii* [42].

Spoilage by *Dekkera/Brettanomyces* spp. characteristically results in dense clouds of yeasts cells and thick sediments. Surface films and particulates may also be formed [227]. Acetic acid is frequently formed from sugars via glycolysis [59], but to a lesser extent by

D. (*Brettanomyces*) *naardenensis* [42]. In addition, phenolic off-flavours may be formed [74], described in wine as "Brett" [134] giving a taste of farm-yard, or sweaty leather. These yeasts are fermentative, particularly in more aerobic conditions [235], however, growth is often slow or delayed and spoilage may not become apparent for a number of weeks. Vitamins, thiamine and biotin, are required for growth of *D. naardenensis* and *D. bruxellensis*, but to a lesser extent for *D. anomala* [17, 42]. *Brettanomyces/Dekkera* species are resistant to both acetic acid and sulphite [73], but show only moderate resistance to sorbic acid and are sensitive to benzoic acid (H. STEELS, unpubl. observ.). Resistance by *Brettanomyces/Dekkera* species to CO_2 in carbonated beverages is exceptionally high, exceeding all other yeasts [82, 87]. This undoubtedly contributes to the role of these yeasts in the spoilage of carbonated soft drinks such as cola, tonic and clear lemon [15, 17, 52, 162].

12.2.2.2 *Candida davenportii* and species of the *Starmerella* clade

A new yeast species was recently isolated from a dead wasp found in a soft drinks factory and named *Candida davenportii* in honour of the late Professor BOB DAVENPORT [199]. This yeast grew well in fruit-containing soft drinks, cola-type beverages and synthetic soft drinks. *C. davenportii* is most closely related to the osmophilic spoilage yeasts *C. lactis-condensii* and *C. stellata* [41, 141, 217, 218]. One closely related species, *C. bombicola* (Fig. 12.2-3), has been induced to sporulate and has been named *Starmerella bombicola*. [152]. For the purposes of this review, yeasts belonging to the same taxonomic group as *C. davenportii* and *St. bombicola* are referred to as the *Starmerella* clade (Fig. 12.2-3). Many of the species of the *Starmerella* clade have been implicated in spoilage of foods, particularly sugary, low a_w foods. Spoilage of high sugar commodities has been reported by *C. apicola* – blackcurrant drink, sugar syrups [70, 169], *C. bombicola* – concentrated grape juice, high sugar vegetables [128, 183], *C. etchellsii* – concentrated citrus juice [147], *C. lactis-condensii* – condensed milk, sugar syrups [72, 169], *C. stellata* – soft drinks, fruit juices and concentrates, tomato sauce [41, 141, 161, 162, 184], and *C. magnoliae* – concentrated fruit juice [41]. Spoilage by these yeasts is relatively uncommon and in the context of spoilage significance, probably Group 2, opportunist spoilage microbes [38]. PITT and HOCKING [140] do not list these yeasts as responsible for spoilage of foods processed and packaged according to normal standards of good manufacturing practice.

The *Starmerella* yeasts are fermentative, osmotolerant (i. e. growth in > 50 % w/v sugar), but sensitive to preservatives. Many of these species have been found to be associated with insects, specifically bees, bumblebees and leafcutter bees. The hypothesis was recently proposed [199] that *C. davenportii* and other yeasts belonging to the *Starmerella* clade, are primarily associated with Aculeates (bees and wasps) which are attracted by sugary residues present in foods such as soft drinks, fruit juices and concentrates. Bees and wasps thus represent the likely source of infection of these yeasts [199].

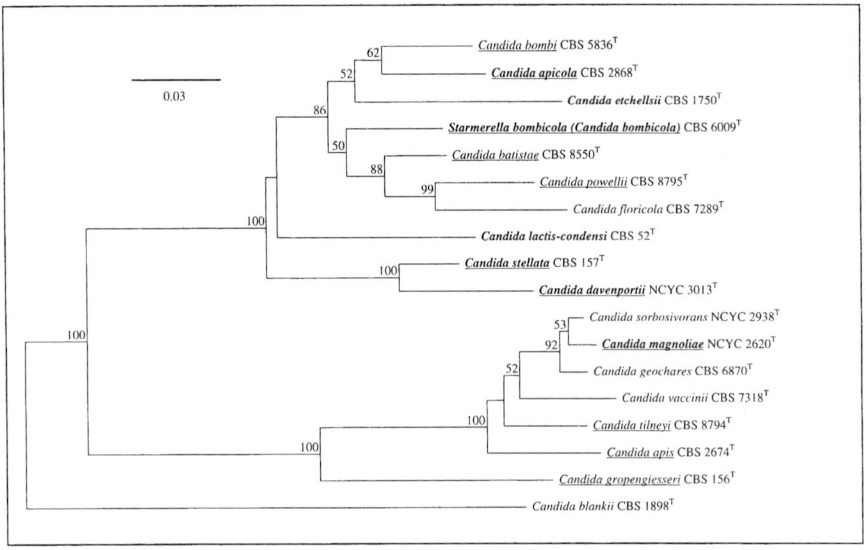

Fig. 12.2-3 Dendrogram showing the phylogenetic relationships of species within the *Starmerella* clade based on 26S rDNA D1/D2 sequences. The tree was constructed using the neighbor-joining method, and bootstrap values, percentages of 100 replications, are given at branch points (only values >50 % are shown). Scale bar, three estimated base substitutions per 100 nucleotides. Species in bold have been reported to be associated with spoilage of high-sugar products and those reported associated with insects, bees and wasps, are underlined.

12.2.2.3 *Candida parapsilosis* and *Lodderomyces elongisporus*

Candida parapsilosis is an asexual (i. e. anamorphic) non-ascospore-forming spoilage species. It was proposed that the ascomycete *L. elongisporus* was teleomorphic with *C. parapsilosis* [71]. However, small subunit rRNA sequencing has revealed that, while these yeasts are indeed closely related to one another, they are nevertheless distinct species [83]. Also closely related are the pathogenic species *C. albicans* and *C. tropicalis*. The close relationship between *C. parapsilosis* and *L. elongisporus* does raise the possibility of mis-identification when these yeasts are isolated from spoiled foods.

Candida parapsilosis is very much an opportunist spoilage yeast, frequently isolated in low numbers from a variety of sources and causing occasional spoilage in a wide variety of materials, ranging from soft drinks, [163] to shampoo. *Lodderomyces elongisporus* similarly causes occasional spoilage of both soft drinks and juice concentrates [41, 143, 158, 166]. *Candida tropicalis* has also been isolated from soft drinks, fruit juices and bottling plants

[18, 143, 163]. *Candida parapsilosis* and *L. elongisporus* share the ability to grow at temperatures above 37 °C, along with *C. tropicalis* and *C. albicans* [17, 140, 226]. *Candida parapsilosis* is frequently isolated from, and causes occasional spoilage in pickles, mayonnaise, high-fat and salt-preserved foods [30, 33, 140] This is in accord with the observed salt resistance of more than 20%, in *C. parapsilosis* [21, 33]. *Candida parapsilosis* strains are osmotolerant, occurring in orange concentrates [167], fermentative [140] and moderately resistant to preservatives [107], at a level comparable to *Saccharomyces cerevisiae* (H. Steels, unpubl. observ.). The observed association of *C. parapsilosis* with *Drosophila* spp. [34, 96] again suggests insect vectors as possible sources of yeast infection.

12.2.2.4 *Issatchenkia orientalis* (teleomorph of *Candida krusei*)

Issatchenkia orientalis is an ascospore-forming species, and is the teleomorph (i. e. sexual form) of *C. krusei*. It is widespread, being commonly found in fresh water [178], as well as on many fruits and in juices [35, 64, 144, 167, 193, 228], cereals and cassava flour [27, 129],

Fig.12.2-4 Surface film formation by *Issatchenkia orientalis* during spoilage of a soft drink. The surface film contains numerous small bubbles generated by fermentation of this yeast.

sourdough breads [62], in sorghum beer [126], tea fungus [114] and oriental fermented foods [203]. In addition to spoilage of soft drinks [221], *I. orientalis* has been reported to be associated with fermentative spoilage in figs [116], cocoa beans [108], olives [123], tomato sauce [141], and flavoured yoghurt [67, 202].

The most obvious sign of spoilage by *I. orientalis* is the presence of a prominent surface film. As this species is also highly fermentative, the surface film may be supported by, or contain, large numbers of gas bubbles (Fig. 12.2-4). Strains of *I. orientalis* are preservative-resistant, although not to the same degree as *Z. bailii*. Resistance to sorbic acid, benzoic acid, acetic acid, and SO_2 has been reported [139, 230, 232]. *I. orientalis* is also resistant to ethanol (H. STEELS, unpubl. observ.). Reports of *I. orientalis* isolated from salt brines [42, 216] are supported by observations that this species is moderately osmotolerant, to both salt [21] and sugar (H. STEELS, unpubl. observ.). *I. orientalis* is unusually resistant to extremes of acidity [139] and to pasteurization [51]. The association of this yeast with *Drosophila* spp. [34] may aid its introduction into soft drinks environments.

12.2.2.5 *Pichia membranifaciens* (teleomorph of *Candida valida*)

As the name suggests, *Pichia membranifaciens*, and its anamorph *C. valida*, are spoilage yeasts that characteristically form a surface film. These are widespread and fairly common yeasts, which have been reported on fruit and in fruit juices [36, 121, 135, 136], in sorghum beer [126], palm wine [12], cane sugar molasses and raisins [5, 133], corn, grass and silage [50]. This species may also be present at low levels on meat and fish [16, 84, 153]. Spoilage by *P. membranifaciens* has been reported in citrus fruit [136, 184], beer [14, 101], wine in bulk or bottle [117, 118, 227], brine preserves such as olives [109], salad dressings [24, 25, 180], cheese [222], and as white spot in bread [185].

The most unusual attribute of *P. membranifaciens* as a spoilage yeast is its ability to grow at low temperature, spoiling chilled foods [69]. This species has a T_{max} for growth of 32–37 °C [226] and was reported sensitive to dry heat [172], but heat-resistant among yeasts, comparable to *Saccharomyces cerevisiae* [22]. *P. membranifaciens* is fast growing under aerobic conditions and non-fermentative [17]. It is resistant to salt [99], SO_2, sorbic and benzoic acids [53], and acetic acid [141]. Association with insect vectors, such as fruit flies [138] may aid infection of foods by *P. membranifaciens*.

12.2.2.6 *Saccharomyces cerevisiae* and *Saccharomyces bayanus* (syn. *Saccharomyces uvarum*)

The taxonomy of the *Saccharomyces* sensu stricto yeasts has been the subject of much debate and revision over the last 20 years [223, 224]. At present 7 species are recognized as belonging to this species complex, namely *S. bayanus* (= S. *uvarum*), *S. cariocanus*, *S. cerevisiae*, *S. kudriavzevii*, *S. mikatae*, *S. paradoxus* and *S. pastorianus* (= *S. carlsber-*

gensis). Interbreeding between these closely related species appears possible [173], with *S. pastorianus* considered to represent a natural hybrid of *S. bayanus* and *S. cerevisiae*. In terms of soft drinks spoilage, *S. bayanus* and *S. cerevisiae* are without doubt responsible for the vast majority of such incidences caused by *Saccharomyces* sensu stricto yeasts, particularly as *S. cariocanus*, *S. kudriavzevii* and *S. mikatae* have not as yet been isolated from foods or beverages.

Saccharomyces cerevisiae is essentially a domesticated organism, widespread due to intensive cultivation in the brewing, winemaking and bread-making industries and subsequent mechanized transportation. As a result, *S. cerevisiae* is found distributed in a wide variety of different foods [17] and is the most frequently isolated yeast in low a_w and low pH products [42]. Together with *S. bayanus*, *S. cerevisiae* is the most common cause of soft drinks spoilage [15, 115, 124]. Brewer's yeast is the main contaminant in facilities producing both beer and soft drinks [115]. Interestingly, the yeast flora from Iraqi soft drinks has been noted [161] for the absence of *S. cerevisiae* and *S. uvarum* (*S. bayanus* [17]), perhaps due to the lack of brewing in this area.

Saccharomyces cerevisiae strains are fast growing, with optimal doubling times ranging from 75 minutes in haploid strains, to 2 hours in diploid/polyploid strains. *S. cerevisiae* strains are highly fermentative, and spoilage is characterized by "blown" bottles and ethanol taints. Heat resistance in *S. cerevisiae* is greater than in most yeasts [22]. The presence of ascospores in this yeast has been reported to greatly increase heat resistance [142, 186]. *S. cerevisiae* is moderately osmotolerant to both salt and sugar. Preservative resistance and vitamin-requirement in *S. cerevisiae* has been reported to be strain variable [17, 38, 140]. While most strains of *S. cerevisiae* display only moderate resistance to sorbic and benzoic acids [139], atypical strains are significantly preservative-resistant and therefore a greater spoilage threat [38, 140]. Possible sources of infection of *S. cerevisiae* in soft drinks production includes local brewing/baking operations, under-washed bottles, and insect vectors [96, 115, 124].

12.2.2.7 *Saccharomyces exiguus* (teleomorph of *Candida holmii*)

Saccharomyces exiguus is an ascospore-forming species, and is the teleomorphic form of *C. holmii* (syn. *Torulopsis holmii*), to which it displays 99.9% 18S rDNA sequence identity [83]. Spoilage characteristics of this species superficially resemble those of *S. cerevisiae*, being fast growing and vigorously fermentative [140]. *Saccharomyces exiguus* has been implicated in spoilage of soft drinks and fruit juices [124], including carbonated beverages and cordials [141], as well as vinegar-and-salt preserved mayonnaise and salads [24, 25, 57].

This yeast is only moderately osmotolerant to salt [21, 99] and sugar (H. STEELS, pers. commun.). *Saccharomyces exiguus* has been reported as preservative resistant to sorbic and benzoic acids [139, 141], and to acetic acid [200]. Using strains authenticated by 26S rDNA D1/D2 sequencing, it was found that preservative resistance was very much strain-variable, with certain strains showing extreme preservative resistance and the ability to grow at pH 4 in the presence of 500 ppm and 800 ppm, sorbic acid and benzoic acid respectively (H. STEELS, unpubl. oberv.). Unlike *S. cerevisiae*, *S. exiguus* does not grow at temperatures above 37 °C [17, 140, 226], is less resistant to ethanol, and shows growth in extremely acidic conditions [21, 139].

12.2.2.8 *Schizosaccharomyces pombe*

Spoilage by the fission yeast *Schizosaccharomyces pombe* is found only occasionally in soft drinks, probably due to the infrequent distribution of this species. When spoilage does occur, the characteristics are those of a Group 1 spoilage yeast, namely excess gas production, preservative resistance, vitamin requirement and osmotolerance [38]. *Sch. pombe* has been isolated from grapes, apples, grape juice, wine and palm wine [12, 17, 93, 174]. It is also found in high-sugar products such as molasses, raisins, dried dates, cane sugar and sugar-syrups, and juice concentrates [133, 140, 217, 218].

Sch. pombe is predictably osmotolerant to sugar [217, 218], but much less so to salt [46, 217]. Like *S. cerevisiae*, it is facultatively fermentative and Crabtree positive [43], and requires B-group vitamins and adenine for growth [17, 42]. It is resistant to preservatives, namely sulphite [230], benzoic acid [231, 232], and acetic acid [140], but only moderately resistant to sorbic acid (H. STEELS, unpubl. observ.). *Sch. pombe* is relatively heat-resistant amongst yeasts, and shows strong growth at 37 °C [140]. Growth of *Sch. pombe* is unusually slow, with a doubling time circa 4 hours, which can result in delayed spoilage. However, it is vigorously fermentative and may produce high levels of H_2S off-flavours if sulphite is present [145]. Where soft drinks spoilage is caused by *Sch. pombe*, sugar syrups and juice concentrates are a likely source of infection.

12.3 Benefits of yeasts in non-alcoholic beverages

Yeast species best able to proliferate in soft drinks are highly fermentative, generating acetate, ethanol or lactate as possible products of fermentation [127]. However, "ethanol is the major end-product of fermentation for most fungi" [85] and if formed in substantial quantities, the product is no longer a non-alcoholic beverage. It is difficult therefore, to envisage a positive role for yeasts in non-alcoholic beverages. There are however, a number of remarkable fermented tea-based beverages where the ethanol formed by yeast action is removed by *Acetobacter* spp. and converted to acetate/acetic acid. Kombucha is one of a number of tea-based beverages of Asian origin, fermented by mixed cultures of bacteria and yeast (and occasionally mould), together forming a surface mat or pellicle described as "the tea fungus". Other synonyms of kombucha and similar beverages include haipao [104], kocha kinoko [210], hongo [20], ishizuchi-kurocha [88], suancha or takezutsu-sancha [206]; the suffix "cha" relating to the tea component of the beverage. Kombucha is typically made from 1.5–5 g/l black tea, infused with boiling water; to which is added 50–100 g/l sucrose before being inoculated with a portion of "the tea fungus" [23, 148, 192]. The taste of kombucha changes during fermentation from sour, fruity and lightly sparkling after a few days, to a mild vinegary taste with prolonged incubation [23, 148, 176]. Kombucha is becoming increasingly popular in Europe and the USA, promoted by claims of health benefits and longevity.

It is probable that kombucha-type beverages have been home-fermented and consumed across Eastern Asia for several millennia, which may account for the variation reported in the micro-organisms comprising the "tea fungus". The acidic nature of these beverages (pH 2.0 – pH 3.6 [23, 148, 176]) limits the bacterial species to acid-tolerant acetic-acid bacteria (*Acetobacter* spp.) and lactic-acid bacteria (*Lactobacillus* spp.) [104, 148, 206]. Yeast species reported in tea fungus include *B. bruxellensis*, *B. lambicus* and *B. custersii* (all now recognized as *D. bruxellensis* [17]), *Debaryomyces hansenii* (= *C. famata*), *Pichia guilliermondii* (= *C. guilliermondii*), *Clavispora lusitaniae* (= *C. obtusa*), *Kloeckera apiculata*, *P. membranifaciens*, *Sch. pombe*, *Saccharomycodes ludwigii*, *S. cerevisiae*, *Z. bailii* and *Z. rouxii* [75, 91, 104, 148]. An examination of two commercial and 32 tea fungus cultures from private households in Germany [114] revealed a very variable composition of yeasts and no defined symbiosis of yeasts and *Acetobacter* spp. The predominant yeast species identified in this study were *B.* (*Dekkera*), *Zygosaccharomyces* and *Saccharomyces* spp. In kombucha the role of yeasts is to invert sucrose and form ethanol, that *Acetobacter* spp. then convert to acetic acid [236].

A recent study by KURTZMAN and colleagues [94], using 26S rDNA D1/D2 sequencing to identify yeasts isolated from kombucha [75], revealed one strain as *P. fluxuum*, and two strains of a hitherto unknown species, which they named *Z. kombuchaensis*. Two further strains from this taxon were isolated in kombucha from the USA. 18S and 26S D1/D2 rDNA sequences both showed *Z. kombuchaensis* to be most closely related to *Z. lentus*, and, more distantly to *Z. bailii* and *Z. bisporus*. All three are recognized as spoilage yeasts [188, 189]. All of these species and *Z. kombuchaensis* are very resistant to acetic acid [191]. It is possible that the acetic acid resistance of many of the yeast species found in kombucha, including *Z. kombuchaensis* [191] might assist their survival in this vinegary environment.

12.4 Physiological background to yeasts in non-alcoholic beverages

As discussed previously, DAVENPORT [38] classified all yeasts found in soft drinks environments as belonging to one of four groups. Group 1 spoilage yeasts were normally absent from soft drinks factories, whereas Group 2 yeasts were opportunistic organisms, taking advantage of mistakes in manufacturing practice to cause spoilage. Group 3 yeasts, widespread in the soft drinks environment, were unable to cause spoilage in soft drinks, even if inoculated into soft drinks in large numbers, and Group 4 were aliens [38, 39, 40].

It is therefore logical to ask the following questions:

1. What are the properties of Group 1 species, (lacking in Group 3 yeasts) which enable them to proliferate so readily in soft drinks?
2. What are the environmental conditions present in soft drinks that prevent or restrict the growth of so many others yeast species? (i. e. Group 2 and 3)
3. What factors prevent growth by Group 2 yeasts in soft drinks and what are the manufacturing errors that permit their growth?

Relatively little physiological research has been carried out into the properties of Group 2 and 3 yeasts. However, by direct observation, the characteristics of Group 1 yeasts of soft drinks noted by DAVENPORT [10, 38, 39] are: 1) a high degree of fermentation; 2) osmotolerance; 3) resistance to weak-acid preservatives such as acetic, sorbic and benzoic acids; 4) vitamin requirement, particularly for B-group vitamins. These properties may be entirely coincidental, or may be associated with a particular taxonomic group of yeast species, or have implications in aiding yeasts to survive and grow in soft drinks.

12.4.1 High degree of fermentation

The highly fermentative property is probably of greatest benefit to Group 1 yeasts in enabling cells to proliferate in soft drinks. A survey of the yeast species listed by BARNETT et al. [17] reveals that of 678 species, 206 are capable of fermentation under semi-anaerobic conditions. Glucose is, by far, the sugar most commonly fermented. In addition, some 129 species are recorded as showing weak, delayed or variable fermentation, with a further 341 species listed as non-fermenting. This suggests that only some one-in-three yeast species are capable of significant fermentation. The majority of yeast species could thus be described as respiratory, requiring oxygen to fully degrade sugars, via the TCA cycle and efficiently generate ATP. Such yeasts are commonly found in very aerobic environments such as on the surfaces of fruit, vegetables and flowers. Large quantities of oxygen are required in respiration, and packages of soft drinks, whether in glass or PET bottles, cans, or tetrapaks are, at best, microaerobic environments and as such, unsuitable for proliferation of respiring yeast species.

Fermentative yeasts, in contrast, possess the ability to ferment sugars without the requirement for oxygen. Carbon dioxide and ethanol are by far the most common products of fungal fermentation [85]. Such a property is likely to be essential for growth and causing spoilage in the near-anaerobic environment of soft drinks containers. Some highly fermentative species, such as *S. cerevisiae*, will continue to ferment sugars even in the presence of oxygen [100] This is an energetically inefficient process, but it does enable fermentative yeasts to become facultative anaerobes, at least as far as energy metabolism is concerned. Oxygen, at low level, may also be required for other metabolic processes. *S. cerevisiae* has been shown to require oxygen for the synthesis of unsaturated fatty acids and sterols, and only shows true anaerobic growth if these are provided [6, 7]. *Z. bailii* has been reported to re-

Fig. 12.4-1 Excess gas generation during soft drinks spoilage following deliberate infection by *Zygosaccharomyces bailii* NCYC 1766. All cans were distended by gas pressure; many ruptured at the ring-pull after several weeks, while others retained their integrity. It is not uncommon to find no viable yeast cells in spoiled soft drinks after prolonged incubation.

quire certain unknown nutritional components, other than fatty acids/sterols, in order to grow anaerobically [149]. It appears possible that other as yet undiscovered nutrients might aid spoilage yeasts to grow anaerobically in soft drinks.

It is also possible that the highly fermentative yeast species are intrinsically resistant to the by-products of fermentation, namely ethanol and carbon dioxide. It is not uncommon after long storage, for obviously "blown" cans of soft drink (Fig. 12.4-1) to contain no viable organisms, with the spoilage yeasts killed either by carbon dioxide or ethanol, or lack of nutrients. Group 1 yeasts such as *S. cerevisiae* and *Z. bailii* are known to be tolerant to high levels of ethanol [196, 212], while some Group 2 yeasts (e. g., *C. parapsilosis* and *Hanseniaspora uvarum*) are ethanol sensitive and Group 3 yeasts (e. g., *Rhodotorula glutinis)* are highly ethanol sensitive (H. STEELS, unpubl. observ.).

Soft drinks are liquids in which most yeast cells sink at a rate of 5mm/hr [194]. Growth therefore occurs in submerged culture (with the exception of film-forming yeasts), at a depth at least 10 to 30 cms of liquid depending on the bottle dimensions. In addition to lack of oxygen, this environment is under slight pressure and likely to contain near-saturated CO_2 from metabolic activity. It is possible that such an environment is hostile to certain species. Carbon dioxide, while inhibitory to most microbes, has relatively little effect on fermentative spoilage yeasts such as *Dekkera/Brettanomyces* spp. and *Z. bailii* [82], thus aiding their ability to cause spoilage. In addition, it is probable that a high degree of fermentation will draw attention to spoilage by distorting packaging, rupturing containers etc. (Fig. 12.4-1), and result in a higher proportion of complaints concerning spoilage by these species [57].

12.4.2 Osmotolerance

Most soft drinks contain only 7–12 % w/w sugar [209] giving a water activity, a_w, as sucrose near 0.995, or as glucose near 0.990 [31]. The great majority of yeast species are able to grow at this water activity; thus 10 % sugar as such does not constitute an osmotically stressful environment for yeasts. Osmophilic yeasts have been defined as able to grow at an a_w of less than 0.85 [214], equivalent to 60 % w/w glucose. Using this definition, the Group 1 spoilage species *Sch. pombe*, *Z. bailii*, *Z. bisporus*, *Z. lentus*, and *Z. rouxii* can be categorized as osmophiles, based on data from BARNETT et al. [17]. On the other hand, *S. exiguus*, *S. cerevisiae*, and *Dekkera/Brettanomyces* spp. are not. Therefore, it would appear that for non-alcoholic beverages, osmotolerance is not an essential requisite for spoilage per se.

The observed association of osmotolerant species with soft drinks spoilage may instead reflect a common route of infection of such yeasts into the soft drinks environment. Most soft drinks are prepared using granulated sugar, sugar syrups or fruit syrups as raw materials. High sugar syrups are a likely source of xerotolerant yeasts including the osmotolerant Group 1 species listed above [213, 214, 215]. If the syrups contain preservatives such as benzoic acid [105], *Zygosaccharomyces* spoilage species may become adapted to growth in preservatives [81, 231], with the syrup thus acting as a source of vigorous, preservative-resistant yeasts, when diluted and added into single strength beverages [80]. It is also possible that osmotolerance may facilitate infection of yeasts via insect vectors. As described previously for yeasts belonging to the *Starmerella* clade (Fig. 12.2-3), it may be speculated that osmotolerant species typically found in fruits and nectar, could be transferred to bees, wasps or fruit flies and carried into the fruit and sugar processing areas within soft drinks facilities.

12.4.3 Preservative resistance

It has been observed that many of the most notorious and damaging spoilage yeasts of soft drinks show very substantial resistance to the preservatives commonly used in these beverages. Resistance to sorbic and benzoic acids is greatest in *Z. bailii*, *Z. bisporus* and *Z. lentus* [81, 125, 139, 187, 189, 229, 232], and high in *S. exiguus* [139, 141], atypical strains of *S. cerevisiae* [38, 140], *Sch. pombe* [231, 232], and *Z. rouxii* [140]. Furthermore, growth of spoilage yeasts in sub-inhibitory levels of preservatives has been shown to result in adapted cells, able to grow in far higher concentrations of preservative [81, 197, 229, 231]. Adaptation to preservatives is a significant possibility in factories, where soft drinks containing preservatives are often splashed. Splashes may be diluted with water and washed into cracks in floors or walls, where they act as substrates for growth and adaptation of preservative-resistant spoilage yeasts. Given that the concentrations of preservatives required to inhibit Group 1 spoilage yeasts are higher than those legally permitted in many parts of the world, spoilage is best prevented by total eradication of such species from the factory and not permitting conditions to exist for adaptation. Details of the mechanisms of preservative resistance by yeasts are discussed in Chapter 6.

12.4.4 Vitamin requirement

Group 1 spoilage yeasts (Table 12.2-1) have been reported to require B-group vitamins, such as biotin [38]. To a large extent, the data of Barnett et al. [17] confirm this observation. With the possible exception of a few individual strains of *D. anomala*, *S. cerevisiae* and *Z. rouxii*, Group 1 spoilage yeasts do not grow in the absence of vitamins, and are thus unlikely to spoil soft drinks when these are absent. Indeed, soft drinks lacking fruit juices have been observed to be less prone to *Zygosaccharomyces* spp. spoilage [52]. The significance of vitamin requirement by spoilage yeasts belonging to several different genera, is at first sight, unexpected and difficult to rationalize. It may possibly indicate that these yeasts have occupied a vitamin-rich environment over a prolonged period, which has acted as a selective pressure, thereby permitting a gradual loss of the ability to synthesize vitamins. The limited information concerning the ecology of Group 1 yeasts, suggests that such a habitat may be based on fruit, particularly dried and mummified fruits [212, 215].

12.5 Quality control and quality assurance

Quality Control and Quality Assurance are two distinct operations in microbiology, which are frequently confused by the public. Quality Control involves testing the final product for numbers of microbes, whereas Quality Assurance involves putting systems in place to ensure the absence of microbes in the final product.

Quality control is routinely carried out in soft drinks production. This involves predominantly end-product testing to determine the total viable cells per millilitre of beverage. Bottles are randomly removed from the production line, aseptically opened and a 20–100 ml sample is membrane filtered, and the filter placed on agar to recover any viable cells. Filtration of larger volumes is rarely possible due to the difficulties of passing soft drinks through filters. The final numbers (i. e. viable cells/ml) gained by quality control are a good indication of general factory hygiene. However, Quality Control is of little use in spoilage

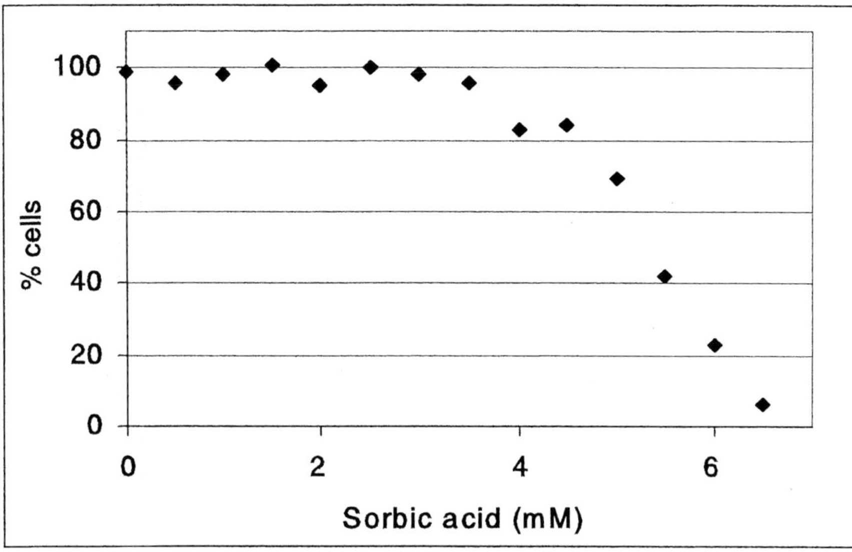

Fig. 12.5-1 Population diversity, expressed as percentage cells in the population, and the inoculum effect demonstrated in *Zygosaccharomyces bailii* NCYC 1766. Only a small proportion of the population of individual cells is capable of growth at high levels of sorbic acid. As a result, substantially greater quantities of preservatives are required to prevent growth of large yeast inocula.

prevention, when microbes are present in very small number, non-uniformly distributed, or spoilage organisms form a very small proportion of the inoculum. Presence of low numbers of yeasts do not necessarily mean lack of spoilage, particularly when *Z. bailii* can cause spoilage from 1 cell/ container [38, 52]. Likewise, high numbers of yeasts may indicate a higher level of risk, although if they are all Group 3 yeasts, then no spoilage will occur. Furthermore, due to the time required for detection of yeasts in filtered samples (2–5 days), the suspect batch of bottles will have been sent to trade unless an expensive quarantine procedure has been set in place. The acceptable level of yeast in the final product varies, depending on the individual type of soft drink. A sensitive fruit-containing soft drink manufactured in a well-run factory may contain less than 1 cell/ litre, whereas in more robust formulations, inoculum levels of 10–100 cells/ml may be acceptable [105].

Quality Assurance, in contrast, describes a process of "safety by design". Within the soft drinks industry there is an increasing trend towards use of quality management systems, such as Hazard Analysis Critical Control Point (HACCP). This process identifies the critical control points during manufacture, and puts systems in place to minimize the risk from these points [113]. For example, if the raw materials (e. g., fruit juice or sugar) are identified as the major source of spoilage yeasts, measures such as pasteurization can be put in place to minimize infection from this source. Using quality assurance measures, the risk of yeast spoilage of soft drinks can be minimized, but never eliminated entirely. Even when using very effective treatments, such as pasteurization, a low level of spoilage will inevitably occur. Pasteurization measures aimed at killing even 10^6 yeast cells/ml will statistically fail occasionally and allow the occasional cell to survive. Higher temperatures will reduce this risk but may also damage the quality of the soft drink. Increasing the yeast load will of course, greatly increase the risk of failure. A microbial load of 10^6/ml would undoubtedly lead to a high level of spoilage in pasteurized bottles. Similarly, it has been shown that large inocula can overcome increased levels of preservatives [103, 211]. The primary role of good factory hygiene must be to minimize the microbial load found in soft drinks. While it is theoretically possible for a single viable yeast cell to cause spoilage (as reported for *Z. bailii* [52]), in practice, it appears that a moderate yeast inoculum is in fact required to cause soft drinks spoilage [115]. Why other yeasts require large inocula has not been investigated, but possibly could involve population diversity effects [190, 201], cell/cell signaling and quorum sensing [146], or sufficient cells to modify the environment before growth can begin.

Overall, a minimal yeast inoculum will act to prevent spoilage in three ways: 1) it will increase the proportion of bottles containing no viable cells, 2) it will decrease the risk of infection by a yeast species able to overcome the preservation systems, and 3) it will decrease the probability of infection by individual cells within a species population, able to cause spoilage. Only a very small proportion of *Z. bailii* cells (Fig. 12.5-1) show the phenomenal preservative-resistance characteristic of this species [190].

12.6 Future prospects and conclusions

At present the role of yeasts in soft drinks is predominantly neutral (albeit undesirable), with most yeasts being unable to grow in this environment. It is possible that in the future, different yeast species may have a beneficial role in soft drinks, but at present this does not appear likely. Conversely, adverse effects caused by the growth of spoilage yeasts are likely to become more prevalent in the future, despite the best efforts of modern preservation technology. The factors impinging on this are: 1) changes in microbial populations, 2) changes in soft drinks formulations, 3) changes in packaging, and 4) changes in preservation systems.

12.6.1 Changes in microbial populations

It is widely known that misuse and over-application of antibiotics has led to the acquisition of multiple drug resistance in pathogenic bacteria. Similarly, medically important yeast species have become increasingly resistant to azole-based therapeutic agents [122]. Despite unsubstantiated claims to the contrary [26], there is no evidence to show that yeasts are becoming increasingly resistant to food preservatives or to pasteurization. Strains of *Z. bailii* isolated many years ago show a broadly similar level of preservative resistance to recently isolated strains (H. STEELS, unpubl. observ.). Yeasts can become "adapted" to preservatives by pre-exposure to low levels of these materials [81]. However, this appears to be short-lived as results from a recent study have shown that after only a few generations growing without preservatives, cell populations had reverted to normal demonstrating that no heritable changes had occurred [190].

However while known spoilage yeast species remain unaltered, it is possible that soft drinks may become exposed to new spoilage yeast species. It is probable that the yeasts discovered to date represent only a small fraction of the true yeast flora existing throughout the world. High-volume modern transportation has increased the risk of accidental exposure of soft drinks to new yeasts present on fruit and vegetables and also on insects. As discussed earlier, insects acts as hosts for many yeasts [96, 97, 98] and as soft drinks are produced in more exotic locations, it is possible that new and unexpected yeast spoilage species may be found, associated with the insect fauna of indigenous fruits and flowers. To date our knowledge of problematic spoilage species has been gained largely empirically, by recording the failures that exploded, or otherwise obviously spoiled. It is to be hoped that a greater scientific understanding of the physiological properties of spoilage yeasts will enable earlier identification of such species.

12.6.2 Changes in soft drink formulations

Over the past two decades the soft drinks industry has been expanding, with a year on year growth rate approaching 10 % [29, 130]. While carbonated soft drinks, which include cola-type products, account for some 50 % of the market, there has been a proliferation of other varieties of soft drink, particularly in the mature markets of the developed world. These beverages are intrinsically more prone to microbiological spoilage, partly due to the absence of carbonation, partly to use of less acidity, and partly to better nutritional status. The expanding markets include sports drinks, health drinks, and drinks made from unusual fruits, as demand for traditional fruits is likely to outstrip supply [130]. Given that some fruit juices are known to spoil easily, due to their richer nutrient status (e. g. blackcurrant [52, 164]), it is likely that soft drinks prepared from certain exotic fruit juices may also be more prone to yeast spoilage.

12.6.3 Changes in packaging

Traditionally soft drinks have been packaged in glass bottles. More recently, steel and aluminium cans have been used, lacquered to prevent corrosion by the acidic contents. Both glass bottles and metal cans are effectively gas impermeable. However, the recent trend towards use of lighter forms of plastic packaging [170] has resulted in increased gas permeability of the packaging, allowing oxygen ingress [150]. These new packaging and formats include plastics of several varieties, metal-foil, cardboard laminates, sleeves, pouches and bag-in-boxes. If this trend to use lighter packaging continues, it is possible that new spoilage yeasts will emerge, enabled to proliferate by the increased availability of oxygen permeating through the packaging into beverages.

Increased use of recycled bottles in the future could also potentially increase spoilage. Recycled bottles, either glass or plastic, in themselves present no hazard; whereas returned bottles form a very potent source of microbial infection, particularly in tropical environments [29]. Empty bottles, often without caps, usually contain a few drops of soft drink residue, in which aerobic environment yeasts can proliferate to 10^6–10^8 cells/ml and adapt to any preservatives that may be present. Return of such sources of infection to factories presents a greatly increased spoilage risk, unless these bottles are segregated, and stringently washed before re-entering production areas.

12.6.4 Changes in preservation

Over the years, formulations of soft drinks have tended to reflect the preservation measures available at the time. In the nineteenth century, soft drinks were very robust, low pH and highly carbonated, even if the taste left much to be desired (e. g., the famous Codd's Wallop [209]). The development of pasteurization and later, use of preservatives such as benzoic acid and sorbic acid, enabled removal of carbonation and the production of still drinks. At present, the spoilage organisms of greatest concern in soft drinks are those largely immune to the current preservation strategies. *Z. bailii* is primarily a problem in soft drinks due to its phenomenal resistance to sorbic acid and benzoic acid. If different preservation systems are applied in the future, it is quite possible that different yeasts may emerge as significant new spoilage species.

The key drivers toward new technologies in preservation are milder processing to give improved taste, the public perception of preservatives and legal constraints to their use. Throughout the developed world, the public is becoming more aware of the presence of chemical residues in foods. While this may have originated in response to pesticide residues in foods, there is increasing public antagonism to all chemical additions to foods, including preservatives. In parallel, there is a move towards more "natural" products, often based on aromatic herbs or essential oils [55]. Legislative changes tend to reflect public concern, and in many countries, there are legal limitations to the use of preservatives in soft drinks. As an example, sorbic acid has GRAS status in the USA, but its use is limited to 300 ppm within the European community [9]. Any further move towards the use of lower levels of preservatives will undoubtedly lead to an increase in yeast spoilage.

New preservative technologies include ultra high pressure, irradiation by ultraviolet or optical wavelengths, rapid aseptic filling, and alternative preservatives such as essential oils or chitosan [2, 44, 79, 112, 132, 137, 151, 154, 168, 175]. The widespread use of such systems in the future is likely to depend upon organoleptic and economic factors, rather than microbiological efficacy in soft drinks.

12.7 References

[1] ADAMS, A.M.: Airborne yeasts from horticultural sites. Can. J. Microbiol. **10** (1964) 641–646.

[2] ADEGOKE, G.O.; IWAHASHI, H.; KOMATSU, Y.: Inhibition of *Saccharomyces cerevisiae* by combination of hydrostatic pressure and monoterpines. J. Fd Sc. **62** (1997) 404–405.

[3] AMERINE, M.A.: Composition of wines. II Inorganic constituents. Adv. Fd Res. **8** (1958) 146–224.

[4] AMERINE, M.A.; BERG, H.W.; KUNKEE, R.E.; OUGH, C.S.; SINGLETON, V.L.; WEBB, A.D.: The Technology of Wine Making, 4th ed. Westport, U.S.A.: AVI (1980).

[5] ANDERSON, P.J.; MCNEIL, K.E.; WATSON, K.A.: Isolation and identification of thermotolerant yeasts from Australian sugar cane mills. J. Gen. Microbiol. **134** (1988) 1691–1698.

[6] ANDREASEN, A.A.; STIER, T.J.B.: Anaerobic nutrition of *Saccharomyces cerevisiae* I. Ergosterol requirement for growth in a defined medium. J. Cell. Compar. Physiol. **41** (1953) 23–26.

[7] ANDREASEN, A.A.; STIER, T.J.B.: Anaerobic nutrition of *Saccharomyces cerevisiae* II. Unsaturated fatty acid requirement for growth in a defined medium. J. Cell. Compar. Physiol. **54** (1954) 271–281.

[8] ANON: Difco Manual of Dehydrated Culture Media and Reagents for Microbiological and Clinical Laboratory Procedures, 9th ed. Detroit, U.S.A.: Difco Laboratories Inc. (1953).

[9] ANON: EC Directive 95/2/EC. Food additives other than colours and sweeteners. Off. J. Eur. Commun. **L61** (1995) 1–40.

[10] ANON: Case Studies of Biotechnology in Action. London, U.K.: Biotechnology Directorate, The Department of Trade and Industry Bulletin, Issue 4 (1998), BMB Initiative.

[11] ASHURST, P.R: Introduction. In: Chemistry and Technology of Soft Drinks and Fruit Juices, (edited by Ashurst, P.R.). Sheffield, U.K.: Sheffield Academic Press, and Boca Raton, U.S.A.: CRC Press (1998) 1–15.

[12] ATPUTHARAJAH, J.D.; WIDANAPATHIRANA, S.; SAMARAJEEWA, U.: Microbiology and biochemistry of natural fermentation of coconut palm sap. Fd Microbiol. **3** (1986) 273–280.

[13] BABSKY, N.E.; TORIBIO, J.L.; LOZANO, J.E.: Influence of storage on the composition of clarified apple juice concentrate. J. Fd Sc. **51** (1986) 564–567.

[14] BACK, W.: Nachweis und Identifizierung von Fremdhefen in der Brauerei. Brauwelt **127** (1987) 735–737.

[15] BACK, W.; ANTHES, S.: Taxonomische Untersuchungen an limonadenschädlichen Hefen. Brauwissenschaft **3236** (1979) 145–154.

[16] BANKS, J.G.; BOARD, R.G.: Some factors influencing the recovery of yeasts and moulds from chilled foods. Int. J. Fd Microbiol. **4** (1987) 197–206.

[17] BARNETT, J.A.; PAYNE, R.W.; YARROW, D.: Yeasts: characteristics and identification. 3rd ed. Cambridge, U.K.: Cambridge University Press (2000).

[18] BEECH, F.W.; CARR, J.G.: Cider and Perry: In: Alcoholic Beverages, Economic Microbiology, Vol. 1 (edited by Rose, A.H.) London, U.K., New York, U.S.A.: Academic Press (1977) 139–313.

[19] BEESLEY, P.M.: Carbohydrate sugars. In: Formulation and Production of Carbonated Soft Drinks, (edited by Mitchell, A.J.). Glasgow and London, U.K.: Blackie and Son Ltd (1990) 37–55.

[20] BENK, E.: The tea fungus fermented drink. Verbraucherdienst **33** (1988) 213–214.

[21] BETTS, G.D.; LINTON, P.; BETTERIDGE, R.J.: Food spoilage yeasts: effects of pH, NaCl and temperature on growth. Fd Contr. **10** (1999) 27–33.

[22] BEUCHAT, L.R.: Synergistic effects of potassium sorbate and sodium benzoate on thermal inactivation of yeasts. J. Fd Sc. **49** (1981) 771–777.

[23] BLANC, P.J.: Characterization of tea fungus metabolites. Biotechnol. Lett. **18** (1996) 139–142.

[24] BROCKLEHURST, T.M.; LUND, B.M.: Yeasts as a cause of spoilage of chilled delicatessen salads. J. Appl. Bacteriol. **59** (6) Spec. Sect. (1985) iv.

[25] BROCKLEHURST, T.F.; WHITE, C.A.; DENNIS, C.: The microflora of stored coleslaw and factors affecting the growth of spoilage yeasts in coleslaw. J. Appl. Bacteriol. **55** (1983) 57–63.

[26] BRUL, S.; COOTE, P.J.: Preservative agents in foods – mode of action and microbial resistance mechanisms. Int. J. Fd Microbiol. **50** (1999) 1–17.

[27] BURMEISTER, H.R.; HARTMAN, P.A.: Yeasts in ensiled high-moisture corn. Appl. Microbiol. **14** (1966) 35–38.

[28] BURROUGHS, L.F.: Analytische Zusammensetzung englischer Apfelsäfte. Flüssiges Obst **51** (1984) 370–372.

[29] BYRNE, M.: The beverage boom continues. Fd Engin. Int. **19** (1994) 52–57.

[30] CANTONI, C; COMI, G.: I lieviti ei prodotti di gastronomia. Ind. Aliment. **27** (1988) 649–651.

[31] CHRISTIAN, J.H.B.: Reduced water activity. In: Microbial Ecology of Foods, Vol. 1, Factors Affecting Life and Death of Microorganisms (edited by the ICMSF, International Commission on Microbiological Specifications for Foods). New York, U.S.A.: Academic Press (1980) 70–91.

[32] CLARK, D.S.; TAKACS, J.: Gases as preservatives. In: Microbial Ecology of Foods, Vol. 1, Factors Affecting Life and Death of Microorganisms (edited by the ICMSF, International Commission on Microbiological Specifications for Foods). New York, U.S.A.: Academic Press (1980) 170–192.

[33] COMI, G.; DENOZZA, D.; CANTONI, C.: Sul deposito di lieviti in bibite analcliche. Ind. Bevande **55** (1981) 346–348.

[34] COOPER, D.M.: Food preferences of larval and adult *Drosophila*. Evolution **14** (1960) 41–55.

[35] DAVENPORT, R.R.: The distribution of yeasts and yeast-like organisms in an English vineyard. Ph.D. thesis, University of Bristol, Bristol, U.K. (1975).

[36] DAVENPORT, R.R.: The distribution of yeasts and yeast-like organisms from aerial surfaces of developing apples and grapes. In: Microbiology of aerial plant surfaces (edited by Nickerson, C.H.; Preece, T.F.) London, U.K.: Academic Press (1976) 325–359.

[37] DAVENPORT, R.R.: Yeasts and Yeast-like Organisms. In: Smith's Introduction to Industrial Mycology, 7th ed. (edited by Onions, A.H.S.; Allsopp, D.; Eggins, H.O.W.). London, U.K.: Edward Arnold (Publishers) Ltd. (1981) 65–92.

[38] DAVENPORT, R.R.: Forensic microbiology for soft drinks business. Soft Drinks Manag. Int. (April 1996) 34–35.

[39] DAVENPORT, R.R.: Forensic microbiology II. Case book investigations. Soft Drinks International (April 1997) 26–30.

[40] DAVENPORT, R.R.: Microbiology of soft drinks. In: The Chemistry and Technology of Soft Drinks and Fruit Juices (edited by Ashurst, P.R.). Sheffield, U.K.: Sheffield Academic Press, and Boca Raton, U.S.A.: CRC Press (1998) 197–216.

[41] DEAK, T.; BEUCHAT, L.R.: Yeasts associated with fruit juice concentrates. J. Fd Prot. **56** (1993) 777–782.

[42] DEAK, T.; BEUCHAT, L.R.: Handbook of Food Spoilage, Yeasts. Boca Raton, U.S.A.: CRC Press Inc.(1996).

[43] VAN DIJKEN, J.P.; WESTHUIS, R.A.; PRONK, J.T.: Kinetics of growth and sugar consumption in yeasts. Antonie van Leeuwenhoek **63** (1993) 343–352.

[44] DORMAN, H.J.D.; DEANS, S.G.: Antimicrobial agents from plants: antibacterial activity of plant volatile oils. J. Appl. Microbiol. **88** (2000) 308–316.

[45] DRAGONI, I.; COMI, G.: Presenza di muffe e lieviti in succhi di frutta prodotti industrialmente. Ind. Bevande **14** (1985) 599–606.

[46] VAN ECK, J.H.; PRIOR, B.A.; BRANDT, E.V.: The water relations of growth and polyhydroxy alcohol production by ascomycetous yeasts. J. Gen. Microbiol. **139** (1993) 1047–1054.

[47] EKLUND, T.: The effect of carbon dioxide on bacterial growth and on uptake processes in bacterial membrane vesicles. Int. J. Fd Microbiol. **1** (1984) 179–185.

[48] ELLINGER, R.H.: Phosphates in food processing. In: Handbook of Food Additives (edited by Furia, T.E.) Cleveland, U.S.A.: CRC Press (1972) 617–671.

[49] ENFORS, S.O.; MOLIN, G.: The influence of temperature on the growth inhibitory effect of carbon dioxide on *Pseudomonas fragi* and *Bacillus cereus*. Can. J. Microbiol. **27** (1981) 15–19.

[50] ENGEL, G.: Hefen in Silagen und Rohmilch. Milchwissensch. **41** (1986) 633–637.

[51] ENGEL, G.; ROSCH, N.; HELLER, K.J.: Heat tolerance of yeasts. Kieler Milchwirtsch. Forschungsber. **46** (1994) 81–90.

[52] VAN ESCH, F.: Yeasts in soft drinks and fruit juice concentrates. De Ware(n) Chemicus **17** (1987) 20–31.

[53] ETHIRAJ, S.; SURESH, E.R.: *Pichia membranaefaciens*: a benzoate resistant yeast from spoiled mango pulp. J. Fd Sc. Technol. **25** (1988) 63–66.

[54] EYTON-JONES, R.: The effect of CO_2 on cell size of *Saccharomyces cerevisiae*. MSc Thesis, Heriot-Watt University, Edinburgh, U.K. (1987).

[55] FARR, S.: 2001, a soft drinks odyssey. Fd Manufact. **69** (1994) 29–30.

[56] FENG, M.; LALOR, B.; HU, S.; HUBER, A.; KIDBY, D.; HOLBEIN, B.: Inhibition of yeast growth in grape juice through removal of iron and other metals. Int. J. Fd Sc. Technol. **32** (1997) 21–28.

[57] FLEET, G.: Spoilage yeasts. Crit. Rev. Biotechnol. **12** (1992) 1–44.

[58] FRY, J.: Authentication of orange juice. In: Production and Packaging of Non-Carbonated Fruit Juices and Fruit Beverages (edited by Hicks, D.). Glasgow and London, U.K.: Blackie and Son Ltd (1990) 68–113.

[59] GEROS, H.; AZEVADO, M.-M.; CASSIO, F.: Biochemical studies on the production of acetic acid by the yeast *Dekkera anomala*. Fd Technol. Biotechnol. **38** (2000) 59–62.

[60] GIGIOLI, I.: Insects and yeasts. Nature **56** (1897) 575–577.

[61] GIMENO, C.J.; LJUNGDAHL, P.O.; STYLES, C.A.; FINK, G.R.: Unipolar cell divisions in the yeast *Saccharomyces cerevisiae* lead to filamentous growth: regulation by starvation and RAS. Cell **68** (1992) 1077–1090.

[62] GOBBETTI, M.: The sourdough microflora: interactions of lactic acid bacteria and yeasts. Trends Fd Sc. Technol. **9** (1998) 267–274.

[63] GOSWELL, R.W.: Microbiology of table wines. Dev. Fd Microbiol. **2** (1986) 21–65.

[64] GOTO, S.; YOKOTSUKA, I.: Wild yeast population in fresh grape must of different harvest time. J. Ferment. Technol. **55** (1977) 417–422.

[65] GOVERD, K.A.; BEECH, F.W.; HOBBS, R.P.; SHANNON, R.: The occurrence and survival of coliforms and salmonellae in apple juice and cider. J. Appl. Bacteriol. **46** (1978) 521–530.

[66] GRAHAM, A.F.; LUND, B.M.: The effect of citric acid on the growth of proteolytic strains of *Clostridium botulinum*. J. Appl. Bacteriol. **61** (1986) 39–49.

[67] GREEN, M.D.; IBE, S.N.: Yeasts as primary contaminants in yogurts produced commercially in Lagos, Nigeria. J. Fd Prot. **50** (1987) 193–198.

[68] GRINBAUM, A.; ASHKENAZI, I.; TREISTER, G.; GOLDSCHMIED-REOUVEN, A.; BLOCK, C.S.: Exploding bottles: eye injury due to yeast fermentation of an uncarbonated soft drink. Brit. J. Ophthalmol. **78** (1994) 883.

[69] GUERZONI, M.E.; LANCIOTTI, R.; MARCHETTI, R.: Survey of the physiological properties of the most frequent yeasts associated with commercial chilled foods. Int. J. Fd Microbiol. **17** (1993) 329–341.

[70] HAJSIG, M.: *Torulopsis apicola* nov. spec., new isolates from bees. Antonie van Leeuwenhoek **24** (1958) 18–22.

[71] HAMAJIMA, K.; NISHIKAWA, A.; SHINODA, T.; FUKAZAWA, Y.: Deoxyribonucleic acid base composition and it homology between two forms of *Candida parapsilosis* and *Lodderomyces elongisporus*. J. Gen. Appl. Microbiol. **33** (1987) 299–302.

[72] HAMMER, B.W.: Studies on the formation of gas in sweetened condensed milk. Iowa Agricult. Exp. Stat. Res. Bull. **54** (1919) 211–220.

[73] HAMMOND, S.M.; CARR, J.G.: The antimicrobial activity of SO_2-with particular reference to fermented and non-fermented fruit juices. In: Inhibition and inactivation of vegetative microbes, (edited by Skinner, F.A.; Hugo, W.B.) London, U.K.: Academic Press (1976) 89–110.

[74] HERESZTYN, T.: Metabolism of volatile phenolic compounds from hydroxycinnamic acids by *Brettanomyces* yeast. Arch. Microbiol. **146** (1986) 96–98.

[75] HESSELTINE, C.W.: A millenium of fungi, food and fermentation. Mycologia **57** (1965) 149–197.

[76] HICKS, D.: Non-carbonated fruit-containing beverages. In: Production and packaging of non-carbonated fruit juices and fruit beverages (edited by Hicks, D.). Glasgow and London, U.K.: Blackie (1990) 264–306.

[77] HIPKIN, C.R.: Nitrate assimilation in yeast. In: Molecular and Genetic Aspects of Nitrate Assimilation (edited by Wray, J.L.; Kinghorn J.R.). Oxford, U.K.: Oxford Science Publications (1989) 51–68.

[78] HUANG, Z.; OUGH, C.S.: Effect of vineyard locations, varieties, and rootstocks on the juice amino acid composition of several cultivars. Am. J. Enol. Viticult. **40** (1989) 135–139.

[79] HULIN, V.; MATHOT, A.G.; MAFART, P.; DUFOSSE, L.: Antimicrobial properties of essential oils and flavour compounds. Sc. Alim. **18** (1998) 563–582.

[80] INGRAM, M.: Fermentation in concentrated orange juice. Fd Manufact. **24** (1949) 77–81.

[81] INGRAM, M.: Studies on benzoate-resistant yeasts. Acta Microbiologica **7** (1960) 95–105.

[82] ISON, R.W.; GUTTERIDGE, C.S.: Determination of the carbonation tolerance of yeasts. Lett. Appl. Microbiol. **5** (1987) 11–13.

[83] JAMES, S.A.; COLLINS, M.D.; ROBERTS, I.N.: The genetic relationship of *Lodderomyces elongisporus* to other ascomycete yeast species as revealed by small-subunit rRNA gene sequences. Lett. Appl. Microbiol. **19** (1994) 308–311.

[84] JAY, J.M.: Meats, poultry and seafoods. In: Food and beverage mycology, 2nd ed. (edited by Beuchat, L.R.). New York, U.S.A.: van Nostrand Reinhold (1987) 155–173.

[85] JENNINGS D.H.: The Physiology of Fungal Nutrition. Cambridge, U.K.: Cambridge University Press (1995).

[86] JONES, M.V.: Modified atmospheres. In: Mechanisms of action of food preservation procedures (edited by Gould, G.W.). London, U.K., New York, U.S.A.: Elsevier Applied Science Press (1989) 247–284.

[87] JONES, R.P.; GREENFIELD, P.F.: Effect of carbon dioxide on yeast growth and fermentation. Enz. Microb. Technol. **4** (1982) 210–223.

[88] KATO, M.; TAMURA, A.; SAITOU, H.; OMORI, M.; NANBA, A.; MIYAGAWA, M.: Changes in flavour during manufacturing process of Japanese fermented tea (Ishizuchi-kurocha) and its characteristics. Nippon Kasei Gakkaishi **46** (1995) 525–530.

[89] DE KEERSMAECKER, J.: The mystery of Lambic beer. Sc. Am. **275** (August 1996) 56–63.

[90] KNATCHBULL, F.B.; SLAUGHTER, J.C.: The effect of low CO_2 pressures on the absorption of amino acids and production of flavour-active volatiles by yeast. J. Inst. Brew. **93** (1987) 420–424.

[91] KOZAKI, M.; KOIZUMI, A.; KITAHARA, K.: Microorganisms of zoogloeal mats formed in tea decoction. J. Fd Hyg. Soc. (Japan) **13** (1972) 89–96.

[92] KUNKEE, R.E.; BISSON, L.F.: Wine-making yeasts. In: The yeasts Vol. 5 2nd ed., Yeast technology (edited by Rose, A.H.; Harrison, J.S.). London, U.K.: Academic Press (1993) 69–127.

[93] KUNKEE, R.E.; GOSWELL, R.W.: Table wines. In: Alcoholic beverages, economic microbiology Vol.1 (edited by Rose, A.H.). London, U.K.: Academic Press (1977) 315–385.

[94] KURTZMAN, C.P.; ROBNETT, C.J.; BASEHOAR-POWER, E.: *Zygosaccharomyces kombuchaensis*, a new ascosporogenous yeast from "kombucha tea". FEMS Yeast Res. **1** (2001) 133–138.

[95] LACHANCE, M.-A.: Yeast biodiversity: how much and how many? In: The rising power of yeasts in science and industry, 10th Int. Symp. Yeasts (edited by van Dijken, J.P.; Scheffers W.A.). Delft, The Netherlands: Delft University Press (2000) 103–104.

[96] LACHANCE, M.-A.; GILBERT, D.G.; STARMER, W.T.: Yeast communities associated with *Drosophila* species and related flies in an eastern oak-pine forest: a comparison with western communities. J. Ind. Microbiol. **14** (1995) 484–494.

[97] LACHANCE, M.-A.; BOWLES, J.M.; DIAZ M.M.C.; JANZEN, D.H.: *Candida cleridarum, Candida tilneyi,* and *Candida powellii*, three new yeast species isolated from insects associated with flowers. Int. J. Syst. Evol. Microbiol. **51** (2001) 1201–1207.

[98] LACHANCE, M.A.; STARMER, W.T.; ROSA C.A.; BOWLES, J.M.; BARKER, J.S.F.; JANZEN, D.H.: Biogeography of the yeasts of ephemeral flowers and their insects. FEMS Yeast Res. **1** (2001) 1–13.

[99] LAGES, F.; SILVA-GRACA, M.; LUCAS, C.: Active glycerol uptake is a mechanism underlying halotolerance in yeasts: a study of 42 species. Microbiology **145** (1999) 2577–2585.

[100] LAGUNAS, R.: Misconceptions about the energy metabolism of *Saccharomyces cerevisiae*. Yeast **2** (1986) 221–228.

[101] LAWRENCE, D.R.: Spoilage organisms in beer. In: Developments in food microbiology, Vol. 3 (edited by Robinson, R.K.). London, U.K.: Elsevier (1988) 1–48.

[102] LEA, A.G.H.: Apple juice. In: Production and packaging of non-carbonated fruit juices and fruit beverages (edited by Hicks, D.). Glasgow and London, U.K.: Blackie (1990) 183–225.

[103] LIEWEN, M.B.; MARTH, E.H.: Growth and inhibition of microorganisms in the presence of sorbic acid: A review. J. Fd Prot. **48** (1985) 364–375.

[104] LIU, C.H.; HSU, W.H.; LEE, F.L.; LIAO, C.C.: The isolation and identification of microbes from a fermented tea beverage, Haipao, and their interactions during Haipao fermentation. Fd Microbiol. **13** (1996) 407–415.

[105] LLOYD, A.C.: Preservation of comminuted orange products. J. Fd Technol. **10** (1975) 565–574.

[106] LUMSDEN, W.B.; DUFFUS, J.H.; SLAUGHTER, J.C.: Effects of CO_2 on budding and fission yeasts. J. Gen. Microbiol. **133** (1987) 877–881.

[107] MAIMER, E.; BUSSE, M.: Growth properties and gas formation by yeasts isolated from processed fruits in media with various brix values and sorbic acid contents. J. Fd Prot. **55** (1992) 192–197.

[108] MARAVALHAS, N.: Mycological deterioration of cocoa beans during fermentation and storage in Bahia. Int. Chocolate Rev. **21** (1966) 375–378.

[109] MARQUINA, D.; PERES, C.; CALDAS, F.V.; MARQUES, J.F.; PEINADO, J.M.; SPENCER-MARTINS, I.: Characterization of the yeast population in olive brines. Lett. Appl. Microbiol. **14** (1992) 279–283.

[110] MARTELL, A.E.; SMITH, R.M.: Critical stability constants, Vol. 1: Amino acids. New York, U.S.A. and London, U.K.: Plenum Press (1974).

[111] MARTELL, A.E.; SMITH, R.M.: Critical stability constants, Vol. 3, Other Organic Ligands. New York, U.S.A., London, U.K.: Plenum Press (1977).

[112] MARUZZELLA, J.C.; Liguori, L.: The *in vitro* antifungal activity of essential oils. J. Am. Pharm. Assoc. **45** (1958) 250–254.

[113] MAYES, T.: HACCP training. Fd Contr. **5** (1994) 190–195.

[114] MAYSER, P.; FROMME, S.; LEITZMAN, C.; GRUNDER, K.: The yeast spectrum of the "tea fungus kombucha". Mycoses **38** (1995) 289–295.

[115] MEYER, O.: Erfahrungen aus mikrobiellen Vorkommnissen in alkoholfreie Getränke herstellenden Betrieben. Lebensmittelindustrie **28** (1981) 308–311.

[116] MILLER, M.W.; PHAFF, H.J.: Successive microbial populations on Calimyrna figs. Appl. Microbiol. **10** (1962) 394–400.

[117] MINARIK, E.: Zur Ökologie von Hefen und hefeartigen Mikroorganismen abgefüllter Weine. Wein Wissensch. **36** (1981) 280–285.

[118] MINARIK, E.: Levures de contamination des vins embuteillees. Bull. O.I.V. **56** (1983) 414–419.

[119] MITCHELL, A.J.: Carbonation and filling. In: Formulation and production of carbonated soft drinks. Glasgow and London, U.K.: Blackie and Son Ltd. (1990) 203–242.

[120] MONCH, D.; KRUGER, E.; STAHL, U.: Wirkung von Streß auf Brauereihefen. Monatsschr. Brauwissensch. **48** (1995) 288–299.

[121] MORA, J.; ROSELLO, C.: The growth and survival of *Pichia membranaefaciens* during fermentation of grape juice. Am. J. Enol. Viticult. **43** (1992) 329–332.

[122] MORAN, G.P.; SANGLARD, D.; DONNELLY, S.M.; SHANLEY, D.B.; SULLIVAN, D.J.; COLEMAN, D.C.: Identification and expression of multidrug transporters responsible for fluconazole resistance in *Candida dubliniensis*. Antimicrob. Agents Chemother. **42** (1998) 1819–1830.

[123] MRAK, E.M.; VAUGHN, R.H.; MILLER, M.W.; PHAFF, H.J.: Yeasts occurring in brines during the fermentation and storage of green olives. Fd Technol. **10** (1956) 416–419.

[124] MUZIKAR, V.: Incidence and significance of yeasts in non-alcoholic beverages. In: Microbial associations and interactions in foods (edited by Kiss, I.; Deak, T.; Incze, K.). Dordrecht, The Netherlands, Boston, U.S.A., Lancaster, U.S.A.: D. Reidel Publishing Company (1984) 247–253.

[125] NEVES, L.; PAMPULHA, M.E.; LOUREIRO-DIAS, M.C.: Resistance of food spoilage yeasts to sorbic acid. Lett. Appl. Microbiol. **19** (1994) 8–11.

[126] NOUVELLIE, L.; DE SCHAEPDRIJVER, P.: Modern developments in traditional African beers. Progr. Ind. Microbiol. **23** (1986) 73–157.

[127] O'FALLON, J.V.; WRIGHT, R.H.; CALZA, R.E.: Glucose metabolic pathways in the anaerobic rumen fungus *Neocallismatix fontalis*. Biochemical J. **274** (1991) 595–600.

[128] OK, T.; HASHINAGA F.: Identification of sugar-tolerant yeasts from high-sugar fermented vegetable extracts. J. Gen. Appl. Microbiol. **43** (1997) 39–47.

[129] OKAGBUE, R.N.: Identification of yeasts and anaerobic spore-forming bacteria from cassava flour. Fd Microbiol. **7** (1990) 27–32.

[130] OLESEN, R.K.: Fruit juice, the major markets. Soft Drinks Manag. Int. (March 1992) 16–17.

[131] PALMER, Q.: Special topics. In: Chemistry and technology of soft drinks and fruit juices (edited by Ashurst, P.R.). Sheffield, U.K.: Sheffield Academic Press and Boca Raton, U.S.A.: CRC Press (1998) 217–249.

[132] PALOU, E.; LOPEZ-MALO, A.; BARBOSA-CANOVAS, G.V.; WELTI-CHANES, J.; SWANSON, B.J.: High hydrostatic pressure as a hurdle for *Zygosaccharomyces bailii* inactivation. J. Fd Sc. **62** (1997) 855–857.

[133] PARFAIT, A.; SABIN, G.: Yeasts isolated from cane sugar and molasses. Ind. Aliment. Agric. **92** (1975) 27–34.

[134] PARKER, R.: Wines of the Rhone valley; revised and expanded edition. London: Dorling Kindersley (1997).

[135] PARISH, M.E.; CAROLL, D.E.: Indigenous yeasts associated with Muscatine (*Vitis rotundifolia*) grapes and musts. Am. J. Enol. Viticult. **36** (1985) 165–169.

[136] PARISH, M.E.; HIGGINS, D.P.: Yeasts and molds isolated from spoiling citrus products and by-products. J. Fd Prot. **52** (1989) 261–263.

[137] PASTER, N.; JUVEN, B.J.; SHAAYA, E.; MENASHEROV, M.; NITZAN, R.; WEISSLOWICZ, H.; RAVID, U.: Inhibitory effect of oregano and thyme essential oils on moulds and foodborne bacteria. Lett. Appl. Microbiol. **11** (1990) 33–37.

[138] PIGNAL, M.-C.; LACHAISE, D.; COUTURIER, G.: Les levures des figues et des drosophiles associees en foret de tai (Cote d'Ivoire). Acta Oecol. **6** (1985) 223–233.

[139] PITT, J.I.: Resistance of some food spoilage yeast to preservatives. Fd Technol. Austr., **6** (1974) 238–241.

[140] PITT, J.I.; HOCKING, A.D.: Fungi and food spoilage, 2nd ed. London, U.K.: Blackie Academic and Professional (1997).

[141] PITT, J.I.; RICHARDSON, K.C.: Spoilage by preservative-resistant yeasts. CSIRO Fd Res. Quart. **33** (1973) 80–85.

[142] PUT, H.M.C.; DE JONG, J.: The heat resistance of ascospores of four *Saccharomyces* spp. isolated from spoiled heat-processed soft drinks and fruit products. J. Appl. Bacteriol. **52** (1982) 235–243.

[143] PUT, H.M.C.; DE JONG, J.; SAND, F.E.M.J.; VAN GRINSVEN, A.M.: Heat resistance studies on yeast spp. causing spoilage in soft drinks. J. Appl. Bacteriol. **40** (1976) 135–152.

[144] RALE, B.V.; VAKIL, J.R.: A note on an improved molybdate agar for the selective isolation of yeasts from tropical fruits. J. Appl. Bacteriol. **56** (1984) 409–413.

[145] RANKINE, B.C.: Hydrogen sulphide production by yeasts. J. Sc. Fd Agricult. **15** (1964) 872–877.

[146] RASMUSSEN, L.; CHRISTENSEN, S.T.; SCHOUSBOE, P.; HEATLEY, D.N.: Cell survival and multiplication. The overriding need for signals: from unicellular to multicellular systems. FEMS Microbiol. Lett. **137** (1996) 123–128.

[147] RECCA, J.; MRAK, E.M.: Yeasts occurring in citrus products. Fd Technol. **6** (1952) 450–454.

[148] REISS, J.: Influence of different sugars on the metabolism of the tea fungus. Zeitschr. Lebensmittel-Untersuch. Forsch. **198** (1994) 258–261.

[149] RODRIGUES, F.; CORTE-REAL, M.; LEAO, C.; DIJKEN, J.P.; PRONK, J.T.: Oxygen requirements of the food spoilage yeast *Zygosaccharomyces bailii* in synthetic and complex media. Appl. Environ. Microbiol. **67** (2001) 2123–2128.

[150] RODRIGUEZ, J.H.; COUSIN, M.A.; NELSON, P.E.: Oxygen requirements of *Bacillus licheniformis* and *Bacillus subtilis* in tomato juice; ability to grow in aseptic packages. J. Fd Sc. **57** (1992) 973–976.

[151] ROLLER, S.; COVILL, N.: The antifungal properties of chitosan in laboratory media and apple juice. Int. J. Fd Microbiol. **47** (1999) 67–77.

[152] ROSA, C.A.; LACHANCE, M.-A.: The yeast genus *Starmerella* gen. nov. and *Starmerella bombicola* sp. nov., the teleomorph of *Candida bombicola* (Spencer, Gorin, & Tullock) Meyer & Yarrow. Int. J. Syst. Bacteriol. **48** (1998) 1413–1417.

[153] ROSS, S.S.: MORRIS, E.O.: An investigation of the yeast flora of marine fish from Scottish coastal waters and a fishing ground off Iceland. J. Appl. Bacteriol. **28** (1965) 224–234.

[154] SAGOO, S.K.; BOARD, R.; ROLLER, S.: Chitosan potentiates the antimicrobial action of sodium benzoate on spoilage yeasts. Lett. Appl. Microbiol. **34** (2002) 168–172.

[155] SAND, F.E.M.J.: Zur Hefeflora von Erfrischungsgetränken. Brauwelt **110** (1970) 225–236.

[156] SAND, F.E.M.J.: Zur Bakterienflora von Erfrischungsgetränken. Brauwelt **111** (1971) 252–264.

[157] SAND, F.E.M.J.: Zur Hygiene von Abfüllkolonnen der Erfrischungsgetränkeindustrie. Brauwelt **111** (1971) 1788–1800.

[158] SAND, F.E.M.J.: Recent investigations on the microbiology of fruit juice concentrates. In: Technology of fruit juices concentrates – chemical composition of fruit juices (edited by Scientific-Technical Commission, XIII). Wien, Austria: Int. Fed. Fruit Juice Producers (1973) 185–216.

[159] SAND, F.E.M.J.: Gluconobacter, Kohlensäurefreies Getränk und Kunststoffverpackung. Das Erfrischungsgetränk Mineralwasser-Ztg **29** (1976) 1–2.

[160] SAND, F.E.M.J.: Gluconobacter, still drinks and plastic containers. Soft Drinks Trade J. (1977) 5–9.

[161] SAND, F.E.M.J.; VAN GRINSVEN, A.M.: Comparison between the yeast flora of Middle Eastern and Western European soft drinks. Antonie van Leeuwenhoek **42** (1976) 523–532.

[162] SAND, F.E.M.J.; VAN GRINSVEN, A.M.: Investigation of yeast strains isolated from Scandinavian soft drinks. Brauwissensch. **29** (1976) 353–355.

[163] SAND, F.E.M.J.; KOLFSCHOTEN, G.A.: Taxonomische und ökologische Untersuchung von einigen aus Colagetränken isolierten Hefen. Brauwissensch. **22** (1969) 129–138.

[164] SAND, F.E.M.J.; KOLFSCHOTEN, G.A.: Mikrobiologische Aspekte bei Feindosierpumpen in der Erfrischungsgetränkeindustrie. Brauwelt **111** (1971) 1524–1532.

[165] SAND, F.E.M.J.; KOLFSCHOTEN, G.A.: Technische Konstruktion und mikrobiologische Qualität. Das Erfrischungsgetränk Mineralwasser Ztg **24** (1971) 1–8.

[166] SAND, F.E.M.J.; KOLFSCHOTEN, G.A.; VAN GRINSVEN, A.M.: Yeasts isolated from proportioning pumps employed in soft drinks plants. Brauwissensch. **29** (1976) 294–298.

[167] SAND, F.E.M.J.; VAN DEN BROEK, W.C.M.; VAN GRINSVEN, A.M.: Yeasts isolated from concentrated orange juice. Proc. 5th Int. Spec. Symp. Yeasts (1977) 121–122.

[168] SATO, M.; KOBORI, H.; ISHIJIMA, S.A.; FENG, Z.H.; HAMADA, K.; SHIMADA, S.; OSUMI, M.: *Schizosaccharomyces pombe* is more sensitive to pressure stress than *Saccharomyces cerevisiae*. Cell Struct. Funct. **21** (1996) 167–174.

[169] SCARR, M.P.; ROSE, D.: Study of osmophilic yeasts producing invertase. J. Gen. Microbiol. **45** (1966) 9–16.

[170] SCHAT, J.J.; RIJN, M.E.: Dutch soft drinks industry in 1991. Soft Drinks Manag. Int. (1992) 25–29.

[171] SCHMIDT, H.J.: Modern microbiology in soft drinks operations. Brauwelt Int. **13** (1995) 220–222.

[172] SCOTT, V.N.; BERNARD, D.T.: Resistance of yeast to dry heat. J. Fd Sc. **50** (1985) 1754–1755.

[173] SEBASTIANI, F.; BARBERIO, C.; CASALONE, E.; CAVALIERI, D.; POLSINELLI, M: Crosses between *Saccharomyces cerevisiae* and *Saccharomyces bayanus* generate fertile hybrids. Res. Microbiol. **153** (2002) 53–58.

[174] SHAMALA, T.R.; SREEKANTIAH, K.R.: Microbiological and biochemical studies on traditional Indian palm-wine fermentation. Fd Microbiol. **5** (1988) 157–162.

[175] SHELEF, L.A.: Antimicrobial effects of spices. J. Fd Safety **6** (1983) 29–44.

[176] SIEVERS, M.; LANINI, C.; WEBER, A.; SCHULER-SCHMID, U.; TEUBER, M.: Microbiology and fermentation balance in a kombucha beverage obtained from a tea fungus fermentation. Syst. Appl. Microbiol. **18** (1995) 590–594.

[177] SLAUGHTER, J.C.: The effects of carbon dioxide on yeasts. In: Biotechnology applications in beverage production (edited by Cantarelli, C.; Lanzarini, G.). London, U.K., New York, U.S.A.: Elsevier Applied Science (1989) 49–64.

[178] SLAVIKOVA, E.; VADKERTIOVA, R.; KOCKOVA-KRATOCHVILOVA, A.: Yeasts isolated from artificial lake water. Can. J. Microbiol. **38** (1992) 1206–1209.

[179] SMITH, M.T.; VAN GRINSVEN, A.M.: *Dekkera anomala* sp.nov., the teleomorph of *Brettanomyces anomala*, recovered from spoiled soft drinks. Antonie van Leeuwenhoek **50** (1984) 143–148.

[180] SMITTLE, R.B.: Microbiology of mayonnaise and salad dressing: a review. J. Fd Prot. **40** (1977) 415–422.

[181] SOMOGYI, L.P.: Direct food additives in fruit processing. In: Processing fruits: science and technology, Vol. 1. Biology, principles, and applications (edited by Somogyi, L.P.; Ramaswamy, H.S.; Hui, Y.H.). Lancaster, U.S.A., Pennsylvania, U.S.A., Basel, Switzerland: Technomic Publishing Co. Inc. (1996) 293–361.

[182] SOUTHGATE, D.A.T.; JOHNSON, I.T.; FENWICK, G.R.: Nutritional value and safety of processed fruit juices. In: Production and packaging of non-carbonated fruit juices and fruit beverages (edited by Hicks, D.). Glasgow and London, U.K.: Blackie (1990) 305–329.

[183] SPENCER, J.F.T.; GORIN, P.A.J.; TULLOCK, A.P.: *Torulopsis bombicola* sp.n. Antonie van Leeuwenhoek **36** (1970) 129–133.

[184] SPENCER, D.M.; SPENCER, J.F.T.; DE FIGUEROA, L.I.C.: Heluane, H.: Yeasts associated with rotting citrus fruit in Tucaman, Argentina. Mycol. Res. **96** (1992) 891–892.

[185] SPICHER, G.: Neue Erkenntnisse über die Erreger der "Kreidekrankheit" des Brotes und Möglichkeiten zur Wachstumsverhinderung. Brot Backwaren **34** (1986) 208–213.

[186] SPLITTSTOESSER, D.F.; LEASOR, S.B.; SWANSON, K.M.J.: Effect of food composition on the heat resistance of yeast ascospores. J. Fd Sc. **51** (1986) 1265–1267.

[187] STEAD, D.: The effect of hydroxycinnamic acids and potassium sorbate on the growth of 11 strains of spoilage yeasts. J. Appl. Bacteriol. **78** (1995) 82–87.

[188] STEELS, H.; STRATFORD, M.; BOND, C.J.; JAMES, S.A.: *Zygosaccharomyces lentus* sp.nov., a new member of the yeast genus *Zygosaccharomyces* Barker. Int. J. Syst. Bacteriol. **49** (1999) 319–327.

[189] STEELS, H.; JAMES, S.A.; ROBERTS, I.N.; STRATFORD, M.: *Zygosaccharomyces lentus*: a significant new osmophilic, preservative-resistant spoilage yeast, capable of growth at low temperature. J. Appl. Microbiol. **87** (1999) 520–527.

[190] STEELS H.; JAMES S.A.; ROBERTS, I.N.; STRATFORD, M.: Sorbic acid resistance: the inoculum effect. Yeast **16** (2000) 1173–1183.

[191] STEELS, H.; JAMES, S.A.; BOND, C.J.; ROBERTS, I.N., STRATFORD, M.: *Zygosaccharomyces kombuchaensis*: the physiology of a new species related to the spoilage yeasts *Zygosaccharomyces lentus* and *Zygosaccharomyces bailii*. FEMS Yeast Res. **2** (2002) in the press.

[192] STEINKRAUS, K.H.; SHAPIRO, K.B.; HOTCHKISS, J.H.; MORTLOCK, R.P.: Investigations into the antibiotic activity of tea fungus/kombucha beverage. Acta Biotechnol. **16** (1996) 199–205.

[193] STOLLARAVA, V.: The presence of yeasts and yeast-like microorganisms on cherry fruits (*Cerasus avium*/L/Moench). Biologia (Bratislava) **37** (1982) 1115–1120.

[194] STRATFORD, M.: Yeast flocculation: a new perspective. Adv. Microb. Physiol. **33** (1992) 1–72.

[195] STRATFORD, M.: Traditional preservatives – organic acids. In: The encyclopedia of food microbiology (edited by Robinson, R.; Batt, C.; Patel, P.). San Diego, U.S.A.: Academic Press (1999) 1729–1737.

[196] STRATFORD, M.; ANSLOW, P.A.: Evidence that sorbic acid does not inhibit yeast as a classic "weak-acid preservative". Lett. Appl. Microbiol. **27** (1998) 203–206.

[197] STRATFORD, M.; LAMBERT, R.J.W.: Weak-acid preservatives: Mechanisms of adaptation and resistance by yeasts. Fd Austr. **51** (1999) 26–29.

[198] STRATFORD, M.; HOFMAN, P.D.; COLE, M.B.: Fruit juices, fruit drinks and soft drinks. In: The microbiological safety and quality of food (edited by Lund, B.M.; Baird-Parker, A.C.; Gould, G.W.). Gaithersburg, U.S.A.: Aspen Publishers, Inc. (2000) 836–869.

[199] STRATFORD, M.; BOND, C.J.; JAMES, S.A.; ROBERTS, I.N.; STEELS, H.: *Candida davenportii* sp. nov., a potential soft-drinks spoilage-yeast associated with wasps. FEMS Yeast Res. **2**, (2002) in the press.

[200] SUIHKO, M.-L.; MAKINEN, V.: Tolerance of acetate, propionate and sorbate by *Saccharomyces cerevisiae* and *Torulopsis holmii*. Fd Microbiol. **1** (1984) 105–110.

[201] SUMNER, E.R.; AVERY, S.V.: Phenotypic heterogeneity: differential stress resistance among individual cells of the yeast *Saccharomyces cerevisiae*. Microbiology **148** (2002) 345–351.

[202] SURIYARACHCHI, V.R.; FLEET, G.H.: Occurrence and growth of yeasts in yogurts. Appl. Environ. Microbiol. **42** (1981) 574–579.

[203] SUZUKI, M.; NAKASE, T.; DAENGSUBHA, W.; CHAONSANGKET, M.; SUYANANDANA, P.; KOMAGATA, K.: Identification of yeasts isolated from fermented foods and related materials in Thailand. J. Gen. Appl. Microbiol. **33** (1987) 205–220.

[204] SUZZI, G.; ROMANO, P.; BENEVELLI, M.: The flocculation of wine yeasts: biochemical and morphological characteristics in *Zygosaccharomyces* – flocculation in *Zygosaccharomyces*. Antonie van Leeuwenhoek **61** (1992) 317–322.

[205] TABAK, H.H.; COOKE, W.B.: The effects of gaseous environments on the growth and metabolism of fungi. Bot. Rev. **34** (1968) 126–252.

[206] TAMURA, A.; KATO, M.; OMORI, M.; NANBA, A.; MIYAGAWA, K.; YANG, C.R.; ZHOU, W.H.: Flavour components and microorganisms isolated from Suancha (sour tea, Takezutsu-sancha in Japanese). Nippon Kasei Gakkaishi **46** (1995) 759–764.

[207] TANNER, H.; DUPERREX, M.: Eine dünnschichtchromatographisch-fluorometrische Mikromethode zur quantitativen Bestimmung von Sorbit, Mannit, Invertzucker, Glycerin und 2,3-Butylenglycol in Getränken. Fruchtsaft-Ind. **13** (1968) 98–114.

[208] TAYLOR, R.B.: Acids, colours, preservatives and other additives. In: Formulation and production of carbonated soft drinks (edited by Mitchell, A.J.). Glasgow, London, U.K.: Blackie and Son Ltd. (1990) 90–107.

[209] TAYLOR, R.B.: Ingredients. In: Chemistry and technology of soft drinks and fruit juices (edited by Ashurst, P.R.). Sheffield, U.K.: Sheffield Academic Press and Boca Raton, U.S.A.: CRC Press (1998) 16–54.

[210] TERADA, S.; NISHIMURA, A.: A study of kocha kinoko. Ichimura Gakuen Tanki Daigaku Shizen Kagaku Kenkyukai Kaishi **10** (1976) 15–18.

[211] TERRELL, F.R.; MORRIS, J.R.; JOHNSON, M.G.; GBUR, E.E.; MAKUS, D.J.: Yeast inhibition in grape juice containing sulphur dioxide, sorbic acid, and dimethyldicarbonate. J. Fd Sc. **58** (1993) 1132–1134.

[212] THOMAS, D.S.; DAVENPORT, R.R.: *Zygosaccharomyces bailii* – a profile of characteristics and spoilage activities. Fd Microbiol. **2** (1985) 157–169.

[213] TILBURY, R.H.: The microbial stability of intermediate moisture foods with respect to yeasts. In: Intermediate moisture foods (edited by Davies, R.; Birch, G.G.; Parker, K.J.). London, U.K.: Applied Science (1976) 138–165.

[214] TILBURY, R.H.: Xerotolerant (osmophilic) yeasts. In: Biology and activity of yeasts, Soc. Appl. Bacteriol. Symp. Ser. No 9 (edited by Skinner, F.A.; Passmore, S.M.; Da-venport, R.R.). London, U.K.: Academic Press (1980) 153–179.

[215] TILBURY, R.H.: Xerotolerant yeasts at high sugar concentrations. In: Microbial growth and survival in extremes of environment, Soc. Appl. Bacteriol. Techn. Ser. 15 (edited by Gould, G.W.; Corry, J.E.L.). London, U.K.: Academic Press (1980) 103–128.

[216] TOKUOKA, K.: Sugar- and salt-tolerant yeasts. J. Appl. Bacteriol. **74** (1993) 101–110.

[217] TOKUOKA, K.; ISHITANI, T.: Minimum water activities for the growth of yeasts isolated from high sugar foods. J. Gen. Appl. Microbiol. **37** (1991) 111–119.

[218] TOKUOKA, K.; ISHITANI, T.; GOTO, S.; KOMAGATA, K.: Identification of yeasts from high sugar foods. J. Gen. Appl. Microbiol. **31** (1985) 411–427.

[219] TUDOR, E.A.: BOARD, R.G.: Food-spoilage yeasts. In: The Yeasts Volume 5, Yeast Technology, 2nd Ed. (edited by Rose, A.H.; Harrison, J.S.). London, U.K.: Academic Press (1993) 435–516.

[220] TURTURA, G.C.; SAMAJA, T.: Alterazioni blastomicetiche delle bevande analcoholiche a base di aromatizzanti naturali – considerazioni sulle possibilita' di prevenzione. Ind. Bevande **4** (1975) 75–79.

[221] TURTURA, G.C.; SAMAJA, T.: Ricerche microbiologiche sulle bevande analcooliche. I. Identificazione di blastomiceti isolati da bibite preparate con aromatizzanti naturali. Ann. Microbiol. **28** (1978) 1–9.

[222] TZANETAKIS, N.; LITOPONEON-TZANETAKI, E.; MANOLKIDIS, K.: Microbiology of kopanisti, a traditional Greek cheese. Fd Microbiol. **4** (1987) 251–256.

[223] VAUGHAN-MARTINI, A.; KURTZMAN, C.P.: Deoxyribonucleic acid relatedness among species of the genus *Saccharomyces* sensu stricto. Int. J. Syst. Bacteriol. **35** (1985) 508–511.

[224] VAUGHAN-MARTINI, A.; MARTINI, A.: Proposal for correct nomenclature of the domesticated species of the genus *Saccharomyces*. In: Biotechnology applications in beverage production (edited by Cantarelli, C.; Lanzarini, G.). London, U.K., New York, U.S.A.: Elsevier Applied Science (1989) 1–16.

[225] VERACHERT, H.; DAWOUD, E.: Yeast in mixed cultures. Louvain Brew. Lett. **3** (1990) 15–40.

[226] VIDAL-LEIRA, M.; BUCKLEY, H.; VAN UDEN, N.: Distribution of maximum temperature for growth among yeasts. Mycologia **71** (1979) 493–501.

[227] VOLLEKOVA, A.; MALIK, F.; VOLLEK, V.; LINCZENYIOVA, K.: Characterization of yeasts isolated from red wine surface film. Folia Microbiol. **41** (1996) 347–352.

[228] WALKER, H.W.; AYRES, J.C.: Yeasts as spoilage organisms. In: The yeasts, Vol. 3, Yeast technology (edited by Rose, A.H.; Harrison, J.S.). London, U.K., New York, U.S.A.: Academic Press (1970) 463–527.

[229] WARTH, A.D.: Mechanism of resistance of *Saccharomyces bailii* to benzoic, sorbic and other weak acids used as food preservatives. J. Appl. Bacteriol. **43** (1977) 215–230.

[230] WARTH A.D.: Resistance of yeast species to benzoic and sorbic acids and sulphur dioxide. J. Fd Prot. **48** (1985) 564–569.

[231] WARTH, A.D.: Effect of benzoic acid on growth yield of yeasts differing in their resistance to preservatives. Appl. Environ. Microbiol. **54** (1988) 2091–2095.

[232] WARTH, A.D.: Relationships between the resistance of yeasts to acetic, propanoic, and benzoic acids and to methyl paraben and pH. Int. J. Fd Microbiol. **8** (1989) 343–349.

[233] WATSON, D.C.: Yeasts in distilled alcoholic-beverage production. In: The yeasts, Vol. 5, 2nd ed., Yeast technology (edited by Rose, A.H.; Harrison, J.S.). London, U.K.: Academic Press (1993) 215–244.

[234] WHITING, G.C.: "Non-fermentable" substance in ciders and perries. Rep. Long Ashton Res. Stat., University of Bristol, Bristol, U.K. (1960) 135–139.

[235] WIJSMAN, M.R.; VAN DIJKEN, J.P.; VAN KLEEFF, B.H.A.; SCHEFFERS, W.A.: Inhibition of fermentation and growth in batch cultures of the yeast *Brettanomyces intermedius* upon a drift from aerobic to anaerobic conditions (Custers effect). Antonie van Leeuwenhoek **50** (1984) 183–192.

[236] YURKEVICH, D.I.; KUTYSHENKO, V.P.: Study of glucose utilization during growth of tea fungus by 1H NMR spectroscopy. Biofizica **43** (1998) 319–322.

[237] ZAAKE, S.: Nachweis und Bedeutung getränkeschädlicher Hefen. Monatssch. Brauerei **32** (1979) 350–356.

[238] ZHAO, T.; DOYLE, M.P.; BESSER, R.E.: Fate of enterohemorrhagic *Escherichia coli* O157:H7 in apple cider with and without preservatives. Appl. Environ. Microbiol. **59** (1993) 2526–2530.

13 Brewing yeasts

Jean-Pierre Dufour, Kevin Verstrepen and Guy Derdelinckx

13.1 Introduction

The earliest historical records indicate that by 6,000 BC humans knew already how to make bread, beer and wine. About the same time they also discovered how to preserve vegetables by lactic acid fermentation. The humans living in these early civilizations had no concept of either microorganisms or fermentations. Consequently, brewing procedures little changed over periods of thousands of years. Traditional brewing, as it presumably was, can still be observed in Africa where there is a great variety of local beers. These beers are made from indigenous cereals such as sorghum. The product is cloudy, containing suspended solids, and is typically consumed, while still fermenting, using dried gourds or calabashes. The knowledge of the recipes resides in the minds of countless women who learned the production of these beers from their mothers and grandmothers.

Scientists became interested in the brewing process during the late 17th century. In 1680, the Dutch microscopist Antonie van Leeuwenhoek was the first to observe brewer's yeast, which he referred to as "animalcules". Lavoisier, in 1789, discovered that carbon dioxide was produced during fermentation. In 1815, Gay-Lussac described the equation of fermentation where two molecules of sucrose are converted into four molecules of ethanol and four molecules of carbon dioxide. A few years later, Friedrich Woehler and Justus von Liebig (1839) published their solution to the mystery of alcoholic fermentation. According to them the production of ethanol from sugar was purely a chemical process dependent on the decomposition of dead yeast. It was not until the 1870's that Louis Pasteur was able to demonstrate that living yeasts were responsible for the fermentation of alcoholic beverages [62]. His studies also enabled him to discover the microorganisms causing deviations and bad taste in beer.

13.2 Yeast biodiversity related to brewing

13.2.1 Taxonomy of brewing yeasts

Yeast has long been recognized to be the key player in the production of beers. With few exceptions, the 250 millions hectoliters of beer and the 10,000 beer brands available around the world are produced by yeasts. Of the worldwide beer production lager beers (produced using lager yeasts) account for 90 %, 5 % for the ale beers (produced using ale yeasts), and the remaining produced through mixed fermentations by yeast and bacteria. It is generally accepted that the yeast strain used in a brewery is a critical component in defining the unique beer characteristics, such as flavour and aroma. For these reasons the choice of yeast is of prime importance to the brewers.

Brewing yeast strains have been the subject to various taxonomic revisions over the years [see e. g., 44]. From a taxonomic point of view, brewing yeast strains belong to the genus *Saccharomyces* and the species *cerevisiae*. Different strains of *S. cerevisiae* relate to more practical information, such as the type of beer or where they have been isolated from. Three other species, *S. paradoxus, S. bayanus* and *S. pastorianus* are closely related to *S. cerevisiae. Saccharomyces cerevisiae* and *S. paradoxus* show maximum growth temperatures of 37 °C or greater, contrary to *S. bayanus* and *S. pastorianus* which grow at 34 °C or lower. They are unable to use melibiose [56] and transport fructose via a facilitated transport mechanism, whereas *S. bayanus* and *S. pastorianus* transport fructose via active transport [69]. The taxonomical identification of yeast species requires an array of tests (see Chapters 1 and 3): morphological (e. g., colony characteristics, microscopic appearance), nutritional (e. g., assimilation of various carbon sources), serological, genetic (e. g., homology of DNAs), molecular biological (e. g., DNA fingerprinting), and enzyme polymorphisms. These tests do not provide the brewers with sufficient information about the characteristics of the brewing yeasts. For them the brewing yeasts are divided between ale (top-fermenting) and lager (bottom-fermenting) yeasts, namely *S. carlsbergensis* and *S. cerevisiae*, respectively. This classification, is taxonomically incorrect: *S. carlsbergensis* (= *S. cerevisiae*) was consolidated within *S. uvarum* (= *S. bayanus*), which itself was repositioned as *S. cerevisiae,* later renamed *S. pastorianus* and now being considered as *S. bayanus* [5–7]. Because of the industrial importance of *S. carlsbergensis* and *S. cerevisiae* and because the two names are routinely used by the brewers, they will be used in this chapter to refer to lager and ale yeasts, respectively.

The use of single-chromosome transfer for the genetic analysis of chromosomes from industrial lager yeast revealed that the lager brewing yeast is alloploid, a hybrid between *S. cerevisiae* and another *Saccharomyces* species [3]. The use of modern molecular biological techniques has shown that the lager *S. cerevisiae*-like genome in general differs by less than 1% from the published *S. cerevisiae* sequence. The other lager yeast genomes show nucleotide sequence divergences typically between 10 to 20 % in coding regions and higher outside coding regions [3]. It has been suggested that this latter lager yeast genome is closely related to the type strain of *S. bayanus.* However there is no agreement on this topic yet, as PEDERSEN [63] suggested that the type strain of *S. monacensis* (= *S. pastorianus*) might be a better candidate. Another study, using polymerase chain reaction (PCR) analysis and genetic analysis of single chromosomes, indicated that neither the *S. bayanus* or the *S. monacensis* genome is identical to that of the non-*S. cerevisiae* parent of lager brewing yeast [3]. This scientific debate thus remains open, but is of little practical interest to the brewers. For the brewers, ale and lager yeasts are fundamentally different from an industrial and physiological point of view. The bottom-fermenting yeast, *S. carlsbergensis*, is exclusive to lager fermentation and the top-fermenting yeast, *S. cerevisiae,* is used in ale fermentation.

13.2.2 Diversity and differences between brewing yeasts: ale and lager yeasts

From the brewers' perspective, there are several phenotypic differences between ale and lager yeasts that warrant maintaining these two types of yeasts, as separate entities. Top fermentation produces beers that are more fruity, estery, and sometimes malty, whereas lager fermentation gives beers with a characteristic sulphurous aroma. Yeast properties of direct relevance to the brewer and that taxonomists do not take into account include fermentative ability and rate of sugar utilization (maltose and maltotriose), oxygen requirements, tolerance to ethanol and temperature, osmotolerance, flocculation characteristics, level and profile of esters formed, reducing activity towards diacetyl, fermentation rate, ability to ferment at low temperature, retention of viability during storage, and the level of sulphite produced. A good brewing yeast should be effective in removing the required nutrients from the wort, able to tolerate the prevailing environmental conditions (e. g., osmotolerance and ethanol tolerance for high gravity brewing) and produce the desired flavour in the beer. Furthermore, once the fermentation is completed the yeast cells should be able to be removed effectively from the beer by flocculation, centrifugation or filtration [74]. A brewing yeast should also have a good growth rate, usually showing a 3–5-fold increase during fermentation. Low growth means that insufficient yeast is available to pitch in the successive fermentations. Too much yeast is not acceptable either, because in this case an excessive amount of carbohydrate will be diverted towards the production of biomass instead of alcohol.

The main characteristics that differentiate lager and ale yeasts are:

- Lager yeasts can utilize melibiose due to the production of an extracellular enzyme, α-galactosidase (melibiase, EC 3.2.1.22), while ale yeasts can not use this compound [56].
- Lager yeasts appear to produce more sulphite than ale yeasts during fermentation [17].
- Lager yeasts are more efficient in assimilating maltotriose [75].
- Lager yeasts transport fructose via active transport (proton symport) [69], while ale yeast transport fructose via facilitated diffusion.
- Lager yeasts are bottom-fermenting, whereas ale yeasts are top-fermenting. At the end of fermentation lager yeast flocculates and settles to the bottom of the fermenter, whereas ale yeast forms loose clumps that adsorb to the carbon dioxide bubbles and is carried to the surface of the medium where the yeast can be skimmed and recovered for re-use. This characteristic, however, tends to disappear in the modern brewing industry. Following the introduction of large cylindroconical tanks, the ale yeast has been harvested from the bottom of the fermenter and has become a bottom-fermenting yeast through selection.
- Lager yeasts can ferment at low temperature (7–15 °C). However, it is wrong to think that this is the preferred temperature for the yeast. The temperature range from 7 to 15 °C used for lager fermentation is a requirement to develop specific flavour characteristics

of the lager beer. This view, however, has been challenged (J.P. DUFOUR, unpubl. observ.). The optimum growth temperatures are below 30 °C for lager yeasts and above 30 °C for ale yeasts [32]. Consequently, lager yeasts grow much more quickly at low temperature than ale yeasts [84], which makes lager yeasts more suitable for fermentation at low temperatures. Most ale yeasts will not growth below 15 °C in industrial fermentations (G. DERDELINCKX, unpubl. observ.).

- The maximum growth temperature of lager yeasts ranges from 31.6 °C to 34.0 °C and from 37.5 °C to 39.8 °C for ale yeast [84]. This criterion can be used to differentiate between lager yeasts and ale yeasts as lager yeasts, never grow above 34 °C.
- Lager and ale yeasts show a different response towards the effect of glucose on the uptake of maltose and maltotriose (glucose effect). It is well established that glucose influences yeast metabolism, and it has been suggested that glucose acts as a metabolic signal. However, the mechanisms of this regulation have yet to be elucidated completely. At the cellular level, metabolic regulation under glucose could be due to the induction or repression of enzyme synthesis (carbon catabolite repression) or the rapid inactivation of enzymes (carbon catabolite inhibition). A systematic analysis of the repression and induction of the maltose and maltotriose permeases in a number of strains of ale and lager yeasts suggested that in nearly all the examined lager yeasts, glucose did not only repress the synthesis of the maltose and maltotriose permeases, but also rapidly inactivated the enzymes, leading to a rapid de-adaptation of the yeasts to utilize maltose and maltotriose [48, 68].
- On sporulation medium, ale yeasts give 1 to 10 % viable ascospores whereas none are obtained with lager yeasts (G. DERDELINCKX, Cl. RAMOS-JEUNEHOMME and C. MASSCHELEIN, unpubl. observ.).

Brewing practices in ale production (top cropping of the yeasts, characteristics of the medium) have likely contributed to the large diversity observed among the ale yeasts. Interestingly, chromosomal fingerprinting analysis performed on numerous ale and lager yeasts around the world indicated a lack of common origin between the ale strains, but identified only two basic fingerprints for the lager strains, namely the Tuborg and the Carlsberg type lager yeast [12]. However, CASEY [12] did identify genuine differences between the lager strains. The strain diversity reported for the lager yeasts were induced likely in response to environmental pressure or 'spontaneous' rearrangement of chromosomes.

13.2.3 *Saccharomyces cerevisiae* laboratory strains and brewing strains

The main characteristics that differentiate laboratory strains and brewing strains are summarized in Table 13.2-1.

Tab. 13.2-1 Main characteristics of *S. cerevisiae* laboratory and brewing strains

Characteristic	Laboratory strains	Brewing strains
Chromosome copy number (ploidy)	Haploid, diploid (seldom triploid or tetraploid)	Polyploid (3n, 4n), aneuploid (e. g., $3<n<4$)
Sporulation	Yes	Very poor
Production of viable spores	Yes	Absent (lager yeasts) to very rare (ale yeasts)
Mating type	Two mating types, a and α	Very rare
Sexual reproduction (formation of zygote)	Yes	Very rare, and highly strain dependent
Asexual (vegetative) reproduction	Yes, multilateral budding	Yes, multilateral budding

Laboratory strains of *S. cerevisiae* are mainly haploid (n) or diploid (2n), whereas brewing strains are polyploid, triploid (3n) or tetraploid (4n). The number of copies of chromosomes is, however, not necessarily the same for all the chromosomes, thus making the brewing strains aneuploid (e. g., a triploid strain may carry three copies for some chromosomes and a different copy number, e. g., two and four of others). For example, the Carlsberg lager production strain 244 has a DNA content corresponding to that of a tetraploid *S. cerevisiae* strain, but a study on the copy number of chromosome III showed that the yeast was pentosomic (5 copies) or had a higher ploidy for that chromosome [39]. Using flow cytometry to determine the cellular DNA content, CODON et al. [13] showed that the ploidy of brewing strains ranged between 1.6 and 2.2n. A direct consequence of the polyploid-aneuploid characteristic is that the brewing strains do not have a mating type, they sporulate poorly and, when sporulation occurs, the spores are non-viable. This lack of sporulation or viable spores means that the traditional genetic improvement methods, such as hybridization, mutation and selection, and spheroplast fusion can not be applied. From a brewer's perspective, the polyploid or aneuploid nature could be advantageous. The more copies of a gene are present, the less likely a mutation will be effective as this would require a mutation in all the copies of the gene. Following the same reasoning, aneuploidy could be one of the mechanisms for the cell to adjust some cellular activities. The more copies of a gene are present, the higher the amount of products produced in the cell, such as enzymes and cell wall components. CODON et al. [13] suggested that the amplification of the genes is an adaptive mechanism conferring a better fitness of the strains in their specific industrial environment.

This apparent genetic stability is however somewhat misleading. Analysis of the genetic properties of brewing strains using DNA fingerprinting techniques, such as chromosomal DNA banding patterns (karyotype) and Restriction Fragment Length Polymorphism (RFLP) of the mitochondrial DNA, have provided evidence that the brewing genome is labile and prone to rearrangements. To obtain consistent fermentation performance and constant taste and flavour of the beer, the genetic stability of brewing yeast is a major concern for brewers.

Based on industrial experience, the changes in yeast characteristics resulting from mutations that are of utmost importance to the brewing fermentation are: 1. a change in flocculation behavior (i. e., a mutation from flocculent to non-flocculent yeast), 2. the loss of the ability to assimilate maltotriose and 3. the presence of petite yeast cells. Genetic fingerprinting suggested that these changes in yeast characteristics have been the results of genuine genetic changes, and are not simply the effect of changes in phenotypic expression [42, 72, 85, 86].

The yeast petite mutation (respiratory deficient yeast) is well known in the brewing industry. It arises spontaneously due to deletions in the mitochondrial DNA. The petite mutation is easily identified by the small size of the yeast colony on agar plates and by the absence of growth on oxidative substrate, such as ethanol or glycerol. It is important to stress that, although the yeast doesn't respire during the brewing fermentation, the yeast cell contains promitochondria (undeveloped mitochondria) [16], whose functions are vital for cellular metabolism [59]. The frequency of the petite mutation is usually low, ranging between 0.5 and 5 %, but frequencies as high as 50 % have been reported [19, 53]. Higher concentrations of ethanol increase the frequency of the petite mutation. A low percentage of petite cells has no impact on the fermentation and characteristics of the beer. Higher proportions of petite yeast cells, however, will be responsible for sluggish fermentation rates, reduced yeast growth, and beer flavour deviations, such as higher levels of diacetyl and an increase in the formation of higher alcohols [29]. Fortunately, petite yeasts are less flocculent and under normal brewing practice, in which the yeast is collected at the bottom of the fermenter, most of the petite cells will remain in suspension in the beer and be removed during beer filtration.

Genetic and physiological instability of brewing yeast strains (see below), complicates the maintenance of brewing yeast in the industry. Chromosomal stability is now one of the most important tests for quality control of brewing yeast in the industry with periodic examinations to check the uniformity of the brewing yeast. Single colony isolation from a yeast slurry and its use as a starter for a new mother culture is not recommended. Single colony isolation carried out from a yeast slurry, almost certainly increases the risk of changes of both the chromosomal and mitochondrial DNA types [85]. If selection of a single colony is required, a complete genetic and physiological assessment of this yeast should be performed before use in commercial production. As genetic and physiological instabilities of brewing yeast are inevitable, the understanding of the diversity of the yeast strain used in the brewery is an essential part of efficient process control and product uniformity. On the other side, we have to realize that the genetic instability of the brewing yeast has contributed, and is still contributing, to the immense variety of beers produced around the world.

13.2.4 *Saccharomyces* and non-*Saccharomyces* wild yeasts

For PASTEUR, the use of a pure ferment, free from foreign ferment of 'an evil character' (such as bacteria) was the guarantee for the production of a good beer [62]. In 1883, at Carlsberg Laboratories in Denmark, EMIL HANSEN was the first who realized that beer yeast should not be contaminated with bacteria, but should also be free of wild yeast cells. He subsequently established which species produced good beer and which did not. In the strict sense, any unwanted yeast is 'wild' for the brewer, as their presence will often result in changes in fermentation behavior and beer flavour characteristics. Wild yeast has been defined as any yeast not deliberately used and under full control [31]. Consequently, this can be a brewing yeast (e. g., an ale yeast contaminated with a lager yeast) or a non-brewing yeast. When several yeasts are used in a brewery, the pitching yeast is constantly at risk of cross-contamination.

The yeasts found in breweries can been classified into domesticated yeasts (brewing yeasts) and wild yeasts (contaminating yeasts) (Table 13.3-1). Domesticated yeasts produce beer with suitable characteristics under well defined and controlled processing conditions. Contaminating yeasts (i. e., the true wild yeasts) originate from infection and are not desirable. The presence of wild yeasts usually generates noticeable defects in the beer characteristics (e. g., phenolic off-flavour, turbidity). *Saccharomyces cerevisiae* killer strains will kill the pitching yeast (see below). To prevent these problems, breweries systematically renew the yeast after a given number of fermentation cycles (usually between 5 to 15 cycles, and this number is a function of the beer produced and fermentation conditions). This renewal of the yeast is one of the basics to guarantee consistent fermentations and the absence of off-flavours. The use of a pure yeast culture will prevent the development and taking over of the fermenting wort by contaminating wild yeast. It also reduces the risk of undesired mutations that happen during successive industrial fermentations, as well as the risk of contamination with bacteria.

13.3 Beneficial aspects of brewing yeasts

The unique flavour characteristics of beers are endowed by small amounts of substances which originate from the raw materials, or which are produced through chemical and biochemical reactions. Raw materials, such as malt and hops, have long been recognized as important factors in defining the character of the beer. Nevertheless, the compounds that exert the greatest impact on the palate and on the sense of smell are mostly produced by the yeast during the fermentation. The yeast not only converts the fermentable sugar of the wort into ethanol and carbon dioxide, but also produces a variety of volatile and non-volatile compounds. These compounds, most of them occurring in trace amounts, contribute to the flavour characteristics of the beer. Consequently the fermentation process is highly critical for the final quality of the beer.

Tab. 13.3-1 Brewing yeasts and wild yeasts found in breweries.

Domesticated yeast (brewing yeast)	Wild yeast (contaminating yeast)
Saccharomyces spp. *Saccharomyces carlsbergensis* (lager yeasts; = *Saccharomyces cerevisiae*) *Saccharomyces cerevisiae* (ale yeasts) *Saccharomyces cerevisiae* var. *chevalieri* (= *Saccharomyces cerevisiae*)	*Saccharomyces* spp. *Saccharomyces cerevisiae* var. *diastaticus* (= *Saccharomyces cerevisiae*) *Saccharomyces bayanus* *Saccharomyces cerevisiae* killer strains
Non-*Saccharomyces* spp. *Brettanomyces bruxellensis* (= *Dekkera bruxellensis*) *Brettanomyces intermedius* (= *Dekkera intermedia*) *Brettanomyces anomalus* (= *Dekkera anomala*)	Non-*Saccharomyces* spp. *Candida* spp. *Pichia* spp. (including *Hansenula* spp.) (e. g., *P. membranifaciens*) *Debaryomyces* spp. *Hanseniaspora* spp. *Issatchenkia* spp. (e. g., *orientalis*) *Kluyveromyces* spp. (e. g., *marxianus*) *Torulaspora* spp. (e. g., *delbrueckii*) *Kloeckera* spp. *Cryptococcus* spp. *Rhodotorula* spp. *Zygosaccharomyces* spp.

The driving force of the main yeast-mediated biochemical events in a brewery fermentation is the growth and multiplication of yeast cells during which the yeast will produce the flavour compounds that determine the beer sensory characteristics. All these flavour compounds are by-products of carbohydrate and/or amino acid metabolism. For the consumer, beer is a delightful beverage, but from the yeast perspective beer is a mixture of its waste products!

While the sensory implications of yeast metabolism are well documented, many questions still remain regarding the biochemical reaction regulatory mechanisms and yeast physiology. The important flavour active compounds produced by yeast can be classified into five categories: alcohols, esters, organic acids, carbonyl compounds (aldehydes and ketones) and sulphur-containing compounds. In some cases these compounds arise as a result of minor biochemical transformations originating from pyruvate (sugar metabolism), in other cases they are the by-products of nitrogen metabolism (Fig. 13.3-1). A description of all the biochemical reactions involved is beyond the scope of this chapter (for more details, see [9, 24, 27, 36, 78]).

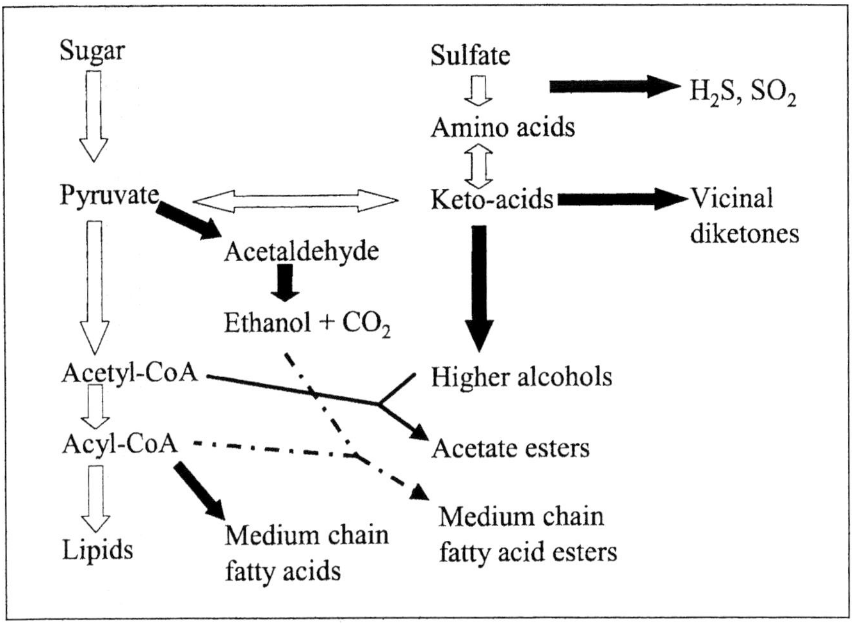

Fig. 13.3-1 **Formation of the major flavour groups during fermentation: alcohols (ethanol, higher alcohols), esters (acetate esters, medium chain fatty acid esters), carbonyl compounds (acetaldehyde, vicinal diketones), sulphurous compounds (hydrogen sulfide, sulphur dioxide), and organic acids (medium chain fatty acids).**

13.3.1 Higher alcohols

The bulk of the volatile constituents formed as by-products of yeast fermentation are the higher alcohols. These compounds, which are less volatile than ethanol, are sometimes referred to as fusel oil. Some 45 alcohols have been reported in beer. The levels of the main higher alcohols in beer are presented in Table 13.3-2. Aliphatic alcohols (e. g., amyl alcohol, isoamyl alcohol, isobutanol) impart an 'alcoholic' or 'solvent-like' aroma to beer, and produce a warming effect on the palate. The aromatic higher alcohols with a fragrant aroma, such as 2-phenylethanol, may contribute positively to the flavour of the beer, whereas tryptophol (bitter, chemical) and tyrosol (almond, solvent) flavour characteristics are generally regarded as undesirable. The alcohols that play the most important role in beer flavour are amyl alcohol, isoamyl alcohol and 2-phenylethanol.

The higher alcohols are formed during fermentation by two routes, the catabolic Ehrlich pathway and the anabolic Genevois pathway. In the catabolic pathway, wort amino acids are taken up by yeast. After transfer of their amino group to α-keto-glutarate (the amino group acceptor), they give the corresponding α-keto-acids. The excess α-keto-acids are subsequently decarboxylated and reduced, which results in the production of alcohols with one carbon atom less than the original amino acids. These alcohols may also be derived anabolically from the carbon source (e. g., pyruvate and acetyl-CoA for the synthesis of valine). The penultimate reaction in the biosynthesis of each amino acid is the production of the corresponding α-keto-acid, which is subsequently transaminated to form the respective amino acid. As with the Ehrlich pathway, it is the decarboxylation and reduction of these α-keto-acids that are responsible for the formation of higher alcohols. Both pathways operate in brewer's yeast, but the relative contributions of the two pathways depend greatly on the level of amino acids present in the fermentation medium. At low levels of amino acids, the biosynthetic pathway predominates, whereas at high levels of amino acids, the Ehrlich pathway becomes dominant, as a result of feedback inhibition and/or repression of key enzymes in the biosynthetic pathway.

The formation of higher alcohols occurs during the main fermentation and is related to yeast growth. The production of higher alcohols depends on the yeast characteristics, medium composition (i. e., wort original extract, amount of zinc, lipids, dissolved oxygen) and fermentation parameters (i. e., temperature, stirring, pressure, fermenter design). As a guideline, conditions that stimulate yeast growth will increase the production of higher alcohols during fermentation.

13.3.2 Esters

Esters represent the largest group of flavour active compounds in beer. Although some ester formation may occur during long maturation of special beers, the yeast forms the most common esters during the fermentation. The major esters in beer are: ethyl acetate (light fruity, solvent-like aroma), isoamyl acetate (banana, peardrop aroma), ethyl caproate (= ethyl hexanoate) (apple-like with a note of aniseed aroma), ethyl caprylate (= ethyl octanoate) (apple-like aroma) and 2-phenylethyl acetate (rose, honey, apple, sweetish). The concentration range and the threshold values for these esters in European lagers are given in Table 13.3-2.

Isoamyl acetate (and its isomer 2-methyl-1-butyl acetate) has an important influence on the flavour of beer. As shown in Table 13.3-2, the mean concentration of isoamyl acetate is the only one above the threshold value. Many technological parameters affect the production of esters in industrial fermentations, which can be divided into three groups. Firstly, those related to the yeast characteristics (e. g., strain, physiological state, pitching rate), secondly those influenced by the medium composition (e. g., lipids, oxygen, sugar levels, amino acid content, zinc concentration, suspended solids) and, thirdly, those affected by the fermentation parameters (e. g., temperature, pressure, stirring, fermenter design, fermentation method) [24]. As previously discussed, most of the higher alcohol synthesis (approx. 80 %) occurs during the growth phase, with the remaining synthesis occurring during the later stage of the industrial fermentation. The synthesis of esters, however, is delayed, as it requires at least the presence of the higher alcohols to proceed. It has been reported that a significant fraction of beer esters (up to 40 %) is formed when yeast growth ceases.

Tab. 13.3-2 Concentration ranges and thresholds of the major volatiles in lager beers.

Compound	Threshold (mg/L) [50, 51]	Concentration (mg/L)[23, 25] Range (n =48)	Mean (n = 48)
n-Propanol	800	7–19	12
Isobutanol	200	4–20	12
Amyl alcohol	70	9–25	15
Isoamyl alcohol	65	25–75	46
2-Phenylethanol	125	11–51	28
Tyrosol	200	5–32 *	16*
Tryptophol	> 400 (limit of solubility)	0.2–3.5 *	1.2*
Ethyl acetate	30	8–32	18.4
Isoamyl acetate	1.2	0.3–3.8	1.72
Ethyl caproate	0.21	0.05–0.3	0.14
Ethyl caprylate	0.9	0.04–0.53	0.17
2-Phenylethyl acetate	3.8	0.10–0.73	0.54
Caproic acid	8	0.7–2.9	1.5
Caprylic acid	14	2.1–7.4	4.5
Capric acid	10	0.1–2.4	0.7

*Range and mean in sixteen lager beers

The synthesis of esters and their regulation have been recently reviewed [24, 46]. Ester synthesis requires 2 substrates, namely ethanol or a higher alcohol and acyl-CoenzymeA (acyl-CoA). Ester synthesis is an intracellular process, which utilizes the energy provided by the thioester linkage of the acyl-CoA co-substrate, and is catalyzed by an acyltransferase (EC 2.3.1) or ester synthase. Acetate ester synthesis during fermentation will mainly depend on yeast ester synthesis potential, i. e., the amount of available acetyl-CoA and the level of ester synthase activity. The evolution of acetate esters synthesis versus yeast growth can be viewed as a bell-shaped curve. As the enzyme is only synthesized during the growth phase, under poor growth conditions (e. g., limiting amount of oxygen, low amino acid levels), it can be considered that ester synthase activity will be the limiting factor. Under excessive growth conditions, the level of available acetyl-CoA can be considered to be limiting. Between these two extremes of yeast growth (i. e., poor growth and excessive growth), the level of esters should reach a maximum level. The net effect will depend on the balance between the two inputs, namely the level of acetyl-CoA and enzyme activities. It has also to be remembered that ester synthesis is closely related to the synthesis of the corresponding alcohol. To reach the optimal conditions one has to take into account the yeast characteristics, the medium composition and the fermentation conditions. Table 13.3-3 summarizes the general guidelines for the control of acetate esters synthesis during fermentation.

Tab. 13.3-3 General guidelines for the control of acetate esters production during fermentation

Parameter	Effect	Net effect on beer esters*
Wort oxygenation	Stimulation of yeast growth Repression of ester synthase	Decrease
Wort unsaturated lipids	Stimulation of yeast growth Repression of ester synthase	Decrease
Wort sugar extract	Increase of alcohol/higher alcohols levels Decrease of oxygen solubility	Increase
Wort amino acids	Stimulation of yeast growth	Variable
Wort zinc	Stimulation of higher alcohol production	Increase
Temperature	Stimulation of yeast growth Increase of ester synthase activity	Increase
Pressure	Product inhibition of decarboxylation reaction Reduction of yeast growth	Decrease
Stirring	Stimulation of yeast growth	Decrease

* the normal lager brewing practice is taken as reference.

13.3.3 Organic acids

Over a hundred organic acids (non-volatile, low-volatile and volatile compounds) have been reported in beer [27]. These compounds, which are derived from both raw materials (malt and hops) and from yeast metabolism, are important in several respects. Firstly, they contribute to lowering the pH during fermentation and, secondly, they influence the sourness attribute, although some acids have their own flavour and aroma characteristics. The major organic acids secreted by yeast, namely pyruvic acid, acetic acid, lactic acid, Krebs cycle acids, and α-keto-acids originate from the amino acid biosynthetic pathways and the carbohydrate metabolism. Their accumulation depends upon a rapid vigorous fermentation.

The most important organic acids are the medium chain length fatty acids (MCFAs), caproic (= hexanoic), caprylic (= octanoic) and capric (= decanoic) acids. These acids can account for 85–90 % of the total fatty acids in beer. Typical concentrations and flavour thresholds of these acids in beer are shown in Table 13.3-2. As their aroma contributions are additive, the flavour thresholds are within the range of concentrations of these compounds in beer. The production of MCFAs above threshold values during fermentation is associated with goaty, cheesy or sweaty off-flavours, whereas their production during maturation is associated with yeasty or autolytic flavour. As lager yeasts have a tendency to produce greater quantities of these acids compare to ale yeasts, the MCFA off-flavour is observed more frequently in lager beers. It has been demonstrated that the MCFAs that are excreted during fermentation are formed *de novo* by fatty acid synthesis and are not a result of β-oxidation of wort or yeast long chain fatty acids. The key enzyme in the regulation of fatty acid biosynthesis is acetyl-CoA carboxylase (EC 6.4.1.2). Because of the close relationship between ester formation and fatty acid biosynthesis (both pathways use acyl-CoA as a substrate), it is likely that the mechanisms operating to control ester levels will also control MCFAs synthesis in beer.

Regarding the autolytic (yeasty) flavour, the activation of hydrolytic systems is responsible for the breakdown of cell constituents and the subsequent release of these acids into beer from the yeast intra-cellular pool. Therefore, it can be expected that the risk of autolytic (yeasty) off-flavour formation is higher at elevated maturation temperatures, where the effect of yeast concentration and contact time on beer flavour will be more critical as well [47].

13.3.4 Carbonyl compounds

Carbonyl compounds of wort and beer are well known off-flavours (i. e., papery, buttery, green, aldehydic) with low thresholds. This property makes them an important group of beer volatiles. Depending on their origin, beer carbonyls may be divided into several groups: sugar metabolism (acetaldehyde), amino acid metabolism (diacetyl, branched aldehydes), Strecker degradation (branched aldehydes) and lipid oxidation (linear carbonyl compounds).

The reduction of these carbonyl compounds by yeast is now widely accepted as a determinant process in the removal of these off-flavours from the beer. The conversion of the carbonyl compounds into their corresponding flavourless alcohols, leading to the elimination of the undesired flavour notes, depends, among others, on the activity of several yeast reductases.

Among the carbonyl compounds, one compound has received particular attention, namely diacetyl causing a buttery off-flavour (threshold, 100 μg/L). Diacetyl (2,3-butanedione) and 2,3-pentanedione, also called the vicinal diketones or VDK, originate from the chemical decomposition of two acids, α-acetolactate and α-acetohydroxybutyrate, respectively. It can be expected that the higher the production of the α-acetohydroxyacids, the higher the levels of VDK. Both acids are intermediates in the synthesis of valine and isoleucine, respectively. The amounts and profile of α-acetohydroxyacids produced during fermentation are influenced by the yeast strain, the medium composition, and the fermentation conditions. High levels of α-acetohydroxyacids will be observed under conditions favoring yeast growth, such as higher dissolved oxygen contents, higher lipid levels, or higher fermentation temperatures, or alternatively if the wort contains low concentrations of amino acids [55]. Fortunately, the yeast transforms the vicinal diketones into the corresponding, much less flavour active, alcohols. These reactions occur mainly at the end of the fermentation, and during the secondary fermentation and maturation of the beer. The efficiency of the yeast in removing the diacetyl during those stages depends strongly on the number of yeast cells in suspension and their reducing activity towards diacetyl.

13.3.5 Sulphur-containing compounds

Many sulphur-containing compounds found in beer derive directly from the raw materials, malt and hops, but some are produced through yeast metabolism. The most important are sulphite (pungent: threshold, 10 mg/L), hydrogen sulfide (rotten egg: threshold, 8 µg/L) and dimethyl sulfide (cooked cabbage: threshold, 30 µg/L).

The brewer wants to keep the production of sulphite under control. Although sulphite is an antioxidant (i. e., it reacts with active oxygen) and a flavour stabilizer (i. e., it forms non-volatile adducts with stale carbonyls), excess amounts of sulphite cause an undesirable taste in beer. Another important consideration is that some countries have a maximum legal limit of sulphite in beer. Sulphite and hydrogen sulfide are intermediates in the biosynthesis of the sulphur-containing amino acids methionine and cysteine. The production of hydrogen sulfide and sulphite during fermentation is strain dependent. Hydrogen sulfide accumulates at the start of fermentation, which is followed by a decrease due to stripping with carbon dioxide in the later stages of fermentation. Excessive production of hydrogen sulfide by yeast has been associated with poor pitching yeast quality and sluggish fermentations. Sulphite production is also greatly influenced by the yeast physiological state. Starving the yeast before pitching results in a significant increase in the production of sulphite during fermentation. Medium composition and fermentation conditions that stimulate yeast growth (i. e., increase in wort oxygenation or lipids, higher temperature) result in beers with low levels of sulphite [20]. The excretion of sulphite occurs mainly after the growth ceases, as long as there is still energy available to the cells. Consequently, the more extract that is fermented after the growth phase, the more sulphite is excreted into the medium. In this regard, there is a positive correlation between the amount of sulphite in the beer and the original extract of the wort.

Dimethyl sulfide (DMS) is an important organic sulphur compound typical of lager beer flavour when present at concentrations between 30 µg/L and 60 µg/L. The two main routes leading to the formation of DMS in beer are firstly the thermal degradation of S-methylmethionine during the hot stages of the brewing process (i. e., wort boiling and wort clarification) and, secondly, the reduction of dimethylsulfoxide (DMSO) by the yeast during fermentation. Because DMS has a low boiling point of 38 °C, the final DMS level in beer depends on the DMS amounts present in the pitching wort, the DMS formed by yeast during fermentation, and the DMS removed with the evolving carbon dioxide. Although it has been recognized that the main source of DMS in beer arises from the breakdown of S-methylmethionine, in some instances it has been shown that the reduction of DMSO by the yeast determines the DMS level. All yeast strains are capable of reducing DMSO. On average, the yeasts reduce 25 % of wort DMSO. The extent of DMSO conversion depends on the yeast strain, the wort extract (higher amount of fermentable extract, higher conversion of DMSO) and the wort composition (e. g., the conversion of DMSO decreases when the level of amino acids increases or when the level of DMSO decreases). The production of DMS occurs at the end of the growth phase and, like sulphite, the production of DMS stops when fermentation is completed.

In addition to DMS, many other sulphur-containing compounds have been reported in beer [27] at much lower levels, like thiols (e. g., methanethiol), thioesters (e. g., methylthioacetate), sulfides and polysulfides. These compounds are highly flavour-active and their presence in trace quantities has a profound effect on the sensory characteristics of beer. Unlike ale beers, lager beers contain significant levels of hydrogen sulfide, methanethiol and methylthioacetate [83]. Thiols and thioesters are mainly fermentation-derived, while sulfides and polysulfides are mainly wort-derived. The yeast strain used has been shown to have a significant impact on the levels of thiols and thioesters in beer after fermentation. Wort oxygenation is also critical for the production of thiols and thioesters (the lower the dissolved oxygen content, the higher the level of sulphur compounds).

13.4 Detrimental aspects of yeasts found in breweries

Non-*Saccharomyces* yeasts are not adapted to survive the stressful environment and conditions of fermentation and beer. Consequently, they pose less of a general threat to product quality than the *Saccharomyces* wild yeasts. The most common non-*Saccharomyces* yeasts found in breweries are *Candida* and *Pichia* (including *Hansenula*) species. These yeasts grow poorly or not at all under anaerobic conditions, and are unable to ferment or are very selective in the sugars they ferment. For these reasons they do not represent a major threat to beer quality. Nevertheless, their presence should be a concern to the brewer as their presence most likely indicates problems in brewing plant hygiene [9]. Several types of yeast are known to cause adverse effects on beer quality, namely *Candida, Saccharomyces, Dekkera, Hanseniaspora, Kloeckera, Rhodotorula, Torulaspora, Brettanomyces* and *Pichia* species (Table 13.3-1). Known beer defects caused by these contaminants are changes in beer flavour (e. g., dry, harsh, washy, thin, bitter, fruity, spicy, medicinal, herbal, sour or sharp), the production of hazy beer, film formation, and gushing.

13.4.1 The POF (phenolic off-flavour) yeasts

Some *Saccharomyces* yeasts, generally ale yeasts used in the production of specialty beers, or *Saccharomyces* wild yeasts (e. g., *Saccharomyces cerevisiae* var. *diastaticus,* killer yeasts) contain an active POF gene coding for an enzyme that decarboxylates wort phenolic acids into flavour active phenols. For instance ferulic acid is converted into 4-vinyl guiacol (clove/spicy aroma), and cinnamic acid into styrene (medicinal aroma). The resulting flavours constitute a flavour defect. In some wheat beers ('Berliner' Weissbier and Weizenbier) and smoked beers (Rauchbier), however, the phenolic or clove-like flavours are part of the beer specific flavour characteristics.

13.4.2 Film forming yeast /particles

Aerobic yeasts (e. g., *Pichia membranifaciens, P. anomala, Candida mycoderma* (= *Pichia fluxuum*)) require oxygen to grow in beer. These yeasts form a film at the surface of the beer which upon breaking results in flaky particles or a deposit in the beers. An unusual high level of esters (e. g., ethyl acetate) is characteristic of a beer spoiled with *Pichia* spp. Significant growth of aerobic wild yeasts will result in loss of ethanol with the concomitant production of acetic acid causing a sauerkraut off-flavour.

13.4.3 Non-finable yeast (hazy beer)

Wild yeasts, such as *Kloeckera* spp. and *Rhodotorula* spp., sediment very slowly and fail to be removed by finings (isinglass) due to differences in their surface charges, because they have a lower negative charge than brewing yeasts.

13.4.4 Super-attenuating yeast (dry beer)

During fermentation, brewing yeasts ferment simple sugars from the wort (glucose, fructose and sucrose), as well as maltose and maltotriose, but they are unable to utilize dextrins (e. g., maltotetraose). Unlike brewing strains, *Saccharomyces cerevisiae* var. *diastaticus* (= *Saccharomyces cerevisiae*) secretes a glucoamylase (glucan 1,4-α-glucosidase, EC 3.2.1.3) enabling it to ferment starch and dextrins. Consequently, this yeast may attack the residual dextrins of beer, thus causing excessive attenuation of the beers. Contamination of bottled beer will result in CO_2 supersaturation, haze and gushing problems.

13.4.5 Killer yeasts

Contamination of the pitching yeast with killer yeasts is a major threat as they will eventually cause the death of sensitive brewing yeast strains. As little of 1 % of killer yeast can completely wipe out the brewing strain from the fermenter [71]. The killer yeasts secrete toxins (exotoxins or zymocins) that bind to cell wall components, making the cell either leaky or causing an arrest of the cell cycle [88]. Killer strains have been found in the genera *Pichia* (the highest frequency), *Candida* and *Saccharomyces* (the lowest frequency and most of them were laboratory strains). Because only a few killer yeasts are connected with brewing, only few cases have been reported in the brewing world. If present, the heavy flocculence of the killer yeast facilitates its transfer to the successive fermentations when bottom cropping is the method of yeast harvest.

13.4.6 Flavour taints

As already stressed the presence of wild yeasts will be frequently associated with unwanted flavours in the beer (e. g., phenolic, medicinal, nail polish). Contamination of the brewing yeast by non-*Saccharomyces* yeasts, *Brettanomyces* spp. and *Dekkera* spp. will result in beer with burnt plastic, wet leather/wet animal (horse) and acetic acid off-flavours in lager and ale beers. However, these yeasts contribute positively to the specific flavour characteristics of specialty beers, such as Belgian Lambic beer, Gueuze and some high gravity beers ([79]; J.P. DUFOUR and R. WIERDA, unpubl. observ.).

13.5 Physiological background of brewing yeast

The uptake of both metabolites and catabolites in and from the yeast cells influences directly the way the brewer will conduct the fermentation. It can be stated that the yeast metabolism is 'driving' the brewing technology, as for example the fermenter is designed to suit the beer characteristics and yeast properties. Moreover, the balance between the assimilation of wort nutrients and the release and removal of compounds by the cells during fermentation (primary fermentation) and maturation (secondary fermentation) determines the quality of the beer.

During the primary fermentation (i. e., the true fermentation), the yeast cells will go through a period of adaptation to the new environment (lag phase), an active cell division phase (exponential phase of growth) and a 'closing down' phase (resting or stationary phase). During the corresponding periods, the wort will be converted to beer involving the catabolism of fermentable sugars to carbon dioxide and ethanol, the assimilation and metabolism of amino acids and lipids, the production of flavour compounds and a fall of pH.

The profiles of the main features of a typical fermentation, such as extract, cell in suspension, free amino nitrogen (FAN), total vicinal dicetones (VDK), sulphite and temperature are illustrated in Figure 13.5-1. The lag phase lasts for several hours during which the dissolved oxygen drops to zero and there is little or no significant changes in other wort characteristics. Although the yeast does not show any visible changes (except a slight decrease in suspended yeast cells, which may often be observed due to sedimentation), very important biochemical events are taking place during this time (see below). Once fermentation

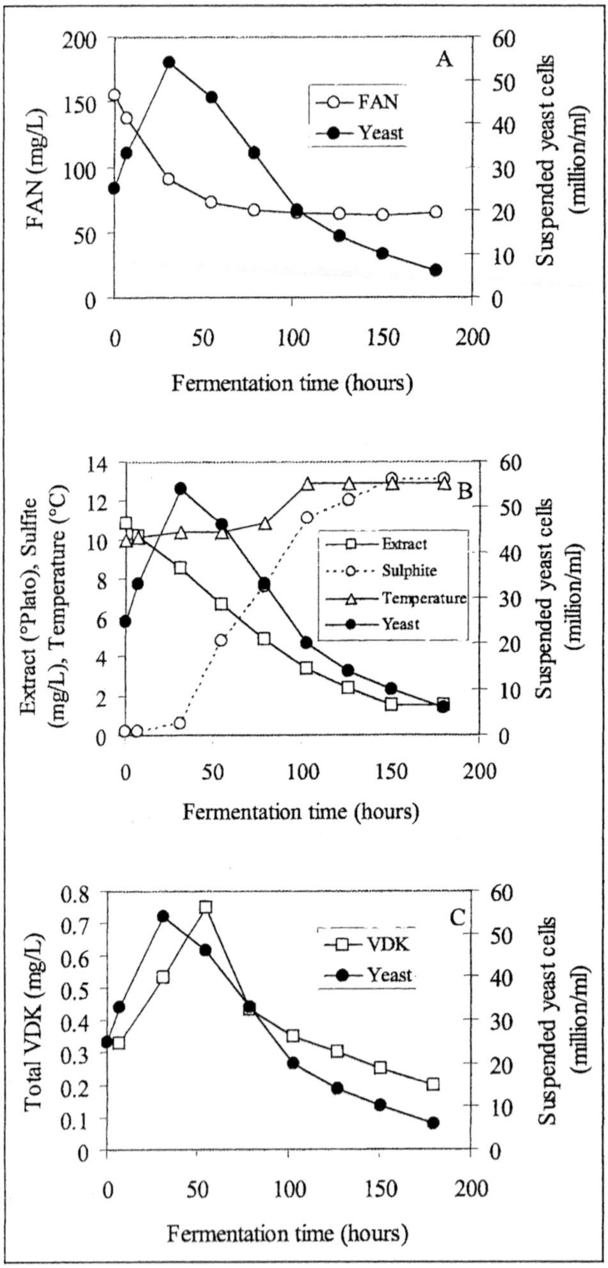

Fig. 13.5-1
Evolution of suspended yeast cells, extract, free amino nitrogen (FAN), sulphite, total vicinal diketones (VDK) and temperature during industrial lager fermentation using a cylindro-conical tank.
A. Suspended yeast cells, FAN.
B. Suspended yeast cells, extract, sulphite, temperature.
C. Suspended yeast cells, VDK
(J.L. VAN HAECHT and J.P. DUFOUR, unpubl. observ.).

starts there is a rapid decline in the level of free amino nitrogen (mainly the amino acids), accompanied by a rapid decline in pH. The fermentation rate gradually increases and reaches a maximum during the exponential phase of growth, then declines once the yeast enters the resting stage. The decrease in fermentation rate originates from changes in the yeast metabolism, such as cessation of yeast growth and slowing down of sugar metabolism, and from the fall in the number of yeast cells in suspension due to flocculation that may occur at that time. The primary fermentation usually ends when all fermentable sugars have been utilized. The pH reaches a minimum ranging from 3.8 to 4.4, and at this stage the beer is called the green beer.

The overall equation of a brewery fermentation may be written as follows:

$$\begin{array}{lcl} \text{sugar (100g/L)} & \rightarrow & \text{ethanol (48.3 g/L)} \\ + \text{ amino acids (0.5 g/L)} & & + \ CO_2 \text{ (46.2 g/L)} \\ + \text{ yeast (1 g dry matter/L)} & & + \text{ by-products (flavour/aroma) (2 g/L)} \\ + \text{ oxygen (8 mg/L)} & & + \text{ yeast (5 g dry matter/L)} + 50 \text{ Kcal} \end{array}$$

The secondary fermentation follows the primary fermentation. During this stage, yeast will continue to ferment slowly the residual maltose/maltotriose. Important processes are carried out, such as the removal of diacetyl, the buttery off-flavour of green beer, and in some cases supply the CO_2 for the final carbonation of the beer. Nowadays, with the use of cylindro-conical tanks, the primary and secondary fermentations processes are combined. Once completed, the fermentation is followed by the maturation during which yeast will complete the removal of the diacetyl, and excrete compounds of importance to the mouthfeel of beer (fullness).

The production of lagers and ales utilizes lager yeast strains (*Saccharomyces carlsbergensis*) and ale yeast strains (*Saccharomyces cerevisiae*) (see above), respectively. Traditional lager fermentations are conducted at temperatures ranging from 7–15 °C. The duration of fermentation usually ranges from 8–20 days, but with the higher temperatures used in modern brewing practice, fermentation times may be reduced to 7 days [54]. Ale fermentations occur more rapidly, and may take typically 3 to 5 days, because the temperatures tend to be relatively high, namely over 20 °C (15–22 °C for pitching, and 19–28 °C for fermentation).

The smooth running of the fermentation, meaning a fast and complete conversion of wort fermentable sugars into ethanol, and the production of beer with the desired sensory characteristics strongly depend on the growth of yeast. The extent of yeast growth has to be set in agreement with the flavour profile specifications of the beer. Excess formation of yeast biomass is an economic loss for the brewer, as this implies the formation of less alcohol and the formation of unsuitable beer flavours.

In breweries, the general practice is to use the yeast harvested at the end of the previous fermentation as the pitching yeast. Pitching yeast cell counts range from 5 to 20 million cells/ml. Lower pitching rates are preferred for ale fermentation and higher pitching rates for lager fermentation. The re-use of the yeast requires proper handling to ensure constant and suitable performances. A weakening of its physiological state during the number of recycles is called 'degeneration' of the yeast. One of the causes underlying this phenomenon is the deterioration of properties of the plasma membrane, such as integrity and fluidity, due to qualitative and quantitative changes in its lipid components. Yeast cells take up and concentrate nutrients from the wort, a cellular event that requires the transport of molecules across the cellular membrane. The integrity of this natural barrier is of prime importance for the various selective exchanges between the wort and the cellular compartment. To maintain functionality, the yeast needs to find or synthesize the essential lipidic constituents, sterols and unsaturated fatty acids [10, 57]. As a direct consequence of the anaerobic brewing fermentation process, the yeast lipid level decreases by 50 % [1]. At the end of fermentation, lipid content accounts for 3 % of yeast cell dry matter, which is less than half the value observed during the active growth. This deterioration of the cellular membrane properties could be responsible for the reduction in uptake of sugar at the end of the fermentation, which causes hanging or tailing fermentation and leads to the production of off-flavours in the beer.

13.5.1 Brewing yeast behavior in aerated wort

To recover both the original assimilation efficiency of wort nutrients and a 'normal' growing activity, the lipid content of the yeast cells needs to be regenerated. This can be achieved in two different ways, namely the use of conditions that favors *de-novo* synthesis of the essential lipids of the yeast cells, such as sterol and unsaturated fatty acids, or the use of wort containing the essential lipids. PASTEUR [62] was the first who realize that his earlier statement "fermentation is life without air" needed modification to take into account the small amount of air required by yeast for satisfactory growth. He also observed that at the time of complete disappearance of the oxygen in solution "the cells of yeast had assumed a younger and fuller appearance than they had at first; but they had not multiplied at all up to that time, nor were there even any buds then visible on them". Without knowing the underlying fundamental justification, PASTEUR noticed the considerable influence that oxygen had on the activity and development of yeast, and, consequently, on the progress of fermentation. Nowadays, the role of oxygen has been elucidated. Although some mitochondrial functions are essential for good fermentation performances [59], the yeast is unable to respire due to the existence of the Crabtree effect (reverse-Pasteur effect or carbon catabolite repression) that is observed for most yeast strains growing in media containing more than 0.4 % (w/v) of sugars. Oxygen is required for the synthesis of essential yeast plasma membrane lipids. These are synthesized at the start of fermentation and their amount is determined by the level of dissolved oxygen into the wort. As such, oxygen may rightly be viewed as a nutrient. Both sterols and unsaturated fatty acids play a vital role in maintaining the structure and function of cell membranes, particularly the plasma membrane. The synthesized sterols will determine the biomass yield and, as a direct consequence, the fermentation rate. The more dissolved oxygen is present in the wort, the better the growth of the yeast, and the faster the fermentation rate. Consequently, the regulation of dissolved oxygen in wort is an integral part of the control of yeast growth.

At the end of the fermentation, the sterol content of the yeast cells is approximately 0.1 % (on yeast dry matter) which prevents any further growth. As soon as anaerobic pitching yeast is inoculated into the aerated wort, *de-novo* synthesis of sterols occurs. Squalene (approx. 1.0–1.5 % on dry matter in anaerobic yeast), the immediate precursor of sterols, undergoes rapid cyclization in the presence of oxygen, finally resulting in ergosterol. More than 70 % of the newly formed sterols are esterified and constitute a sterol pool that will be used later during the growth of yeast under anaerobic conditions. The maximum level of total sterols in aerobically grown yeast (i. e., excess of oxygen) is around 5 % (on yeast dry matter), which includes all intermediates between lanosterol and ergosterol. Under normal fermentation conditions however, a 1.0–1.5 % (on yeast dry matter) upper limit is observed at the end of the aerobic phase. Once the medium is depleted of oxygen, the total level of sterols in the culture remains almost unchanged. In practice, the sterols present in the cell at this time determine the potential yeast biomass, assuming that other nutrients are not limiting. The parent cells will share their sterols with their progeny until a lower limit is reached, because a critical level of sterols is required for survival. Simultaneously to sterols synthesis, there is also a rapid synthesis of long chain unsaturated fatty acids (palmitoleic and oleic acids) in aerated wort, which are incorporated into phospholipids and triglycerides.

The overall synthesis of sterols and unsaturated fatty acids is dependent on the initial lipid composition of the pitching yeast and the amount of oxygen available. For industrial fermentation using aerated wort (about 8 mg/L of dissolved oxygen using air), the oxygen disappears from the wort in 2 to 24 hours depending on the fermentation temperature. Consequently, the time available for the yeast to synthesize its essential lipids is rather short, often less than 10 % of the overall fermentation length. Therefore, it is not surprising that levels and distribution of lipids are very different between cells harvested at the end of industrial fermentation and cells from an aerobic culture. Although the synthesis of lipids is limited in industrial wort, it is widely recognized that it plays an essential role in relation to the performance of yeast in brewery fermentation and beer sensory properties.

13.5.2 Brewing yeast growth and metabolic changes during primary fermentation

During the early phase of the fermentation and before any growth occurs, the yeast cells show a large and rapid increase in sterol and unsaturated fatty acid contents (see above). When the sterols level in the yeast reaches about 0.25 % (on yeast dry matter), yeast growth will start. This occurs typically after 12–24 hours and depends on the fermentation conditions. Growth proceeds to the point where the yeast sterol content decreases to the critical limit of 0.1 % (on yeast dry matter). The harvest of yeast (i. e., growing yield) is a function of the absolute amount of sterols synthesized during the lag phase. The higher the dissolved oxygen concentration in the wort, the higher the levels of sterols synthesized, and the better will be the yeast growth. Additional oxygen may be supplied, but it is well established that it is difficult to dissolve oxygen in fermenting wort as it is quickly washed out by the carbon dioxide produced by the fermenting yeast. Alternatives include aeration of the yeast prior to pitching, or the use of oxygen gas to oxygenate the wort. Up to 30 mg/L of dissolved oxygen can be obtained using pure oxygen. As the quantity of oxygen required is yeast strain dependent [41, 43], the brewer has to evaluate the specific oxygen requirements for each particular strain. In addition to oxygen, wort should also contain an appropriate level of lipids, varying between 10 mg/L and 20 mg/L (measured as total long chain fatty acids). Amounts of lipids below 5 mg/L have been shown to be detrimental when the yeast is recycled [11]. Ideally, every brewer needs to determine the yeast-wort lipid dependence using their specific fermentation conditions (e. g., dissolved oxygen, wort gravity, temperature).

A few hours after pitching, once the cellular membrane properties have been restored, the cells start multiplying as evidenced by the budding of the yeast. Once again it is worth stressing that alcoholic fermentation and growth proceed under anaerobic conditions. The growing phase will last until one or several essential nutrients become limiting. It is now well known that in most cases the arrest of growth results from a change in the cellular membrane properties. As the extent of growth directly affects the level of flavour active compounds, such as e. g., higher alcohols, esters, and diacetyl, it is of prime importance to keep yeast growth under control. Parameters controlling yeast growth include fermentation conditions (e. g., the contact time with oxygen, the wort lipid level, temperature, stirring), the number of times the yeast is recycled, and the yeast storage conditions (temperature, time). Yeast obtained from a propagation vessel is characterized by much better cellular membranes than cells harvested after an eight fermentation cycle of a wort poor in lipids (i. e., containing less than 5 mg/L). Yeast carrying out successive fermentations under these conditions will have a sterol content that will progressively fall down below 0.1% (on yeast dry matter). This will eventually lead to a reduction of the fermenting capacity and, ultimately, will result in an uncompleted fermentation during which fermentable sugars are being left in the medium. Theoretically, the higher the fermentation temperature is, the lower the requirement for sterols and unsaturated fatty acids is, which may explain why ale yeasts are less sensitive to degeneration.

At the end of the growth phase, yeast cells enter into the most stressing phase of the fermentation cycle, where most of the cellular activities slow down. The importance of this stage to the quality of the beer should not be underestimated. The fermentation capacity, especially towards maltotriose, slowly decreases to a level where unfermented sugars are left in the beer. During this secondary fermentation phase, the reducing activity of the yeast results in the removal of diacetyl.

13.5.3 Sugar and amino acid metabolisms

The most important quantitative change in the wort during the brewing fermentation is the conversion of the sugars into ethanol. In batch fermentation, yeast assimilates wort fermentable sugars, such as sucrose, fructose, glucose, maltose and maltotriose in a sequential fashion. Sucrose, fructose and glucose are used first, followed by maltose, the major wort sugar (45–65 %, w/w), and maltotriose. Maltotriose can be utilized simultaneously to maltose or after the completion of assimilation of maltose. In the latter case the maltotriose is often left unfermented in industrial fermentations. Of relevance to the brewing fermentation is that utilization of maltose and maltotriose is inhibited in the presence of glucose. This is known as carbon catabolite repression and catabolite inhibition and is described above (for a review, see [45]). The global effect of the presence of exogenous glucose on the yeast metabolism depends on the genotype of the yeasts. Yeasts showing carbon catabolite inhibition have their fermentation performance strongly slowed down or, in some cases, even stopped at the beginning of maltose utilization, if the glucose concentration is too high (e. g., 30 % of total fermentable sugars). Once inside the cells, the sugars are converted into ethanol and carbon dioxide by the glycolytic pathway. Under normal fermentation conditions the conversion of the extract of the medium into alcohol and other volatile by-products, which is called attenuation of the wort, ranges between 80 and 84 %. The residual beer extract consists of dextrins, peptides and proteins, and limited amount of unused nutrients, such as amino acids and unfermented sugars.

The uptake of amino acids is also highly regulated. The twenty amino acids enter the cells in a highly organized fashion, and the activity of the permeases is being modulated by the spectrum and concentration of the amino acids present in the wort (i. e., nitrogen catabolite repression) [34]. Amino acid uptake is governed by the timing of the synthesis of the permeases, their turnover (V_{max}), their affinity for the transported amino acid(s) (K_m), and the competitive binding between different amino acids. Based on their uptake profiles during fermentation, the wort amino acids have been divided into four groups [28, 65]. Group I amino acids (arginine, aspartic and glutamic acids, asparagine, glutamine, lysine, threonine and serine) are characterized by a fast uptake and complete adsorption. Group II amino acids (valine, methionine, isoleucine, leucine, histidine) have a slow uptake rate at the start of the fermentation, but this accelerates once a significant proportion of group I has been used by the yeast. Under traditional fermentation conditions, adsorption of group II amino acids is only partial in lager fermentations, but complete in ale fermentations. The uptake of group III amino acids (glycine, alanine, tyrosine, phenylalanine, tryptophan) shows a significant delay, and adsorption starts after the complete removal of amino acids of group I. The adsorption is only significant in ale fermentations or when using wort of low amino acid content. Proline, the major amino acid in wort (approx. 1/3 of the total amino acids) and the only amino acid in group IV, is only slightly adsorbed during brewing fermentation.

Amino acids are required for the growth of yeast, and lysine is an essential one. The amino acids profile in wort is relatively constant, but the level of amino acids may vary significantly depending on the amount of adjunct (e. g., rice, corn, starch, sugar) used [22]. A minimum of 150 mg (measured as mg of NH_2) per liter of wort (12° Plato – 12 %, w/w) is recommended for satisfactory fermentation performance during successive fermentations [26]. Parameters that affect growth of yeast, such as temperature, lipids, dissolved oxygen, and stirring, will affect the rate of amino acid uptake as well as the amount adsorbed [11, 21].

As described earlier, metabolism of amino acids is essential for the synthesis of beer flavour compounds. For the brewer the important aspects of nitrogen metabolism are: 1. the formation of diacetyl (the buttery off-flavour) as a by-product of valine synthesis, 2. the production of volatile sulphurous compounds, hydrogen sulfide (rotten egg) and sulphite, as by-products of the synthesis of methionine and cysteine, and 3. the production of higher alcohols as overflow products of amino acid catabolism (degradation of amino acids from wort – Ehrlich pathway) and anabolism (*de-novo* synthesis of amino acids by the yeast – Genevois pathway).

13.5.4 Secondary fermentation: bottle-conditioned beers

Refermentation in bottles was originally developed for the carbonation of the beer similarly as is performed for sparkling wines. DOM PERIGNON, at the abbey of Hauvillers-France, was the first to describe the technique of bottle conditioned white wines, called 'champagne'. From a sensory perspective, carbonation contributes to the refreshing characteristic of the beverage. In traditional lagering, beer with a residual fermentable extract of 1 % (w/w) is transferred from the fermenter to the lagering tank and carbonation of beer is obtained during maturation in a cellar (or chilled tank) maintained at low temperature (less than 2 °C) or in pressure resistant containers, such as bottles. The contribution of bottle conditioning to the final quality of the beer are twofold: firstly, the carbonation of the beer (6.0 to 9.0 g of CO_2/L) and, secondly the removal of traces of oxygen that have dissolved in the beer during the filling process. Secondary fermentations may last from a few days up to several weeks. Refermentation in bottles eliminates the negative impact of oxygen, which may have been picked up during filling, on the early staling of beer.

Bottle conditioning is mainly applied to ale beers with high levels of ethanol (6 to 11 %, v/v) that are produced in Belgium and the North of France. Other popular bottle conditioned beers are the German and Belgian wheat and white beers (less than 6 % alcohol, v/v).

An ale strain (*S. cerevisiae*) is routinely used in bottle conditioning, but this yeast strain can be different from the strain used in the primary fermentation.The selection of the yeast strain and yeast handling are the most important determinants for successful bottle conditioning. In particular, the following properties are especially relevant to the yeast strain used in bottle conditioning: 1. tolerance to high levels of ethanol, 2. resistance to inhibitory substances (melanoidins) from Maillard reactions from specialty malts (e. g., roasted malts), 3. absence of effect on beer foam (some *S. cerevisiae* strains show proteolytic activity toward the proteins of the foam), 4. an appropriate flocculent character and 5. a good adhesion of the yeast to the glass surface. Great care is required to maintain the microbial quality of the yeast. For example, if the yeast is contaminated with *Lactobacillus* spp. or *Pediococcus* spp., granular structures of yeast cells and bacteria (0.1 mm to 1 mm in diameter) are formed that do not adhere to the glass.

The source of the yeast can be different. This may be 1. the active phase of growth at the maximum fermentation rate during the primary fermentation, which is usually at the second or third day of the fermentation, 2. the end of the fermentation, 3. a freshly cultured yeast in a specially designed yeast propagator, 4. a pressed brewing or even baking yeast, or 5. a dried yeast. The advantages and disadvantages of each yeast preparation are presented in Table 13.5-1. Typically the beer will be inoculated with 250,000 to 300,000 cells/ml, but the values may range from 250,000 to 2 millions cells/ml, depending on the alcohol level (G. DERDELINCKX and B. VANDERHASSELT, unpubl. observ.).

Tab. 13.5-1 Main characteristics of the various yeast sources used for bottle conditioning

Yeast source	Advantages	Drawbacks
Fermenting yeast	Easy to use High fermenting capacity High reducing power No investment needed Reduced microbiological risk	Temperature stress Increasing risk of colloidal and sensory defects due to the addition of fermenting wort Limited number of suitable yeast strains
Yeast cultured in specially designed propagator	Large selection of yeast strains Conditioning of the yeast strain at the suitable temperature High fermenting capacity High reducing power Reduce microbiological risk	High capital costs Labor intensive unless automated
Yeast taken from the end of fermentation or from yeast storage tank	Easy to use High level of cellular glycogen No investment needed	Risk of microbial infection Risk of fast autolysis of the yeast Very often, the re-fermentation process is slow, associated with a low reducing power
Dried yeast	Easy to use No investment needed Reduce microbiological risk	Limited number of suitable yeast strains Very often, the re-fermentation process is slow Need to dissolve and condition the yeast
Pressed yeast	Easy to use No investment needed Fast re-fermentation	Available strains included bakery strains and lager strains Need to dissolve and condition the yeast

The type of sugars added to the bottle includes glucose (syrup or solid), maltose, sucrose, invert sugar or candi sugar. The added amounts may range from 5 to 10 g/L depending on the level of carbon dioxide in beer at bottling and the level of carbon dioxide saturation required. Sugars can be added to the beer in a buffer tank prior to filling or directly injected in-line during the filling process. The yeasts also utilize a small amount of amino acids, as indicated by the production of diacetyl. In some instances trace levels of sulphurous compounds have been detected, but this is strain dependent. Levels of higher alcohols and esters do not show significant changes during bottle conditioning. At most there is an increased sensory perception of short chain fatty acid esters, such as ethyl caproate, ethyl caprylate and ethyl caprate. It has been suggested that this could be the result of the high saturation in carbon dioxide (8 to 11 g/L) of these beers that influences the overall bouquet of the beer that reaches the olfactory epithelium.

The yeast also contributes to the removal of carbonyl compounds (Fig. 13.5-2). It is still believed by some brewers that oxygen dissolved during the filling process or present in the head-space is beneficial to re-fermentation. The apparent positive effect of oxygen follows on the use of yeast in a poor physiological state. This can be remedied by using properly

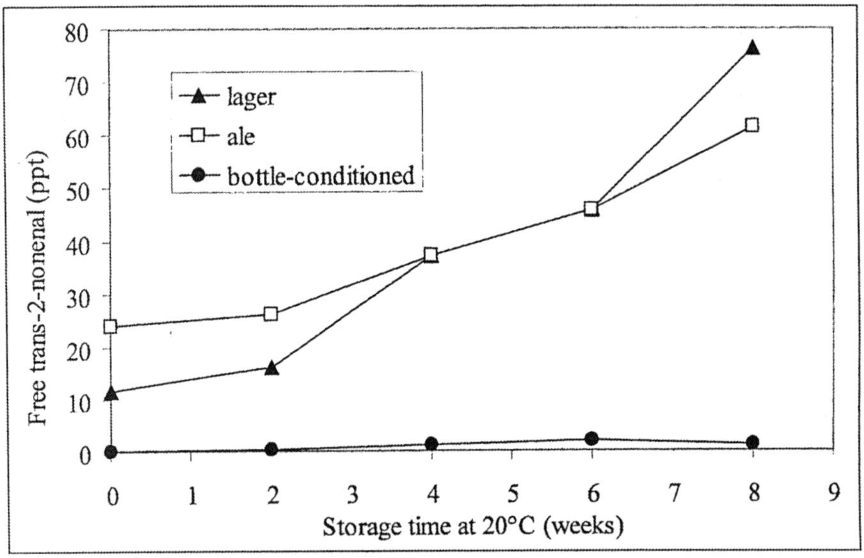

Fig. 13.5-2 Evolution of free trans-2-nonenal during storage at 20 °C. Trans-2-nonenal, the carbonyl compound responsible for the papery off-flavor in aged beer, was analyzed in a lager, an ale and a bottle conditioned beer using a dynamic headspace-gas chromatography-mass spectrometry technique. In the lager and ale beers, free trans-2-nonenal increased with aging whereas in the bottle-conditioned beer, the free trans-2-nonenal concentration was kept close to zero. This low level was due to the reducing activity of the yeast (S. Bonte, S. Dupire and L. Degelin, unpubl. observ.).

prepared yeast (see Table 13.5-1). In general, dissolved oxygen should be kept to a minimum as it increases the risk of flavour staling and the formation of haze. Bottled beer is incubated for a period of 10–20 days at 18–24 °C depending on the beer characteristics (G. DERDELINCKX, unpubl. observ.).

In some bottle-conditioned beers, the beer ester level decreases with time as a result of yeast esterase activities. The effect of esterases from brewing yeasts to the final level of esters in bottle-conditioned beers was investigated by NEVEN et al. [58]. During fermentation and lagering, ester-hydrolyzing enzyme activities are released into the beer and remain active in the finished non-pasteurized product, thus decreasing the levels of measured esters, such as isoamyl acetate and ethyl caproate, during the storage of beer to a level corresponding to the chemical equilibrium with the alcohol and acid.

Bottle conditioning may also involve non-*Saccharomyces* yeasts and bacteria. Yeasts of the genus *Brettanomyces* also show esterase activity towards a large number of esters. Such activities have been identified by the presence of ethyl acetate and ethyl lactate in the production of specialty beers (Lambic, Gueuze, ale beer) and the hydrolysis of isoamyl acetate ([73]; J.P. DUFOUR and R.WIERDA, unpubl. observ.) (Fig. 13.5-3). In some cases, bacteria are inoculated together with the yeast (see below). Their combined action during bottle con-

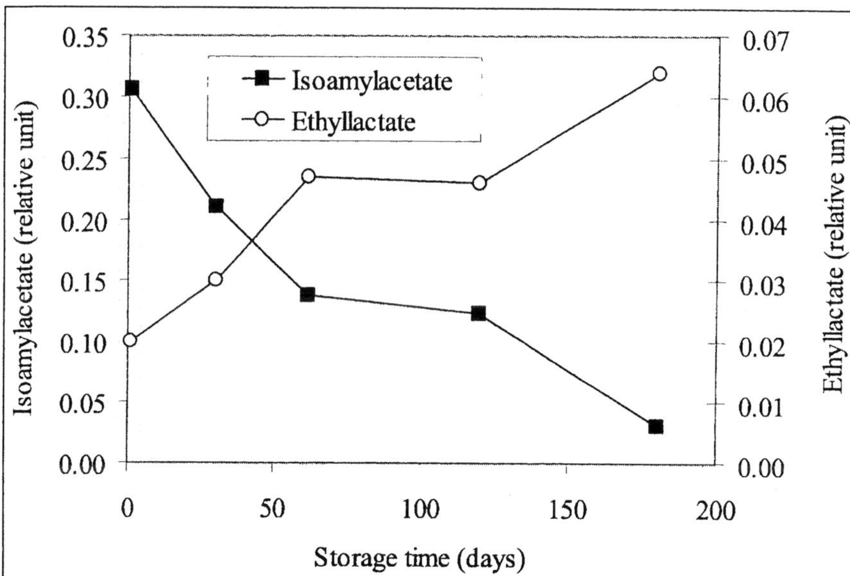

Fig. 13.5-3 **Evolution of isoamylacetate and ethyllacetate during storage of a bottle-conditioned ale using *Brettanomyces* (J.P. DUFOUR and R. WIERDA, unpubl. observ.).**

Tab. 13.5-2 **Main characteristics (microorganisms and typical flavour compounds) of various specialty bottle-conditioned beers**

Microorganisms	**Berliner Weissbier (Germany)**	**Bavarian Weizenbier (Germany)**	**Brussels's acid beers (Lambic, gueuze, fruit beers) (Belgium)**	**White beers (Belgium)**	**Abbey and strong pale ales (Belgium)**	**Trappist beers (Belgium, The Netherlands)**
Yeast	*S. cerevisiae*, *Brettanomyces* spp. (= *Dekkera* spp.), *Candida* spp.	*S. cerevisiae* (POF strains)	*Saccharomyces* spp., non-*Saccharomyces* spp., *Brettanomyces* spp. (= *Dekkera* spp.), *Candida* spp.	*S. cerevisiae*	*S. cerevisiae*	*S. cerevisiae*, *S. cerevisiae* with *Brettanomyces* spp. (= *Dekkera* spp.)
Bacteria	*Lactobacillus* spp., *Pediococcus* spp.		*Enterobacter* spp., *Acetobacter* spp., *Acetomonas* spp., *Lactobacillus* spp., *Pediococcus* spp.			
Characteristic flavour compounds (mg/L)						
Ethylacetate	6–87	24–41	25–150	20–40	40–100	15–25
Isoamylacetate	0.04–0.47	2.7–5.6		1.7–3.2	1.5–15	0.1–15
Ethyllactate	8.0–345	~ 20	40–300			
Ethylcaproate	0.05–0.11	~ 0.1		0.20–0.70	0.1–3.0	
Ethylcaprylate	0.05–0.18	~ 0.1		0.30–1.0	0.1–3.0	
Acetic acid	65–730	30–400	1,000–2,500			
Lactic acid	1,100–4, 600	60–125	1,500–6,000	600–900[b]		
Isovaleric acid	0.03–1.3	1.5–3.5				0.5–3.5[c]
Isoamyl/amyl alcohols	~ 100	60–85		70–100	60–250	120–250
Phenylethanol	10–50	10–60			25–70	25–70
4-Vinylguaiacol[d]	0.070–0.150	0.2–4.4	0.0–0.4	0.16–1.33	0.20–1.08	
4-Vinylphenol	0.025–0.090	0.02–2.7	0.0–0.125	0.035–0.75	0.12– 0.76	
Fruit aromas			Cherries, raspberries			
Geranial and geranyl acetate						0.005–0.015, in dry hopped beer
Linalool (coriander)				Above threshold		
Dimethylsulfide		0.05–0.1[a]	0.05–0.20			

[a]From specialty malt. [b]From addition of lactic acid. [c]From hops. [d]Flavour threshold, > 0.5 mg/L

ditioning produce very characteristic beer flavours, such as solvent/nail polish (ethyl acetate), fruity (isoamyl acetate, ethyl caproate and ethyl caprylate), acidic (lactic acid) and phenolic (4-vinylguiacol, 4-ethylgaiacol) flavours. The main characteristics of various specialty bottle-conditioned beers are presented in Table 13.5-2.

13.5.5 Mixed fermentations: yeast and bacteria

Mixed cultures occur fairly commonly in fermented foods worldwide (see Chapter 17). Mixed culture fermentations can be initiated spontaneously, e. g., from air or equipment, or

Tab. 13.5-3 Typical beers produced using mixed fermentation

Beer type: Microorganisms	Substrates	Country
Acid/sour ales: Brewing and wild yeasts, lactic acid bacteria (*Lactobacillus* spp.)	Malted barley, specialty malt	Belgium
Lambic, Gueuze: *Saccharomyces* spp. and non-*Saccharomyces* spp. and wild yeast (*Brettanomyces* spp., *Dekkera* spp., *Candida* spp.) *Enterobacter* spp., acetic acid bacteria (*Acetobacter* spp., *Acetomonas* spp.) and lactic acid bacteria (*Lactobacillus* spp., *Pediococcus* spp.)	Malted barley, wheat, oat	Belgium
White beer (Berliner Weissbier): *S. cerevisiae* and wild yeasts (*Brettanomyces bruxellensis* – *Dekkera bruxellensis*), *Candida* spp. Lactic acid bacteria (*Lactobacillus* spp., *Pediococcus* spp.)	Malted barley and malted wheat	Germany
Rice beer: Wild yeasts (*Hansenula* spp., *Endomycopsis* spp.)	Rice	Indian subcontinent, South East and East Asia
Sorghum beer: *Saccharomyces* spp., *Candida* spp., *Geotrichum candidum* Lactic acid bacteria (*Leuconostoc* spp., *Lactobacillus* spp.)	Malted and raw cereals (sorghum, millet), maize, green banana (plantain)	Africa, Middle East, Europe, Indian subontinent, East Asia, South America
Pulque: *S. cerevisiae, Pichia* spp., *Torulopsis* spp. *Kloeckera* spp. Lactic acid bacteria (*Lactobacillus brevis, Leuconostoc* (*mesenteroides, dextranicum)*, *Lactobacillus plantarum*) *Zymomonas* spp.	Agave spp., Opuntia spp.	South America

through the use of inocula containing the different organisms involved. The latter can be inoculated either simultaneously or sequentially. Sequential inoculation is typical of the production of kaffir beer. Examples of simultaneous inoculation are the production of acid beers from barley or wheat, which rely on the inoculation of *Saccharomyces* yeasts with lactic acid bacteria. A list of beers obtained by mixed fermentation, either in the primary and/ or the secondary fermentation is presented in Table 13.5-3. Mixed fermentation presents a definite advantage when the brewer is aiming at special flavour characteristics of a beer, for instance acid beers. In some cases, a stable association between yeasts and bacteria is found, such as yeasts and lactic acid bacteria in Berlin 'Weissbier' [82] or the Belgian acid ale Rodenbach. Mixed fermentations involve complex types of interactions. In the production of Belgian Lambic and Gueuze beers, the succession of the microflora has been studied in detail [79, 80]. It has been shown that the spontaneous fermentation starts with a rapid development of Enterobacteria that die off completely after 30 to 40 days, followed by the main yeast fermentation that last a few months. This phase is followed by a period of strong development of lactic acid bacteria (*Pediococcus* spp.). Finally in the last phase, lactic acid bacteria and the yeast *Brettanomyces* spp. are the predominant microorganisms.

13.5.6 Continuous fermentation systems

Brewing fermentation is traditionally carried out as a batch process in what has become the standard fermentation vessel, namely the cylindroconical tank. In this system, the fermentation and maturation steps are combined. Since the introduction of the first continuous fermentation system in the early 20th century, numerous continuous systems have been described (for a review see [9]), but most of them did not progress beyond the laboratory or the pilot scale. The main reason for this is that the brewing fermentation is much more than just ethanol production, and therefore the simple chemostat theory can not be applied. Beer flavour characteristics encompass numerous reactions, and most of them occur in the yeast as a response to a complex sequence of physiological events.

The minimum requirements for a continuous fermentation are the continuous supply of wort and the continuous handling of green beer. The requirements that are critical to the success of a continuous system were reviewed by PORTNO [66]. The advantages of continuous fermentation are manifold, namely 1. a greater efficiency in the utilization of fermentable sugars (e. g., low yeast growth and high ethanol yield) and equipment (e. g., faster fermentation, less downtime in filling, emptying and cleaning the fermentation vessels), 2. less beer losses, 3. an improved consistency of beer quality due to a better control of yeast physiology, 4. space savings, 5. lower running costs because of less cleaning, 6. no need for pitching yeast storage, 7. simpler carbon dioxide collection, 8. use of a smaller yeast crop, and 9. a better utilization of hop, because less adsorption of hop resins occurs on the yeast. One of the main disadvantages of the continuous fermentation system is the increased risk of microbial infection. Infection will cause the shut down of the entire system very quickly. The genetic stability of the yeast is another important issue. Continuous operations can lead to the selection of yeast variants, which may have potentially disastrous consequences on the beer quality. Other disadvantages of continuous fermentation include the lack of flexibility (e. g., limited number of products and a narrow range of product output rate), the more sophisticated fermentation process and equipments, and the very long start-up time, which may take two weeks or more.

It is difficult to match the flavour characteristics of beer produced with a batch system using a continuous system. The process should aim at the production of a beer with suitable flavour characteristics that meet consumer expectations. All the successful continuous systems have adopted a multistage design. Several continuous systems implemented in commercial breweries in Canada, U.S.A. and the U.K., ran for a limited period of time in the 1960's and 1970's. Today, continuous fermentation is only applied in New Zealand. This has been developed in New Zealand by COUTTS [15], and is made of a cascade of three vessels, followed by a yeast separation and a maturation vessel. The main reason for the success of the multistage system is that it mimics the successive stages of the batch fermentation process, i. e., it includes the three key stages of yeast aeration, yeast growth, and the yeast resting phase. Because of the use of three vessels, each stage is physically separated. Such systems have been running at commercial scale for over 50 years at DB Breweries Limited of New Zealand (new plant commissioned in 1993) [76] and New Zealand Breweries Limited (continuous process discontinued in the early 90s) [18]. Operating conditions, such as residence time, wort gravity, wort oxygenation, yeast concentration, temperature, and stirring, are selected to achieve optimal transformation in each step. The total volume of the first three vessels (yeast aeration, growth and resting phase) is rather small, less than 250 hl, with an hourly production averaging 15 hl. Fermentation time can vary from 40 to 120 hours depending on the production requirements.

13.5.7 Yeast immobilized systems

Another important progress in brewing fermentation technology has been the development of commercial immobilized yeast reactors. The scope and limitations for immobilized cell systems in the brewing industry were reviewed by MCMURROUGH [49]. Immobilized systems offer similar advantages as the conventional continuous systems (see above) or perform better. For example, the immobilization of a high yeast concentration contributes to a much faster fermentation rate and a reduced risk of microbial infection. These reactors are also relatively small. The main disadvantages are the requirement to use bright wort to avoid clogging of the reactor, the interference of the evolving carbon dioxide with the mass transfer of nutrients, and its role in the mechanical breakage of fragile support (e. g., alginate beads).

Yeasts are immobilized by entrapment into gels (carragheenan, alginates, gelatin) or binding to a solid carrier (e. g., porous glass beads, DEAE-cellulose, silicon carbide). In an immobilized reactor, yeast growth is strongly impeded because the high density of cells has only limited access to the dissolved oxygen due to restricted mass transfer. Although this is an advantage with respect to the efficiency of conversion of sugars to ethanol, the reduced growth leads to major flavour deviations, such as low levels of esters. Therefore, it is not surprising that the first successful industrial applications of an immobilized yeast reactor were obtained for processes that did not require yeast growth, such as the maturation and the production of non-alcoholic beer. The immobilized yeast applications in the brewing industry were reviewed in 1995 during a European Brewery Convention symposium. The first industrial system was installed at the Sinebrychoff brewery in Finland, where immobilized yeast reactors are used for beer maturation, resulting in the shortening of the fermentation time from two weeks to two hours [61]. The technology is presently used in several breweries around the world for beer maturation and the production of alcohol-free beer as well [52]. For alcohol-free beer, the process only requires a contact between yeast and the wort to remove the undesired worty off-flavours [14, 77].

Immobilized yeast reactors have also been investigated at the pilot scale to carry out the primary fermentation [4, 87]. However, one should not underestimate the importance of this challenge. Many important beer flavour compounds are produced in amounts that are directly related to yeast growth, and therefore, the production of beer using an immobilized yeast reactor matching the quality of conventionally produced beer is challenging. Research on the development of an industrial process using an immobilized yeast reactor for the primary fermentation is continuing.

13.6 Genetic improvement of brewing yeasts

Optimization of brewing yeast performance is difficult as beer is not a single-fermentation product, and emphasis has been placed on the flavour characteristics of the beer. With the completion of the yeast genome sequence [33], and the availability of a range of very powerful molecular biological techniques (for a review see [38]), yeast geneticists now have a tremendous battery of tools and a source of information for the genetic improvement of process organisms. The complexity of yeast physiology during fermentation, however, makes it difficult to alter one metabolic step or yeast characteristic without affecting other functions or characteristics. Attempts to improve yeast performance fall into four broad categories: 1. reduction of material costs (e. g., improved fermentation efficiency, utilization of non-fermentable dextrins, production of chill-proof beer, and degradation of β-glucans), 2. increase of the production efficiency (e. g., reduced production of diacetyl, reduced production of H_2S, modified flocculation properties, and resistance to contamination), 3. increase in the production of SO_2 and flavours, such acetate esters, and 4. improvement of the microbiological stability by increasing the resistance to contamination. Progresses on producing genetically modified brewing yeasts have been reviewed by HAMMOND [37] and VERSTREPEN et al. [81].

Constraining factors to the successful exploitation of genetically engineered yeasts are the undesirable effects on yeast physiology, the specificity of the desired change, the stability of the introduced attribute, and the regulatory approval and consumer acceptance. The latter topic is today by far the major constraint to the utilization of transformed organisms in many countries around the world. Beer is consumed with the expectation that natural ingredients and traditional methods are being used. Researchers at the Brewing Research Foundation International (BRFI), now named Brewing Research International (BRI), in Nutfield (U.K.) produced an amylolytic brewing yeast, successfully transformed with glucoamylase genes from *Saccharomyces cerevisiae* var. *diastaticus* (= *Saccharomyces cerevisiae*), which imparts hydrolytic activity toward α-1,4 bonds with the enzyme being fully secreted by the yeast. In 1994, BRFI was eventually granted permission to use the yeast 'commercially'. This beer, Nutfield Lyte, was the first and only beer in the world made from a genetically-modified yeast to be approved for production and general tasting. Nevertheless, because of the increasing pressure of consumer organizations and their negative perception of the use of genetically modified organisms in the food industries, the brewers worldwide have adopted a very clear-cut no-go attitude.

13.7 Typing of brewing yeasts

Each brewery ideally needs to ascertain the identity of their yeast. Practical experience has shown that it is much easier to differentiate between brewing yeasts when they are used on a production scale than when they are compared using laboratory tests. Consequently, a plethora of laboratory methods have been developed to help brewing microbiologists to differentiate between brewing strains (for a review, see [9]). Methods to differentiate brewing yeast strains can be divided into traditional and modern methods.

Traditional methods do not give unequivocal identification, but the tests are simple and give a practical answer to the brewer. They rely on morphological (e. g., colony size and shape), physiological (e. g., resistance of wild yeasts to the antibiotic actidione and flocculation test) and biochemical (e. g., utilization of melibiose by lager yeasts) differences between *S. cerevisiae* yeast strains. As already mentioned, brewers are more interested in identifying their yeast strains using parameters directly relevant to the fermentation performance and the quality of the beer, such as rate of fermentation, extent of fermentation, flocculation, and the production of volatiles. This is not an easy task as laboratory conditions simulate only with great difficulty the industrial process.

Modern methods can give an unequivocal identification and allow the differentiation of individual brewing strains thanks to the development of molecular biological techniques. Karyotyping (determination of chromosomal size and number) has been the most successfully used technique for the differentiation of brewing yeast strains. Recently, the Amplified Fragment Length Polymorphism (AFLP) DNA fingerprinting technique has been applied with some success to the identification of brewing strains [64].

13.8 Yeast quality control

The brewing practice of re-using the yeast for successive fermentations requires an assessment of the quality of the yeast before the yeast is pitched in the wort. Yeast quality can be examined at two levels, namely fermentation performance and microbiological purity.

13.8.1 Fermentation performance

The repeatability of the fermentation performance is vital for the production of beer of consistent quality. This strongly relies on the use of yeast of appropriate physiological state. Viability and vitality measurements are routinely performed by the brewers in view of maintaining and improving the fermentation performance. Predicting the fermentation behavior of the pitching yeast is difficult as it relies on many different aspects of yeast metabolism. Therefore, it is not surprising that numerous metabolic tests have been developed. Unfortunately, none can be applied universally, and none is perfect. The metabolic activity tests available to the brewers and their ability to predict fermentation performance have been reviewed by BENDIAK [8]. These tests have been grouped into seven categories: 1. energy level tests (e. g., ATP levels, $NAD^+/NADH_2$ reducing power), 2. cellular component tests (e. g., O_2 uptake, CO_2 evolution), 3. fermentation capacity tests (e. g., acidification power tests, intracellular pH), 4. cell surface tests (e. g., hydrophobicity measurements, surface charge zeta potential), 5. replication tests (e. g., slide culture, yeast growth and budding index), 6. flow cytometry tests (use of dyes to measure specific activity parameters), 7. yeast capacitance tests and 8. stress indicator tests (e. g., levels of glutathione, trehalose).

Brewers prefer simple, rapid, cheap and reproducible vitality tests. Ideally the tests should allow them to decide the appropriate level of oxygenation and amount of pitching yeast. Classical microbiological methods, such as the replication tests, take too long to meet the needs of the production. Moreover, they do not inform the brewer about growth and fermentation performances of the yeast. Measuring some aspects of metabolic activity may appear a better approach to assess yeast physiology that is directly relevant to fermentation performance. The most obvious one is to run a laboratory fermentation test that will provide information about parameters related directly to the industrial fermentation. Unfortunately, this test again is too time consuming for routine application. A practical approach is to identify a couple of tests that can be implemented in the brewery environment and to perform them in a consistent manner. This will allow the collection of data that can be integrated with those of the industrial fermentations, and, which eventually can be used by the brewer to identify trends in the performance of the yeast. Such analysis will serve to set series of criteria to decide to use or to reject a specific yeast batch. The acidification power test or the measurement of the specific oxygen uptake rate are rapid, and can be used with readily available equipment. A typical example of specific oxygen uptake rate and the corresponding maximal rate of fermentation for successive fermentations is presented in Fig. 13.8-1 (J.L. VAN HAECHT and J.P. DUFOUR, unpubl. observ.).

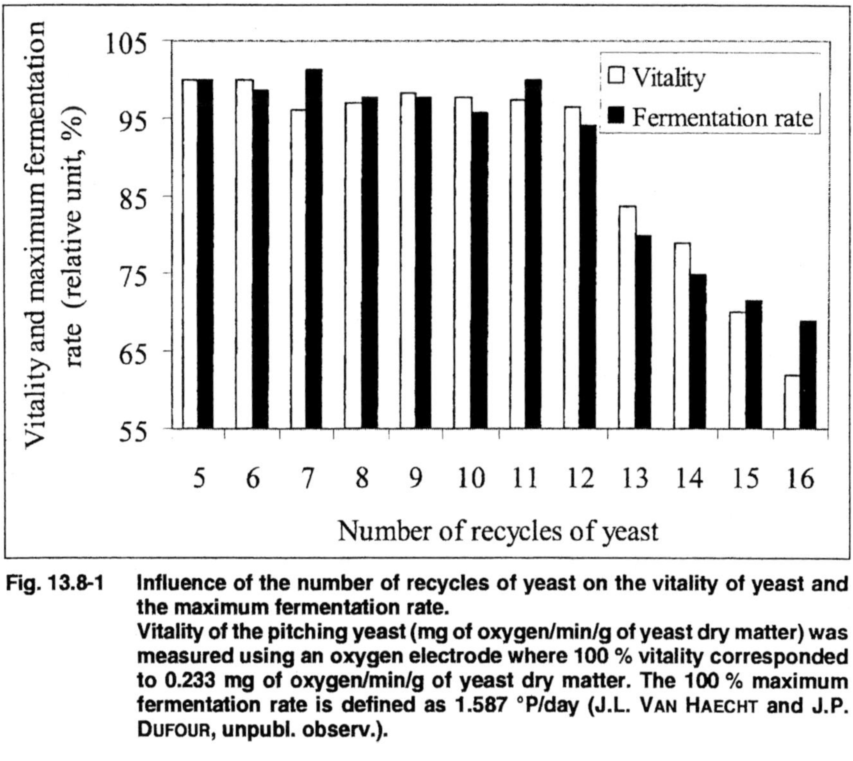

Fig. 13.8-1 Influence of the number of recycles of yeast on the vitality of yeast and the maximum fermentation rate.
Vitality of the pitching yeast (mg of oxygen/min/g of yeast dry matter) was measured using an oxygen electrode where 100 % vitality corresponded to 0.233 mg of oxygen/min/g of yeast dry matter. The 100 % maximum fermentation rate is defined as 1.587 °P/day (J.L. VAN HAECHT and J.P. DUFOUR, unpubl. observ.).

13.8.2 Microbial contamination

The microbiological threats during the brewing process are contamination with bacteria and contamination with a brewing (e. g., an ale yeast contaminated with a lager yeast) or a wild yeast. The presence of contaminants can potentially affect yeast fermentation performances and give a product with unacceptable flavours. Contamination of the fermentation with wild yeasts could also be a potential cause of failure of pasteurization. Unlike brewing yeasts, many wild yeasts form ascospores (e. g., *Dekkera* and *Hanseniaspora* spp.), which are more heat resistant than the vegetative cells. The latter cells are the target of the routine pasteurization programs. Consequently, the ascospore-forming contaminants may survive and creates haze and off-flavours in the beer during storage.

The microbiological control can not be done without looking at the entire brewing process. An excellent review of all aspects of brewing microbiology has been published by PRIEST and CAMPBELL [67]. A series of recommended sampling and microbiological methods have been published for routine quality control [2, 30].

For the purpose of this chapter, the main critical points are the yeast propagation, the fermentation and yeast handling. As mentioned before, the pitching yeast must be examined before use to ensure good viability and vitality. However, it is as important to check for the absence of contaminating bacteria and wild yeasts. The use of closed fermenters has considerably reduced the risk of pitching yeast contamination, but it still exists. Traditionally, detection of a contaminant is based on the inoculation of a sample to a specific medium under defined culture conditions, i. e., temperature, and length of incubation. Unfortunately, there is no single medium that allows the growth of all wild yeasts, while suppressing the growth of the brewing strains. Media routinely used are the synthetic medium using lysine as the sole source of nitrogen for the detection of non-*Saccharomyces* strains, a medium containing dextrins or starch as a sole source of carbon for the detection of *S. cerevisiae* var. *diastaticus* (= *Saccharomyces cerevisiae*), and a medium containing actidione to trace non-*Saccharomyces* yeasts. INGLEDEW and CASEY [40] have provided a comprehensive survey of the culture media for detection and isolation of wild *Saccharomyces* yeasts. In most instances, however, the detection of wild *Saccharomyces* contaminants remains a serious problem. Immunological methods using immunofluorescence could be a solution when no selective medium exists, provided the appropriate antiserum is available.

Recent developments in molecular biology and DNA technology have led to new methods for the detection and identification of specific contaminants (for reviews, see [35, 70], see also Chapters 2, 3 and 4). One of the disadvantages of the methods based on DNA analysis, such as the polymerase chain reaction (PCR), is their inability to distinguish between cells that are dead and alive.

Prevention of contamination is the keystone to efficiently control the microbiological quality of yeast. The implementation of very strict hazard analysis and critical control points (HACCP) programs in the breweries today have largely contributed to the consistency of the pitching yeast and consequently of the product quality.

13.9 Conclusions

Fermentation remains one of the most fascinating and challenging processes in the production of beer. Nowadays, the sensory implications of yeast metabolism are well documented. The biochemical reactions and metabolic pathways related to the conversion of wort into beer have been widely studied and described. The mechanisms leading to the formation and/or removal of flavour-active compounds are more or less elucidated. Consequently, one could think that the brewer is in a better position to adjust the process parameters in order to reach the desired levels of flavour compounds in the beer at the end of fermentation and maturation. However, many questions still remain opened regarding most of the regulatory mechanisms involved and yeast physiology. Because of the gaps in our knowledge, the brewer needs to constantly adjust to the changing mood of the yeast.

The current brewing industry drivers for yeast research have been recently reviewed by PAJUNEN [60]. They focus on quality, consistency, costs, control and development. Quality issues include all aspects of fermentation related to beer flavour and require a more detailed understanding of yeast growth, fermentation and energy metabolism. Consistency of the beer quality should be obtained from batch to batch. Consequently, an understanding of how the yeast physiological state degenerates is required to help the brewer to implement procedures that guarantee a consistent performance from one fermentation to another. Improving the fermentation rates or shortening the lag phase at the start of the fermentation are key issues in terms of fermentation costs, due to a better efficiency in the utilization of the equipment. Process control is very critical to produce consistent quality at the lowest cost possible. This requires the availability of tools to control yeast vitality, such as metabolic markers. As stated by PAJUNEN [60]: 'what you cannot measure and monitor, you cannot control'. Finally, brewers are always attentive to technological developments that could lead to improved consistency and quality. In an era of major scientific achievements in yeast knowledge (genomics and proteomics), one has all the tools to develop yeast strains with ideal properties for a given brewing process and product. Developments in the use of genetically modified yeasts with improved properties, however, has now been put on hold because of the consumer's attitude. We think it is very unlikely that modified yeast will find their way into the brewing process in the near future.

13.10 References

[1] AHVENAINEN, J.: Lipid composition of aerobic and anaerobically propagated brewer's bottom yeast. J. Inst. Brew. **88** (1982) 367–370.

[2] American Society of Brewing Chemists, Methods of Analysis, 8th ed. Amer. Soc. Brew. Chem., St. Paul, U.S.A. (1992).

[3] ANDERSEN, T.H.; HOFFMAN, L.; GRIFONE, R.; NILSSON-TILLGREN, T.; KIELLAND-BRANDT, M.C.: Brewing yeast genetics. In: Yeast physiology – a new era of opportunity, Eur. Brew. Conv. Symp., Monograph 28. Fachverlag Hans Carl, Nürnberg, Germany (2000) 140–147.

[4] ANDRIES, M.; VAN BEVEREN, P.C.; GOFFIN, O.; RAJOTTE, P.; MASSCHELEIN, C.A.: First results on semi-industrial continuous top fermentation with the Meura-Delta immobilised yeast reactor. Techn. Quart. Master Brew Assoc. Amer. **34** (1997) 119–122.

[5] BARNETT, J.A.: The taxonomy of the genus *Saccharomyces* Meyen ex Rees: a short review for non taxonomists. Yeast **8** (1992) 1–23.

[6] BARNETT, J.A.; PAYNE, R.W.; YARROW, D.: Yeasts, characteristics and identification, 2nd ed. Cambridge, U.K.: Cambridge University Press (1990).

[7] BARNETT, J.A.; PAYNE, R.W.; YARROW, D.: Yeasts, characteristics and identification, 3rd ed. Cambridge, U.K.: Cambridge University Press (2000).

[8] BENDIAK, D.: Review of metabolic activity tests and their ability to predict fermentation performance. In: Brewing yeast performance (edited by Smart, K.). Oxford, U.K.: Blackwell Science (2000) 34–45.

[9] BOULTON, C.; QUAIN, D.: Brewing yeast and fermentation. Oxford, U.K.: Blackwell Science (2001).

[10] BRENNER, R.R.: Effect of unsaturated acids on membrane structure and enzyme kinetics. Progr. Lipid Res. **23** (1984) 69–96.

[11] CARPENTIER, B.; VAN HAECHT, J.L.; DUFOUR, J.P.: Influence of the trub content of the pitching wort on yeast by-products synthesis. In: Proc. 3rd Scient. Technic. Symp., Inst. Brewing, Central and Southern African section (Victoria Falls, Zimbabwe), Institute of Brewing, Central and Southern African section, Sandton, South Africa (1991) 144–149.

[12] CASEY, G.P.: Practical application of pulse field electrophoresis and yeast chromosome fingerprinting in brewing QA and R&D. Techn. Quart. Masters Brew. Assoc. Amer. **33** (1996) 1–10.

[13] CODON, A.C.; BENITEZ, T.; KORHOLA, M.: Chromosomal polymorphism and adaptation to specific industrial environments of *Saccharomyces* strains. Appl. Microbiol. Biotechnol. **49** (1998) 154–163.

[14] COLIN, S.; MONTESINOS, M.; MEERSMAN, E.; SWINKELS, W.; DUFOUR, J.P.: Yeast dehydrogenase activities in relation to carbonyl compounds removal from wort and beer. In: Proc. 23rd Congr. Eur. Brew. Conv. (Lisbon, Portugal). Oxford, U.K.: Oxford University Press (1991) 409–416.

[15] COUTTS, M.W.: Process for the manufacture of beer, ale and the like. New Zealand: United States Patent 3,234,026 (1956).

[16] CRIDDLE, R.S.; SCHATZ, G.: Promitochondria of anaerobically grown yeast. I. Isolation and biochemical properties. Biochem. **8** (1969) 322–334.

[17] CRUMPLEN, R.; D'AMORE, T.; SLAUGHTER, C.; STEWART, G.G.: Novel differences between ale and lager brewing yeasts. In: Proc. 24th Congr. Eur. Brew. Conv. (Oslo, Norway), Oxford, U.K.: Oxford University Press (1993) 267–274.

[18] DAVIES, A.W.: Continuous fermentation – 30 years on. In: Proc. 20th Conv. Inst. Brew., Australia and New Zealand section,Brooklyn Park, South Australia (1988) 159–165.

[19] DONNELLY, D.; HURLEY, J.: Yeast monitoring: the Guinness experience. Ferment **9** (1996) 283–286.

[20] DUFOUR, J.P.: Influence of industrial brewing and fermentation working conditions on beer sulphur dioxide level and flavour stability. In: Proc. 23rd Eur. Brew. Conv. (Lisbon, Portugal). Oxford, U.K.: Oxford University Press (1991) 209–216.

[21] DUFOUR, J.P.; CARPENTIER, B.; KULAKUMBA, M.; VANHAECHT, J.L.; DEVREUX, A.: Alteration of SO_2 production during fermentation. In: Proc. 22nd Eur. Brew. Conv. (Zurich, Switzerland). Oxford, U.K.: Oxford University Press (1989) 331–338.

[22] DUFOUR, J.P.; DEVREUX, A.: The use of amino acids analysis as a tool to control malting and brewing processes. In: Wort production, Eur. Brew. Conv. Symp., Monograph XI. Nürnberg, Germany: Verlag Hans Carl (1986) 227–249.

[23] DUFOUR, J.P.; MALCORPS, P.: Ester synthesis during fermentation: enzyme characterization and modulation mechanism. In: Proc. 4th Aviemore Confer. on Malting, Brewing and Distilling (edited by CAMPBELL, I.; PRIEST, F.G.). London, U.K.: Institute of Brewing (1995) 137–151.

[24] DUFOUR, J.P.; MALCORPS, P.; SILCOCK, P.: Control of ester synthesis during brewery fermentation. In: Brewing yeast fermentation performance (edited by Smart, K.). Oxford, U.K.: Blackwell Science, in press.

[25] DUFOUR, J.P.; WIERDA, R.; LEUS, M.; LISSENS, G.; DELVAUX, F.; DERDELINCKX, G.; LARSEN, D.: Quantitative analysis of beer aromatic alcohols using stable isotope dilution assay. J. Amer. Soc. Brew. Chem. **60** (2002) 88–96.

[26] ENARI, T.M.: Amino acids, peptides and proteins of wort. In: Eur. Brew. Conv. Symp., Monograph I (Zeist, The Netherlands), European Brewery Convention, Rotterdam, The Netherlands (1974) 73–89.

[27] ENGAN, S.: Beer composition: volatile substances. In: Brewing science, Vol 2. (edited by POLLOCK, J.R.A.). London, U.K.: Academic Press (1981) 93–165.

[28] ENGASSSER, J.M.; MARC, I.; MOLL, M.; DUTEURTE, B.: Kinetics modeling of beer fermentation. In: Proc. 18th Eur. Brew. Conv. (Copenhagen, Denmark), IRL Press, London, U.K. (1981) 579–586.

[29] ERNANDES, J.R.; WILLIAMS, J.W.; RUSSEL, I.; STEWART, G.G.: Respiratory deficiency in brewing yeasts – effects on fermentation, flocculation and beer flavour components. J. Amer. Soc. Brew. Chem. **51** (1993) 16–20.

[30] European Brewery Convention, Analytica-Microbiologica-EBC, CD-Rom. Nürnberg, Germany: Fachverlag Hans Carl (2001).

[31] GILLILAND, R.B.: Wild yeast spoilage. Brewers' Guardian **96** (1967) 37–45.

[32] GIUDICI, P.; CAGGIA, C.; PULVIRENTI, A.; RAINIERI, S.: Karyotyping of *Saccharomyces* strains with different temperature profiles. J. Appl. Microbiol. **84** (1998) 811–819.

[33] GOFFEAU, A.; BARRELL, B.G.; BUSSEY, H.; DAVIS, R.W.; DUJON, B.; FELDMANN, H.; GALIBERT, F.; HOHEISEL, J.D.; JACQ, C.; JOHNSTON, M.; LOUIS, E.J.; MEWES, H.W.; MURAKAMI, Y.; PHILIPPSEN, P.; TETTELIN, H.; OLIVER, S.G.: Life with 6000 genes. Science **274** (1996) 546, 563–567.

[34] GRENSON, M.: Amino acid transporters in yeast: structure, function and regulation. New Compr. Biochem. **21** (1992) 219–245.

[35] GUTTERIDGE, C.S.; PRIEST, F.G.: Methods for the rapid identification of microorganisms. In: Brewing microbiology (edited by PRIEST, F.G.: CAMPBELL, I.). London, U.K.: Chapman and Hall (1996) 237–270.

[36] HAMMOND, R.M.: Brewer's yeast. In: The yeasts, Vol. 5 (edited by ROSE, A.H.: HARRISON, J.S.). New York, U.S.A.: Academic Press (1993) 7–67.

[37] HAMMOND, R.M.: Genetically-modified brewing yeasts for the 21st century. Progress to date. Yeast **11** (1995) 1613–1627.

[38] HAMMOND, R.M.: Yeast genetics. In: Brewing microbiology (edited by PRIEST, F.G.; CAMPBELL, I.). London, U.K.: Chapman and Hall (1996) 43–82.

[39] HOFFMAN, L.: The defective sporulation of lager brewing yeast. Ph.D. thesis, Copenhagen, Denmark: University of Copenhagen (1999).

[40] INGLEDEW, W.M.; CASEY, G.P.: The use and understanding of media used in brewing mycology. I. Media for wild yeast. Brewers Digest **57** (1982) 18–20, 22.

[41] JAKOBSEN, M.; THORNE, R.S.W.: Oxygen requirements of brewing strains of *Saccharomyces uvarum* (*carlsbergensis*)-bottom fermentation yeast. J. Inst. Brew. **86** (1980) 284–287.

[42] JIBIKI, M.; ISHIBIKI, T.; YUUKI, T.; KAGAMI, N.: Application of polymerase chain reaction to determine the flocculation properties of brewer's lager yeast. J. Amer. Soc. Brew. Chem. **59** (2001) 107–110.

[43] KIRSOP, B.H.: Oxygen in brewery fermentation. J. Inst. Brew. **80** (1974) 252–259.

[44] KURTZMAN, C.P.; FELL, J.W.: The yeasts – A taxonomic study, 4th ed. (edited by KURTZMAN, C.P.: FELL, J.W.). Amsterdam, The Netherlands: Elsevier (1998).

[45] LAGUNAS, R.: Sugar transport in *Saccharomyces cerevisiae*. FEMS Microbiol. Rev. **104** (1993) 229–242.

[46] MASON, A.B.; DUFOUR, J.P.: Alcohol acetyltransferases and the significance of ester synthesis in yeast. Yeast **16** (1999) 1287–1298.

[47] MASSCHELEIN, C.A.: Role of fatty acids in beer flavour. In: Flavour, Eur. Brew. Conv. Symp., Monograph VII. Nürnberg, Germany: Brauwelt-Verlag (1982) 211–236.

[48] MASSCHELEIN, M.: Intrinsic characterization of brewing yeast. Techn. Quart. Masters Brew. Assoc. Amer. **4** (1967) 159–167.

[49] MCMURROUGH, I.: Scope and limitations for immobilized cell systems in the brewing industry. In: Immobilized yeast applications in the brewing industry. Eur. Brew. Conv. Symp., Monograph 24. Nürnberg, Germany: Verlag Hans Carl Getränke-Fachverlag (1996) 2–12.

[50] MEILGAARD, M.C.: Flavour chemistry of beer. I. Flavour interaction between principal volatiles. Techn. Quart. Masters Brew. Assoc. Amer. **12** (1975) 107–17.

[51] MEILGAARD, M.C.: Flavour chemistry of beer. Part II: Flavour and threshold of 239 aroma volatiles. Techn. Quart. Masters Brew. Assoc. Amer. **12** (1975) 151–173.

[52] MIETH, H.O.: Immobilised yeast plants for alcohol free beer production and rapid maturation. In: Proc. 5th Congr. Inst. Brew., Central and Southern African section (Victoria Falls, Zimbabwe), Institute of Brewing, Central and Southern African section, Hout Bay, South Africa (1995) 166–172.

[53] MORRISON, K.B.; SUGGETT, A.: Yeast handling, petite mutants, and lager flavour. In: Proc. 19th Congr. Eur. Brew. Conv. (London, U.K.). Oxford, U.K.: Oxford University Press (1983) 489–496.

[54] MUNROE, J.: Fermentation. In: Handbook of brewing (edited by HARDWICK, W.A.). New York, U.S.A.: Marcel Dekker (1995) 323–353.

[55] NAKATANI, K.; TAKAHASHI, T.; NAGAMI, K.; KUMADA, J.: Kinetic studies of vicinal diketones in brewing. 2. Theoretical aspects for the formation of total vicinal diketones. Techn. Quart. Masters Brew. Ass. Amer. **21** (1984) 175–183.

[56] NAUMOV, G.I.: Genetic identification of biological species in the *Saccharomyces* sensu stricto complex. J. Ind. Microbiol. Biotechnol. **17** (1996) 295–302.

[57] NES, W.D.; JANSSEN, G.G.; CRUMLEY, F.G.; KALINOWSKA, M.; AKIHISA, T.: The structural requirements of sterols for membrane function in *Saccharomyces cerevisiae*. Arch. Biochem. Biophys. **300** (1993) 724–733.

[58] NEVEN, H.; DELVAUX, F.; DERDELINCKX, G.S.: Flavour evolution of top fermented beers. Techn. Quart. Master Brew. Ass. Amer. **34** (1997) 115–118.

[59] O'CONNOR-COX, E.S.C.; LODOLO, E.J.; AXCELL, B.C.: Mitochondrial relevance to yeast fermentative performance: A review. J. Inst. Brew. **102** (1996) 19–25.

[60] PAJUNEN, E.: Current brewing industry drivers for yeast research. In: Yeast Physiology – a new era of opportunity, Eur. Brew. Conv. Symp., Monograph 28. Nürnberg, Germany: Fachverlag Hans Carl (2000) 12–14.

[61] PAJUNEN, E.; GRONQVIST, A.: Immobilized yeast lager beer maturation: DEAE-cellulose at Synebrychoff. In: Immobilized yeast applications in the brewing industry, Eur. Brew. Conv. Symp. Monograph 24. Nürnberg, Germany: Verlag Hans Carl Getränke-Fachverlag (1996) 24–40.

[62] PASTEUR, L.: Etudes sur la bière, ses maladies, causes qui les provoquent, procédé pour la rendre inaltérable, avec une théorie nouvelle de la fermentation. Paris, France: Gauthier-Villars (1876).

[63] PEDERSEN, M.B.: DNA sequence polymorphism in the genus *Saccharomyces* III. Restriction endonuclease fragment patterns of chromosomal regions in brewing and other yeast strains. Carlsberg Res. Commun. **51** (1986) 163–183.

[64] PERPETE, P.; CUTSEM, P.V.; BOUTTE, C.; COLSON-CORBISIER, A.M.; COLLIN, S.: Amplified fragment-length polymorphism, a new method for the analysis of brewer's yeast DNA polymorphism. J. Amer. Soc. Brew. Chem. **59** (2001) 195–200.

[65] PIERCE, J.S.: The role of nitrogen in brewing. J. Inst. Brew. **93** (1987) 378–381.

[66] PORTNO, A.D.: Continuous fermentation in the brewing industry – the future outlook. In: Fermentation and storage, Eur. Brew. Conv. Symp., Monograph V. Zouterwoude, The Netherlands: European Brewery Convention (1978) 145–154.

[67] PRIEST, F.G.; CAMPBELL, I.: Brewing microbiology, 2nd ed. (edited by PRIEST, F.G.: CAMPBELL, I.). London, U.K.: Chapman and Hall (1996).

[68] RAMOS-JEUNEHOMME, C.I.; HABOUCHA, J.; DEVREUX, A.; MASSCHELEIN, C.A.: Inhibition catabolique chez les levures et consequences industrielles. In: Proc. 14th Congr. Eur. Brew. Conv. (Salzburg, Austria). Amsterdam, The Netherlands: Elsevier Scientific Publishing Company (1974) 111–124.

[69] RODRIGUES DE SOUSA, H.; MADEIRA-LOPES, A.; SPENCER-MARTINS, I.: The significance of active fructose transport and maximum temperature for growth in the taxonomy of *Saccharomyces sensu stricto*. Syst. Appl. Microbiol. **18** (1995) 44–51.

[70] RUSSELL, I.; DOWHANICK, T.: Rapid detection of microbial spoilage. In: Brewing microbiology (edited by PRIEST, F.G.: CAMPBELL, I.). London, U.K.: Chapman and Hall (1996) 209–235.

[71] RUSSELL, I.; STEWART, G.G.: Valuable techniques in the genetic manipulation of industrial yeast strains. J. Amer. Soc. Brew. Chem. **43** (1985) 84–90.

[72] SATO, M.; WATARI, J.; SHINOTSUKA, K.: Genetic instability in flocculation of bottom-fermenting yeast. J. Amer. Soc. Brew. Chem. **59** (2001) 130–134.

[73] SPAEPEN, M.; VERACHTERT, H.: Esterase activity in the genus *Brettanomyces*. J. Inst. Brew. **88** (1982) 11–17.

[74] STEWART, G.G.; RUSSEL, I.: Brewer's yeast. In: Food biotechnology (edited by HUI, Y.H.: KHACHATOURIANS, G.G.). New York, U.S.A.: VCH Publishers (1995) 847–871.

[75] STEWART, G.G.; ZHENG, X.; RUSSEL, I.: Wort sugar uptake and metabolism – the influence of genetic and environmental factors. In: Proc. 25th Congr. Eur. Brew. Conv. (Brussels, Belgium). Oxford, U.K.: Oxford University Press (1995) 403–410.

[76] STRATTON, M.K.; CAMPBELL, S.J.; BANKS, D.J.: Process developments in the continuous fermentation of beer. In: Proc. 23rd Inst. Brew. Conv., Asia Pacific section. Sydney, Australia: (1994) 96–100.

[77] VAN DIEREN, B.: Yeast metabolism and the production of alcohol-free beer. In: Immobilized yeast applications in the brewery industry. Eur. Brew. Conv. Symp., Monograph 24. Nürnberg, Germany: Verlag Hans Carl Getränke-Fachverlag (1996) 66–76.

[78] VAN HAECHT, J.L.; DUFOUR, J.P.: The production of sulphur compounds by brewing yeast: A review. Cerevisia **20** (1995) 51–64.

[79] VAN OEVELEN, D.; SPAEPEN, M.; TIMMERMANS, P.; VERACHTERT, H.: Microbiological aspects of spontaneous wort fermentation in the production of Lambic and Gueuze. J. Inst. Brew. **83** (1977) 356–60.

[80] VERACHTERT, H.; DAWOUD, E.: Yeast in mixed cultures. Louvain Brew. Lett. **3** (1990) 15–40.

[81] VERSTREPEN, K.J.; BAUER, F.F.; WINDERICKX, J.; DERDELINCKX, G.; DUFOUR, J.P.; THEVELEIN, J.M.; PRETORIUS, I.S.; DELVAUX, F.R.: Genetic modification of *Saccharomyces cerevisiae*: fitting the modern brewer's needs. Cerevisia **26** (2001) 89–97.

[82] WACKERBAUER, K.; METHNER, F.-J.: The microorganisms of "Berliner Weissbier" and their influence on the beer flavour. Brauwelt Int. (1988) 382–388.

[83] WALKER, M.D.; SIMPSON, W.J.: Production of volatile sulphur compounds by ale and lager brewing strains of *Saccharomyces cerevisiae*. Lett. Appl. Microbiol. **16** (1993) 40–43.

[84] WALSH, R.M.; MARTIN, P.A.: Growth of *Saccharomyces cerevisiae* and *Saccharomyces uvarum* in a temperature gradient incubator. J. Inst. Brew. **83** (1977) 169–172.

[85] WATARI, J.; SATO, M.; OGAWA, M.; SHINOTSUKA, K.: Genetic and physiological instability of brewing yeast. In: Yeast physiology – a new era of opportunity. Eur. Brew. Conv. Symp., Monograph 28. Nürnberg, Germany: Fachverlag Hans Carl (2000) 148–158.

[86] WIGHTMAN, P.; QUAIN, D.E.; MEADEN, P.G.: Analysis of production brewing strains of yeast by DNA fingerprinting. Lett. Appl. Microbiol. **22** (1996) 90–94.

[87] YAMAUCHI, Y.; KASHIHARA, T.; MURAYAMA, H.; NAGARA, A.; OKAMOTO, T.; MAWATARI, M.: Scale-up of immobilised bioreactor for continuous fermentation of beer. Techn. Quart. Master Brew. Assoc. Amer. **31** (1994) 90–94.

[88] YOUNG, T.W.: Killer yeasts. In: The yeasts, Vol. 2 (edited by Rose, A.H.: Harrison, J.S.). London, U.K.: Academic Press (1987).

14 Wine yeasts

SYLVIE DEQUIN, JEAN-MICHEL SALMON, HUU-VANG NGUYEN and BRUNO BLONDIN

14.1 Introduction

Yeasts play a key role in wine making in performing the alcoholic fermentation of grape musts sugars. During this fermentation not only hexoses are converted to ethanol and carbon dioxide, but many compounds are removed from the medium and a large set of by-products are formed that influence the sensory properties of wines. In classical wine production (dry wines) a good fermentation leads to a complete fermentation of the sugars in a reasonable time without the production of off-flavours. Because of its strong fermentation capacity in grape musts and its ability to produce wines with a pleasant "fermentation bouquet", *Saccharomyces cerevisiae* is the preferred yeast in winemaking and is often designed as the "wine yeast". In spontaneous wine fermentations other yeast species contribute to the process and have a variable and poorly predictable impact on the quality of the product. In the past 30 years most of the wine industry has moved from spontaneous to inoculated fermentations that are more reliable and facilitate wine processing. *Saccharomyces cerevisiae* yeast starters, which are available as active dried yeasts, represent a significant market where more than 150 strains are used for a total volume of about 1500 tons per year.

Grape must fermentations are characterized by a high sugar content (140–260 g l^{-1}), a low pH (3.0–3.5), anaerobiosis, the presence of added sulphites (40–80 mg l^{-1}) and often limiting amounts of nitrogen and lipids (reviewed in [31, 44, 80]). The reliability of alcoholic fermentation is still strongly limited by variation in grape juice composition. Nutrient disequilibria may affect the fermentation capacity of the yeast and lead to sluggish or stuck fermentations or to an excessive release of unsuitable compounds detrimental for the wine quality [12, 31]. Our knowledge on the physiology of yeasts under such conditions and their response to various nutritional and environmental factors has made considerable progress in the last 20 years. This permits a better management of the alcoholic fermentations, but further developments are required to both improve the processes and better specify targets for strains selection or genetic engineering.

Wine yeasts have been primarily selected on the bases of strong growth and fermentation capacity, ethanol tolerance and limited production of undesirable compounds, such as H_2S and acetate. As a result of a better understanding of the interactions between yeasts and the fermenting medium, as well as the impact of yeasts on the quality of the wine, new criteria for yeast selection have emerged and there is a quest for better adapted strains to a given wine style or wine processing condition. While some of these objectives could be reached by classical strain selection, the transfer of some specific attributes will only be obtained by gene transfer technology. Several major achievements have demonstrated the potential of these technologies to improve the properties of wine yeasts. This chapter reviews the role of yeasts in winemaking with emphasis on the physiology of *S. cerevisiae* during alcoholic fermentation and on the progress in strain improvement by genetic engineering.

14.2 Yeast biodiversity related to grapes and wines fermentations

Traditional wine fermentation relied on the activity of indigenous wine yeast's present in grape-musts. The "wine yeast" par excellence, *S. cerevisiae* and related species, was early shown to play a major role in the fermentation of grape-musts sugars. In the course of the 20th century a series of ecological surveys performed in various winemaking areas have established that next to *S. cerevisiae* many other yeast species contributed to the spontaneous microflora of the must [21, 32, 44, 62]. Yeasts of the genera *Kloeckera, Hanseniaspora, Rhodotorula, Hansenula, Candida, Metschnikowia* and *Debaryomyces*, were frequently isolated from grapes or musts.

In the vineyard, yeasts may be transported from the soil to the grapes by various insects (e. g., *Drosophila* spp., honey bees and wasps) carrying them on their body or by the wind. The yeast microflora of grapes is dominated by the presence of oxidative or poorly fermentative yeasts. *Kloeckera apiculata,* or its ascospore forming equivalent *Hanseniaspora uvarum,* often account for more than 60 % of the total yeast cell population. Surprisingly, *S. cerevisiae* occurs rarely on grapes and this led some authors to the provocative suggestion that *S. cerevisiae* may not be present on grapes [51]. Because of their low number *S. cerevisiae* strains are difficult to isolate from grapes by direct plating, but are regularly found if enrichment procedures are applied. The origin of *S. cerevisiae* responsible for spontaneous industrial fermentations is still a question in debate. Due to their scarcity on grapes, some authors consider that the wineries are more likely the natural store than the vineyard. Indeed, *S. cerevisiae* has been isolated from cellar walls and from the surfaces of equipment [62]. On the other hand, MORTIMER and POLSINELLI [53] have recently observed that damaged grape berries are rich depositories of *S. cerevisiae* and may be sufficient to provide inocula of 10^2–10^3 cells/ml of must. This could account for a vineyard origin of *S. cerevisiae*. Depending on enological situations and practices the importance of each yeast source may vary considerably. The harvesting method (i. e., hand picking or mechanical), the temperature of the grapes, the transportation time, the amount of SO_2, and the handling method of the must, have all an influence on the dynamic of the yeast microflora on the grapes. In any case, a few days after the beginning of the harvest of the grapes, *S. cerevisiae* has colonized the winery equipment and all sugar-coated surfaces so that inoculation by the winery material is important. Indeed winemakers have often observed that spontaneous fermentations start much more quickly several days after harvest than at the start.

Grape-musts are strongly selective conditions for microorganisms because of the low pH (2.9 to 3.7), the high sugar content, the presence of sulphites (40–80 mg l^{-1}) and anaerobiosis. Under these conditions yeasts efficiently compete with other micro-organisms present in musts, such as moulds, and lactic and acetic acid bacteria. Spontaneous fermentations of grape-musts are initially carried out by the abundant apiculate yeast species *Hanseniaspora uvarum* (= *Kloeckera apiculata*). Yeasts of the species *Metschnikowia pulcherrima* (= *Candida pulcherrima*), *Debaryomyces hansenii* (= *Candida famata*), *Candida stellata* or *Pichia* spp. may also contribute to a significant extent at the beginning of the fermentation. The apiculate yeasts usually dominate the fermentation until the ethanol level reaches 3–5 % when they are outgrown by *S. cerevisiae*. Some enological practices, such as low-temperature fermentations, may increase the contribution of non-*Saccharomyces* yeasts [32]. Depending on the extent of their development the non-*Saccharomyces* yeasts can influence the chemical composition of wines and their sensory quality. The impact of these yeasts can be detrimental for the quality of the wine because of their frequent ability to produce high amounts of acetic acid, ethyl acetate ester or other unsuitable compounds. However, some winemakers still use spontaneous fermentations because they consider that these "wild" yeasts may add aromatic complexity to the wines.

Although *S. cerevisiae* ultimately dominates almost all spontaneous fermentations, it was recently shown that some low-temperature fermentations were mainly caused by the related species *S. bayanus*, in particular the variety *uvarum* (*S. uvarum*) [54]. The contribution of *S. uvarum* (= *S. bayanus*) is probably restricted to cool winemaking areas and practices, and is obviously connected to the cryotolerance of this species. Since such yeasts display some specific aromatic properties their involvement is expected to give distinct sensory profiles to the wines.

One prevailing idea in the wine industry was that a given area could harbour a dominant yeast strain with specific attributes responsible for the quality of the corresponding wines. Several ecological surveys have established that the populations of *S. cerevisiae* in the vineyard were essentially polyclonal. It has also been observed that wineries could harbour a dominant strain for a while, but that this dominance was temporary [52]. Therefore, "territorial" yeasts do not seem to exist.

When inoculating grape musts with selected yeasts, one has to take into account the existence of a natural but variable yeast microflora. The implantation of an exogenous yeast strain in a must is not guaranteed, since it has to compete with the natural microflora which can be numerically important (10^3 to 10^6 cells ml^{-1}). The success depends both on the vigour of the inoculated yeast and on the numerical ratio between exogenous and indigenous yeasts. Labelled strains, which harbour resistance markers to mitochondrial inhibitors (mitochondrial, spontaneous or induced mutants), can be easily monitored and are helpful to control the efficiency of inoculation [92].

14.3 Beneficial aspects of wine yeasts

Complete fermentation of grape juice by *S. cerevisiae* leads to the production of 8 to 15 % (v/v) ethanol, low amounts (< 1 g l^{-1}) of several fermentation by-products such as glycerol, organic acids (acetate and succinate), and to trace amounts of alcohols and esters. Beneficial aspects of wine yeasts are mainly due to this release of molecules by the yeasts during the fermentation. Glycerol, which is the major by-product of fermentation (5 to 8 g l^{-1}), is thought to contribute to the smoothness, consistency and overall body of the wine. More than one hundred organic acids are produced by yeast cells during alcoholic fermentation. These compounds may represent up to 0.3 to 0.5 % of the fermented sugars and exert some impact on the overall organoleptic attributes of wines [60]. Succinate is one of the most abundantly formed acids, and its concentration in the final product may reach less than 1 g l^{-1}. Some of these organic acids, i. e., ketonic acids, are side products from amino acids and higher alcohols synthesis pathways. Contents of higher alcohols (n-propanol, phenylethanol, isobutyl, isoamyl, and active amyl alcohols or "fusel oils") may vary from 50 to 300 mg l^{-1} in wines [68]. Although elevated concentrations of higher alcohols are undesirable, they are thought to have a positive contribution to the global sensorial attributes of wine when present in limited amounts. One exception might be phenetyl alcohol, which has a pleasant rose flavour and whose enhanced contribution can be desired. The use of *S. uvarum* (= *S. bayanus*), which releases high amounts of this alcohol, provides a means to increase the concentration of phenetyl alcohol in wines. Esters are the most abundant aromatic compounds produced by yeasts during fermentation and are the main contributors to the bouquet of young wines. Isoamyl acetate, hexyl acetate, and ethyl caproate are thought to be the major contributors to the fruity flavour. They are produced by esterification of free alcohols by fatty acid derivatives of Coenzyme A. During fermentation, the production of acetate esters is directly linked to the availability of the alcoholic precursors and to the level of the alcohol-acetyl transferase activity within the cell [36]. On the contrary, the presence of unsaturated fatty acids in the must lowers the production of short chain fatty acids and of the correspondent esters [88].

Yeasts can also affect the aromatic properties of wines by their action on grape must compounds. The release of aromatic components from non-aromatic grape precursors has received little attention. Two heavy mercaptans involved in the sauvignon flavour, namely 4-mercapto-4-methylpentan-2-one and 3-mercaptohexan-1-ol, are liberated from S-conjugated cysteines must precursors by yeast [89]. In addition it was observed that yeast strains displayed variable abilities to release these compounds.

A last class of compounds, mannoproteins, which have no aromatic properties, are released by yeasts and are considered to have a positive effect on the quality of the wine by increasing both wine sensorial properties and its physico-chemical stability. These molecules originate from cell wall proteins and are released in small amounts during the actual fermentation, but in higher amount during wine aging on yeast lees [30].

14.4 Detrimental effect of wine yeasts

Two types of detrimental effects are associated to yeasts in wine making, namely the production of off-flavours by the fermenting yeast during alcoholic fermentation, and the alteration of finished wines due to yeast growth. During fermentations yeasts are able to produce various sulphur compounds that are detrimental to the quality of the wine. This is the case of hydrogen sulphide (H_2S) because of its characteristic rotten egg flavour, but also of various mercaptans and thioesters, which are unsuitable in wine [69]. *Saccharomyces cerevisiae* is able to assimilate most of the sulphur compounds from grapes (sulphates, sulphur amino acids, glutathione, thiamine, biotin) and to produce new sulphur compounds. The main source of sulphur in musts is sulphate (SO_4^-), which is reduced to sulphide before its incorporation in a carbon skeleton to fuel organic sulphur synthesis such as that of methionine. Sulphite added to musts is also a source of sulphur directly available for the sulphite reductase.

H_2S production depends on the availability of O-Ac-serine or O-Ac-homoserine since they are the H_2S acceptors used to synthesize cysteine or homocysteine rather then methionine. Consequently H_2S production is raised when these sequestering molecules are not present in sufficient amounts. This occurs mainly when musts are deficient in assimilable nitrogen (amino acids and ammonia), which may be overcome by the addition of nitrogen compounds [29, 38, 39]. Yeast strains vary widely in their susceptibility to produce H_2S, and a low H_2S production is a primary criterion for the selection of wine yeast strains. Yeasts can also produce hydrogen sulphide from molecular sulphur, which is frequently used as antifungal treatment in vineyards [69].

Excessive acetic acid production (> 700 mg l^{-1}) by yeasts can be considered as being problematic. Selected wine strains release usually low amounts of acetate (100–400 mg l^{-1}) during alcoholic fermentations. However, fermentation conditions influence the release of acetic acid. In particular, strongly clarified white musts are known to enhance its production. This is connected to the deficiency in lipids associated with must clarification. *S. cerevisiae* strains have a variable ability to form acetic acid. High levels of acetate may also originate from the activity of the wild yeast microflora, since some non-*Saccharomyces* species (*Kloeckera* and *Hansenula* spp.) can be strong producers. Ethyl acetate gives a vinegary character to wines and is considered as a default when its concentration exceeds 200 mg l^{-1}. Unlike *S. cerevisiae strains,* which usually are low producers, non-*Saccharomyces* yeasts can display a strong capacity to produce this ester.

Various phenolic compounds, which are usually undesirable in wine, can be formed from hydroxycinnamic acids (p-coumaric acid, ferulic acid) present in musts. *S. cerevisiae* can release 4-vinylphenol 4-vinylguaiacol by decarboxylation of the acids, but is unable to reduce them to 4-ethylphenol or 4-ethylguaiacol. These compounds give odours described as 'horsey', 'wet dog' or 'pharmaceutical' and are mainly released by *Brettanomyces/Dekkera* yeasts during storage of the wines [24]. Such alterations of wines caused by *Brettanomyces* species seem to occur rather frequently. These yeasts are also responsible of the off-flavour which results from the synthesis of various pyridines. Other spoilages of wines, such as yeast growth and sugar re-fermentation in bottles, can occur during wine storage. This can be caused by yeast species such as *S. cerevisiae* or the sulphite-resistant *Zygosaccharomyces* species.

14.5 Physiological background of wine yeasts

Wine fermentations are typically carried out at temperatures ranging between 16 and 20 °C for white wines and from 24 to 30 °C for red wines. When grape must is inoculated with active dried yeasts, growth is restricted to 5–7 generations resulting in 50–200 x 10^6 cells ml^{-1}. This corresponds to a final yeast biomass of about 1.5–6 g l^{-1} (dry weight). The growth of the cells depends on the nutrients available in the grape must, in particular on the presence of assimilable nitrogen [9], and vitamins, in particular thiamin [7, 58]. The typical evolution of the yeast cell population and fermentation rate during a wine fermentation cycle is depicted in figure 14.5-1. It is important to note that during wine fermentations stationary-phase cells ferment most of the sugars. As shown in figure 14.5-1, the fermentation rate decreases progressively throughout this phase as a result of physiological regulations and inhibition by ethanol. During alcoholic fermentations about 92–94% of the sugars are converted to ethanol and CO_2 and the remaining carbons are used for biomass and the formation of by products such as glycerol, organic acids and higher alcohols [17].

14.5.1 Sugar transport and metabolism

Grape must hexoses (glucose and fructose in equimolar amount) enter yeast cells by facilitated diffusion (carrier-mediated). Yeast can express high or low affinity transporters but

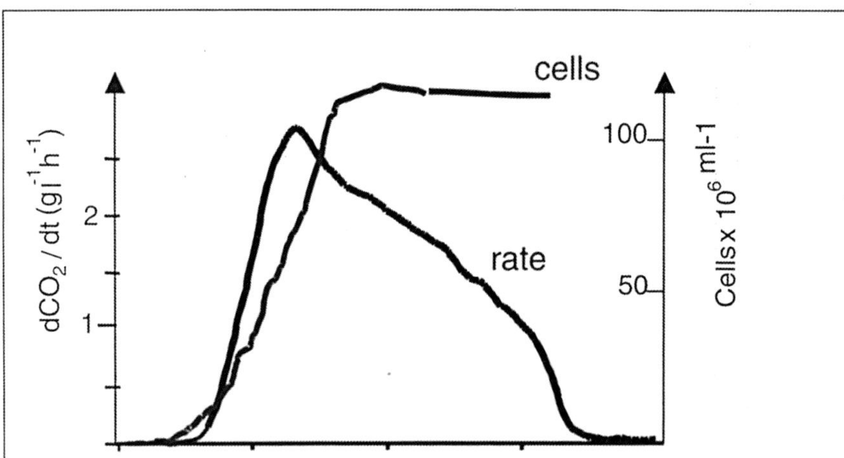

Fig. 14.5-1 Evolution of the yeast cell population and fermentation rate during a typical fermentation cycle.

due to the high amount of hexoses in grapes musts, the low affinity carriers were expected to play a key role under enological conditions [13, 61]. The kinetic properties of these carriers were characterized by REIFENBERGER et al. and indicated low affinity for hexoses (Km glucose 50–100 mM; Km fructose 100–300 mM) [70]. Amongst the multigenic family of 20 genes potentially implicated in hexose transport in yeast, only few have been identified as susceptible and playing a significant role in sugar transport under enological conditions. Recent results revealed that the low-affinity transporter Hxt3 plays a major role throughout wine fermentation, while the second major low-affinity transporter Hxt1 seems to be active only at the beginning of the fermentation [49]. In the same work, it has been shown that the high-affinity transporters Hxt7 and Hxt6 could play an important role in the completion of sugar utilization at the end of the fermentation, while the Hxt2 carrier is involved in growth initiation. The high affinity carriers are probably much more effective at the end of the fermentation than low affinity carriers, since only low amount of fructose is present at this stage and these transporters display a higher affinity for this sugar (Km about 2 mM). Several laboratories have identified that hexose uptake is the main point of control of the glycolytic flux during stationary phase, since hexose transport activity is submitted to a catabolic inactivation process when the rate of protein synthesis decreases in stationary phase after growth on glucose [18]. Under enological conditions, this assertion has been verified [83] because protein synthesis activity decreased before the maximal biomass was reached. The rate of decrease has been determined by the availability of assimilable nitrogen in the medium [84]. Whether phosphorylation of hexoses plays a role in the control of sugar utilization during fermentation is unclear. The relative affinities of the kinases for fructose and glucose are different and a shift from Hxk2 to Hxk1, which has a higher affinity for fructose, may favour fructose assimilation at the end of the fermentation. During wine fermentation, sugar phosphates are metabolized to pyruvate by the standard glycolytic pathway. It has been clearly demonstrated that overproduction of the glycolytic enzymes has no effect on the rate of ethanol production, thus suggesting that the set up of glycolytic enzyme is not rate limiting under anaerobic conditions [85]. Recent work indicates that, under enological stationary phase, the fermentation rate is directly correlated to the amount of ethanol in the medium [3]. Given that the known target of ethanol is the plasma membrane and that sugar transport is probably the rate limiting step of the carbon flux, it has been suggested that ethanol inhibition of sugar transport might be critical under such conditions.

14.5.2 Formation of by-products

During wine fermentation, glycerol production may reach 5 to 11 g l^{-1} [67]. Its concentration may vary depending on environmental factors and cultivation conditions, but the predominant factor is the yeast strain involved [71]. Glycerol is formed in two enzymatic steps from dihydroxyacetone phosphate (DHAP) involving the glycerol-3-phosphate dehydrogenase (GPDH) (EC 1.1.1.8) and a glycerol-3-phosphatase (EC 3.1.3.-). As a non-ionized molecule, this polyol can cross the plasma membrane by passive diffusion or be transported by facilitated diffusion through the MIP (Major Intrinsic Protein) protein channel Fps1p [48]. In yeast, glycerol production has a dual role in redox-balancing and in osmoregulation [40]. During alcoholic fermentation of grape must, glycerol production is essential to convert the excess of NADH, generated during biomass formation and the associated anabolism, to NAD^+. In addition, Gpd1p, which plays an important role in the response to osmotic stress, is the principal GPDH isoform expressed during wine fermentation, suggesting that glycerol is also important for combating osmotic stress in grape musts [72].

The most abundant organic acids derived from yeast metabolism are pyruvate, acetate, succinate and malate. Acetate is produced from pyruvate via the pyruvate dehydrogenase (PDH) bypass, which is an alternative route to the PDH reaction for the conversion of pyruvate to acetyl coenzyme A (CoA). Its main role is to supply the cytosol with acetyl-CoA, a precursor of lipids. A key enzyme of this bypass is the acetaldehyde dehydrogenase (ACDH) (EC 1.2.1.3), which comprises five isoforms located in different cellular compartments. While the five corresponding genes are all expressed under enological conditions, the main isoform is Ald6p, which is located in the cytosol. The expression level of *ALD6* has been shown to correlate with the level of acetate produced [71]. Succinate, malate, and other minor acids (α-ketoglutarate and citrate) derive from the tricarboxylic acid (TCA) cycle. During enological fermentation, most enzymes of the TCA cycle are subjected to glucose repression, but a limited activity remains necessary to accomplish biosynthetic demands. Under laboratory anaerobic conditions, it has been shown that the TCA cycle is operating in a branched fashion [35]. During the fermentation of the grape must, the TCA route operates as an oxydative branch leading to α-ketoglutarate and a reductive branch producing malate and leading to succinate. The succinate dehydrogenase is not functional. Succinate is produced via a fumarase reductase (EC 1.3.99.1) which has an essential role in redox balancing [20]. Other organic acids (isovaleric and isobutyric acids) are side products from amino acids and higher alcohols synthesis pathways. Several short-chain fatty acids (mainly octanoic and decanoic) are produced by *S. cerevisiae* during fermentation. These fatty acids have been suspected to exert a toxic effect on yeast strains during fermentation as they accumulate within the medium [45, 87].

Higher alcohols are produced by yeasts through transamination of normal and branched amino acids, and by decarboxylation of the correspondent ketonic acids in aldehydes. Each aldehyde may then be reduced by an alcohol dehydrogenase activity leading to the correspondent final higher alcohol. From a chemical point of view, a direct relationship between a specific amino acid and the correspondent higher alcohol exists. However, several studies have shown that no direct relation exist between the amino acids content of a must and the formation of higher alcohols by yeast during fermentation [25, 68]. Their synthesis pathway seems to play a role in the fermentative metabolism by re-oxidizing NADH, especially at the beginning of the growth [57, 90].

14.5.3 Factors affecting the fermentation capacity of the yeast

14.5.3.1 Oxygen

During fermentation, oxygen may be added in order to improve biomass synthesis, and consequently the fermentation rate, when slow fermentation is suspected [78]. Such oxygen addition is only efficient at the end of the cell growth phase [76, 77]. This molecular oxygen requirement is low and has been estimated to be 5 to 10 mg l^{-1} during enological fermentations [77]. As a matter of fact, yeast growth under strict anaerobiosis normally requires the addition of oxygen to favour the synthesis of sterols and unsaturated fatty acids. Unsaturated fatty acids are formed from saturated fatty acids by desaturation reactions involving molecular oxygen. However, during enological fermentations the use of oxygen for fatty acid desaturation is questionable since deletion of the desaturase encoding gene *OLE1* was shown to have no effect on oxygen consumption [82]. In addition, during enological fermentations, the situation is more complex since oxygen consumption by yeast cells has been attributed to the partial functioning of several mitochondrial and microsomal alternative pathways [82].

14.5.3.2 Nitrogen uptake and metabolism

Grape must contains a great variety of nitrogen compounds that represent potential nitrogen sources for yeast growth: ammonium ions ($NH4^+$), amino acids, peptides and small polypeptides. The total nitrogen content may vary between 60 and 2400 mg N l^{-1}, with 19 to 240 mg l^{-1} as $NH4^+$ depending on the grape variety and vineyard management [38]. Ammonium and amino acids enter yeast cells by the way of several different permeases [38, 80]. Few studies have been done on the regulation of amino acid transport during enological conditions. The pattern of these transport systems during enological fermentation have led various authors to classify nitrogen substrates in groups corresponding to their assimilation rank [42, 50]. It is important to notice that this classification order is different depending on the growth phase where the nitrogen is added (exponential phase [42], or stationary phase [50]). Ammonium is the preferred nitrogen source and severely represses genes involved in the uptake and catabolism of poorly utilized nitrogen sources [80, 81]. All the nitrogen compounds assimilated by yeast cells during fermentation are either incorporated into proteins or degraded into ammonium or glutamic acid [17, 46]. Both compounds can be rapidly exchanged into the cell, as they represent the main precursors of nitrogen compounds synthesis within the cell. Nitrogen and oxygen additions at the beginning of the stationary phase have been shown to be particularly effective on fermentation kinetics when sluggish fermentations are suspected [14].

Neo-synthesis of amino acids in *S. cerevisiae* requires the existence of the correspondent carbon skeleton within the cell, which is the reason for the existence of a direct relationship between intermediary compounds in the carbon metabolism and amino acids. Carbon skeletons released after degradation of amino acids and incorporated into the cell are mainly excreted into the fermentation medium after decarboxylation and/or reduction as higher alcohols [57].

14.6 Genetic improvement of wine yeasts

In the past 10 years, the demand for new, specialized wine yeast strains, that may improve the winemaking process, the quality of wine and even subtly influence the style of the wine, has been growing. The number of commercialized strains has increased from 20 to about 150. At the same time, research has been emphasized on the genetic improvement of industrial wine yeast strains. During the 1980's, genetic improvement of wine yeasts relied on classical genetic techniques (e. g., mutagenesis, hybridization, protoplast fusion, cytoduction) followed by the selection for broad traits such as fermentation performance, ethanol tolerance, absence of off-flavours. The impressive advances in yeast genetics and – genomics [34] during the past decade have opened the way to the development of approaches based on recombinant DNA technology. A new generation of specialized wine yeast strains has been developed, to improve fermentation performance and process efficiency, wine sensory quality and health benefits for the consumers [6, 15, 19, 26, 63, 64, 66].

14.6.1 Fermentation processes

Malolactic fermentation (MLF), the decarboxylation of malate to lactate, plays an essential role in the de-acidification and stabilization of wine. However, due to the poor development of lactic acid bacteria in wine, MLF remains unreliable in numerous situations, leading to scheduling problems in cellars and increased risks of wine alteration. Increasing the reliability of MLF could be achieved by using wine yeast strains able to degrade malic acid completely into lactic acid and CO_2, instead of lactic acid bacteria. Considerable progress has been made in this area, with as its key achievement the cloning of the gene of the *Lactococcus lactis* malolactic enzyme [1]. *S. cerevisiae* strains able to degrade malic acid completely into lactic acid and CO_2 have been constructed by introducing a new malate degradation pathway, composed of the malolactic gene and the malate permease from *Schizosaccharomyces pombe* [2, 16, 93]. The recombinant strains fully degraded up to 7g l^{-1} of malate in four days, simultaneously with the alcoholic fermentation and without affecting the growth properties and fermentation rate [16].

Another target of strain improvement that would be of great interest to reduce the riddling operation (remuage) in the elaboration of sparkling wines, is the introduction of the flocculent character in wine yeast strains. Flocculation is an asexual, calcium-dependent, reversible aggregation of cells into flocs. Significant progress has been made to understand the molecular and the biochemical bases of this phenomenon. It has been shown that non-flocculent strains possess inactive structural flocculation genes distributed on different chromosomes. The flocculent character has been transferred to non-flocculent wine yeasts by placing the dominant *FLO1* gene, which codes for a cell wall protein containing a lectin domain, under the control of a strong and constitutive promoter [10].

A significant amount of work has also been performed to construct yeast strains expressing a wide variety of heterologous pectinases, glucanases, xylanases, or a combination of these enzymes (reviewed by PRETORIUS [63]). This set of engineered strains expressing specific enzymes may be used to facilitate wine clarification, to improve liquefaction of grapes, thereby increasing the juice yield, or to enhance the liberation of various compounds trapped in grape skins. Consequently the bouquet and colour of the wine will improve.

14.6.2 Wine sensory quality

In this area, metabolic engineering approaches have been successfully applied to wine yeast to redirect the carbon flux towards appropriate levels of by-products. A new generation of strains that could be used in specific situations has been generated. For example, a correct balance between sugar and acidity is sometimes difficult to achieve for wines produced in hot regions. Insufficiently acidic wines are unstable, and exhibit problems of taste and colour. This can be solved using a lactic acid producing strain of *S. cerevisiae*. To this end, the lactate dehydrogenase (LDH) gene from *Lactobacillus casei* [28] (EC 1.1.1.27) has been expressed in *S. cerevisiae*. Wine produced by the recombinant strain at 5 g l^{-1} of lactic acid showed a decrease in pH of 0.2 to 0.3 units [27]. Consistent with the diversion of sugars towards lactate, the ethanol content of the wines obtained is slightly lower.

Another example is the construction of glycerol overproducing strains that may improve wine quality by providing sweetness and fullness. In addition, re-routing the carbon flux towards glycerol is expected to decrease ethanol yield. Wine yeast strains over-expressing *GPD1* encoding the *S. cerevisiae* glycerol 3-phosphate dehydrogenase (EC 1.1.1.8) on a multi-copy plasmid under the control of a *ADH1* promoter produced between 12 and 18 g l^{-1} glycerol and about 1 % (v/v) less ethanol [73]. To compensate the redox imbalance, these strains exhibited increased production of by-products, mainly acetate, 2,3 butanediol and succinate. Since the high production of acetate was a major disadvantage, an additional modification was required to properly adjust the flux. Deletion of *ALD6*, which encodes the $NADP^+$-dependent Mg^{2+}-activated cytosolic acetaldehyde dehydrogenase isoform (EC 1.2.1.3) (Fig. 14.6-1), combined to *GPD1* over-expression led to the production of high amounts of glycerol without increasing acetate formation. The redox balance was equilibrated by increased formation of succinate and 2,3-butanediol [72]. Deleting *ALD6* alone may be useful to control the level of acetate produced [71]. Although the concentration of acetic acid in wines is usually approximately 0.5 g l^{-1}, higher levels can sometimes be produced during alcoholic fermentation, depending on the yeast strain, the must composition, cultivation conditions and the winemaking process (e. g., excessive clarification) used. Inactivation *ALD6* in a wine yeast led to a two-fold reduction in the amount of acetate pro-

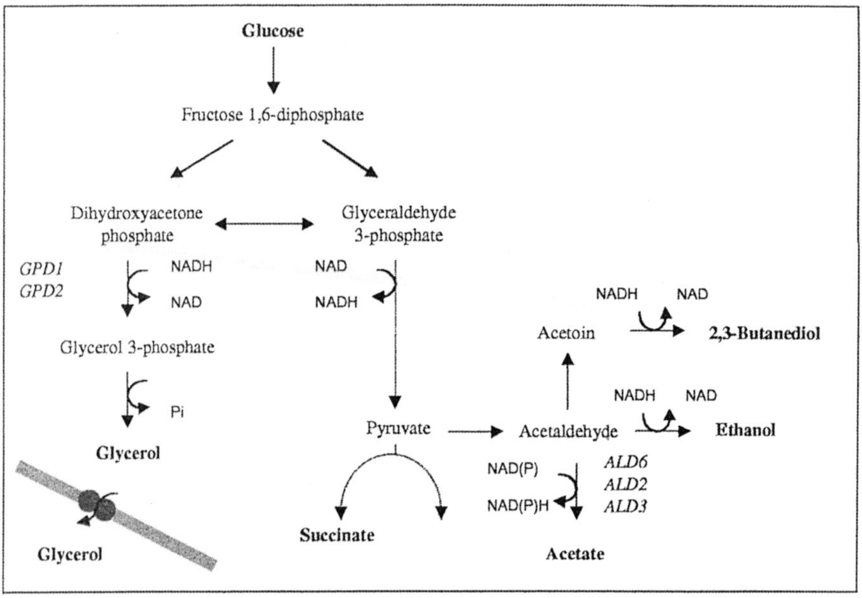

Fig. 14.6-1 Metabolic pathways of main by products involved in the redox metabolism

duced during wine fermentation. As a consequence of the resulting redox imbalance, the production of glycerol, succinate and 2,3-butanedediol is slightly increased.

Several attempts have been made to better control the production of aromas. Isoamyl acetate, produced from isoamyl alcohol, which itself is a by-product of leucine synthesis, is responsible for the banana-like aroma characteristic of young wines. Recently, LILLY et al. [47] have increased the expression level of the *ATF1* gene, which codes for the alcohol acetyltransferase (EC 2.3.1.84) that catalyses the formation of esters from acetyl CoA and the relevant alcohols. This resulted in increased esters production (ethyl acetate, isoamyl acetate and 2-phenyl acetate) affecting the flavour profile of wines. A limitation of this approach is that the production of various esters is increased, including those, such as ethyl acetate, which can be undesirable in wine above critical levels. Another approach, developed on sake yeast strains, has been to disrupt the *EST2* gene, coding for the major esterase isoamyl acetate [33]. This has resulted in a two-fold decrease in isoamyl acetate production.

Since a significant part of the wine aroma results from the hydrolyzation of non-aromatic precursors contained in grape must, attempts have been made to construct wine yeast strains with the ability to liberate a variety of aromas. For example, terpenols (e. g., linalol, geraniol) can be released from terpenyl-glycosides by a β-glucosidase cleaving the 1,6 osidic linkage. During wine fermentation, the β-glucosidases present in grapes and in yeast are not very efficient because of glucose inhibition and instability at low pH. Efforts have been focused on the characterization of more active enzymes, such a highly glucose-tolerant β-glucosidase purified from *Candida peltata* [79], or a β-glucosidase purified from *Aspergillus oryzae* that is highly resistant to inhibition by glucose and is stable at low pH [74]. The latter enzyme has a broad-specificity, because it can hydrolyze 1,3-, 1,4-, and 1,6-β-diglycosidase and can release flavour compounds such as geraniol, nerol and linalol from the corresponding monoglucosides in a glucose rich medium at pH 2.9. A completely different approach relied on the isolation of yeast mutants able to produce monoterpenes. Strains mutated in the sterol metabolic pathway and producing geraniol, citronelol and linalol similar to those of the floral grape cultivars have been developed [23, 41].

14.6.3 Safety and health benefits

Attempts to reduce the concentration of ethyl carbamate in wine have been performed, since this compound is a suspected carcinogen. Ethyl carbamate is mainly formed by a spontaneous chemical reaction of ethanol and urea at elevated temperatures in acidic media. Although not present in measurable levels in young wines, it can be detected in aged wines or in wines stored at elevated temperatures [59]. On the other hand, since urea arises mainly from the cleavage by arginase (EC 3.5.3.1) of arginine, wines obtained from arginine rich grape musts may contain ethyl carbamate in amounts exceeding the authorized concentrations. To reduce the formation of urea, the two copies of the *CAR1* gene coding for arginase were disrupted from industrial sake yeast, resulting in the elimination of urea and ethyl carbamate formation during sake brewing [43]. However, the amount of nitrogen that can be assimilated by the recombinant strain is reduced, thereby limiting its commercial use.

Another concern is to reduce the amount of SO_2 added to grape must and wine as a preservative. To decrease the risk of bacterial contamination during wine fermentation, bactericidal wine yeast strains have been recently developed by expression of genes encoding a pediocin and a leucocin gene from *Pediococcus acidilactici* and *Leuconostoc carnosum*, respectively [86]. Perspectives rely on the development of yeast strains with a larger spectrum, which could be useful for the production of wine with reduced levels of sulphur dioxide and other chemical preservatives. The addition of limited amounts of SO_2 will, however, remain necessary because of its antioxidant properties.

14.7 Typing of wine yeasts

Typing methods aim to recognize yeasts at the strain level. This means that yeast strains submitted for typing have been previously identified at the species level. However, the classical methods of yeast identification frequently give ambiguous results at the species level, thus making them inconvenient for typing purposes. New molecular techniques are of great interest for this purpose.

14.7.1 Taxonomy of wine yeasts

Wine yeasts generally belong to the *Saccharomyces* sensu stricto species complex. Molecular techniques allowed to recognize three species and one hybrid: *S. cerevisiae, S. bayanus, S. paradoxus* and *S. pastorianus*. Classical identification allowed to class these four species into either one group, *S. cerevisiae,* or two groups, namely *S. cerevisiae/S. paradoxus* on one hand, and *S. bayanus/S. pastorianus* on the other [5, 94]. From these four species only *S. cerevisiae* and, to a lesser extent, *S. bayanus* are used in the production of wine.

Until recently, the distinction between *S. cerevisiae* and *S. bayanus* was based on only one physiological character. *S. cerevisiae* is able to ferment and assimilate galactose (Gal+), while *S. bayanus* cannot use this compound (Gal-). However, this differentiation is hampered by the fact that galactose is metabolized via the glycolysis in two mutable steps. This leads to the misidentification of *S. cerevisiae* Gal- strains. *S. uvarum*, a species classified as a synonym of *S. bayanus,* is Gal+ and Mel+. *S. uvarum* strains ferment melibiose and raffinose completely, whereas *S. bayanus* is unable to ferment melibiose. The taxonomic confusion in the species *S. bayanus* is caused by the type strain (CBS 380), which is Gal- and Mel-. Other strains belonging to the species *S. uvarum* (e. g., CBS 395) are Gal+ and Mel+, and named *S. bayanus* by taxonomic definition. If these two physiological characters are considered alone, *S. uvarum* cannot be distinguished from *S. cerevisiae,* because this latter species is Gal+, and Mel+ strains of *S. cerevisiae* do exist. Fortunately, *S. uvarum* strains cannot grow at temperature higher than 37 °C and *S. cerevisiae* strains can grow at 40 °C or higher.

The confused taxonomic situation in the *S. bayanus* complex has been clarified recently. Using PCR amplification of the NTS2 (Non Transcribed Spacer = Intergenic Spacer (IGS)) of the ribosomal DNA (rDNA) with specific primers, followed by restriction fragment length polymorphism analysis (RFLP), we have reinvestigated a large number of strains identified as *S. bayanus* by other methods. The results showed that the majority of strains identified as *S. bayanus* and isolated from wine, cider and apple juice, are identical to *S. uvarum* (CBS 395). The genetic material of the type strain of *S. bayanus* (CBS 380) is mainly identical to that of *S. uvarum,* but seems partly derived from *S. cerevisiae* as well. Therefore, we consider strain CBS 380 as a partial hybrid between *S. uvarum* and *S. cerevisiae* [55, 56]. Consequently, yeast strains isolated from wine or related products, such as grape juice, belong mainly to *S. cerevisiae* or *S. uvarum.* They may be identified by a combination of two tests, namely fermentation of melibiose and growth at 37 °C. *S. uvarum* is Mel+ and unable to grow at 37 °C, whereas *S. cerevisiae* is usually Mel- and is able to grow at 37 °C.

14.7.2 Typing of *S. cerevisiae* and *S. uvarum* strains

Various techniques of typing wine yeast strains are in use. These include pulsed field gel electrophoresis (PFGE or karyotyping), restriction fragment length polymorphism (RFLP) of the mitochondrial DNA (mtDNA) and, in the case of *S. cerevisiae,* fingerprinting of nuclear DNA using Y′ subtelomeric sequences [65, 91].

Karyotypes obtained from *S. cerevisiae* wine strains are generally more complex than those of *S. cerevisiae* standard strains. This may be due either to higher ploidy, multiple chromosomal rearrangements or their hybrid nature [11, 91]. Among 40 *S. cerevisiae* strains from the "Collection de Levures d'Intérêt Biotechnologique" (CLIB), only two showed identical karyotypes, and the others presented enough polymorphism for differentiation. When the amount of chromosomal length polymorphism is sufficient for a pairwise comparison, PFGE is the typing method of choice. The karyotypes of *S. uvarum* are different from those of *S. cerevisiae*, and can be recognized immediately. In many cases, and like in *S. cerevisiae*, the polymorphisms present in the chromosomal patterns of *S. uvarum* are sufficient for strain differentiation (Fig. 14.7-1A).

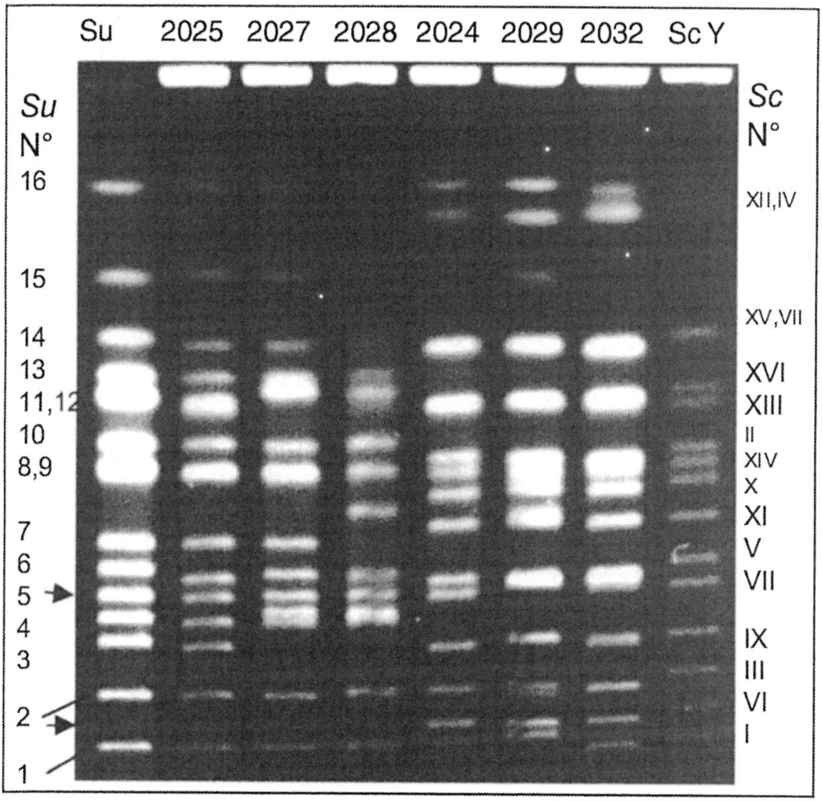

Fig. 14.7-1A Typing of *Saccharomyces* wine yeasts. Karyotypes obtained from *S. uvarum* and *S. cerevisiae* strains. Su: *S. uvarum* strain CLIB 111, 2025-2028: deposited *S. uvarum* wine strains CLIB 2025, CLIB 2027, CLIB 2028; 2024-2032: deposited *S. cerevisiae* wine strains CLIB 2024, CLIB 2029, CLIB 2032. ScY: *S. cerevisiae* standard strain YNN295. Arrow heads indicate the characteristic pattern of *S. uvarum* small chromosomes.

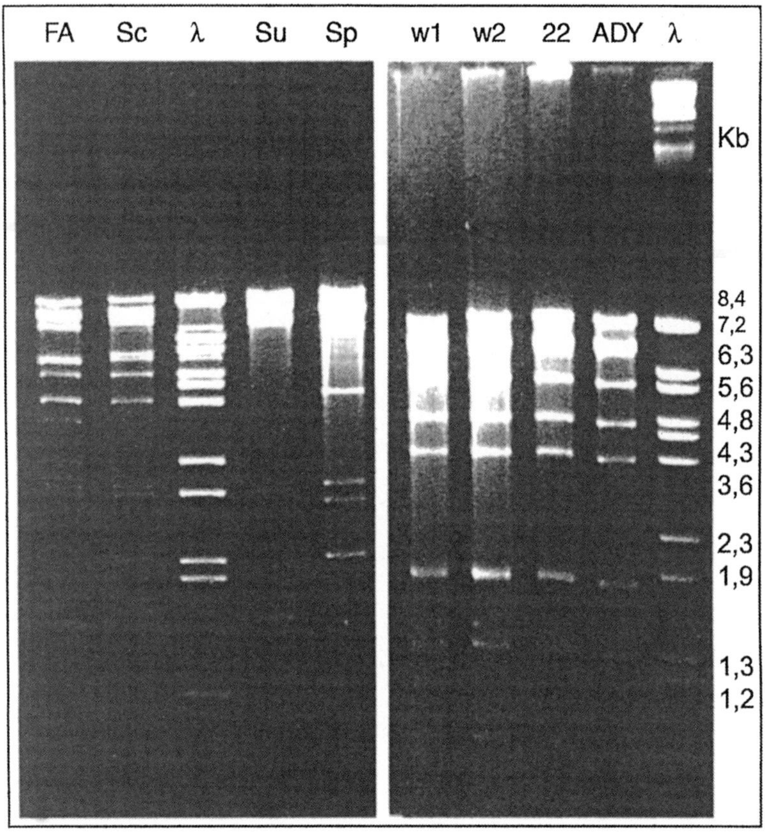

Fig. 14.7-1B Typing of *Saccharomyces* wine yeasts. EcoRV restriction patterns of mtDNA from *Saccharomyces*. (Left panel) FA: *S. cerevisiae* Cognac wine strain FA; Sc: *S. cerevisiae*; Su: *S. uvarum*; Sp: *S. pastorianus*. (Right panel) *S. cerevisiae* strain deposit (22: CLIB 2022) Active Dry Yeast (ADY) produced from it and clones isolated from wine fermented with this ADY (w1, w2). λ: λ *Bst*EII marker.

Comparison of the mitochondrial genome may be used to differentiate between strains of *S. cerevisiae*. *Eco*RV restriction of the mtDNA generates strain specific RFLP patterns. Comparison of RFLP patterns of 40 strains from the CLIB collection showed that the amount of polymorphism is sufficient to differentiate each strain (Fig.14.7-1B). The mitochondrial genome of *S. uvarum* (50 kb) is smaller than that of *S. cerevisiae* (75 kb), and, consequently, the *Eco*RV restriction patterns of the mtDNA differ between these two species. Unfortunately, little polymorphism is observed between the *Eco*RV restriction patterns

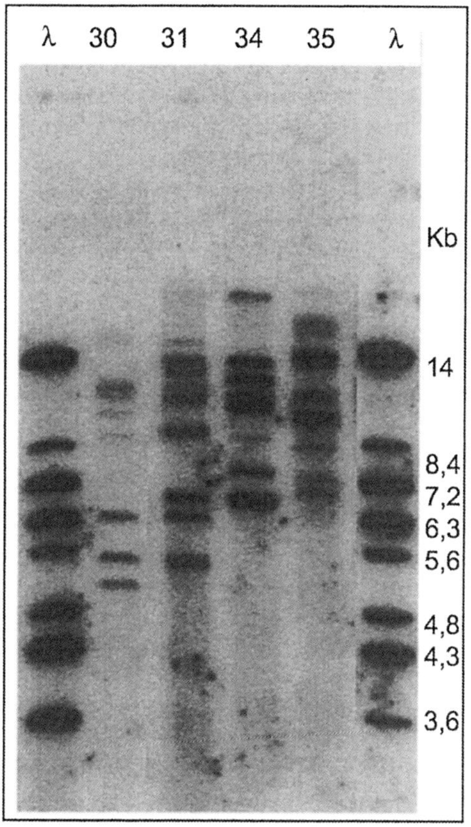

Fig. 14.7-1C Typing of *Saccharomyces* wine yeast Fingerprinting of *S. cerevisiae* wine strains CLIB 2030, CLIB 2031, CLIB 2034, CLIB 2035. Total DNA was digested with *XhoI* and plasmid pJY harboring *S. cerevisiae* Y' sequence was used as probe; λ: λ *BstEII* marker.

of the mtDNA of strains of *S. uvarum* and, therefore, examination of karyotypes is required to differentiate these strains.

Fingerprinting of *S. cerevisiae* wine strains was performed by digesting the total DNA with *XhoI* and by Southern hybridisation with labelled Y′ subtelomeric fragment (Fig. 14.7-1C). This proved to be highly discriminatory and strains with similar karyotypes could be differentiated with this method. This method cannot be applied to strains of *S. uvarum*, because they do not harbour Y′ subtelomeric sequences.

14.8 Conclusion and future prospect

The use of pure and selected *S. cerevisiae* strains in alcoholic wine fermentations has strongly improved the reliability of the fermentation process and the quality of wines by restricting the impact of yeasts with unsuitable fermentation performances or aromatic properties. However, in the last decade a renewed interest in non-*Saccharomyces* strains appeared with the aim to gain some benefits of the aromatic properties produced by these yeasts under controlled inoculations. The exploitation of some specific metabolic capacities of these yeasts may represent a promising way to influence the flavour profiles of the wines. Progress in this field will require a thorough characterization of these yeasts and their physiological role in wine fermentations. Despite a large amount of available data, our understanding of the impact of yeasts (both *Saccharomyces* and *non-Saccharomyces* yeasts) on the flavour of wines is still limited. Progress in the understanding of the biochemical mechanisms by which yeasts influence the sensorial quality of wines will be of key importance in the future to select or genetically construct yeasts that are better adapted to the production of wines fulfilling the consumers demands.

Thanks to the wealth of knowledge acquired on *S. cerevisiae* in enological fermentations, some main factors influencing the performance of yeasts have been specified. This has permitted a better management of the alcoholic fermentations by an appropriate temperature control or by supplying the required nutrients at the optimum fermentation moment. However, because of the high variability of the composition of musts, many factors can occasionally influence the yeast fermentation capacity and give delayed, sluggish or stuck fermentations [12]. Progress in the prediction of problematic fermentations will rely on a better knowledge of the physiology of yeasts under enological conditions, such as, for instance, non-proliferating cells in the presence of stressing amounts of ethanol in a nutrient depleted medium. Data already available at the molecular level on the response of yeast to stress, represent a sound starting point to address this problem. The acquisition of data using wine yeasts under wine fermentation conditions will be a prerequisite to integrate this knowledge in the appropriate genetic and physiological context [8]. The availability of new powerful methods such as DNA microarrays, which allow genome-wide analysis of gene expression, should help to decipher the regulatory networks underlying wine yeast response in fermentation. Global genomic approaches represent undoubtedly a breakthrough in the manner we can address the properties of wine yeasts as already demonstrated by some studies [4, 22, 37, 75].

Progress in the development of genetically modified wine strains has permitted the construction of yeast strains possessing optimized and new characteristics. The availability of these strains in the future will increase the possibilities to achieve a better control of the wine making process, to increase wine quality and even to improve the wine style. However, while the great potential benefits of these strains are well established, none of these strains has been commercialized yet, principally because of public acceptance considerations. In relation with public and winemakers concerns, it remains necessary to increase our knowledge concerning the potential risks associated with the use of genetically modified

(GM) yeasts in the wine production chain. Global genomic tools, such as DNA microarrays and 2D protein gels, will be helpful to assess the impact triggered by a genetic modification at a genome wide scale. To increase the winemakers and consumers confidence and acceptability of genetically modified wine yeasts, it will also be necessary to get additional knowledge on the impact of the release of GM wine yeast strains in the environment.

14.9 References

[1] ANSANAY, V.; DEQUIN, S.; BLONDIN, B.; BARRE, P.: Cloning, sequence and expression of the gene encoding the malolactic enzyme from *Lactococcus lactis*. FEBS Lett. **332** (1993) 74–80.

[2] ANSANAY, V.; DEQUIN, S.; CAMARASA, C.; SCHAEFFER, V.; GRIVET, J.P.; BLONDIN, B.; SALMON, J.M.; BARRE, P.: Malolactic fermentation by engineered *Saccharomyces cerevisiae* as compared with engineered *Schizosaccharomyces pombe*. Yeast **12** (1996) 215–225.

[3] ANSANAY-GALEOTE, V.; BLONDIN, B.; DEQUIN, S.; SABLAYROLLES, J.M.: Stress effect of ethanol on fermentation kinetics by stationary-phase cells of *Saccharomyces cerevisiae*. Biotechnol. Lett. **23** (2001) 677–681.

[4] BACKHUS, L.E.; DERISI, J.; BROWN, P.; BISSON, L.: Functional genomic analysis of a commercial wine yeast strain of *Saccharomyces cerevisiae* under differing nitrogen conditions. FEMS Yeast Res. **1** (2001) 1–15.

[5] BARNETT, J.A.: Yeast identification program, Version 4 (2000).

[6] BARRE, P.; VEZINHET, F.; DEQUIN, S.; BLONDIN, B.: Genetic improvement of wine yeasts. In: Wine microbiology and biotechnology (edited by Fleet, G.H.). Chur, Switzerland: Harwood Academic Publishers (1993) 265–287.

[7] BATAILLON, M.; RICO, A.; SABLAYROLLES, J.M.; SALMON, J.M.; BARRE, P.: Early thiamin assimilation by yeasts in enological conditions: impact on alcoholic fermentation kinetics. J. Ferment. Bioeng. **82** (1996) 101–106.

[8] BAUER, F.F.; PRETORIUS, I.S.: Yeast stress response and fermentation efficiency: How to survive the making of wine – a review. S. Afr. J. Enol. **21** (2000) 27–51.

[9] BELY, M.; SABLAYROLLES, J.M.; BARRE, P.: Automatic detection of assimilable nitrogen deficiencies during alcoholic fermentation in enological conditions. J. Ferm. Bioeng. **4** (1990) 246–252.

[10] BIDARD, F.: Clonage et analyse d'un gène de floculation *FLO5* de *Saccharomyces cerevsiae*. Ph. D. Thesis, Université de Montpellier II, Montpellier, France (1993).

[11] BIDENNE, C.; BLONDIN, B.; DEQUIN, S.; VEZINHET, F.: Analysis of the chromosomal DNA polymorphism of wine strains of *Saccharomyces cerevisae*. Curr. Genet. **22** (1992) 1–7.

[12] BISSON, L.F.: Stuck and sluggish fermentations. Am. J. Enol. Vitic. **50** (1999) 107–119.

[13] BISSON, L.F.; COONS, D.M.; KRUCKEBERG, A.L.; LEWIS, D.A.: Yeast sugar transporters. Crit. Rev. Biochem. Mol. Biol. **28** (1993) 259–308.

[14] BLAYTERON, L.; SABLAYROLLES, J.M.: Stuck and slow fermentations in enology: statistical study of causes and effectiveness of combined additions of oxygen and diammonium phosphate. J. Biosc. Bioeng. **91** (2001) 184–189.

[15] BLONDIN, B.; DEQUIN, S.: Perspectives dans l'amélioration des levures en œnologie. In: Oenologie – fondements scientifiques et technologiques, Technique et Documentation (edited by Flanzy, C.). Paris, France: Lavoisier (1998) 415–444.

[16] BONY, M.; BIDARD, F.; CAMARASA, C.; DULAU, L.; BARRE, P.; DEQUIN, S.: Complete malolactic fermentation by *S. cerevisiae* strains expressing heterologous genes for malolactic enzyme and malate transport. FEBS Lett. **410** (1997) 452–456.

[17] BOULTON, R.B.; SINGLETON, V.L.; BISSON, L.F.; KUNKEE, R.E.: Yeast and biochemistry of yeast fermentation. In: Principle and practices of winemaking (edited by Boulton, R.B.; Singleton, V.L.; Bisson, L.; Kunkee, R.E.). New York, U.S.A.: Chapman & Hall (1996) 139–172.

[18] BUSTURIA, A.; LAGUNAS, R.: Catabolite inactivation of the glucose transport system in *Saccharomyces cerevisiae*. J. Gen. Microbiol. **132** (1986) 379–385.

[19] BUTZKE, C.E.; BISSON, L.F.: Genetic engineering of yeast for wine production. Agro-Food-Ind. Hi-Tech. Jul/Aug (1996) 26–30.

[20] CAMARASA, C.; TRIHAN, F.; DEQUIN, S.: Production of succinate during alcoholic fermentation by *Saccharomyces cerevisiae*. In: Abstr. 10th int. symp. yeasts (ISY 2000) (edited by van Dijken, J.P.; Scheffers, W.A.). Papendal, The Netherlands (2000) 248.

[21] CASTELLI, T.: Aspetto biologico della fermentazionne vinaria nelle zone calde. Atti. Acad. Ital. Vite Vino. **4** (1952) 284–306.

[22] CAVALIERI, D.; TOWNSEND, J.P.; HARTL, D.L.: Manifold anomalies in gene expression in a vineyard isolate of *Saccharomyces cerevisiae* revealed by DNA microarrays analysis. Proc. Natl. Acad. Sci. USA **97** (2000) 12369–12374.

[23] CHAMBON, C.; LADEVEZE, V.; OULMOUDEN, A.; SERVOUSE, M.; KARST, F.: Isolation and properties of yeast mutants affected in farnesyl diphosphate synthetase. Curr. Genet. **18** (1990) 41–46.

[24] CHATONNET, P.; DUBOURDIEU, D.; BOIDRON, J.: The influence of *Dekkerra/Brettanomyces* and lactic acid bacteria on the ethylphenol content of red wines. Am. J. Enol. Vitic. **46** (1995) 463–467.

[25] CHEN, E.C.H.: The relative contribution of Ehrlich and biosynthetic pathways to the formation of fusel alcohols. J. Am. Soc. Brew. Chem. **36** (1978) 39–43.

[26] DEQUIN, S.: The potential of genetic engineering for improving brewing, wine-making and baking yeasts. Appl. Microbiol. Biotechnol. **56** (2001) 577–588.

[27] DEQUIN, S.; BAPTISTA, E.; BARRE, P.: Acidification of grape musts by *S. cerevisiae* wine yeast strains genetically engineered to produce lactate. Am. J. Enol. Vitic. **50** (1999) 45–50.

[28] DEQUIN, S.; BARRE, P.: Mixed lactic acid-alcoholic fermentation by *Saccharomyces cerevisiae* expressing the *Lactobacillus casei* L(+)-LDH. Bio/Technol. **12** (1994) 173–177.

[29] ESCHENBRUCH, R.: Sulphite and sulphide formation during wine making. Am. J. Enol. Vitic. **25** (1974) 157–161.

[30] FEUILLAT, M.: Autolyse des levures. In: Oenologie – fondements scientifiques et technologiques (edited by Flanzy, C.). Paris, France: Lavoisier (1998) 444–454.

[31] FLEET, G.H.; HEARD, G.M.: Wine microbiology and biotechnology. Chur, Switzerland: Harwood Academic (1993).

[32] FLEET, G.H.; HEARD, G.M.: Yeast-growth during fermentation. In: Wine microbiology and biotechnology (edited by Fleet, G.H.). Chur, Switzerland: Harwood Academic (1993) 27–54.

[33] FUKUDA, K.; YAMAMOTO, N.; KIYOKAWA, Y.; YANAGIUCHI, T.; WAKAI, Y.; KITAMOTO, K.; INOUE, Y.; KIMURA, A.: Brewing properties of sake yeast whose *EST2* gene encoding isoamyl acetate-hydrolysing esterase was disrupted. J. Ferm. Bioeng. **85** (1998) 101–106.

[34] GOFFEAU, A.; BARRELL, B.G.; BUSSEY, H.; DAVIS, R.W.; DUJON, B.; FELDMANN, H.; GALIBERT, F.; HOHEISEL, J.D.; JACQ, C.; JOHNSTON, M.; LOUIS, E.J.; MEWES, H.W.; MURAKAMI, Y.; PHILIPPSEN, P.; TETTELIN, H.; OLIVER, S.G.: Life with 6000 genes.: Science **274** (1996) 563–567.

[35] GOMBERT, A.K.; DOS SANTOS, M.M.; CHRISTENSEN B.; NIELSEN J.: Network identification and flux quantification in the central metabolism of *Saccharomyces cerevisiae* under different conditions of glucose repression. J. Bacteriol. **183** (2001) 1441–1451.

[36] HAMMOND, J.R.M.; PYE, J.M.: Acetate ester synthesis: the role of alcohol acetyl transferase. Yeast Newsletter **45** (1996) 12–13.

[37] HAUSER, N.C.; FELLENBERG, K.; GIL, R.; BASTUCK, S.; HOHEISEL, J.D.; PEREZ-ORTIN, J.E.: Whole genome analysis of a wine yeast strain. Comp. Funct. Genom. **2** (2001) 69–79.

[38] HENSCKE, P.; JIRANEK, V.: Yeasts – metabolism of nitrogen compounds. In: Wine microbiology and biotechnology (edited by Fleet, G.H.). Chur, Switzerland: Harwood Academic (1993) 77–164.

[39] HENSCKE, P.A.; JIRANEK, V.: H_2S formation during fermentation: effect of composition in model grape musts. In: Int. symp. nitrogen in grapes and wines (edited by Rantz, J.A.). Davis, U.S.A.: American Society of Enology and Viticulture (1991) 172–184.

[40] HOHMANN, S.: Shaping up: the response of yeast to osmotic stress. In: Yeast stress response (edited by Hohmann, S.: Mager, W.H.). Austin, U.S.A.: R.G. Landes (1997) 101–145.

[41] JAVELOT, C.; KARST, F.; LADEVEZE, V.; CHAMBON, C.; VLADESCU, B.: Production of monoterpenes by yeast mutants defective in sterol biosynthesis. In: Microbiology application in food biotechnology (edited by Nga, B.H.; Lee, Y.K.). London, U.K.: Elsevier Applied Science (1990) 101–122.

[42] JIRANEK, V.; LANGRIDGE, P.; HENSCHKE, P.A.: Nitrogen requirements of yeast during wine fermentation. In: Proc. 7th Austr. wine industry techn. conference (Adelaïde, 1989) (edited by Williams, P.J.; Davidson, D.M.; Lee, T.H.). Adelaïde, Australia: Australian Industrial Publishers S.A. (1990) 166–171.

[43] KITAMOTO, K.; ODA, K.; GOMI, K.; TAKASHI, K.: Genetic engineering of a sake yeast producing no urea by successive disruption of arginase gene. Appl. Envir. Microbiol. **57** (1991) 2568–2575.

[44] KUNKEE, R.M.; AMERINE, M.A.: Yeast in wine making. In: The yeasts, Vol. 3, Yeast technology (edited by Rose, A.H.). London, U.K.: Academic Press (1970) 5–71.

[45] LAFONT-LAFOURCADE, S.; GENEIX, C.; RIBÉREAU-GAYON, P.: Inhibition of alcoholic fermentation of grape musts by fatty acids produced by yeasts and their elimination by yeast ghosts. Appl. Envir. Microbiol. **47** (1984) 1246–1249.

[46] LARGE, P.J.: Degradation of organic nitrogen by yeasts. Yeast **2** (1986) 1–34.

[47] LILLY, M.; LAMBRECHTS, M.G.; PRETORIUS, I.S.: Effect of increased yeast alcohol acetyltransferase activity on flavour profiles of wine and distillates. Appl. Envir. Microbiol. **66** (2000) 744–53.

[48] LUYTEN, K.; ALBERTYN, J.; SKIBLE, F.; PRIOR, B.A.; RAMOS, J.; THEVELEIN, J.M.; HOHMANN, S.: FPS1, a yeast member of the MIP family of channel proteins, is a facilitator for glycerol uptake and efflux and is inactive under osmotic stress. EMBO J. **14** (1995) 1360–1371.

[49] LUYTEN, K.; RIOU, C.; BLONDIN, B.: The hexose transporters of S*accharomyces cerevisiae* play different roles during wine fermentation. Yeast (2002) in press.

[50] MANGINOT, C.; ROUSTAN, J.L.; SABLAYROLLES, J.M.: Use of constant rate alcoholic fermentations to compare the effectiveness of different nitrogen sources during the stationary phase. Enzyme Microb. Technol. **23** (1998) 511–517.

[51] MARTINI, A.; MARTINI, V.: Grape must fermentation: past and present. In: Yeast technology (edited by Spencer, J.F.T.; Spencer, D.M.). Berlin, Germany: Springer Verlag (1990) 105–123.

[52] MASNEUF, I.: Recherches sur l'identification génétique des levures de vinification, applications œnologiques. Ph. D. Thesis, Université de Bordeaux II, Bordeaux, France (1996).

[53] MORTIMER, M.; POLSINELLI, M.: On the origin of wine yeast. Res. Microbiol. **150** (1999) 199–204.

[54] NAUMOV, G.I.; MASNEUF, I.; NAUMOVA, E.S.; AIGLE, M.; DUBOURDIEU, D.: Association of *Saccharomyces bayanus* var. *uvarum* with some French wines: genetic analysis of yeast populations. Res. Microbiol. **151** (2000) 83–691.

[55] NGUYEN, H.-V.; GAILLARDIN, C.: Two subgroups within the *Saccharomyces bayanus* species evidenced by PCR amplification and restriction polymorphism of the non-transcribed spacer 2 in the ribosomal DNA unit. Syst. Appl. Microbiol. **20** (1997) 286–294.

[56] NGUYEN, H.-V.; LÉPINGLE, A.; GAILLARDIN, C.: Molecular typing demonstrates homogeneity of *Saccharomyces uvarum* strains and reveals the existence of hybrids between *S. uvarum* and *S. cerevisiae*, including the *S. bayanus* type strain CBS 380. Syst. Appl. Microbiol. **23** (2000) 71–85.

[57] NYKÄNEN, L.: Formation and occurrence of flavour compounds in wine and distilled alcoholic beverages. Am. J. Enol. Vitic. **37** (1986) 84–96.

[58] OUGH, C.S.; DAVENPORT, M.; JOSEPH, K.: Effects of certain vitamins on growth and fermentation rate of several commercial active dry wine yeasts. Am. J. Enol. Vitic. **40** (1989) 208–213.

[59] OUGH, C.S.; CROWELL, E.A.; MOONEY, L.A.: Formation of ethyl carbamate precursors during grape juice (Chardonnay) fermentation. I. Addition of amino acids, urea and ammonia: effects of fortification on intracellular and extracellular precursors. Am. J. Enol. Vitic. **39** (1988) 243–249.

[60] OURA, E.: Reaction products of yeast fermentations. Process Biochem. **12** (1977) 19–21.

[61] OZCAN, S.; JOHNSTON, M.: Three different regulatory mechanisms enable yeast hexose transporter (HXT) genes to be induced by different levels of glucose. Mol. Cell. Biol. **15** (1995) 1564–1572.

[62] PEYNAUD, E.; DOMERCQ, M.C.: Etude des levures de la Gironde. Ann. Inst. Nat. Rech. Agr. **4** (1953) 265–300.

[63] PRETORIUS, I.S.: Tailoring wine yeast for the new millennium: novel approaches to the ancient art of winemaking. Yeast **16** (2000) 675–729.

[64] PRETORIUS, I.S.; VAN DER WESTHUIZEN, T.J.: The impact of yeast genetics and recombinant DNA technology on the wine yeast industry. A review. S. Afr. J. Enol. Vit. **12** (1991) 3–31.

[65] QUEROL, A.; BARRIO, E.; RAMON, D.: A comparative study of different methods of yeast strains characterization. Syst. Appl. Microbiol. **15** (1992) 439–446.

[66] QUEROL, A.; RAMON, D.: The application of molecular techniques in wine microbiology. Trends Food Sci. Technol. **7** (1996) 73–78.

[67] RANKINE, B.C.; BRIDSON, D.A.: Glycerol in Australian wines and factors influencing its formation. Am. J. Enol. Vitic. **22** (1971) 6–12.

[68] RAPP, A.; VERSINI, G.: Influence of nitrogen compounds in grapes on aroma compounds of wine. In: Proc. int. symp. on nitrogen in grapes and wines (Seattle, WA, 1991) (edited by Rantz, J.). Davis, U.S.A.: American Society for Enology and Viticulture (1991) 156–164.

[69] RAUHUT, D.: Yeasts – Production of sulphur compounds. In: Wine microbiology and biotechnology (edited by Fleet, G.H.). Chur, Switzerland: Harwood Academic (1993) 183–223.

[70] REIFENBERGER, E.; BOLES, E.; CIRIACY, M.: Kinetic characterization of individual hexose transporters of *Saccharomyces cerevisiae* and their relation of the triggering mechanism of glucose repression. Eur. J. Biochem. **245** (1997) 324–333.

[71] REMIZE, F.; ANDRIEU, E.; DEQUIN, S.: Engineering of the pyruvate dehydrogenase by-pass in *S. cerevisiae* – Role of the cytosolic Mg2+ and mitochondrial K+ acetaldehyde dehydrogenases Ald6p and Ald4p in acetate formation during alcoholic fermentation. Appl. Envir. Microbiol. **66** (2000) 3151–3159.

[72] REMIZE, F.; DEQUIN, S.: Engineering of glycerol metabolism in wine yeast. Proc. EC framework IV symp. yeast as a cell factory. Vlaardingen, The Netherlands (1998) 30–31.

[73] REMIZE, F.; ROUSTAN, J.L.; SABLAYROLLES, J.M.; BARRE, P.; DEQUIN, S.: Glycerol overproduction by engineered *Saccharomyces cerevisiae* wine yeast strains leads to substantial changes in by-product formation and to a stimulation of fermentation rate in stationary phase. Appl. Envir. Microbiol. **65** (1999) 143–149.

[74] RIOU, C.; SALMON, J.M.; VALLIER, M.J.; GÜNATA, Z.; BARRE, P.: Purification, characterization, and substrate specificity of a novel glucose-tolerant β-glucosidase from *Aspergillus orizae*. Appl. Envir. Microbiol. **64** (1998) 3607–3614.

[75] ROSSIGNOL, T.; JULIEN, A.; DULAU, L.; BLONDIN, B.: Genome-wide analysis of yeast gene expression during wine fermentation. Yeast **18** (S1) (2001) 298.

[76] SABLAYROLLES, J.M.: Besoins en oxygène lors des fermentations œnologiques. Rev. Fr. Oenol. **124** (1990) 77–79.

[77] SABLAYROLLES, J.M.; BARRE, P.: Evaluation des besoins en oxygène de fermentations alcooliques en conditions oenologiques simulées. Sci. Alim. **6** (1986) 373–383.

[78] SABLAYROLLES, J.M.; SALMON, J.M.; BARRE, P.: Carences nutritionnelles des moûts. Efficacité des ajouts combinés d'oxygène et d'azote ammoniacal. Rev. Fr. Oenol. **159** (1996) 25–32.

[79] SAHA, B.C.; BOTHAST, R.J.: Production, purification and characterisation of a highly glucose-tolerant novel β-glucosidase from *Candida peltata*. Appl. Envir. Microbiol. **62** (1996) 3165–3170.

[80] SALMON, J.M.: Relations levure-milieu. In: Oenologie – fondements scientifiques et technologiques (edited by Flanzy, C.). Paris, France: Lavoisier (1998) 415–444.

[81] SALMON, J.M.; BARRE, P.: Improvement of nitrogen assimilation and fermentation kinetics under enological conditions by derepression of alternative assimilatory pathways in an industrial *Saccharomyces cerevisiae* strain. Appl. Envir. Microbiol. **64** (1998) 3831–3837.

[82] SALMON, J.M.; FORNAIRON, C.; BARRE, P.: Determination of oxygen utilization pathways in an industrial strain of *Saccharomyces cerevisiae* during enological fermentation. J. Ferment. Bioeng. **86** (1998) 154–163.

[83] SALMON, J.M.; VINCENT, O.; MAURICIO, J.C.; BELY, M.; BARRE, P.: Sugar transport inhibition and apparent loss of activity in *Saccharomyces cerevisiae* as a major limiting factor of enological fermentations. Am. J. Enol. Vitic. **44** (1993) 127–133.

[84] SALMON, J.M.: Effect of sugar transport inactivation in *Saccharomyces cerevisiae* on sluggish and stuck enological fermentations. Appl. Envir. Microbiol. **55** (1989) 953–958.

[85] SCHAAF, I.; HEINISCH, J.; ZIMMERMANN, F.K.: Overproduction of glycolytic enzymes in yeast. Yeast **5** (1989) 285–290.

[86] SCHOEMAN, H.; VIVIER, M.A.; DU TOIT, M.; DICKS, L.M.; PRETORIUS, I.S.: The development of bactericidal yeast strains by expressing the *Pediococcus acidilactici* pediocin gene (pedA) in *Saccharomyces cerevisiae.* Yeast **15** (1999) 647–56.

[87] STEVENS, S.; HOFMEYR, J.H.M.: Effects of ethanol, octanoic and decanoic acids on fermentation and the passive influx of protons through the plasma membrane of *Saccharomyces cerevisiae.* Appl. Microbiol. Biotechnol. **38** (1993) 656–663.

[88] THURSTON, P.A.; QUAIN, P.E.; TUBB R.S.: Lipid metabolism and the regulation of volatile ester synthesis in *Saccharomyces cerevisiae*. J. Inst. Brew. **88** (1982) 90–94.

[89] TOMINANGA, T.; BALTENWECK-GUYOT, R.; PEYROT DES GACHONS, C.; DUBOURDIEU, D.: Contribution of volatile thiols to the aroma of white wines made from several *Vitis vinifera* grapes varieties. Am. J. Enol. Vitic. **51** (2000) 178–181 .

[90] VAN DIJKEN, J.P.; SCHEFFERS, W.A.: Redox balances in the metabolism of sugars by yeasts. FEMS Microbiol. Rev. **32** (1986) 199–224.

[91] VÉZINHET, F.; BLONDIN, B.; HALLET J.N.: Chromosomal DNA patterns and mitochondrial DNA polymorphism as tools for identification of enological strains of *Saccharomyces cerevisiae.* Appl. Microbiol. Biotechnol. **32** (1990) 568–571.

[92] VEZINHET, F.; LACROIX, S.: Marquage génétique de levures: outil de contrôle des fermentations. Bull OIV **643, 644** (1984) 759–773.

[93] VOLSCHENK, H.; VILJOEN, M.; GROBLER, J.; BAUER, F.; LONVAUD-FUNEL, A.; DENAYROLLES, M.; SUBDEN, R.E.; VAN VUUREN, H.J.J.: Malolactic fermentation in grape musts by a genetically engineered strain of *Saccharomyces cerevisiae.* Am. J. Enol. Vitic. **48** (1997) 193–197.

[94] YARROW, D.; NAKASE, T.: DNA base composition of species of the genus *Saccharomyces*. Antonie van Leeuwenhoek **41** (1985) 81–88.

15 Yeasts and soy products

YOSHIKI HANYA and TADANOBU NAKADAI

15.1 Introduction

Soy products were invented in China more than 2500 years ago and were later developed in many Asian countries. Soy sauce can be divided into fermented soy sauce and chemical soy sauce, which is made by hydrolysis of vegetable protein (Figure 15.1-1) [9]. Fermented soy sauce can be divided into a Japanese-type, which uses soybeans and wheat, and a Chinese-type, which mostly uses soybeans. Japanese-type soy sauce is based on fermentation by yeasts. Soy sauce has long been used as an all-purpose seasoning in Asia. Nowadays it is also widely used in Western countries.

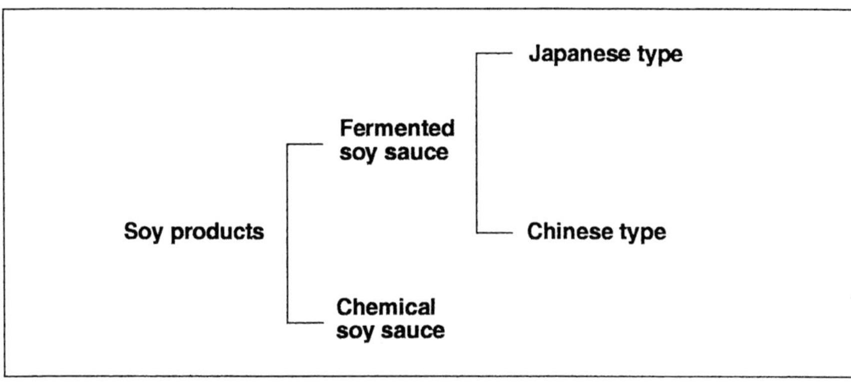

Fig. 15.1-1 Classification of soy products

15.1.1 Production of Japanese-type soy sauce

The procedure for producing Japanese-type soy sauce is shown in Figure 15.1-2 [18].

Soybeans and wheat are cooked to break down the proteins and starches, which are subsequently easily degraded by mold enzymes. After heat treatment, koji mold (*Aspergillus oryzae* or *A. sojae*) is inoculated and cultivated for a few days. During growth, the koji mold produces many kinds of enzymes, including protease, amylase, and peptidase. Next, brine is added to produce moromi mash, and the fermentation starts. The NaCl concentration of the moromi is 16 to 20 %. During fermentation, proteins and starches are dissolved by the koji enzymes and converted into sugars, amino acids, and precursors of flavour compounds.

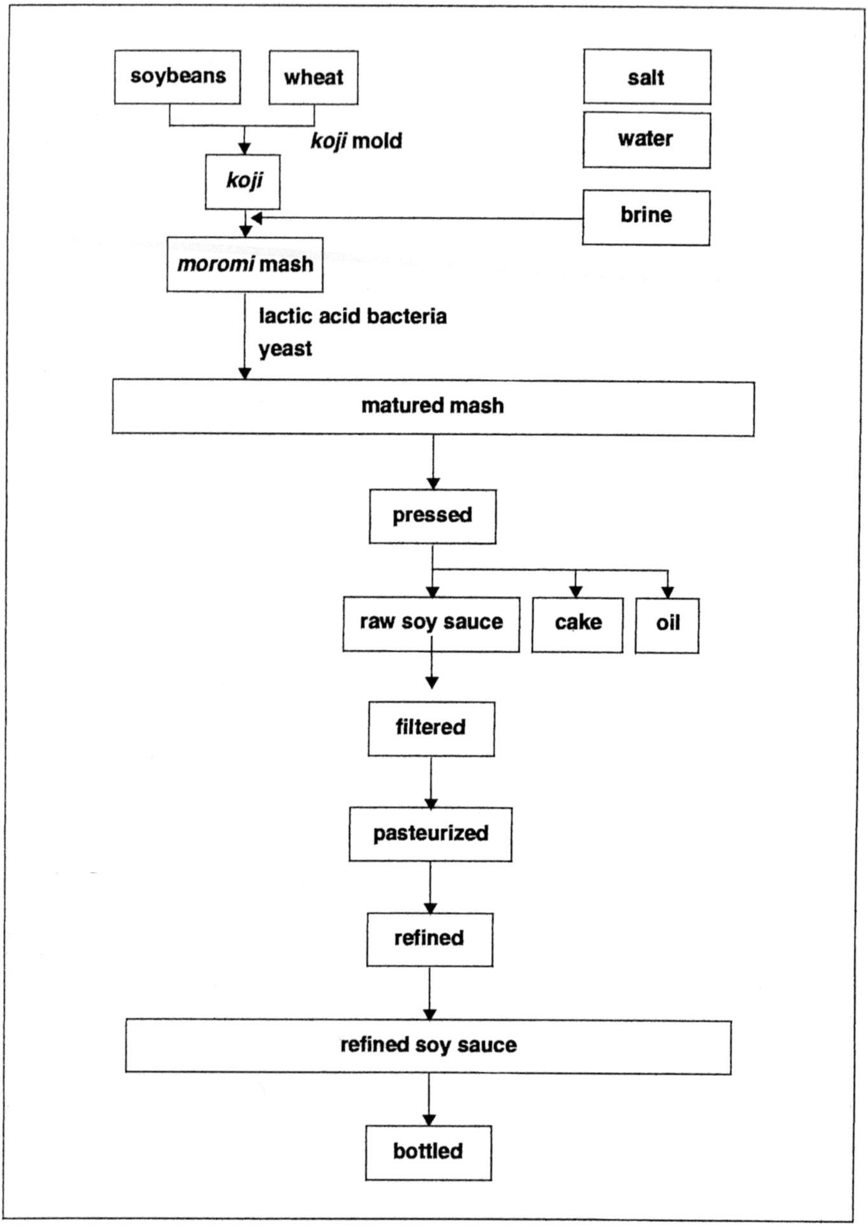

Fig. 15.1-2 Production procedure of Japanese-type soy sauce.

It takes about 6 months for a moromi fermentation and aging. At the beginning of the fermentation, lactic acid bacteria (*Tetragenococcus halophilus*) are added. Lactic acid is produced, which lowers the pH. This produces an optimum environment for fermentation by yeast. In the next phase, soy yeast (*Zygosaccharomyces rouxii*) is added and alcoholic fermentation starts. In some cases *Candida versatilis* or *C. etchellsii* is added to enhance the production of phenolic compounds. During the fermentation many flavour compounds are produced that are important to the quality of the soy sauce. After aging, the moromi is pressed and raw soy sauce is obtained. The raw sauce is heat-treated for pasteurization and enhancing the flavours.

15.2 Yeast biodiversity

Because the production of soy sauce generally occurs in the open air, microbial contamination is difficult to prevent. However, because of the high NaCl concentration, only salt-tolerant microorganisms can grow. Figure 15.2-1 lists the yeast species isolated from koji and moromi [22]. Because most contaminating yeasts in koji are not salt-tolerant, their numbers decrease rapidly in the moromi mash.

The yeast flora of moromi varies according to the phase of fermentation. In the early stage, the pH of the moromi is high (about 6.5), and *Pichia, Debaryomyces*, and some *Candida* spp. predominate. As the pH lowers due to the fermentation by lactic acid bacteria, *Z. rouxii* multiplies and produces ethanol. However, growth of *Z. rouxii* decreases as the alcohol fer-

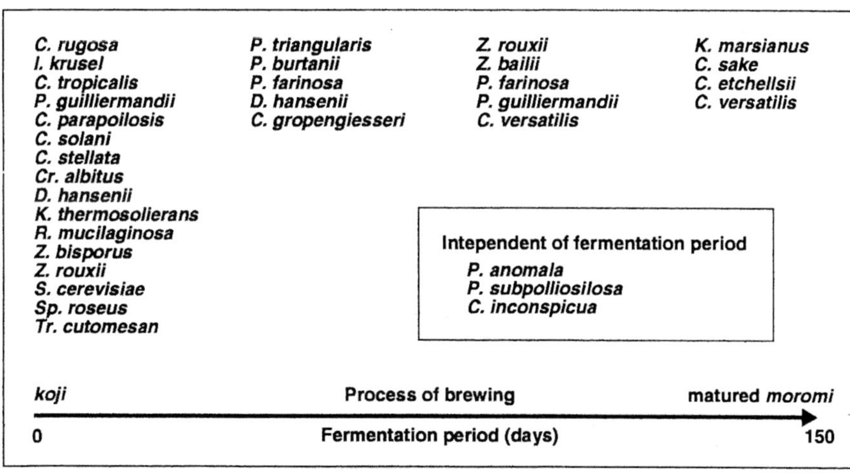

Fig. 15.2-1 Yeasts isolated from soy sauce koji and moromi.
C.: *Candida*; Cr.: *Cryptococcus*; D.: *Debaryomyces*; I.: *Issatchenkia*; K.: *Kluyveromyces*; P.: *Pichia*; R.: *Rhodotorula*; S.: *Saccharomyces*; Sp.: *Sporobolomyces*; Tr.: *Trichosporon*; Z.: *Zygosaccharomyces*.

mentation progresses, because the species is sensitive to alcohol at a concentration of about 4 %. In matured moromi, *C. versatilis* and *C. etchellsii* predominate and these yeasts produce phenolic compounds.

15.3 Beneficial aspects of yeasts in fermented soy products

About 300 kinds of flavour compounds have so far been found in Japanese-type soy sauce [24]. They are classified according to their origin, e. g., materials (soybeans and wheat), koji molds, lactic acid bacteria, yeasts, and chemical reaction products. The flavour compounds produced by fermentation include alcohols, esters, phenols, and furanones. Table 15.3-1 lists the most important flavour compounds produced by yeasts.

Table 15.3-1 Flavour compounds of soy sauce produced by yeasts

Flavour compounds	Yeast	Concentration (ppm)
ethanol	*Z. rouxii*	30 000
glycerol	*Z. rouxii*	10 000
4-hydroxy-5-methyl-3(2H)-furanone (HMMF)	*Z. rouxii*	256
4-hydroxy-2(or 5)-ethyl-5(or 2)-methyl-3(2H)-furanone (HEMF)	*Z. rouxii, C. versatilis, C. etchellsii*	200 12
isobutyl alcohol	*Z. rouxii*	10
4-hydroxy-2,5-dimethyl-3(2H)-furanone (HDMF)	*Z. rouxii*	10
isoamyl alcohol	*Z. rouxii*	4
2-phenylethanol	*Z. rouxii*	3.65
methionol	*Z. rouxii*	3
4-ethylguaiacol (4-EG)	*C. versatilis*	3
methionol	*Z. rouxii*	2
f-Abutyllactone	*Z. rouxii*	0.3
4-ethylphenol (4-EP)	*C. versatilis*	

'Z: Zygosaccharomyce'
C: Candida

15.3.1 4-Hydroxy-2(or 5)-ethyl-5(or 2)-methyl-3-furanone (HEMF)

Among the furanones, HEMF gives Japanese-type soy sauce its characteristic flavour. The compound is produced by *Z. rouxii* and *Candida* spp. The concentration of HEMF in soy sauce is contained around 200 ppm, as the threshold value of HEMF is low (0.04 ppb). HEMF is a characteristic impact flavour compound of Japanese-type soy sauce. HEMF has antioxidative and antitumor activities as well [23, 17, 19].

The biosynthetic pathway of HEMF production in yeasts is unknown. SASAKI and coworkers suggested that sugar phosphates, such as d-ribulose 5-phosphate and d-sedoheptulose 7-phosphate, are suitable precursors for HEMF and that the pentose-phosphate cycle is essential for the production of HEMF by yeasts [30]. However, the low permeability of the cell membrane makes it difficult for yeasts to take up sugar phosphates directly. Moreover, HEMF is found in heat reaction mixtures of sugars and amino acids without any fermentation [5]. Based on these observations, some researchers proposed that HEMF is biosynthesized by yeasts from intermediates of the amino-carbonyl reaction. HAYASHIDA and coworkers reported that aging increased the concentration of HEMF in the moromi [11]. They inoculated *Z. rouxii* into YPD (yeast extract, peptone and dextrose) medium and added intermediates of the amino-carbonyl reaction (a heat-treated solution containing d-ribose and sodium l-glutamate), and measured the production of HEMF [10]. Their results suggested that intermediates of the amino-carbonyl reaction are precursors of HEMF. The same result was found in miso fermentation [34].

15.3.2 Phenolic compounds

4-Ethylguaiacol (4-EG) and 4-ethylphenol (4-EP) are phenolic compounds produced by *C. versatilis* [33]. Ferulic acid, a precursor of 4-EG, is released from plant cell wall polysaccharides by feruloyl esterase (EC 3.1.1.73.), produced by the koji molds. Cinnamate decarboxylase converts ferulic acid into 4-vinylguaiacol (4-VG), which is converted into 4-EG by vinylphenol reductase. In the same way 4-EP is produced from p-coumaric acid as a precursor. *Candida etchellsii* does not produce these compounds, because it lacks vinylphenol reductase [32].

15.3.3 Higher alcohols (fusel alcohols)

Higher alcohols, such as isobutyl alcohol, isoamyl alcohol, and 2-phenyl ethanol, produced by *Z. rouxii* are important flavour compounds. It is assumed that *S. cerevisiae* produces these higher alcohols from corresponding alpha-keto acids by decarboxylation and reduction [28]. These alpha-keto acids are derived from branched-chain amino acids that result from deamination of extracellular amino acids or from an amino acid biosynthetic pathway (Fig. 15.3-1).

AOKI and coworkers isolated mutants of *Z. rouxii* that were deficient in amino acid uptake and measured the mutants' ability to produce higher alcohols [9]. They found that higher alcohols in soy sauce are derived from extracellular amino acids, and that decreasing the uptake of leucine and phenylalanine resulted in a decreased production of 2-phenyl ethanol and isoamyl alcohol, respectively.

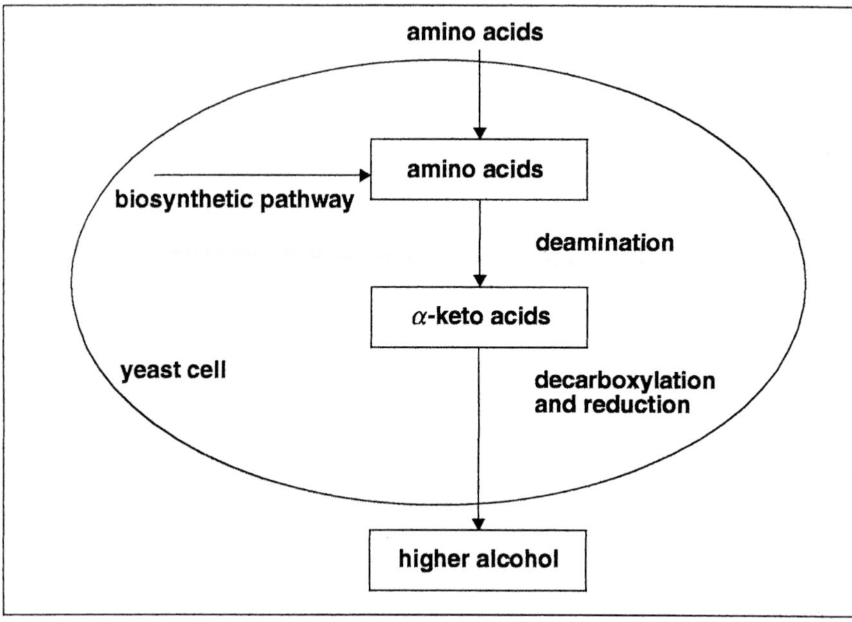

Fig. 15.3-1 Production of higher alcohol in yeasts.

15.3.3.1 2-Phenyl ethanol

In *S. cerevisiae,* 2-phenyl ethanol is produced from glucose or from phenylalanine [28]. AOKI and coworkers isolated a mutant of *Z. rouxii* resistant to p-fluoro-phenylalanine and studied the production of higher alcohols by this isolate [2]. This mutant, which could not metabolize phenylalanine, produced a large amount of 2-phenyl ethanol in moromi, but no isoamyl alcohol, isobutyl alcohol, or methionol. The authors suggested that 2-phenyl ethanol is produced from glucose through the phenylalanine biosynthetic pathway, and not from extracellular precursors as in *Z. rouxii.* The activity of prephenate dehydrogenase (EC 4.2.1.51) was decreased in this mutant, and this decrease may reduce the intracellular concentration of tyrosine and derepress the biosynthesis of prephenate. As a result of the conversion of prephenate to phenylpyruvate, which is a precursor of 2-phenyl ethanol, 2-phenyl ethanol was over produced.

15.3.3.2 Isoamylalcohol

To produce a large amount of isoamyl alcohol, YOSHIKAWA and coworkers isolated mutants resistant to 5,5,5-trifluoro-DL-leucine, an analogue of L-leucine [48]. One of these mutants produced about three times as much isoamyl alcohol as the parental strain. The activity of alpha-isopropylmalate synthase (EC 4.1.3.12) was not inhibited by L-leucine, a feedback inhibitor. This result suggested that the accumulation of alpha-isopropylmalate, a precursor of isoamyl alcohol, is due to the overproduction of isoamyl alcohol in the mutant.

15.3.3.3 3-(Methylthio)-1-propanol (Methionol)

Methionol is usually produced from methionine in yeasts. However, a mutant of *Z. rouxii* produced a large amount of methionol in a medium without methionine [3]. This mutant was derived from a mutant resistant to L-ethionine, an analogue of methionine. The authors observed a decrease in the concentration of S-adenosylmethionine due to a reduction of S-adenosylmethionine synthase (EC 2.5.1.6), and an intracellular accumulation of methionine. It was assumed that overproduction of methionol in this strain resulted in the accumulation of methionine.

15.3.3.4 Polyol

Polyols, such as glycerol and arabitol, are produced by *Z. rouxii*. These compounds are involved in the salt-tolerance of *Z. rouxii*. Polyol may contribute to the mild taste of soy sauce. The mechanism of polyol production is discussed under "Salt-tolerance", below.

15.4 Detrimental aspects of yeasts in fermented soy products

Sometimes moromi is spoiled by film-forming yeasts, mainly belonging to the genera *Zygosaccharomyces*, *Hansenula*, and *Pichia*. Film forming yeasts cover the surface of the moromi and produce undesirable flavours, such as n-butyric acid. The use of sanitary fermentation tanks decreased spoilage by film-forming yeasts.

15.5 Salt tolerance of yeasts in soy fermentation

Soy yeasts can grow in the presence of a high concentration of NaCl (~ 4 M). The mechanisms of salt tolerance have been investigated in *Z. rouxii*. In a medium with a high salt concentration, polyols accumulate [7], and the sterol-ester and free fatty acid content of membrane lipids increase [44, 43]. Based on these observations, several approaches have been used to elucidate the mechanisms of salt tolerance.

15.5.1 Accumulation of polyols

Zygosaccharomyces rouxii accumulates intracellular glycerol in response to an increased concentration of NaCl [7]. Glycerol regulates and protects the cells against osmotic pressure. The production and accumulation of glycerol starts immediately after initiation of salt stress, and the intracellular osmotic pressure is maintained similarly to the external osmotic pressure [47]. In the non salt-tolerant species *S. cerevisiae* glycerol accumulation is induced by osmotic shocks [7]. Recently, the osmosensing signal transduction pathway, named the high-osmolarity glycerol response (HOG) pathway, has been elucidated in *S. cerevisiae* [6, 29, 13]. Under highly osmotic conditions, osmotic stress is sensed by SLN1 and SHO1, sensor proteins located in the membrane. SLN1 forms a phosphorelay system with YPD1 and SSK1. This system transmits the signal to a pair of SSK2 and SSK22 (MAP kinase kinase kinase) and further to PBS2 (MAP kinase kinase). SHO1 transmits the signal to STE11 (MAP kinase kinase kinase) and further to the PBS2 (MAP kinase kinase). PBS2 phosphorylates HOG1 (MAP kinase) and HOG1 enhances the transcription of the GPD1 gene, which encodes for glycerol-3-phosphate dehydrogenase (EC 1.1.99.5). At the same time, glucose influx increases, and the glucose is metabolized to dihydroxyacetone phosphate. This is converted to glycerol-3-phosphate, which subsequently is altered into glycerol (Fig. 15.5-1).

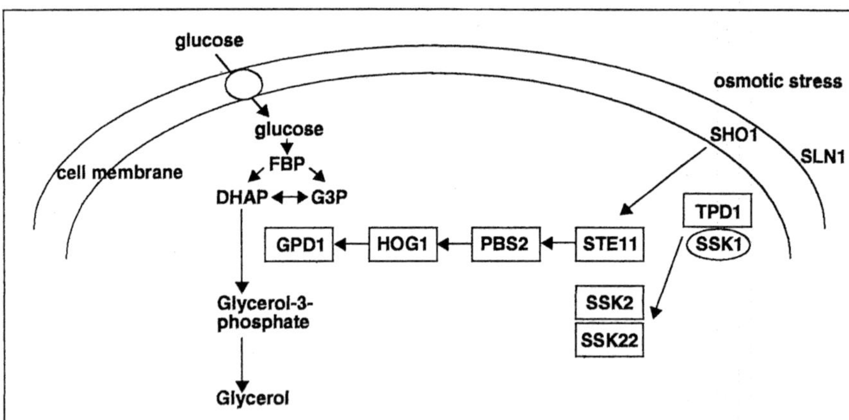

Fig. 15.5-1 The osmosensing signal transduction pathway in *S. cerevisiae*.
FBP: fructose-1,6-biphosphate; G3P: glyceraldehydes-3-phosphate; DHAP: dihydroxyacetone lactoylglutathione; GPD1: glyceraldehydes-3-phosphate dehydrogenase; SHO1: osmosensor; SLN1: histidine kinase; YPD1: inter-mediate protein between SLN and SSK1; SSK1: enzyme activator; STE11: MAP kinase kinase kinase; SSK2: MAP kinase kinase kinase; SSK22:
MAP kinase kinase kinase; PD32: MAP kinase kinase: HOG1: MAP kinase; MAPK: mitogen-activated protein kinase.

Two putative mitogen-activated protein (MAP) kinase genes have been cloned from *Z. rouxii*, namely ZrHOG1 and ZrHOG2, which are homologous to the *S. cerevisiae* HOG1 gene [15]. The deduced amino acid sequences of these genes show a high homology to the sequence of HOG1 and contain a TGY motif for phosphorylation by MAP kinase kinase. These genes can complement the hog1-delta-null mutant of *S. cerevisiae*. In disruption experiments, disruptants of ZrHOG1 or ZrHOG2 in Z. rouxii could grow only at a concentration of 2.5 M NaCl, even though the parental strain could grow at 3 M NaCl. Consequently, the authors suggested that the HOG pathway of *Z. rouxii* is similar to that of *S. cerevisiae*.

15.5.2 Alteration of membrane lipid composition

Alterrations in membrane lipid composition have been investigated in media with various NaCl concentrations [44, 43]. Sterol-ester, free fatty acids, and oleic acid increased, and triacylglycerol and linoleic acid decreased as the NaCl concentration increased. Polyene antibiotics, such as nystatin, amphotericin B, and filipin, bind to ergosterol in the cell membrane and change the membrane permeability. USHIO and coworkers studied the relationship between sterol and salt tolerance in nystatin-resistant mutants of *Z. rouxii* [38]. A mutant with a decreased ergosterol content, showed delayed growth at a high NaCl concentration. A second mutant with an altered sterol composition, could not grow in medium containing more than 8 % NaCl. The accumulation of glycerol in these mutants was not changed, but they leaked extracellularly glycerol. Moreover, the fluidity of the membrane lipid bilayer was increased. These results suggest that alterations in the fluidity of the membrane lipid bilayer and the permeability of the cell membrane are involved in salt tolerance.

15.5.3 H^+-ATPase and sodium-proton antiporter

The intracellular concentration of Na^+ remained low when *Z. rouxii* was cultivated in a medium with a high concentration of NaCl [26]. In *S. cerevisiae*, it has been suggested that the transfer of Na^+ depends on the sodium-proton antiport mechanism, which uses a proton gradient as driving force [31, 8]. This proton gradient is formed by H^+-ATPase in the plasma membrane and allows the cell to take up amino acids and sugars.

15.5.3.1 H^+-ATPase

WATANABE and coworkers compared the characteristics of membrane located H^+-ATPase in membranes from *Z. rouxii* and *S. cerevisiae* [46]. Characteristics of H^+-ATPase of *Z. rouxii*, such as optimal pH, Km values for ATP, and sensitivity toward ATPase inhibitors, did not depend on the concentration of NaCl. However, the activity of the H^+-ATPase doubled because the addition of NaCl to the medium increased the amount of the enzyme. Furthermore, the salt tolerance of *Z. rouxii* decreased in the presence of H+-ATPase inhibitors. The activity of H^+-ATPase of this species was constitutively higher than that of *S. cerevisiae*, and the H^+-ATPase was activated by the addition of glucose to the medium. Unlike in *S. cerevisiae*, the specific activity of the enzyme from *Z. rouxii* was found to be independent of the growth phase. On the other hand, it has been suggested that mitochondrial ATPase activity is not essential for salt-tolerance, because respiratory-deficient mutants that lacked mitochondrial ATPase activity were still salt tolerant [42]. From these observations, it appears that plasma membrane H^+-ATPase is involved in salt tolerance in *Z. rouxii*. A cloned plasma membrane H^+-ATPase gene from *Z. rouxii* showed a low homology of the amino acid sequences of the N-terminal region when compared with that of *S. cerevisiae* [41]. The relation between salt tolerance and amino acid sequences of the N-terminal region was not clear. Moreover, the plasma membrane H^+-ATPase isolated from *C. versatilis* behaved like that of *Z. rouxii* [45]. As a result of these observations, WATANABE and coworkers assumed that these traits of plasma membrane H^+-ATPase isolated from salt-tolerant yeasts are significant for supplying the energy to drive transport systems, such as sodium-proton antiporters.

15.5.3.2 Sodium-proton antiporter

To elucidate the relationship between sodium-proton antiporters and salt-tolerance, two sodium-proton antiporter genes, namely *Z-SOD2* and *Z-SOD22*, have been isolated from *Z. rouxii* [40, 14]. These genes are homologues of a sodium-proton antiporter gene of *Sch. pombe*. Cells in which *Z-SOD2* was disrupted could not grow in the presence of 3 M NaCl, but were able to grow in the presence of 50 % sorbitol. This proved that *Z-SOD2* is involved in salt-tolerance, but not in osmotolerance. The salt-tolerance and osmotolerance of cells in which *Z-SOD22* was disrupted were unchanged. Thus, *Z-SOD2* encodes a functional product as an antiporter.

15.5.3.3 Other genes

In a study of salt-sensitive mutants, USHIO and coworkers attempted to clone the gene involved in with salt-tolerance by using the gene complementing method [37]. The cloned gene, *STA1*, showed 53.3 % similarity to *UGA43* of *S. cerevisiae*, a negative regulator of the nitrogen assimilation pathway that contains a typical zinc-finger motif to bind DNA. Decreased salt-tolerance and intracellular accumulation of Na^+ were observed in cells in which *STA1* was disrupted. The researchers assumed that *STA1* is essential for NaCl tolerance and that the protein regulates the transcription of some genes involved with the efflux of Na^+.

15.6 Genetic improvement of soy yeasts

There are only few reports on the genetic improvement of *Z. rouxii*. In this section we discuss the genetic engineering and genetic improvement of the species.

15.6.1 Plasmids

Among soy yeasts, only *Z. rouxii* has plasmid DNA. A plasmid resembling the 2 µm plasmid of *S. cerevisiae* DNA was isolated from *Z. rouxii* IFO 1130 [20]. Characteristics of this plasmid, named pSR1, were investigated in detail [39]. The plasmid has the same gross structure as 2 µm plasmid DNA, and is a double-stranded circular DNA molecule of a 6251-bp nucleotide sequence with inverted repeats and two isomeric forms resulting from intramolecular recombination at the inverted repeat region. pSR1 shows no similarity to other circular DNA plasmids in yeasts. The pair of inverted repeats of pSR1 are able to function as autonomously replicating sequence (ARS) in *Z. rouxii* and *S. cerevisiae*. The plasmid encodes 3 open reading frames named P, R, and S loci. The R locus encodes a site-specific recombinase, and the P and R loci encodes trans-acting products to stabilize pSR1 [20,6].

15.6.2 Construction of a host-vector system for *Zygosaccharomyces rouxii*

Using pSR1, researchers constructed a host-vector system for *Z. rouxii*. IMURA and coworkers cloned a glyceraldehyde-3-phosphate dehydrogenase (EC 1.2.1.12) gene from *Z. rouxii* to obtain a promoter that is functional in the species [12]. The same promoter is well known as a high expression promoter in *S. cerevisiae*. They verified the ability to promote gene expression by using *LacZ* as a reporter gene and showed that the promoter has a comparable expression ability to that of *S. cerevisiae*. An *Escherichia coli* – *Z. rouxii* shuttle vector has been constructed with the *ARS* sequence of pSR1 or *ARS1* of *S. cerevisiae* as a replicon in yeast and the *LEU2* gene of *S. cerevisiae* or the *Tn601* gene, encoding the *G418* resistance gene of *E. coli*, as a selectable marker [39]. Several hundred to 2000 transformants were obtained per microgram of plasmid DNA.

15.6.3 Improvement of *Zygosaccharomyces rouxii* using a host-vector system

The secretion of *Aspergillus oryzae* alkaline protease in *Z. rouxii* was investigated to increase the efficiency of soy sauce production [25]. An expression plasmid consisting of the glyceraldehyde-3-phosphate dehydrogenase promoter, the prepro-alkaline protease cDNA of *A. oryzae*, and the whole sequence of pSR1 and the G418 resistance gene has been constructed. The transformants obtained secreted about 300 mg/mL of *A. oryzae* alkaline protease (EC 3.4.24.40) into the medium.

15.6.4 Other reports of genetic engineering

Genetic engineering is a powerful tool to improve strains. However, the technique is not widely accepted in the food industry and its consumers. One of the reasons is that heterogeneous DNAs, which are used to improve the properties of strains and act as a selection marker of the transformants, remained in the host. *LacZ*, *LEU2*, and the *G418* resistance gene are very useful as selectable markers in transformants, but unfortunately, they do not occur in *Z. rouxii*. Food-grade vectors that are acceptable for the food industry and its consumers, need to be based on the use of homogenous selection marker genes. Therefore, a host-vector system using selection marker genes from *Z. rouxii* is needed. Recently, SYCHROVA and coworkers isolated *ADE2* [35] and *HIS3* [36] from *Z. rouxii*, which encode phosphoribosyl-aminoimidazole carboxylase and imidazoleglycerol-phosphate dehydratase (EC 4.2.1.19), respectively. These are very useful as transformant-selectable markers in *S. cerevisiae*. Therefore, the construction of a food-grade host-vector system in *Z. rouxii* is expected to occur in the near future.

The R locus of the plasmid pSR1 encodes a site-specific recombinase [20]. MATSUZAKI and coworkers developed methods to delete or invert a chromosome segment that uses the R gene of the plasmid [21]. At first, they inserted a DNA fragment, bearing a specific recombination site on the inverted repeats of pSR1, at the target site of *S. cerevisiae* using an integrative vector. Then the cells were transformed with a plasmid bearing the R gene of pSR1, which is placed downstream of the *GAL1* promoter. When the transformants were cultivated in galactose medium, the R gene was expressed. A similar method developed for DNA manipulation in plant cells [27] has not been adapted to soy yeasts yet.

15.7 Prospects and conclusions

In the brewing of soy sauce, yeasts play an important role, especially in the production of many flavour compounds. Because most flavour compounds, including higher alcohols and polyols, are derived from the fermentation by *Z. rouxii*, research has concentrated on this species. To enhance the production of higher alcohols, which are derived from branched-chain amino acids, researchers have isolated mutants that are resistant to amino acid analogues and have investigated the mechanisms of overproduction of these compounds. As the brewer's yeast *S. cerevisiae* produces these flavour compounds, and because molecular genetics has been well developed, the mechanisms of higher alcohol production and polyol accumulation are best understood in this species. Comparative studies suggest that the mechanisms in *Z. rouxii* are similar to those of *S. cerevisiae*. To achieve further improvements in the quality of soy sauce it will be necessary to establish methods to control the production of these flavour compounds by *Z. rouxii*.

A *Z. rouxii* host-vector system has been developed, but only few reports are available on the improvement of *Z. rouxii* by genetic engineering. The reason for this may be that it is not sufficient to elucidate just the genes involved in regulating the desirable characteristics in soy sauce brewing. Therefore, the development of a genomics approach to study soy yeasts will be important.

Sodium chloride plays an important role in soy sauce brewing. At high NaCl concentrations, *Z. rouxii* and *C. versatilis* are able to grow, but other microorganisms, which produce undesirable flavours, do not grow. This control of the microorganism flora in moromi determines the quality of soy sauce.

To explain the salt-tolerance of the important microorganisms, researchers have investigated the accumulation of polyols, the increase in sterol-esters and free fatty acids in the membranes, and characteristics of H^+-ATPase and the sodium-proton antiporter in *Z. rouxii*. However, we believe it is important to elucidate the interactions between these mechanisms as well.

15.8 References

[1] AOKI, T.; UCHIDA, K.: Amino acids-uptake deficient mutants of *Zygosaccharomyces rouxii* with altered production of higher alcohols. Agric. Biol. Chem. **55** (1991) 2893–2894.

[2] AOKI, T.; UCHIDA, K.: Enhanced formation of 2-phenyl-ethanol in *Zygosaccharomyces rouxii* due to prephenate dehydrogenase deficiency. Agric. Biol. Chem. **54** (1990) 273–274.

[3] AOKI, T.; UCHIDA, K.: Enhanced formation of 3-(methylthiol)-1-propanol in *Zygosaccharomyces rouxii*, due to deficiency of S-adenosylmethionine synthase. Agric. Biol. Chem. **55** (1991) 2113–2116.

[4] ARAKI, H.; JEARNPIPATKUL, A.; TATSUMI, H.; SAKURAI, T.; USHIO, K.; MUTA, T.; OSHIMA, Y.: Molecular and functional organization of yeast plasmid pSR1. J. Mol. Biol. **182** (1985) 191–203.

[5] BLANK, I.; FAY, L.B.: Formation of 4-hydroxy-2,5-dimethyl-3(2H)-furanone and 4-hydroxy-2(or 5)-methyl-3(2H)-furanone through Maillard reaction based on pentose sugars. J. Agric. Fd Chem. **44** (1996) 531–536.

[6] BREWSTER, J.L.; VALOIR, T.; DWYER, N.D.; WINTER, E.; GUSTIN, M.C.: An osmosensing signal transduction pathway in yeast. Science **259** (1993) 1760–1763.

[7] BROWN, A.D.: Compatible solutes and extreme water stress in eukaryotic micro-organisms. Adv. Microbiol. Physiol. Vol. 17 (edited by Rose, A.H.; Morris, J.G.). London, U.K.: Academic Press (1978) 181–242.

[8] EDDY, A.A.: Mechanisms of solute transport in selected eukaryotic micro-organisms. Adv. Microbiol. Physiol. Vol. 23 (edited by Rose, A.H.; Morris, J.G.). London, U.K.: Academic Press (1982) 1–78.

[9] HAYASHI, K.: Classification of soy products. In: Science and Technology of Soy Sauce (edited by Tochikura, S.). Soy Sauce Res. Inst. (1993), 506–520 (Japanese).

[10] HAYASHIDA, Y.; NISHIMURA, K.; KURIYAMA, H.; SLAUGHTER, J.C.: 4-Hydroxy-3(2H)-furanones, HDMF and HEMF, production during yeast fermentation in YPD medium with heated sugar and amino acid solution. J. Japan Soy Sauce Res. Inst. **25** (1999) 237–242 (Japanese).

[11] HAYASHIDA, Y.; NISHIMURA, K.; SLAUGHTER, J.C.: The influence of mash pre-aging on the development of the flavour-active compound, 4-hydroxy-2(or 5)-ethyl-5(or 2)-methyl-3 (2H)-furanone (HEMF), during soy sauce fermentation. Int. J. Fd Sci. Technol. **32** (1997) 11–14.

[12] IMURA, T.; UTATSU, I.; TOH-E, A.: Glyceraldehyde-3-phosphate dehydrogenase genes of *Zygosaccharomyces rouxii*: The source of a promoter for a host-vector system for *Z. rouxii*. Agric. Biol. Chem. **51** (1987) 1641–1647.

[13] INOUE, Y.; TSUJIMOTO, Y.; KIMURA, A.: Expression of the glyoxalase I gene of *Saccharomyces cerevisiae* is regulated by high osmolarity glycerol mitogen-activated protein kinase pathway in osmotic stress response. J. Biol. Chem. **273** (1998) 2977–2983.

[14] IWAKI, T.; HIGASHIDA, Y.; TUJI, H.; TAMAI, Y.; WATANABE, Y.: Characterization of a second gene (*ZSOD22*) of Na^+/H^+-antiporter from salt-tolerant yeast *Zygosaccharomyces rouxii* and functional expression of *ZSOD2* and *ZSOD22* in *Saccharomyces cerevisiae*. Yeast **14** (1998) 1167–1174.

[15] IWAKI, T.; TAMAI, Y.; WATANABE, Y.: Two putative MAP kinase genes, *ZrHOG1* and *ZrHOG2*, cloned from the salt-tolerant yeast *Zygosaccharomyces rouxii* are functionally homologous to the *Saccharomyces cerevisiae HOG1* gene. Microbiology **145** (1999) 241–248.

[16] JEARNPIPATKUL, A.; ARAKI, H.; OSHIMA, Y.: Factors encoded by and affecting the holding stability of yeast plasmid pSR1. Mol. Gen. Genet. **206** (1987) 88–94.

[17] KATAOKA, S; LIU, W.; ALBRIGHT, K.; STORKSON, J.; PARIZA, M.: Inhibition of benzo[α]pyrene-induced mouse forestomach neoplasia and reduction of H_2O_2 concentration in human polymorphonuclear leucocytes by flavour components of Japanese-style fermented soy sauce. Fd Chem. Toxicol. **35** (1997) 449–457.

[18] KIKKOMAN home page: www.kikkoman-usa.com

[19] KOGA, T.; MORO, K.; MATSUDO, T.: Antioxidative behaviors of 4-hydroxy-2,5-dimethyl-3(2H)- furanone and 4-hydroxy-2(or 5)-ethyl-5(or 2)-methyl-3(2H)-furanone against lipid peroxidation. J. Agric. Food. Chem. **46** (1998) 946–951.

[20] MATSUZAKI, H.; ARAKI, H.; OSHIMA, Y.: Gene conversion associated with site-specific recombination in yeast plasmid pSR1. Mol. Cell Biol. **8** (1988) 955–962 .

[21] MATSUZAKI, H.; NAKAJIMA, R.; NISHIYAMA, J.; ARAKI, H.; OSHIMA, Y.: Chromosome engineering in *Saccharomyces cerevisiae* by using a site-specific recombination system of a yeast plasmid. J. Bacteriol. **172** (1990) 610–618.

[22] MORI, H.: Microorganisms. In: Science and Technology of Soy Sauce (edited by Tochikura, S.). Soy Sauce Res. Inst. (1993) 152–227 (Japanese).

[23] NAGAHARA, A.; BENJAMIN, H.; STORKSON, J.; KREWSON, J.; SHENG, K.; LIU, W.; PARIZA, M.W.: Inhibition of benzo[α]pyrene-induced mouse forestomach neoplasia by a principal flavour component of Japanese-style fermented soy sauce. Cancer Res. **52** (1992) 1754–1756.

[24] NUNOMURA, N.; SASAKI, M.: Japanese soy sauce flavour with emphasis on off-flavours. In: Off-flavours in foods and beverages (edited by Charalambous, G.). Amsterdam, The Netherlands: Elsevier (1992) 287–312.

[25] OGAWA, Y.; TATSUMI, H.; MURAKAMI, S.; ISHIDA, Y.; MURAKAMI, K.; MASAKI, A.; KAWABE, H.; ARIMURA, H.; NAKANA, E.; MOTAI, H.; TOH-E, A.: Secretion of *Aspergillus oryzae* alkaline protease in an osmophilic yeast, *Zygosaccharomyces rouxii*. Agric. Biol. Chem. **54** (1990) 2521–2529.

[26] ONISHI, H.: Osmophilic yeasts. IV. Change in permeability of cell membranes of the osmophilic yeasts and maintenance of their viability in the saline medium. Bull. Agr. Chem. Soc. Japan **23** (1959) 332–339.

[27] ONOUCHI, H.; YOKOI, K.; MACHIDA, C.; MATSUZAKI, H.; OSHIMA, Y.; MATSUOKA, K.; NAKAMURA, K.; MACHIDA, Y.: Operation of an efficient site-specific recombination system of *Zygosaccharomyces rouxii* in tobacco cells. Nucl. Acids Res. 19 (1991) 6373–6378.

[28] RAINBOW, C.: Brewer's yeasts. In: The yeasts, Vol. 3: Yeast technology (edited by Rose A.H.; Harrison, J.S.). London, U.K. and New York, U.S.A.: Academic Press (1970) 147–224.

[29] REYNOLDS, T.B.; HOPKINS, B.D.; LYONS, M.R.; GRAHAM, T.R.: The high osmolarity glycerol response (HOG) MAP kinase pathway controls localization of yeast golgi glycosyltransferase. J. Cell. Biol. **143** (1998) 935–946.

[30] SASAKI, M.; NUNOMURA, N.; MATSUDO, T.: Biosynthesis of 4-hydroxy-2(or 5)-ethyl-5(or 2)-methyl-3(2H)-furanone by yeasts. J. Agric. Fd Chem. **39** (1991) 934–938.

[31] SERRANO, R.: Energy requirements for maltose transport in yeast. Eur. J. Biochem. **80** (1977) 97–102.

[32] SUEZAWA, Y.: Bioconversions of ferulic acid and p-coumaric acid to volatile phenols by halotolerant yeasts. Nippon Nogeikagaku Kaishi **69** (1995) 1587–1596.

[33] SUEZAWA, Y.; YOSHIOKA, N.; MORI, H.: Bioconversions of ferulic acid and p-coumaric acid to volatile phenols by *Aspergillus* spp. and bacteria found in soy sauce koji and mashes. Nippon Nogeikagaku Kaishi **72** (1998) 43–49.

[34] SUGAWARA, E.; SAKURAI, Y.: Effect of media constituents on the formation by halophilic yeast of 4-hydroxy-2(or 5)-ethyl-5(or 2)-methyl-3(2H)-furanone aroma component specific to miso. Biosc. Biotechnol. Biochem. **63** (1999) 749–752.

[35] SYCHROVA, H.; BRAUN, V.; SPUCIET, J.L.: Molecular cloning and sequence analysis of *Zygosaccharomyces rouxii ADE2* gene encoding a phosphoribosyl-aminoimidazole carboxylase. Yeast **15** (1999) 1399–1402.

[36] SYCHROVA, H.; BRAUN, V.; SPUCIET, J.L.: Sequence and organization analysis of a Zygosaccharomyces rouxii DNA fragment containing the *HIS3* gene. Yeast **16** (2000) 581–587.

[37] USHIO, K.; OTSUKA, H.; YOSHIKAWA, S.; TAGUCHI, G.; SHIMOSAKA, M.; MITSUI, N.; OKAZAKI, M.: Cloning of the *SAT1* gene concerned with salt tolerance of the yeast *Zygosaccharomyces rouxii*. J. Ferment. Bioeng. **82** (1996) 16–21.

[38] USHIO, K.; SAWATANI, M.; NAKATA, Y.: Effect of altered sterol composition on the salt tolerance of *Zygosaccharomyces rouxii*. J. Ferment. Bioeng. **71** (1991) 390–396.

[39] USHIO, K.; TATSUMI, H.; ARAKI, H.; TOH-E, A.; OSHIMA, Y.: Construction of a host-vector system in the osmophilic haploid yeast *Zygosaccharomyces rouxii*. J. Ferment. Technol. **66** (1988) 481–488.

[40] WATANABE, Y.; MIWA, S.; TAMAI, Y.: Characterization of Na^+/H^+-antiporter gene closely related to the salt-tolerance of yeast *Zygosaccharomyces rouxii*. Yeast **11** (1995) 829–838.

[41] WATANABE, Y.; SANEMITSU, Y.; TAMAI, Y.: Expression of plasma membrane proton-ATPase gene in salt-tolerant yeast *Zygosaccharomyces rouxii* is induced by sodium chloride. FEMS Microbiol. Lett. **114** (1993) 105–108.

[42] WATANABE, Y.; SHIRAMIZU, M.; SANEMITSU, Y.: Molecular cloning and sequencing of plasma membrane H^+-ATPase gene from the salt-tolerant yeast *Zygosaccharomyces rouxii*. J. Biochem. **110** (1991) 237–240.

[43] WATANABE, Y.; TAKAKUWA, M.: Change of lipid composition of *Zygosaccharomyces rouxii* after transfer to high sodium chloride culture medium. J. Ferment. Technol. **65** (1987) 365–369.

[44] WATANABE, Y.; TAKAKUWA, M.: Effect of sodium chloride on lipid composition of *Saccharomyces rouxii*. Agric. Biol. Chem. **48** (1984) 2415–2422.

[45] WATANABE, Y.; YAMAGUCHI, M.; SAKAMOTO, J.; TAMAI, Y.: Characterization of plasma membrane H^+-ATPase from salt-tolerant yeast *Candida versatilis*. Yeast **9** (1993) 213–220.

[46] WATANABE, Y.; YAMAGUCHI, M.; SANEMITSU, Y.; TAMAI, Y.: Characterization of plasma membrane proton-ATPase from salt-tolerant yeasts *Zygosaccharomyces rouxii* cells. Yeast **7** (1991) 599–606.

[47] YAGI, T.; NOGAMI, A.; NISHI, T.: Salt tolerance and glycerol accumulation of a respiration-deficient mutant isolated from the petite-negative, salt-tolerant yeast *Zygosaccharomyces rouxii*. FEMS Microbiol. Lett. **92** (1992) 289–294.

[48] YOSHIKAWA, S.; OGURI, I.; KONDO, K.; FUKUZAWA, M.; SHIMOSAKA, M.; OKAZAKI, M.: Enhanced formation of isoamyl alcohol in *Zygosaccharomyces rouxii* due to elimination of feedback inhibition of α-isopropylmalate synthase. FEMS Microbiol. Lett. **127** (1995) 139–143.

16 Mixed microbial fermentations of chocolate and coffee

ROSANE FREITAS SCHWAN and ALAN E. WHEALS

16.1 Introduction

Both chocolate and coffee are consumed as the processed products of fruits in which the seeds have been roasted and ground. One of the early steps in processing is to remove the pulp surrounding the seeds and this is done by microbial activity, sometimes together with mechanical action. In the case of chocolate, the microbial fermentation and the roles of yeasts are well understood and are essential to the full development of chocolate flavour. In the case of coffee the level of understanding remains limited and only recently research has started to reveal the level of complexity of the process.

Coffee and cocoa fermentations fulfil the classic description of a microbial fermentation: Microbial action on the mucilaginous pulp producing alcohols and acids which diffuse into the bean causing biochemical reactions that are responsible for the quality of the final product – coffee beverage and chocolate. In both fermentations bacteria, yeasts and filamentous fungi are present. In the case of cocoa, yeast activity is an essential part of an efficient fermentation. In the case of coffee, the data are less secure but pectinolytic activity seems to play an important role. However, in neither case is the yeast or its metabolites consumed. The fermentation, in which the yeasts take part, causes changes in the coffee and cocoa beans that render them suitable for further processing.

16.1.1 Cocoa and chocolate

Cocoa probably originated in Mesoamerica [22] and has been used as a food, a beverage and as medicine for over 2,000 years. It was also used by the Aztecs as a currency – an example of money literally growing on trees! Christopher Columbus first encountered chocolate "almonds" (beans) at Guanaja off the coast of Honduras in 1502, but its first recorded appearance in the Spanish courts was in 1544. The origin of the word chocolate is unclear but it may have come from *chocol* (Mayan for hot) and *atl* (Aztec for water) reflecting its popularity with Europeans as a hot drink sweetened with cane sugar [20, 24]. Its special status in human culture is reflected in its binomial generic name, *Theobroma*, meaning food of the gods. *Theobroma* has been divided into 22 species of which *T. cacao* is the most widely known. The specific name, *cacao*, probably originated as an Olmec word in Mexico [20].

The cocoa tree is a small-sized under-storey tree of the tropics naturally found within ± 10° of the equator. Mature fruits (pods) arise directly from the stem of the tree, are thick walled and contain 30–40 beans (seeds). Each dicotyledonous bean is enveloped in a sweet, white, mucilaginous pulp comprising approximately 40 % of the seed fresh weight. This pulp is a rich medium for microbial growth. It consists of 82–86 % water, 10–14 % sugar (glucose, sucrose and fructose), 2–3 % pentosans, 1–2 % citrate and 1–2 % pectin together with proteins, amino acids, vitamins (mainly vitamin C) and minerals, and has a pH value close to 3.6 [59]. Harvested seeds are immediately allowed to undergo a natural fermentation in heaps of 25 to 2500 kg enclosed by banana or plantain leaves, in baskets or in large, perforated wooden boxes (holding up to 800 kg of beans) that allow pulp to drain away and air to enter. The fermentation process begins with the growth of micro-organisms. Insects, such as *Drosophila* spp., the vinegar-fly, are probably responsible for the transfer of micro-organisms to the heaps of beans. Microbial action on the mucilaginous pulp creates an array of acids and alcohols and liberates heat. Diffusion of these metabolites into the bean initiates the formation of the precursors of cocoa flavour [47]. After about 5 to 6 days, the pulp disappears and the beans develop a purple colour due to anthocyanins. The length of fermentation varies depending on the type of beans: Forastero beans require about 6 days and Criollo beans 2–3 days. The beans are then sun or machine-dried. During this process some acids in excess may volatilize and oxidation occurs and both are beneficial for the development of chocolate flavour. To produce chocolate the beans are roasted and the cracked husks are air-separated from the cotyledons, while a further series of chemical reactions leads to the development of full chocolate flavour.

16.1.2 Coffee

The coffee tree of the genus *Coffea* probably originated in the province of Kaffa in the Ethiopian highlands and was being cultivated in Yemen by the 15th century. The Dutch brought coffee plants to Holland in 1616, to Batavia in Java (Indonesia) in 1699, to the Dutch colony of Surinam in 1718 and shortly after that to French Guyana and to Brazil. In 1730 the British introduced coffee to Jamaica, and by 1825 coffee was planted in Hawaii. Venetian traders first brought coffee to Europe in 1615. Two species dominate the world market – *C. arabica* (yielding coffee variety *arabica*) and *C. canephora* (yielding coffee variety *robusta*). Like cocoa, it was originally an understorey tree but it is now usually grown in the open air. It is mainly cultivated in the tropics, where it is not exposed to frost, and usually at higher altitudes to avoid excessively high temperatures.

Coffee is produced from the central pair of seeds in the berry (botanically a drupe) that have to be separated from the outer layers, which include skins, mucilage and pulp. Two main methods are used to process coffee berries [5, 19, 40, 61]. In Colombia, Central America and Hawaii for example, the wet method is used on arabica coffees. Although details vary, it involves hand-picking of mature berries from bushes, mechanical depulping to separate the pulp and skins from the beans that are still surrounded by a thin mucilaginous layer, and then a relatively short (ideally 24–48 hour) final water wash or fermentation to aid removal of the mucilage. It is during this stage that microbial activity is important. The beans are then machine or sun-dried to 11–12 % moisture content to produce coffee beans suitable for transport and roasting. In Brazil and Ethiopia, and for robusta coffees worldwide, the dry or natural method of fermentation is usually used. When most berries are mature, the bushes are completely hand or machine-stripped yielding varying amounts of green (immature) berries, over-mature dark brown raisins and dry shrivelled berries. They are then spread on patios (earth, platforms, concrete or tarmac) in layers approximately 10 cm thick, heaped at night and re-spread each day. During the course of 10 to 25 days of sun drying, a natural microbial fermentation occurs and secreted enzymes break down the pulp and mucilage [40]. Eventually a dry berry is left, free of mucilage and pulp, but still surrounded by dry skins from which the beans are mechanically separated and stored at 11–12 % moisture content. In this state the beans are transported to their final destination where they are roasted and ground.

The coffee berry consists of six parts: skin (exocarp), pulp, mucilage, parchment (the outer layer surrounding the beans), silver skin (the inner layer) and bean. Coffee pulp represents about 29 % of the dry weight of the whole fruit. The pulp consists of 76 % water, 10 % protein, 2 % fibre, 8 % mineral salts and 4 % of different soluble and insoluble matter such as pectin, tannins, reducing and non reducing sugars, caffeine, chlorogenic and caffeic acids, cellulose, hemicellulose, lignin and amino acids [23]. The mucilage is located between the pulp and the parchment and represents 5 % of dry weight of the berry. The mucilage is a hydrogel system consisting of water, pectic acid, small amounts of arabinose, galactose, xylose, rhamnose [3] and organic acids [23]. The mucilage also contains hydrolytic enzymes [3, 4]. The polysaccharide composition is: pectic substances (30 %), cellulose (8 %), and neutral non-cellulosic polysaccharides (18 %). The pectins contain 60 % uronic acids with a high degree of methyl esterification (62 %) and a moderate degree of acetylation (5 %) [6].

16.2 Importance

World annual production of cocoa is ca. 2.5 M tonnes and the major producers are, in order, the Ivory Coast, Ghana, Indonesia, Brazil, Nigeria, Cameroon, Malaysia and Ecuador, but there are many other smaller producers. The trade of cocoa is complex: farmers produce fermented beans, warehouses store beans, processors turn this into cocoa products, traders ship to mainly North America and Europe and manufacturers convert this into consumable products. The "first" world dominates the commodities market and determines the price of cocoa for "third" world farmers. After reaching a peak of well over US$ 3,000/tonne in 1977, the price of beans has fallen to an average about US$ 1,000/tonne during the last decade.

World export production of coffee in 1999 was 3.7 M tonnes. The biggest producer is Brazil (1.38 M tonnes) followed by Colombia, Vietnam, Indonesia, India, Mexico, Guatemala and Ethiopia and another 40 smaller exporting countries. *Arabica* coffees are worth approximately US$ 2,000 per tonne, which is at least twice that of *robusta* coffees. The trade is similar to that of cocoa and the price is very variable and prone to sudden surges, particularly when frosts occur in Brazil.

These products of tropical countries are a good example of the effects of global markets on agricultural crops from these countries. Over-production and control by global markets that are in the hands of Western countries have led to dramatic falls in income for farmers. What once were important cash crops for farmers are now becoming uneconomic. Only large cocoa and coffee farms are making substantial profits although the largest profits are made after the products leave the country of origin.

16.3 Yeast biodiversity

Coffee and cocoa pulp are natural substrates for the growth of microorganisms. Yeasts, bacteria and filamentous fungi have been implicated in the processing of both fruits. Both cocoa and coffee are the result of natural fermentations without reliance on defined inocula. In principle a large variety of yeasts can be present initially, but natural selection results in the predominance of a restricted number of species, capable of coping with low pH and high sugar content. The biochemical transformations that are necessary for a normal fermentation in both coffee and cocoa can be performed by many different species provided they have the appropriate physiological capacity. It is possible that the different species found in different countries or in different types of fermentation are not important with respect to the fermentation process *per se*. It may be both necessary and sufficient if representatives of each physiological/ecological group are present in order to provide the appropriate transformations during the fermentation. With cocoa, lactic and then acetic acid bacteria supersede the yeasts in importance. In dry processing of coffee the yeasts co-exist with bacteria and moulds. In wet coffee processing, particular microbial groups may dominate the re-used fermentation tanks and this might be either bacteria or yeasts.

16.3.1 Cocoa

Seeds within the ripe pod are microbiologically sterile. On opening, the pulp becomes contaminated with micro-organisms from the hands of workers, knives, unwashed baskets used for transport of seeds, and dried mucilage left on the walls of boxes from previous fermentations. There is a well-defined microbial succession [45, 54, 64, 73]. In the early stages of the fermentation, several species of yeasts proliferate, leading to the production of ethanol and the secretion of pectinolytic enzymes. This is followed by a phase in which bacteria appear, principally lactic acid bacteria and acetic acid bacteria, which is followed by growth of aerobic spore-forming bacteria. Finally, some filamentous fungi may appear on the surface. A schematic figure of a microbial succession (Figure 16.3-1) summarizes the key events during the processes that occur during cocoa fermentation in Bahia, Brazil.

The yeast population increases from 5 x 10^7 up to 2 x 10^8 viable cells/g of pulp during the first 12 h, then remains almost constant for the next 12 h after which there is a dramatic decline of four orders of magnitude over the next day followed by a slower decrease leading

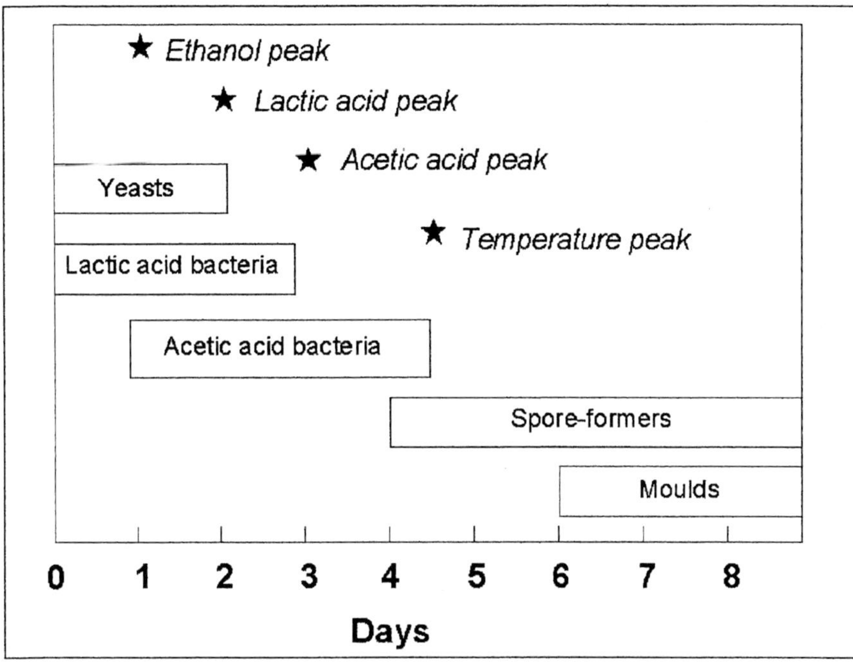

Figure 16.3-1 Schematic representation of a cocoa fermentation from Bahia, Brazil. The boxes indicate the duration of important metabolic activities for the respective groups. The asterisks indicate key transition points.

to a final population density of only 10 viable cells per gram of pulp [73]. Another minor increase in the number of yeasts can be seen in the last day of the fermentation due to growth of thermotolerant yeasts and survivors in the cooler external layers of the fermentation coinciding with an increase in the oxygen content of the fermenting mass.

The amended conditions favour the development of lactic-acid bacteria that reach a peak at the time when the yeast population is in decline. As aeration of the fermenting mass increases and the temperature rises above 37 °C, acetic acid bacteria become the dominant organisms. The exothermic reactions of acetic-acid bacteria raise the temperature of the fermenting mass to 50 °C or higher. The strong acetic acid odour, evident from 48 to 112 h, decreases progressively towards the end of the fermentation.

Yeasts have been isolated from cocoa fermentations by many groups from all the major cocoa-growing areas (Table 16.3-1). It is not possible to determine whether differences in the yeast flora were due to geography or to fermentation practices. In the most comprehensive study [73] frequency with time has been monitored in detail. *Saccharomyces cerevisiae* is the dominant yeast in the cocoa beans taken from boxes immediately after filling. *Kloeckera*

Table 16.3-1 Yeasts isolated from cocoa fermentations in six countries

Country	Yeast genera and species
Brazil [73]	*Candida bombi, C. pelliculosa* (= *Pichia anomala*), *C. rugopelliculosa, C. rugosa, Kloeckera apiculata* (= *Hanseniaspora uvarum*), *Kluyveromyces marxianus, K. thermotolerans, Lodderomyces elongisporus, Pichia fermentans, Saccharomyces cerevisiae, S. cerevisiae* var. *chevalieri (= S. cerevisiae), Torulaspora pretoriensis*
Belize [79]	*Brettanomyces clausenii* (= *Dekkera anomala*), *Candida* spp., *C. boidinii, C. cacaoi* (= Pichia farinosa), *C. guilliermondii* (= *Pichia guilliermondii*), *C. intermedia, C. krusei* (= *Issatchenkia orientalis*), *C. reukaufii* (= *Metschnikowia gruessii*), *Kloeckera apis* (= *Hanseniaspora guilliermondii*), *Pichia membranifaciens, Saccharomyces cerevisiae, S. chevalieri* (= *S. cerevisiae*), *Saccharomycopsis* spp., *Schizosaccharomyces malidevorans* (= *Schizosaccharomyces pombe*), *Schizosaccharomyces* spp.
Malaysia [17]	*Candida* spp., *Debaryomyces* spp., *Hanseniaspora* spp., *Hansenula* spp., *Kloeckera* spp., *Rhodotorula* spp., *Saccharomyces* spp., *Torulopsis* (= *Candida*) spp.
Indonesia [46]	*Candida pelliculosa* (= *Pichia anomala*), *C. tropicalis, Kloeckera apis* (= *Hanseniaspora guilliermondii*), *Saccharomyces cerevisiae*
Java [2]	*Candida, Kloeckera, Pichia, Saccharomyces, Schizosaccharomyces and Trichosporon* spp.
Ivory Coast [32]	*Candida zeylanoides, Pichia membranifaciens, Saccharomyces cerevisiae, Torulopsis candida* (= *Candida saitoana*), *T. castelli* (= *Candida castellii*), *T. holmii* (= *Saccharomyces exiguus*)
Ghana [17]	*Candida* spp., *Hansenula* spp., *Kloeckera* spp., *Pichia* spp., *Saccharomyces* spp., *Saccharomycopsis* spp., *Schizosaccharomyces* spp., *Torulopsis* (= *Candida*) spp.

The species names are those used in the original literature.

apiculata grows during the early phase of fermentation, but declines rapidly in such a way that it could not be isolated after 24 h of fermentation. This probably reflects its intolerance to ethanol at concentrations above 4 % (v/v) [73]. *Kluyveromyces marxianus* grows slowly at the outset of fermentation and then decline gradually. Two different strains of *S. cerevisiae* dominate the alcoholic fermentation phase and survive throughout the fermentation process. Small numbers of *Pichia fermentans* and *Lodderomyces elongisporus* have been isolated, but only during the first few hours of the fermentation. *Candida* spp. increase in numbers after 24 h. *Candida rugosa* is present up to the end of the fermentation when the temperature is ca. 50 °C. *Torulaspora pretoriensis* and *Kluyveromyces thermotolerans* are also found when the temperature of the fermenting mass is ca. 50 °C. The yeast flora is abundant and varies as expected for pulp that contains approximately 14 % sugars. In fresh pods up to 60 % is sucrose and 40 % a mixture of glucose and fructose [59]. All these sugars are fermented by the above species, but *S. cerevisiae* is the most common species of yeast identified in the study, probably because of its rapid growth at pH 3.6 and ethanol-tolerance. The species is also found in high numbers during the first 24 h of cocoa fermentation in Trinidad [54]. The initial reduction in the numbers of the other yeast species is likely to be due to intolerance to low oxygen levels and not to an increasing ethanol concentration. This has been demonstrated both for cider [55] and wine yeasts [38], including *K. thermotolerans*. In cocoa fermentations the oxygen tension decreases from an initial value of 21 % to zero after 36 hours. This latter decline may well be due to a combination of ethanol and high temperature.

16.3.2 Coffee

There have been numerous studies of wet coffee processing [8, 19, 40, 61, 80], but only one comprehensive study of dry processing [75]. In addition, many analyses of the microbial composition of dried beans have been made [5, 80]. A wide variety of microorganisms have been isolated, including fifteen genera of fungi, with a wide array of species capable to produce mycotoxins, as well as 16 genera of bacteria including pectinolytic Gram negatives [2, 82], Gram positives [82] and *Erwinia dissolvens* [29, 80].

16.3.2.1 Wet processing

Entire mature beans are mechanically depulped. Yeasts have been found on the surfaces of berries in trees and during the fermentation process [21, 81] in *robusta* coffee. The discovery of pectinolytic yeasts [2] suggests that they might be responsible for pulp degradation. *Saccharomyces* (= *Kluyveromyces*) *marxianus*, *Saccharomyces bayanus*, *S. cerevisiae* var. *ellipsoideus* (= *S. cerevisiae*) and *Schizosaccharomyces* spp. were identified [2]. *Candida guilliermondii* (= *Pichia guilliermondii*) (including the variety *membranifaciens*), *C. tropicalis*, *C. parapsilosis*, *C. pelliculosa* (= *Pichia anomala*), *Saccharomyces cerevisiae*, *Torulopsis famata* (= *Debaryomyces hansenii* var. *hansenii*), and *Rhodotorula mucilaginosa* have also been identified [81]. One study [8] claimed that yeasts are not involved in mucilage hydrolysis because no pectinolytic yeasts have been isolated from fermentation tanks in Mexico. However, this conclusion derived from the failure to isolate pectinolytic yeasts using selective medium with pectin as the sole carbon source. The ability to hydrolyze pectin has been well established [11, 51], and both *Candida boidinii* [52, 77] and *Geotrichum lactis* [57] can grow on pectin as sole carbon source, but neither have been found in coffee. In contrast, the pectinolytic yeast species that that have been found on coffee include *K. marxianus*, which does not utilize pectin either as sole carbon source or in the presence of glucose [72], and *Saccharomyces cerevisiae* (var. *chevalieri*) that uses the products of pectin hydrolysis only in the presence of non-fermentable carbon sources and, furthermore, high glucose concentrations repress endopolygalacturonase transcription [30, 36]. It is therefore likely that many pectinolytic yeasts were not detected on the selective medium used. Some of the genera of yeasts found on non-selective media include pectinolytic yeasts. It thus remains an open question as to whether yeasts or bacteria are the prime sources of pectinases in wet fermentations.

Usually, a preponderance of bacteria is observed [8, 29, 58, 65], although more yeasts have been described in the mucilage and on the surfaces of coffee berries [2, 81]. One factor that could affect the relative incidence of the different groups of organisms during a wet fermentation may be the pH. An increase in acidity, caused by the release of organic acids, is accompanied by a reduction in the number of bacteria and an increase of that of yeasts [2]. It has been claimed that yeasts can cause degradation of pulp and mucilage in the absence of bacteria [2].

16.3.2.2 Dry processing

Only one study is known on the microbiology of the dry processing method [75]. With few exceptions, bacteria are the most abundant of the three groups of microbes during the process and constitute approximately 80 % of the microorganisms found. The proportions of the filamentous fungi and yeasts vary substantially. Both may be well represented or one group may be almost absent. Approximately 15 % of 754 randomly chosen isolates of microorganisms are yeasts and there is no obvious microbial succession with maturation and fermentation, although there is some indication that the number of microbes increases during the ground fermentation stage. More than double the amount of yeasts has been isolated in a dry year, probably because bacteria generally require a higher water activity to develop than yeasts and filamentous fungi. Only the predominant genera and species should be considered important to understand coffee maturation and processing. *Candida* (22.3 %), *Arxula* (18.9 %) and *Saccharomycopsis* (9.7 %) are the most commonly encountered genera and these three genera appear more often as the fruit dries and ferments. The yeast species identified are *Arxula adeninivorans* (17 isolates), *Blastobotrys proliferans* (1), *Candida auringiensis* (1), *C. glucosophila* (1), *C. incommunis* (3), *C. paludigena* (2), *C. schatavii* (1), *C. vartiovaarae* (2), *Citeromyces matritensis* (1), *Geotrichum fermentans* (1), *Pichia acaciae* (1), *P. anomala* (3), *P. ciferii* (3), *P. jadinii* (2), *P. lynferdii* (4); *P. ofunaensis* (16), *P. sydowiorium* (2), *Saccharomyces cerevisiae* (4), *Saccharomycopsis fermentans* (3), *S. fibuligera* (3), *Schizosaccharomyces pombe* (3), *Sporopachydermia cereana* (1), *Trichosporonoides oedocephales* (1) and *Williopsis saturnus* var. *sargentensis* (1). Almost all these yeasts have been previously isolated from soil, fruits, trees and insect frass [44]. Several species are relatively uncommon, such as *Pichia lynferdii*. Perhaps the biggest surprise is that one quarter of the yeasts isolated belong to *Arxula adeninivorans*, the most common species found at eight out of 12 coffee farms. This is a relatively rare yeast previously reported from soil and silage [41]. Finding *Saccharomycopsis fibuligera*, an amylolytic yeast associated with starchy foods, is unexpected. A study of the natural decomposition of separated coffee waste skins and pulp from the wet process, material which has much in common with that formed during dry processing, has provided isolates of *Rhodotorula*, *Candida* and *Saccharomyces* [14].

16.4 Benefical aspects

Yeasts are an essential part in the fermentation of cocoa because they are the only group of microorganisms simultaneously capable of hydrolyzing pectin and to use ethanol rapidly. In coffee production, it is likely that yeasts can be substituted by bacteria during wet processing. In dry processing of coffee, it has not yet been possible to understand the respective roles of bacteria, yeasts and moulds.

16.4.1 Cocoa

Yeasts have clearly established beneficial and essential effects during the fermentation process. The two most obvious are the utilization of the sugars to produce ethanol and the production of pectinases that degrade the pulp. The crucial role of rapid pectin hydrolysis in the first 24 hours of a cocoa fermentation has been confirmed by the addition of extra enzyme to an early inoculum. This resulted in a more rapid fermentation producing chocolate of high quality after processing (R.F. SCHWAN, unpubl. observ.). It further suggests that pectinase production may be less than optimal during natural fermentations. This may be one of the causes of the occurrence of occasionally slow and unsatisfactory fermentations. Aeration caused by pectin hydrolysis enables the lactic acid bacteria to produce acids rapidly, and it is the combination of heat and acid that kills the beans making them permeable to metabolites that are essential for the development of the chocolate flavour (see below). The hydrolyzed pulp drains from the bottom of the fermenters as sweatings or cacoa honey. This sweet product has a local market distinct from the unfermented cocoa pulp [1, 15].

16.4.2 Coffee

The purpose of dry processing is to yield a bean with no mucilage or pulp but with attached skins that can be mechanically removed. Therefore the microbial consortium involved has to be able to degrade the components. Pectinolytic filamentous fungi, yeasts and bacteria are present during the fermentation. An analysis of the relative importance of each of the groups has not yet been made, but it seems that as long as the pectin is degraded the microbial composition may not be important and may vary from case to case.

The purpose of the water wash during wet processing is to remove the mucilage, a layer that can allow bacterial growth and therefore results in the production of acids that may deteriorate the final product. However, electron microscopy studies suggest that the mucilaginous layer is still present at the end of the fermentation process (quoted in [8]). Furthermore, the chemistry of the mucilage cell wall polysaccharides seems to be relatively little changed and thus the nature of the pectinolysis remains to be elucidated [7].

16.5 Detrimental aspects

Yeasts are not directly responsible for any detrimental effects, but the development of an inappropriate consortium of yeasts can prevent an optimum fermentation. Over-fermentation can lead to taints, but only ethanolic flavours in coffees can be ascribed to yeasts. The predominance of unsuitable yeasts at the beginning of a cocoa fermentation can result in a slower and less efficient fermentation. Rapid and highly efficient yeast fermentation prevents bacteria and fungi from producing deleterious compounds.

16.5.1 Cocoa

Occasionally fermentations take a longer period of time and the quality of the chocolate is sub-optimal. The problem of over-fermentation is caused by filamentous fungal growth at the end of the fermentation resulting in the production of organic acids that will attenuate chocolate flavour. A different consortium of yeasts tends to appear when cocoa pods have been contaminated with pathogenic fungi such as *Crinipellis perniciosa* (witches broom) or *Phytophytora palmivora* (black pod) (R.F. SCHWAN, unpubl. observ.). In these fermentations the genus *Candida* predominates and the fermentation is slow and incomplete after a week.

16.5.2 Coffee

In the wet process, the selection of a uniform population of mature berries and rapid processing seems to be the key to a good quality product [3, 40, 61, 80]. Over-fermentation can lead to spoilage and taints [18, 19, 40]. The fermentative yeasts may be responsible for the ethanol flavour that can be detected in over-fermented beans.

In the dry process over-fermentation or prolonged fermentations due to lack of sun and high humidity will give opportunities for spoilage and mycotoxin-producing organisms to predominate. Ochratoxin A is a mycotoxin that can be found at low levels in ground coffee and seems to be associated with some batches of berries that have been maintained at insufficiently reduced moisture contents for extended periods. A rapid fermentation will reduce the chance of such an occurrence. The formation of butyric and propionic acids from bacterial fermentations causes a loss of quality due to the diffusion of the acids into the beans [3].

16.6 Physiological background

In cocoa processing, yeasts have well established and multiple roles, such as hydrolysis of pectin, and the production of ethanol, organic acid and volatiles. A well-defined microbial succession occurs. This conclusion is based on substantial research and by experimental testing of the hypotheses [68, 73]. In coffee, yeasts have been proposed to be (one of the) pectinase-secreting organisms involved in the degradation of pulp and mucilage. However, the limited amount of studies conducted as well as the simultaneous occurrence and activity of bacteria and fungi have hampered the critical confirmation of this hypothesis.

16.6.1 Roles of yeasts in cocoa fermentation

Five separate roles have been ascribed to yeasts during cocoa fermentations:

a) Ethanol production. The sugar-rich, acidic pulp presents ideal conditions for rapid yeast growth. Conversion of sucrose, glucose and fructose to ethanol and CO_2 is the primary activity of the fermentative yeasts. Measurements of ethanol show clearly how rising concentrations in the pulp eventually penetrate the cotyledons of the beans. However, it is reputedly the acetic acid that kills the beans [28].

b) Breakdown of citric acid. Some of the yeasts, including *Candida* spp. and *Pichia* spp., metabolize citric acid causing an increase of the pH in the pulp, thus allowing growth of bacteria. The loss of citric acid, both in the sweatings and by microbial metabolism, causes an alkaline drift in pH. This, together with the increasing levels of alcohol and aeration, inhibits the yeasts and their activity wanes.

c) Production of organic acids. Several yeast species isolated produce organic acids including acetic, oxalic, phosphoric, succinic and malic acids. These weak organic acids have a buffering capacity reducing fluctuations in pH.

d) Production of volatiles. Yeasts produce a large array of aroma compounds, principally fusel alcohols, fatty acids and fatty acid esters [78], and different species produce different aromas [50]. Volatile compounds are important in the development of full chocolate flavour. The five major yeast species that produce these volatiles (*Kloeckera apiculata* (= *H. uvarum*), *S. cerevisiae*, *S. cerevisiae* var. *chevalieri* (= *S. cerevisiae*), *Candida spp. and Kluyveromyces marxianus*) have been studied individually. *Kloeckera apiculata* (= *H. uvarum*) and *S. cerevisiae* var. *chevalieri* (= *S. cerevisiae*) are the major producers of volatiles, such as isopropyl acetate, ethyl acetate, methanol, 1-propanol, isoamyl alcohol, 2,3 butanediol, diethyl succinate, 2-phenylethanol. Among the yeasts with high fermentative capacity, *S. cerevisiae* var. *chevalieri* (= *S. cerevisiae*) produces large amounts of aroma compounds suggesting that these strains may collaborate in the elaboration of aroma and flavour characteristics (R.F. SCHWAN, unpubl. observ.). Although pure cultures have been used in this study, it provides clues as to which species may be involved in the development of flavour.

e) Production of pectinolytic enzymes. Pectins give pulp its sticky, viscous and cohesive properties. Some yeast strains produce pectinolytic enzymes [32, 64, 67], which break down the cement between the walls of the pulp cells. The resultant juice drains away as sweatings resulting in the formation of void spaces into which air percolates. *Kluyveromyces marxianus* and *S. cerevisiae* var. *chevalieri* (= *S. cerevisiae*) show substantial pectinolytic activity, while *C. rugopelliculosa* and *K. thermotolerans* show a much lower activity [69]. An increasing number of yeast species with pectinolytic activity have been discovered [11, 12]. Pectinolytic yeasts belonging to the genera *Candida*, *Pichia*, *Saccharomyces* and *Zygosaccharomyces* have been found in Java [62]. Yeasts from cocoa fermentations produce various pectinolytic enzymes that aide the maceration of cocoa pulp and the drainage of sweatings [32]. It has been claimed that *S. chevalieri* (= *S. cerevisiae* [9, 44]), *Torulopsis candida* (= *C. saitoana*) and *T. holmii* (*S. exiguus*) produce pectin methyl esterase (PME) and that *S. chevalieri* and *C. zeylanoides* secrete polygalacturonase (PG), although the sequencing of the genome of *S. cerevisiae* has revealed a PG gene but no PME gene. *S. chevalieri* (= *S. cerevisiae*), *C. norvegensis* and *T. candida* (= *C. saitoana*) were the only pectinolytic yeasts isolated from cocoa fermentations in another study [66, 67].

In the only other study where yeast species have been identified, the pectinase-producing variety of *Saccharomyces cerevisiae* var. *chevalieri* was found. In other studies representatives of the genera *Saccharomyces* and *Candida* were found, and both genera include pectinolytic species. It seems likely that all cocoa fermentations contain both strongly fermentative and pectinolytic yeasts. The available data suggest that the anaerobic conditions, occurring rapidly after initiation of natural cocoa fermentations are ideal for the appearance of the enzyme, but that its production becomes self-limiting as the pulp drains away and air percolates through the fermenter.

Pectins are branched heteropolysaccharides of L-rhamnose and α-D-(1,4)-linked residues of D-galactopyranosiduronic acid [27] with partially methanol esterified carboxyl groups of the D-galactopyranosiduronic acid residues. At values less than 65 % esterification, endopolygalacturonase (EC 3.2.1.15) can hydrolyze the backbone. The cocoa pulp pectin has about 60–62 % of methylation [10] and it has been shown that yeast polygalacturonase can hydrolyze the cocoa pectin [70]. Addition of pectin or polygalacturonic acid to the growth medium does not increase enzyme secretion, indicating that PG production is constitutive under these conditions. However, the yeasts are unable to grow on pectin or galacturonic acid as the sole carbon source. The gene for endoPG has been cloned from *Saccharomyces cerevisiae* [13, 37, 39], *S. bayanus* [34, 53] and *K. marxianus* [39, 74]. Although pectin lyase (PL) and PME enzyme activities have been reported from yeasts [31, 51, 56], no genes have yet been detected or cloned.

It has been difficult to understand why the yeasts possess the gene, because it is not induced by pectin and the hydrolyzed products are not assimilated. Some recent work has thrown light on this matter. The non-pectinolytic laboratory strains of *S. cerevisiae* used in genetic studies contain a PG homologue [33]. The endoPG in wild *S. cerevisiae* strains can be induced only under nitrogen starvation with the induction of the pseudohyphal development pathway [48]. This is an invasive method of growth found in plant and animal pathogens. Recent taxonomic work suggests that *S. cerevisiae*, which is naturally found on plant tissues, may be closely related to certain plant pathogens [60]. It is therefore possible that *S. cerevisiae* originated as an opportunistic plant pathogen and is well suited to the unnatural conditions occurring during coffee and cocoa fermentations. Experiments with *S. cerevisiae* have confirmed this potential role as a plant pathogen [35].

16.6.2 Coffee (wet processing)

The purpose of wet processing is to remove the mucilage layer surrounding the bean prior to final drying. It starts with mechanical depulping and the removal of pulp and the outer skin, but it also provides an opportunity for contamination by the processing machines. The subsequent wash is essentially anerobic in water of relatively low temperature. The inner mucilage layer is the only available solid substrate for microbial growth [40]. The role of micro-organisms in this process is to secrete pectinases which break down the pectinaceous mucilage during the fermentation. However, although some pectinolytic yeasts (and bacteria) have been isolated, direct proof of this microbial role is lacking. Indeed it has been reported that the mucilage may remain intact after processing [8]. This implies that other factors are important and a simple bacterial fermentation of 60 % of the low molecular weight saccharides has been suggested as the main purpose of the wash [8].

16.6.3 Coffee (dry processing)

A low pH of 3.6 together with low oxygen levels favour colonization by yeasts [71], which are able to utilize pulp carbohydrates under aerobic and anaerobic conditions. Fermentation of the pectinaceous sugars produces alcohol, acetic, lactic, butyric and other higher carboxylic acids. The presumed function of the yeasts is to ferment the oligosaccharides and produce pectinolytic enzymes, which can degrade the mucilage and pulp. In dry processing, the pulp and outer skins are still present and the substrate is therefore complex. Contamination can occur during re-spreading of the beans on the ground while raining. This also happens when the ground temperature is high and the conditions are largely aerobic. Although pectinolytic yeasts are routinely isolated, the principal pectinolytic organisms that are important in pulp degradation have not yet been identified. Of particular interest is the abundance of an otherwise rarely observed yeast, *Arxula adeninivorans*. Some strains of this species can grow at 45 °C and the species is remarkable for the large array of secreted enzymes, including pectinases, cellobiase I and II, xylosidase, acid phosphatase, RNases and some proteases [41]. This repertoire of enzymes may make the species well suited for coffee fermentations.

16.7 Specific methods to study mixed fermentations

Standard microbiological methods can be used with wet processing of coffee since the medium is aqueous, and individual beans can be isolated and examined. During natural processing, fermenting beans are routinely mixed and re-spread, which tends to make the population more uniform, but the material can cover areas of one hectare. Sampling has to be done from sufficient sites within the sample to be statistically representative. The hydrolyzed pulp that runs off also provides a convenient source for sampling. During the processing of cocoa, sampling from a solid mass of up to one tonne causes difficulties of repeatability. This can be solved by sampling from many places within the mass, at least twice a day and doing experiments in sufficient replication. The counting and sampling of mixed populations of microbes is difficult because most "selective" media are not 100 % efficient. For example, MRS (de Man, Rogosa and Sharpe) medium enriches for lactobacilli, but many filamentous fungi and yeasts can grow on this medium as well. The general solution to this problem is to use a range of media and to check internal consistency in the enumeration.

16.8 Future prospects and conclusions

16.8.1 Starter cultures

From our knowledge of the microorganisms responsible for spontaneous cocoa fermentations and their physiological roles during the process, an attempt has been made to manipulate the fermentation [68]. A defined microbial cocktail has been prepared for use as an inoculum consisting of one pectinolytic yeast species, two lactic acid bacterial species and two acetic acid bacterial species. The yeast *S. cerevisiae* var. *chevalieri* produces pectinase, can ferment all pulp sugars at pH 3.5–4.2, is ethanol tolerant and is present at the beginning of natural fermentations. The lactic-acid bacteria *Lactobacillus lactis* and *L. plantarum* are able to produce lactic acid in the particular conditions in cocoa pulp, namely a pH of 3.5, low oxygen levels and temperatures between 33 °C and 39 °C. The best producers of acetic acid that are also tolerant to temperatures of 45 °C are isolates of *Acetobacter aceti*. *Gluconobacter oxydans* is added to oxidize the ethanol to acetic acid, CO_2 and water.

A cocktail of these microorganisms was inoculated onto cocoa beans immediately after the pod had been broken open and left to ferment in sterilized 200kg wooden boxes for 7 days. The three key metabolites in the pulp, namely ethanol, lactic acid and acetic acid showed similar sequential rises and falls to that found in spontaneous fermentations. Contamination with extraneous microorganisms was kept to a minimum. The beans were dried, roasted and chocolate was produced by the usual means. A taste panel found the product as good as that formed by a natural fermentation, thus suggesting that the fermentation occurs normally. Defined cocktails of microorganisms are more reliably, because of the lack of spoilage organisms. Depulping is another advantageous approach to the course of the fermentation because excess pulp can lead to over-acid fermentation of the beans [46]. The precise design of the depulping machine is important in this process.

16.8.2 Fermenter design

A newly designed stainless steel vessel that can be sterilized, and which is capable of turning a 50kg load of cocoa beans, has been constructed (E.S. FREIRE et al., unpubl. ovserv.). Inoculum, aeration and turn rate can be controlled, the temperature can be monitored and samples can be taken at regular intervals. Early results show that it mimics the natural conditions of fermentation boxes and produces fermented beans in five days. The result is a good quality chocolate. Combined with the use of defined inocula there is the prospect of producing reliably the best quality chocolate in less time. Another approach is to modify a rotary dryer capable of fermenting up to 9 tonnes of beans (A.H. ZAIBUNNISA et al., unpubl. observ.). Initial results show that fermentations using this dryer performs well in comparison with traditional fermentations, despite the fact that the process was stopped after four days. Further research with such fermenters will accelerate the number of variables which can be studied and allow early transmission of this information to farmers.

16.8.3 Identification

A persistent problem for the microbiologist when sampling from natural inocula is the identification of the isolates. When high levels of sugars are present the species can be reasonably well predicted. However, in mixed cultures with bacteria and fungi, the possible consortia are unknown and extensive identification procedures must be applied. Traditional methods using a wide range of biochemical tests are slow, time-consuming, expensive and not fully reliable. The advent of molecular methods is starting to make an impact [42] and from almost all currently accepted yeast species partial ribosomal gene sequences are available [26, 43]. However, when large numbers of yeasts have to be analyzed, sequencing can become prohibitively expensive. Restriction fragment length polymorphisms (RFLPs) can be used in identification [25], but better and more rapid technologies are required, such as species-specific probes for PCR [49] or *in situ* hybridization [76].

16.8.4 Coffee prospects

Coffee fermentation research lags behind cocoa fermentation research. The recent results on both wet [8] and dry [75] processing have laid the groundwork for a better understanding of these processes. The most obvious long-term gains will be in the use of starter cultures and in a more rapid processing. For both wet and dry processing it may prove useful to inoculate the initial fermentation with the appropriate microbial flora but in neither case can the exact composition be specified yet. Experiments have been conducted by one of the major pectinase producing companies to spray coffee beans, harvested for the dry processing method, with pectinolytic enzymes to accelerate the process (R.F. SCHWAN, unpubl. observ.). The fermentation process is accelerated but the cost of such a treatment is prohibitive for farmers. The isolation and study of pectinolytic yeasts with enhanced secretion capacities provide prospects to inoculate dry-processed coffee fermentations with yeasts by spraying.

With coffee there has also been an understandable concentration of effort on spoilage and mycotoxins. No attention has been paid to possible beneficial aspects of the fermentation, for example in improving the quality of under or over-mature beans. Mycotoxins, particularly Ochratoxin A (OTA) can be found in coffee samples. Usually the levels are within acceptable international limits but high values can occur [63]. Analysis of the appearance of OTA [16] suggests that this is a harvesting and post-harvesting problem, when beans are kept at intermediate moisture levels for too long, therefore promoting fungal growth and the synthesis of mycotoxins. These conditions most noticeably occur when green (processed) beans are improperly stored under humid conditions, as well as during slow sun drying of beans during the dry process, particularly when they are not protected from rain. Knowledge of the roles of the micro-organisms involved, should enable more rapid fermentations to be achieved and thus obviate this latter problem.

16.9 References

[1] ADOMAKO, D.; TAKRAMA, J.F.: Large scale collection of cocoa bean pulp juice (sweatings). In: Proc. 12th Int. Cocoa Res. Confer., Salvador, Bahia, Brasil, 1996. Lagos, Nigeria: Cocoa Producers Alliance: (1999) 929–938.

[2] AGATE, A.D.; BHAT, J.V.: Role of pectinolytic yeasts in the degradation of mucilage layer of coffee robusta berries. Appl. Microbiol. **14** (1966) 256–260.

[3] AMORIM, H.V.; AMORIM, V.L.: Coffee enzymes and coffee quality. In: Enzymes in food and beverage processing. ACS Symposium Series, Vol. 47 (edited by Ory, R.L.; Angelo, A.J.St.). Washington, DC, U.S.A.: American Chemical Society (1977) 27–55.

[4] AMORIM, H.V.; MELLO, M.: Significance of enzymes in non alcoholic coffee beverage. In: Food enzymology, Vol. 2 (edited by Fox, P.F.). Amsterdam, The Netherlands: Elsevier (1991) 189–209.

[5] ARUNGA, R.O.: Coffee. In: Fermented foods: economic microbiology, Vol. 7 (edited by Rose, A.H.). London, U.K.: Academic Press (1982) 259–274.

[6] AVALLONE, S.; GUIRAUD, J.P.; GUYOT, B.; OLGUIN, E.; BRILLOUET, J.M.: Polysaccharide constituents of coffee-bean mucilage. J. Fd Sci. **65** (2000) 1308–1311.

[7] AVALLONE, S.; GUIRAUD, J.P.; GUYOT, B.; OLGUIN, E.; BRILLOUET, J.M.: Fate of mucilage cell wall polysaccharides during coffee fermentation. J. Agric. Fd. Chem. **49** (2001) 5556–5559.

[8] AVALLONE, S.; GUYOT, B.; BRILLOUET, J.M.; OLGUIN, E.; GUIRAUD, J.P.: Microbiological and biochemical study of coffee fermentation. Curr. Microbiol. **42** (2001) 252–256.

[9] BARNETT, J.A.; PAYNE, R.W.; YARROW, D.: Yeasts: Classification and identification, 3rd ed. Cambridge, U.K.: Cambridge University Press (2000).

[10] BERBET, P. R. F.: Contribuiçao para o conhecimento dos açúcares componentes da amêndoa e do mel de cacau. Rev. Theobroma **9** (1979) 55–61.

[11] BIELY, P.; SLAVIKOVA, E.: New search for pectolytic yeasts. Fol. Microbiol. **39** (1994) 485–488.

[12] BLANCO, P.; SIEIRO C.; VILLA, T.G.: Production of pectic enzymes in yeasts. FEMS Microbiol. Lett. **175** (1999) 1–9.

[13] BLANCO, P.; SIEIRO, C.; REBOREDO N.M.; VILLA T.G.: Cloning, molecular characterisation, and expression of an endo-polygalacturonase-encoding gene from *Saccharomyces cerevisiae* IM1–8b. FEMS Microbiol. Lett. **164** (1998) 249–255.

[14] BLANDÓN-CASTAÑO, G.; RODRÍGUEZ-VALENCIA, N.; DÁVILA-ARIAS, M.T.: Caracterización microbiológica y físico-química de los subproductos del beneficio del café en proceso de compostaje. Cenicafé **49** (1998) 169–185.

[15] BUAMAH, R.; DZOGBEFIA, V.P.; OLDHAM, J.H.: Pure yeast culture fermentation of cocoa (*Theobroma cacao* L): effect on yield of sweatings and cocoa bean quality. World J. Microbiol. Biotechnol. **13** (1997) 457–462.

[16] BUCHELI, P.; KANCHANOMAI, C.; MEYER, I.; PITTET, A.: Development of ochratoxin A during Robusta (*Coffea canephora*) coffee cherry drying. J. Agric. Fd Chem. **48** (2000) 1358–1362.

[17] CARR, J.G.; DAVIES, P.A.; DOUGAN, J.: Cocoa fermentation in Ghana and Malaysia II. Bristol, London, U.K.: University of Bristol Research Station, Tropical Products Institute (1979).

[18] CHALFOUN, S.M.; CARVALHO, V.D.: Efeito de microrganismos na qualidade da bebida de café. Informe Agropecuário (Belo Horizonte) **18** (1997) 21–26.

[19] CLARKE, R.J.; MACRAE, R.: Coffee Technology, Vol. 2. New York, U.S.A.: Elsevier Applied Science (1987).

[20] COE, S.D.; COE, M.D.: The true history of chocolate. London, U.K.: Thames and Hudson (1996).

[21] DAIVASIKAMANI, S.; KANNAN, N.: Studies on post-harvest mycoflora of coffee berry of *Robusta*. J. Coffee Res. **16** (1986) 102–106.

[22] DE LA CRUZ, M.; WHITKUS, R.; GOMEZ-POMPA, A.; MOTA-BRAVO, L.: Origins of cacao cultivation. Science **375** (1995) 542–543.

[23] ELIAS, L.G.: Composición quimica de la pulpa de café y otros subprodutos. In: Pulpa de café: Composición, tecnologiay utilización (edited by Braham, J.E.; Bressan, R.). Panamá, Panamá: INCAP (1978) 19–29.

[24] ERDMAN, J.W.; WILLS, J.; FINLEY, D'A. (editors): Chocolate: Modern science investigates an ancient medicine. J. Nutrit. **130**, Supplement 8S (2000).

[25] ESTEVE-ZARZOSO, B.; BELLOCH, C.; URUBURU, F.; QUEROL, A.: Identification of yeasts by RFLP analysis of the 5.8S rRNA gene and the two ribosomal internal transcribed spacers. Int. J. Syst. Bacter. **49** (1999) 329–337.

[26] FELL, J.W.; BOEKHOUT, T.; FONSECA, A.; SCORZETTI, G.; STATZELL-TALLMAN, A.: Biodiversity and systematics of basidiomycetous yeasts as determined by large-subunit rDNA D1/D2 domain sequence analysis. Int. J. Syst. Evol. Microbiol. **50** (2000) 1351–1371.

[27] FOGARTY, W.M.; WARD, O.P.: Pectinases and pectic polysaccharides. Progr. Ind. Microbiol. **13** (1996) 59–119.

[28] FORSYTH, W.G.C.; QUESNEL, V. C.: Mechanisms of cocoa curing. Adv. Enzym. **25** (1963) 457–492.

[29] FRANK, H.A.; LUM, N.A.; DELA CRUZ, A.S.: Bacteria responsible for mucilage layer decomposition in Kona coffee berries. Appl. Microbiol. **13** (1965) 201–207.

[30] GAINVORS, A.; BELARBI, A.: Detection methods for polygalacturonase-producing strains of *Saccharomyces cerevisiae*. Yeast **11** (1995) 1493–1499.

[31] GAINVORS, A.; FREZIER, V.; LEMARESQUIER, H.; LEQUART, C.; AIGLE, M.; BEALRBI, A.: Detection of polygalacturonase, pectin-lyase and pectin-esterase activities in a *Saccharomyces cerevisiae* strain. Yeast **10** (1994) 1311–1319.

[32] GAUTHIER, B.; GUIRAUD, J.; VINCENT, J.C.; PORVAIS, J.P.; GALZY, P.: Comments on yeast flora from traditional fermentation of cocoa in the Ivory Coast. Rev. Ferment. Ind. Aliment. **32** (1977) 160–163.

[33] GOFFEAU, A.; BARRELL, B.G.; BUSSEY, H.; DAVIS, R.W.; DUJON, B.; FELDMANN, H.; GALIBERT, F.; HOHEISEL, J.D.; JACQ, C.; JOHNSTON, M.; LOUIS, E.J.; MEWES, H.W.; MURAKAMI, Y.; PHILIPPSEN, P.; TETTELIN, H.; OLIVER, S.G.: Life with 6000 genes. Science **274** (1996): 546–567.

[34] GOGNIES, S.: GAINVORS, A.; AIGLE, M.; BELARBI, A.: Cloning, sequencing analysis and overexpression of a *Saccharomyces cerevisiae* endopolygalacturonase-encoding gene (*PGL1*). Yeast **15** (1999) 11–22.

[35] GOGNIES, S.; BELARBI, A.; BARKA, E.A.: *Saccharomyces cerevisiae*, a potential pathogen towards grapevine, *Vitis vinifera*. FEMS Microbiol. Lett. **1271** (2001) 1–8.

[36] GOGNIES, S.; SIMON, G.; BELARBI, A.: Regulation of the endopolygalacturonase gene *PGU1* in *Saccharomyces*. Yeast **18** (2001) 423–432.

[37] HIROSE, N.; KISHIDA, M.; KAWASAKI, H.; SAKAI, T.: Molecular cloning and expression of a polygalacturonase gene in *Saccharomyces cerevisiae*. J. Ferment Bioengin. **86** (1998) 332–334.

[38] HOLM HANSEN, E.; NISSEN, P.; SOMMER, P.; NIELSEN, J. C.; ARNEBORG, N.: The effect of oxygen on the survival of non-*Saccharomyces* yeasts with mixed culture fermentations of grape juice with *Saccharomyces cerevisiae*. J. Appl. Microbiol. **91** (2001) 541–547.

[39] JIA, J.; WHEALS, A.E.: Endopolygalacturonase genes and enzymes of *Saccharomyces cerevisiae* and *Kluyveromyces marxianus*. Curr. Genet. **38** (2000) 264–270.

[40] JONES, K.L.; JONES, S.E.: Fermentations involved in the production of cocoa, coffee and tea. Progr. Ind. Microbiol. **19** (1984) 411–456.

[41] KUNZE, G.; KUNZE, I.: *Arxula adeninivorans*. In: Nonconventional yeasts in biotechnology: A handbook (edited by Wolf, K.). Berlin, Germany: Springer Verlag (1996) 389–409.

[42] KURTZMAN, C.P.: Yeast systematics – from phenotype to genotype. Fd Technol. Biotechnol. **36** (1998) 261–266.

[43] KURTZMAN, C.P.; ROBNETT, C.J.: Identification and phylogeny of ascomycetous yeasts from analysis of nuclear large subunit (26S) ribosomal DNA partial sequences. Antonie van Leeuwenhoek **73** (1998) 331–371.

[44] KURTZMAN, C.P; FELL, J.W. (editors): The Yeasts: a taxonomic study, 4th ed. Amsterdam, The Netherlands: Elsevier (1998).

[45] LEHRIAN, D.W.; PATTERSON, G.R.: Cocoa fermentation. In: Biotechnology, a comprehensive treatise, Vol. 5 (edited by Reed, G.). Basel, Switzerland: Verlag Chemie (1983) 529–575.

[46] ARDHANA, M.M.: Microbial ecology and biochemistry of cocoa bean fermentations, PhD thesis, University of New South Wales, Sydney, Australia (1990).

[47] LOPEZ, A.S.: Fermentation and organoleptic quality of cacao as affected by partial removal of pulp juices from the beans prior to curing. Rev. Theobroma **9** (1979) 25–37.

[48] LOPEZ, A.S.; DIMICK, P.S.: Cocoa fermentation. In: Biotechnology: A comprehensive treatise, Vol. 9, Enzymes, food and feed, 2nd ed. (edited by Reed, G.; Nagodawithana, T.W.). Weinheim, Germany: VCH (1995) 563–577.

[49] MADHANI, H.D.; GALITSKI, T.; LANDER, E.S.; FINK, G.R.: Effectors of a developmental mitogen-activated protein kinase cascade revealed by expression signatures of signalling mutants. Proc. Nat. Acad. Sc. USA **96** (1999) 12530–12535.

[50] MANNARELLI, B.M.; KURTZMAN, C.P.: Rapid identification of *Candida albicans* and other human pathogenic yeasts by using short oligonucleotides in a PCR. J. Clin. Microb. **36** (1998) 1634–1641.

[51] MATEO, J.J.; JIMENEZ, M.; HUERTA, T.; PASTOR, A.: Contribution of different yeasts isolated from Monastrell grapes to the aroma of wine. Int. J. Fd Microbiol. **14** (1991) 153–10.

[52] MIDDELHOVEN, W.J.: Identity and degradative abilities of yeasts isolated from plants growing in an arid climate. Antonie van Leeuwenhoek **72** (1997) 81–89.

[53] NAKAGAWA, T.; MIYAJI, T.; YURIMOTO, H.; SAKAI, Y.; KATO, N.; TOMIZUKA, N.: A methylotrophic pathway participates in pectin utilization by *Candida boidinii*. Appl. Environ. Microbiol. **66** (2000) 4253–4257.

[54] NAUMOV, G.I.; NAUMOVA, E.S.; AIGLE, M.; MASNEUF, I.; BELARBI, A.: Genetic reidentification of the pectinolytic yeast strain SCPP as *Saccharomyces bayanus* var. *uvarum*. Appl. Microbiol. Technol. **55** (2001) 108–111.

[55] OSTOVAR, K.; KEENEY, P.G.: Isolation and characterization of microorganisms involved in the fermentation of Trinidad's cacao beans. J. Fd Sci. **38** (1973) 611–617.

[56] PANON, G.: Influence of oxygen on fermentation pattern in model media containing mixed or sequential cultures of three cider-producing yeasts: *Saccharomyces cerevisiae, Hanseniaspora valbyensis* and *Metschnikowia pulcherrima*. Sci. Alim. **17** (1997) 193–217.

[57] PANON, G.; MASSIOT, P; DRILLEAU, J.F.: Pectinolytic enzymes production by yeasts in cider fermentation. Sci. Alim. **15** (1995) 31–42.

[58] PARDO, C.; GACTO, M.: Production and properties of an exo-polygalacturonic acid hydrolase from *Geotrichum lactis*. Microbios **71** (1992) 193–202.

[59] PEDERSON, C.S.; BREED, R.S.: Fermentation of coffee. Fd Res. **11** (1946) 99–106.

[60] PETTIPHER, G.L.: Analysis of cocoa pulp and the formulation of a standardised artificial cocoa pulp medium. J. Sci. Fd Agric. **37** (1986) 297–309.

[61] PRILLINGER, H.; SCHWEIGKOFLER, W.; BREITENBACH, M.; BRIZA, P.; STAUDACHER, E.; LOPANDIC, K.; MOLNAR, O.; WEIGANG, F.; IBL, M.; ELLINGER, A.: Phytopathogenic filamentous (*Ashbya, Eremothecium*) and dimorphic fungi (*Holleya, Nematospora*) with needle-shaped ascospores as new members within the Saccharomycetaceae. Yeast **13** (1997) 945–960.

[62] PUERTA, G.I.: Evaluación de la calidad del café colombiano processado por vía seca. Cenicafé, Colombia **47** (1996) 85–90.

[63] Roelofsen, P.A.: Polygalacturonase activity in yeast, *Neurospora* and tomato extract. Biochim. Biophys. Acta **10** (1953) 410–413.

[64] Romani, S.; Sacchetti, G.; Lopez, C.C.; Pinnavaia, G.G.; Dalla Rosa, M.: Screening on the occurrence of ochratoxin A in green coffee beans of different origins and types. J. Agric. Fd Chem. **48** (2000) 3616–3619.

[65] Rombouts, J.E.: Observations on the microflora of fermenting cocoa beans in Trinidad. Proc. Soc. Appl. Bacteriol. **15** (1952) 103–111.

[66] Roussos, S.; Angeles Aquiáhuatl, M.; Trejo-Hernández, M.R.; Gaime Perraud, I.; Favela, E.; Ramakrishna, M.; Raimbault, M.; Viniegra-González, G.: Biotechnological management of coffee pulp-isolation, screening, characterization, selection of caffeine-degrading fungi and natural microflora present in coffee pulp and husk. Appl. Microbiol. Biotechnol. **42** (1995) 756–762.

[67] Sanchez, J.; Daguenet, G.; Guiraud, J.P.; Vincent, J.C.; Galzy, P.: A study of the yeast flora and the effect of pure culture seeding during the fermentation of cocoa beans. Lebensmittel-wiss. Technol. **18** (1985) 69–76.

[68] Sanchez, J.; Guiraud, J.P.; Galzy, P.: A study of the polygalacturonase activity of several yeast strains isolated from cocoa. Appl. Microbiol. Biotechnol. **20** (1984) 262–267.

[69] Schwan, R.F.: Cocoa fermentations conducted with a defined microbial cocktail inoculum. Appl. Environ. Microbiol. **64** (1998) 1477–1483.

[70] Schwan, R.F.; Cooper, R.M.; Wheals, A.E.: Endo-polygalacturonase secretion by *Kluyveromyces marxianus* and other cocoa pulp-degrading yeasts. Enz. Microb. Technol. **21** (1997) 234–244.

[71] Schwan, R.F.; Cooper, R.M.; Wheals, A.E.: Endopolygalacturonase of the yeast *Kluyveromyces marxianus* is constitutive, highly active on native pectin and is the main extracelleular protein. In: Pectins and pectinases (edited by Visser. J.; Voragen, A.G.). Amsterdam, The Netherlands: Elsevier (1996) 861–868.

[72] Schwan, R.F.; Lopez, A.; Silva, D.O.; Vanetti, M.C.D.: Influência da frequência e intervalos de revolvimentos sobre a fermentaçao de cacau e qualidade do chocolate. Rev. Agrotróp. **2** (1990) 22–31.

[73] Schwan, R.F.; Rose, A.H.: Polygalacturonase secretion by *Kluyveromyces marxianus*: effect of medium composition. J. Appl. Bacter. **76** (1994) 62–67.

[74] Schwan, R.F.; Rose, A.H.; Board, R.G.: Microbial fermentation of cocoa beans, with emphasis on enzymatic degradation of the pulp. J. Appl. Bacteriol. Symp. Suppl. **79** (1995) 96S–107S.

[75] Siekštele, R.; Bartkeviciute, D.; Sasnaukas, K.: Cloning, targeted disruption and heterologous expression of the *Kluyveromyces marxianus* endopolygalacturonase gene (*EPG1*). Yeast **15** (1999) 311–322.

[76] Silva, C.F.; Schwan, R.F.; Dias, E.S.; Wheals, A.E.: Microbial diversity during maturation and natural processing of coffee cherries of *Coffea arabica* in Brazil. Int. J. Fd Microbiol. **60** (2000) 251–260.

[77] Stender, H.; Kurtzman, C.P.; Hyldig-Nielsen, J.J.; Sorensen, D.; Broomer, A.; Oliveira, K.; Perry-O'Keefe, H.; Sage, A.; Young, B.; Coull, J.: Identification of *Dekkera bruxellensis* (*Brettanomyces*) from wine by fluorescence *in situ* hybridization using peptide nucleic acid probes. Appl. Environ. Microbiol. **67** (2001) 938–941.

[78] STRATILOVA, E.; BREIEROVA, E.; VADKERTIOVA, R.; MACHOVA, E.; MALOVIKOVA, A.; SLAVIKOVA, E.: The adaptability of the methylotrophic yeast *Candida boidinii* on media containing pectic substances. Can. J. Microbiol. **44** (1998) 116–120.

[79] SUOMALAINEN, H.; LEHTONEN, M.: The production of aroma compounds by yeast. J. Inst. Brew. **85** (1979) 149–156.

[80] THOMPSON, S.S.; MILLER, K.B.; LOPEZ, A.: Cocoa and Coffee. In: Food microbiology fundamentals and frontiers (edited by Doyle, M.P.; Beuchat, L.R.; Montville T.J.). Washington, DC, U.S.A.: ASM Press (1997) 649–661.

[81] VAN PEE, W.; CASTELEIN, J.M.: Study of the pectinolytic microflora, particularly the Enterobacteriaceae, from fermenting coffee in the Congo. J. Fd Sci. **37** (1972) 171–174.

[82] VAN PEE, W.; CASTELEIN, J.M.: The yeast flora of fermenting *robusta* coffee. East Afr. Agric. Forest. J. **26** (1971) 308–310.

[83] VAUGHN, R.H.; CAMARGO, R.DE; FALLANGE, H.; MELLO AYRES, G.; SERGEDELLO, A.: Observations on the microbiology of the coffee fermentation in Brazil. Fd Technol. Suppl. **4** (1958) 12–57.

17 Traditional fermented products from Africa, Latin America and Asia

M.J. ROBERT NOUT

17.1 Introduction

Fermentation is regarded as one of the oldest-known methods of food processing and preservation. Fermented foods and beverages are obtained by the action of micro-organisms (bacteria, yeasts and mycelial fungi) and their enzymes. Traditional fermented foods, also referred to as indigenous fermented foods, are those popular products that are known since early history and that can be prepared in the household or cottage industry using relatively simple techniques and equipment. The world-wide diversity of traditional fermented foods, their preparation methods and safety aspects are the subject of several textbooks and encyclopedia [1, 7, 61]. This chapter will deal with some of the documented foods of Africa, Latin America and Asia. As other chapters deal specifically with shoyu (Chapter 15), cocoa (Chapter 16), coffee (Chapter 16), wine (Chapter 14) and kefyr (Chapter 8), this chapter will not include those products.

Yeasts occur in a wide range of fermented foods, made from ingredients of plant as well as animal origin. When yeasts are present as an abundant group of organisms, they usually have a significant impact on food quality parameters such as taste, texture, odour, as well as nutritive value. There are many products however, in which yeasts take part in the fermentation in small numbers next to the presence of bacteria or mycelial fungi. Such mixed food fermentations sometimes are characterized by synergistic and very stable communities of e. g., yeasts and lactic acid bacteria (LAB), or yeasts and amylolytic mycelial fungi that depend on each other for nutrient supply [38]. On the other hand, the presence of some yeasts in fermented foods can result in undesirable reactions and can thus be considered as spoilage. The present chapter will deal with the occurrence of yeasts (biodiversity), their beneficial effects in fermented foods, as well as some cases of negative effects. Future prospects for development and industrialization will be discussed.

17.2 Yeast biodiversity related to specific fermented products

A large variety of yeasts can be found as functional flora in traditional fermented foods world-wide. Table 17.2-1 summarizes some major catogories of fermented foods and beverages. Some representative examples will be discussed in more detail.

In all continents, yeasts play a predominant role in the preparation of alcoholic beverages made from cereals or sugary juices. Likewise, fermented doughs and batters are encoun-

Table 17.2-1 Major categories of traditional fermented foods involving yeasts

Continent	Alcoholic beverages	Starters for alcoholic fermentation	Doughs and Batters	Miscellaneous
Africa	**made from maize, sorghum or millets, using malt for brewing:** busaa [36, 37], pito [56]), ting [34]. **palmwine** [65]	**natural fermentation (no starters added), or yeast grown on inoculation belt** [56]	**made from cereals (maize, sorghum, millets):** kenkey [50], kisra [42], mawè [23], **or root-crops (cassava):** agbelima [2], fufu [48], lafun [42], **mostly fermented before cooking.**	**fermented milk products:** amasi [14], m'bannick [35], **fermented vegetables:** kawal [18], **non-or low-alcoholic cereal beverages:** togwa [42]
Latin America	**made from sugary juices:** aguardente [33], pulque [8, 60], toddy [30] found world-wide	natural fermentation	**made from maize, cooked before fermentation:** pozol [45]	**non-or low-alcoholic beverages:** sugary kefyr [38], tibicos [3, 8].
Asia	**made from rice, using amylolytic starters for brewing: beers or wines:** brem bali [39], chongju [29], ou [39], sake [29, 37], samsu [30], sato [39], shaohing [39], takju [29], tapuy [39], yakju [39], **or pasty snacks:** khaomak [29], peuyeum (= tape ketella) [29, 37], tapai pulut [29], tape ketan [9].	**yeasts present in mixed fungal amylolytic starters** [42]: bubod [52, 53], koji [39], murcha [63], nuruk [29], ragi-tape [37].	**made mostly from rice and leguminous seed flour (dal), fermented before steaming (some other cereals such as millets may be used):** dhokla [29], dosa [5], idli [43, 58, 66], jalebies [5], nan [39], phool waries [5], punjabi waries [5].	**fermented milk products: kumiss [32], fermented mass of leguminous seeds (mainly soybeans):** hamanatto [37], kinema [55], miso [37], tempe [41, 44], **non-or low-alcoholic beverages:** kombucha [37, 49, 64].

tered in all continents. These doughs and batters may be leavened (i. e., have a spongy texture due to gas produced) or not, but in most cases they are cooked or steamed prior to consumption as a staple food providing starch, energy and some protein. Unlike alcoholic beverages, doughs and batters are fermented by a majority population of bacteria usually dominated by lactic acid bacteria. Nevertheless, yeasts are present in significant numbers and contribute to flavour, texture and nutritive value of the products.

17.2.1 Alcoholic beverages

A distinction will be made on the basis of principles of generating fermentable sugars. Plant or fruit juices containing sufficient levels of glucose or sucrose are the easiest to ferment. They simply need to be inoculated, naturally from the environment or using an enrichment starter or pure culture. Pulque is an example of such a fermentation.

On the other hand, when cereals are used as ingredients, endosperm starch needs to be converted into fermentable sugars (maltose and glucose). This can be achieved using added amylolytic enzymes, sources of which are germinated (sprouted) cereal grains or mixed fungal amylolytic starters. In African beermaking, germinated sorghum and millets are commonly used in brewing; the preparation of pito beer from sorghum is a representative example. In most Asian countries, amylolytic starters are used in the form of starchy tablets containing mixed cultures of starch degrading moulds and yeasts. Such starters are used for the manufacture of beers, wines and pasty snacks from various kinds of rice, sorghum, and cassava. The preparation of takju in Korea, using a starter called nuruk is a representative example.

Mexican pulque (Figure 17.2-1) is made from Agave juice (*Agave atrovirens* or *A. americana*). Essential micro-organisms in the fermentation are *Lactobacillus plantarum*, a heterofermentative *Leuconostoc*, *Zymomonas mobilis* and *Saccharomyces cerevisiae*. Other yeasts include, *Candida parapsilosis*, *C. rugosa*, *C. rugopelliculosa*, *Debaryomyces carsonii*, *Pichia guilliermondii*, *P. membranifaciens* and *Torulaspora delbrueckii* [28]. Although *S. cerevisiae* appears to be the major producer of ethanol, it is *Z. mobilis* that transforms 45 % of the glucose to ethanol (4–6 % v/v in final product) and carbon dioxide [60].

Pito beer (Figure 17.2-2) from Ghana is obtained by mixed activities of lactic acid bacteria and yeasts. It is a yellow to brown coloured sorghum beer that is obtained from previously germinated sorghum which is extracted, boiled and inoculated. Depending on the type of pito, the inoculation is achieved by immersing a woven "inoculation belt" (Figure 17.2-3) which allows entrapment of microbial cells. Other inoculation methods include back-slop-

cut *Agave* inflorescence
↓
collect juice
↓
inoculate with previous juice
↓
ferment 8-30 days (15-30°C)
↓
pulque
(white viscous, sour-alcoholic beverage)

Fig. 17.2-1 Manufacture of pulque (Mexico)

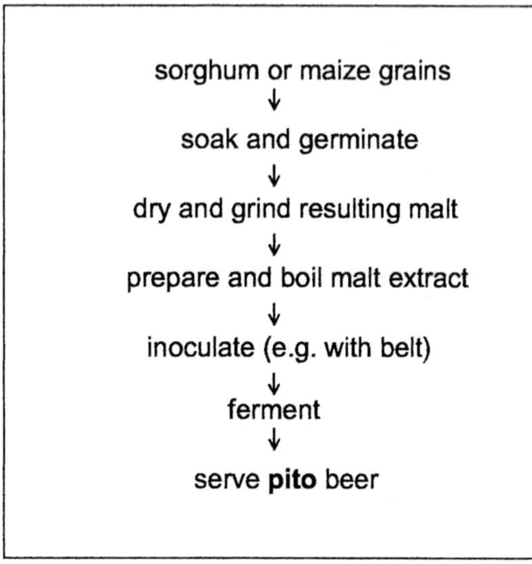

Fig. 17.2-2 Manufacture of pito (Ghana)

ping (addition of previous beer), or adding dried scum (foam) of previous beer [56]. After fermentation it typically contains 1.5–3.5 % v/v ethanol and 0.7–1.0 % w/w lactic acid and a corresponding pH of about 3.5.

Takju (Figure 17.2-4) is a Korean rice beer, which can also be prepared from other cereals. The nuruk starter is made by solid-state fermentation of wheat flour with *Aspergillus usamii.* After approximately 2 months fermentation, nuruk also contains *Rhizopus, Aspergillus niger* and yeasts such as *Debaryomyces hansenii, D. occidentalis, Pichia anomala, P. fabianii* and *Saccharomycopsis fibuligera.* Nuruk contains fungal amylolytic enzymes to saccharify starch (brewing), as well as the yeasts needed for alcoholic fermentation. Takju is a turbid beer with suspended insoluble solids and living yeasts, containing 7–10 % ethanol, approx. 1 % titratable acidity and has pH 4 after 3 days fermentation [29, 39].

17.2.2 Fermented doughs and batters

In Africa, fermented doughs form the basis for a variety of staple foods. These doughs are first fermented, then cooked.

Mawè, an uncooked fermented maize dough from Bénin, is not consumed as such but it is used as an ingredient for the preparation of a wide variety of beverages, cooked and fried meals and snacks. Its fermentation (Figure 17.2-5) is dominated by heterofermentative lactic acid bacteria (> 9 Log CFU/g) but a minority of yeasts (7–8 Log CFU/g) are important for the correct taste [23].

Fig. 17.2-3 Inoculation belt for pito fermentation

polished rice (4 parts)

↓

wash and steep

↓

steam

↓

mix with 1 part of powdered
nuruk and 10 parts of water
in an earthen jar

↓

ferment for 2-3 days

↓

sieve

↓

serve **takju** beer

Fig. 17.2-4 Manufacture of takju (Korea)

Ghanaian Kenkey, a fermented and cooked stiff maize dough is fermented by mixed lactic acid bacteria and yeasts. Although yeasts (*Issatchenkia orientalis* and *S. cerevisiae*) are a minority with about 6–7 Log CFU/g compared to the LAB, they contribute to the taste and odour of kenkey [17, 42].

Nigerian Fufu is a product obtained by submerged cassava fermentation [48]. Yeasts are present in relatively high numbers (8 Log CFU/g) and comprise predominantly *Issatchenkia orientalis*, *Candida tropicalis* and *Zygosaccharomyces bailii*. These yeasts co-exist with lactic acid bacteria such as *Lactobacillus plantarum*. The growth of the latter was reported to be enhanced in the presence of *Issatchenkia orientalis*.

In Mexico, a typical alkaline maize dough (nixtamal) is fermented after having been cooked. Mexican pozol (Figure 17.2-6) is a refreshing beverage prepared from fermented nixtamal, which is a dough made from maize cooked in alkali. In this fermentation, which takes about 12–60 h, yeasts are a minority (2–7 Log CFU/g), and the fermented dough contains a majority of lactic acid bacteria resulting in pH 4.7–5.7 and 0.35–0.75 % titratable acidity. It was reported that 50 % of the yeasts isolated from this product can hydrolyze starch [8].

In Asia, leavened batters of rice and leguminous flour are obtained by fermentation. Subsequently they are steamed. Idli is popular throughout India and Sri Lanka because of its typ-

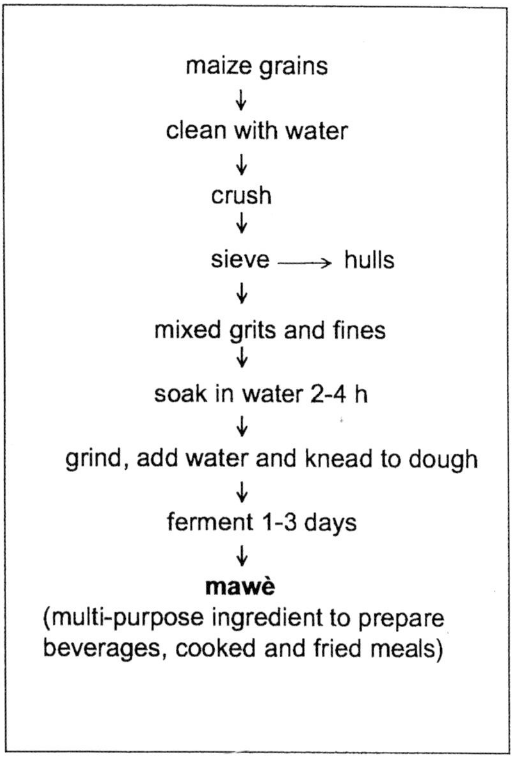

Fig. 17.2-5 Home-scale preparation of mawè (Bénin)

ical sour flavour and spongy texture, nutritional quality and improved digestibility. It is fed to infants as a weaning food, and as a main dish in diets in hospitals [43].

The main ingredients used in the traditional preparation of idli (Figure 17.2-7) are white polished rice (*Oryza sativa* L.) and black gram (*Phaseolus mungo* L.) dal, which are washed, soaked separately in water at room temperature for 5–10 hours before grinding in a stone mortar or other grinders. While rice is coarsely ground, the dal is ground to a fine smooth paste. The rice and dal slurries are mixed and stirred to form a thick batter. Salt is added to taste. The batter, put in a closed container, is kept at a warm place to ferment overnight or longer. The fermentation period must allow a definite leavening of the batter and development of a pleasant acid flavour. The fermented batter is poured in small cups or in a special idli pan having cups (8–10 cm diam), and steamed until the starch is gelatinized and the idli cakes are soft and spongy. The fermented batter is consumed the same day and there is no effort to preserve it.

Idli is a natural fermented food; no inoculum is added generally for fermentation. This is because the essential microorganisms have been found to be naturally present in the ingre-

maize grains
↓
cook in lime water
(about 1 hour)
↓
wash → hulls
↓
alkaline dough "nixtamal"
↓
grind
↓
knead to balls
↓
wrap in banana leaves
↓
ferment 5-8 days
↓
pozol
(mix with water and serve
as porridge or beverage)

Fig. 17.2-6 Preparatio and use of pozol (Mexico)

dients. When the product is made daily, it is often the practice of adding a bit of freshly fermented batter ("backslop") to the newly ground one. In addition to lactic acid bacteria (*Leuconostoc mesenteroides, Enterococcus faecalis, Lactobacillus fermentum* and *Pediococcus cerevisiae)* the slightly acid environment favours the growth and activity of yeasts, mainly *Saccharomyces cerevisiae, Debaromyoces hansenii* var. *hansenii*, *Pichia anomala*, *Candida saitoana* and *Trichosporon cutaneum* var. *cutaneum.*

The major functions of the fermentation include the leavening of the batter and the improvement of taste and nutritional value of idli. *Leuconostoc mesenteroides* is the main species responsible for the production of CO_2 which results in about 2–3 times increase in the original volume of batter [58].

17.2.3 Some other products

In western Sudan, kawal, sigda and furundu are fermented products made by solid-state fermentation from plant leaves and seeds [18]. The fermentation of kawal is dominated by *Ba-*

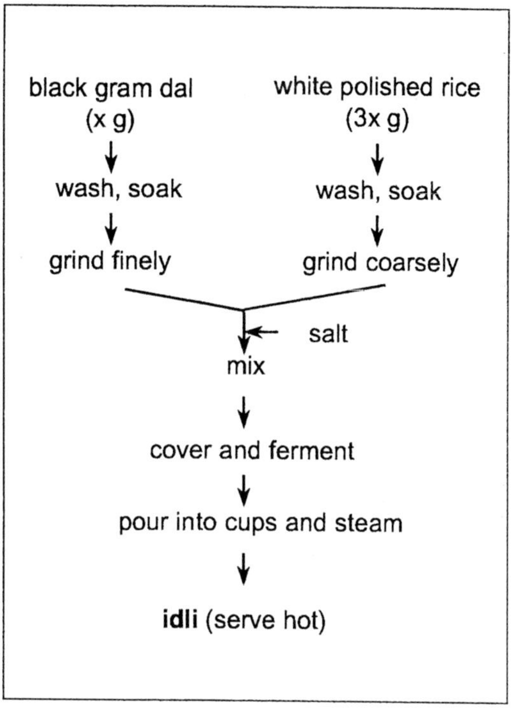

Fig. 17.2-7 Preparation of idli (India, Sri Lanka)

cillus subtilis, *Lactobacillus plantarum*, *Propionibacterium* sp. and *Staphylococcus sciuri*, and two yeasts *Issatchenkia orientalis* and *Saccharomyces* sp. During later stages of the fermentation, *Debaryomyces* and *Candida* spp. are detected in low numbers. The role of the yeasts is to degrade starch into fermentable sugars in the early stages of fermentation, and in later stages they consume lactic acid. It is not clear whether amylolytic yeasts are to be considered desirable or not for this type of fermentation.

Mexican tibicos, also called tibi grains, and probably similar to sugary kefyr, are microbiogleae consisting of dextran with embedded bacteria and yeasts (*Dekkera anomala*, *Pichia guilliermondii*, *P. membranifaciens* var. *membranifaciens*, *Cryptococcus albidus*, *Rhodotorula mucilaginosa* var. *mucilaginosa* and *Saccharomyces cerevisiae*) that live in symbiosis. Tibicos are used to prepare a kind of soft drink from sugar cane juice, containing approximately 3 % sugar, 0.6 % lactic acid and traces of alcohol and acetic acid, after 5 days of fermentation [51].

Kinema from Nepal and north-east India [55] is one of several "alkaline fermented foods" made from protein-rich seeds (e. g., soy beans) with a predominant *Bacillus subtilis* fermentation. During the fermentation a sticky mass is formed of pH 8.5 that is consumed as a flavouring condiment. Yeasts, particularly *Candida parapsilosis* are a minority (<5.2 Log CFU/g) in the microflora and their contribution to the product quality is still uncertain.

Tempe from Indonesia is a fungal solid-state fermented cake-like product, usually made from soy beans [41]. It's principal functional micro-organisms are *Rhizopus* and *Mucor* spp. but various bacteria and yeasts may be present, in relatively high (4–8 Log CFU/g) numbers. The predominant yeasts include *Debaryomyces, Rhodotorula, Candida, Pichia* and *Cryptococcus* spp. (VAN LAARHOVEN, unpubl. observ.).

Kombucha from Central and East Asia is a beverage obtained by fermentation of sweetened boiled tea with a mixed culture of yeasts and acetic acid bacteria [7].

17.3 Beneficial aspects of yeasts in fermentations

Yeasts can have several beneficial effects:

Functional effects, by the production of alcohol, gas, flavour and taste, as well as a contribution to food preservation by scavenging of sugars and other compounds that could otherwise serve as assimilable carbon sources for spoilage-causing micro-organisms. The production of alcohol improves the aroma of the product. In addition, alcohol at a certain concentration makes the substrate unsuitable for spoilage-causing microorganisms. This effect is increased in the presence of organic acids produced by bacteria. However, this preservative effect is not always a guarantee for long shelf-life as was observed in African beers such as Burukutu and Pito made from maize and sorghum [54]. These underwent spoilage within 72 h after the end of fermentation. The spoilage was associated with a decline of the yeast flora, and a concomitant increase and dominance of acetic acid bacteria including *Acetobacter aceti, A. pasteurianus* and *A. hansenii.*

The typical attractive volatiles produced by yeasts in fermentations of cereals such as maize include ethanol, propanol, 2-methyl-1-propanol, 3-methyl-1-butanol (isoamylalcohol), 2,3 butanediol, acetaldehyde, acetoin, diacetyl (2,3 butanedione), acetic acid, and ethyl acetate [10, 17, 40]. In agbelima, a fermented cassava dough [2], the volatile aroma was also attributed to the yeasts present, which produce various alcohols as well as ethylacetate and acetoin. In particular, the ability to assimilate or ferment the available carbohydrates, determines the evolution of attractive odours. These are produced almost exclusively by fermentation [40]. In Table 17.3-1, the ability to ferment glucose is mentioned for all relevant yeast species.

A study of palm wine volatile aroma [65] pointed out that *Saccharomyces* spp. were the major responsible organisms for the attractive odour. About 17 esters, 4 alcohols, 4 terpenes and 2 hydrocarbons were detected in the headspace volatiles.

Nutritional improvement due to increasing the digestibility of foods by the degradation of anti-nutritional factors such as phytate, and by the synthesis of nutrients such as vitamins. For example, *S. cerevisiae* and *Issatchenkia orientalis* occurring in African sorghum

Tab. 17.3-1 Yeast genera and species predominating in fermented foods

Genus	Species	Synonyms reported in literature sources	Fermentation of glucose	Traditional Fermented Foods
Candida	*fennica*	*Trichosporon fennicum*	+	tapuy
Candida	*glabrata*		+	idli, mawè
Candida	*inconspicua*	*Torulopsis inconspicua*	–	yakju
Candida	*intermedia* var. *intermedia*	*Candida intermedia*	+	tempe
Candida	*maltosa*		+	tempe
Candida	*parapsilosis* var. *parapsilosis*	*Candida parapsilosis*	+	bubod, kinema, pulque, tapuy
Candida	*rugopelliculosa*		+	pulque
Candida	*rugosa* var. *rugosa*	*Candida rugosa*	–	pulque
Candida	*saitoana*	*Torulopsis candida*	w	bubod, idli, dosa, dhokla
Candida	*sake*		v	aguardente, idli, tempe
Candida	*sake*	*Torulopsis sake*	v	yakju
Candida	spp.			brem bali, fufu, kombucha (= teekvass), lafun, ou, pito, pozol, sato, toddy
Candida	*stellata*		w	kombucha
Candida	*tropicalis* var. *tropicalis*		+	agbelima, bubod, fufu, idli, pito
Candida	*versatilis*	*Torulopsis versatilis*	w	hamanatto, idli, kombucha, miso
Clavispora	*lusitaniae*	*Candida lusitaniae*	+	amasi, tempe
Cryptococcus	*albidus* var. *albidus*	*Cryptococcus albidus*	–	tibicos
Cryptococcus	*humicola*		–	tempe
Debaryomyces	*carsonii*	*Pichia carsonii*	–	pulque
Debaryomyces	*hansenii* var. *hansenii*	*Candida famata*	v	nuruk, tibicos

Tab. 17.3-1 Continued

Genus	Species	Synonyms reported in literature sources	Fermentation of glucose	Traditional Fermented Foods
Debaryomyces	*hansenii* var. *hansenii*		v	idli, tapuy, tempe
Debaryomyces	*occidentalis* var. *occidentalis*		+	tempe
Debaryomyces	*occidentalis* var. *occidentalis*	*Schwanniomyces occidentalis*	+	nuruk
Debaryomyces	*polymorphus* var. *polymorphus*	*Pichia polymorpha*	v	yakju
Dekkera	*bruxellensis*	*Brettanomyces bruxellensis*	w	kombucha
Dekkera	*anomala*	*Brettanomyces claussenii*	+	tibicos
Hanseniaspora	*uvarum*	*Kloeckera apiculata*	+	pito, sugary kefir
Issatchenkia	*orientalis*	*Candida krusei*	+	busaa, fufu, idli, kawal, kenkey, kisra, mawè, phool waries, punjabi waries, togwa
Kluyveromyces	*africanus*		+	pito
Kluyveromyces	*marxianus* var. *marxianus*	*Candida kefyr*	+	kumiss, mawè
Kluyveromyces	*marxianus* var. *marxianus*	*Candida pseudotropicalis*	+	m'bannick
Kluyveromyces	*marxianus* var. *marxianus*	*Kluyveromyces fragilis*	+	kumiss
Kluyveromyces	*marxianus* var. *marxianus*		+	aguardente, kumiss, pulque
Lodderomyces	*elongisporus*		+	tempe
Pichia	*anomala*		+	bubod, fufu, murcha
Pichia	*anomala*	*Candida javanica*	+	amylolytic starters, idli, kombucha, murcha

Tab. 17.3-1 Continued

Genus	Species	Synonyms reported in literature sources	Fermentation of glucose	Traditional Fermented Foods
Pichia	*anomala*	*Hansenula anomala*	+	bubod, idli, jalebies, koji, murcha, nuruk, pito, tape ketan, saké, yakju
Pichia	*fabianii*	*Candida fabianii*	+	nuruk
Pichia	*fabiani*	*Hansenula fabiani*	+	tempe
Pichia	*fermentans* var. *fermentans*	*Candida lambica*	+	sugary kefir
Pichia	*fluxuum*	*Candida mycoderma*	w	ting
Pichia	*guilliermondii*	*Candida guilliermondii*	w	bubod, pulque, tibicos
Pichia	*membranifaciens* var. *membranifaciens*	*Candida valida*	w	sugary kefir, tibicos
Pichia	*membranifaciens* var. *membranifaciens*	*Pichia membranaefaciens*	w	pulque, ting
Pichia	*pini*		v	tempe
Pichia	spp		+	khaomak, tapai pulut
Pichia	*subpelliculosa*	*Hansenula subpelliculosa*	+	yakju
Rhodotorula	*glutinis* var. *glutinis*		–	tapuy
Rhodotorula	*minuta* var. *minuta*		–	tempe
Rhodotorula	*mucilaginosa* var. *mucilaginosa*		–	tempe
Rhodotorula	*mucilaginosa* var. *mucilaginosa*	*Rhodotorula rubra*	–	tibicos, ting

Tab. 17.3-1 Continued

Genus	Species	Synonyms reported in literature sources	Fermentation of glucose	Traditional Fermented Foods
Saccharomyces	*cerevisiae* var. *cerevisiae*		+	aguardente, amasi, amylolytic starters, bubod, busaa, chongju, fufu, idli, jalebies, kenkey, mawè, nan, palm wine, pito, pulque, punjabi waries, saké, shaohing, sugary kefir, takju, tibicos, ting, yakju
Saccharomyces	*cerevisiae* var. *cerevisiae*	*Saccharomyces chevalieri*	+	palm wine
Saccharomyces	*dairenensis*		+	amasi, tempe
Saccharomyces	*bayanus*	*Saccharomyces globosus*	+	kumiss
Saccharomyces	*exiguus*	*Torulopsis holmii*	+	idli
Saccharomyces	*kluyveri*		+	nan
Saccharomyces	spp		+	brem bali, kawal, khaomak, kombucha (= teekvass), m'bannick, ou, pito, sato, toddy
Saccharomyces	*unisporus*		+	bubod, kumiss
Saccharomycodes	*ludwigii* var. *ludwigii*		+	kombucha
Saccharomycopsis	*fibuligera*	*Candida lactosa*	w	samsu
Saccharomycopsis	*fibuligera*	*Endomycopsis fibuliger*	w	bubod, murcha, nuruk, peuyeum (= tape ketella), ragi-tape, tapai pulut, tapuy
Saturnispora	*saitoi*	*Pichia saitoi*	+	fufu
Schizosaccharomyces	*pombe* var. *pombe*		+	kombucha, pito
Schizosaccharomyces	spp		+	toddy

Tab. 17.3-1 Continued

Genus	Species	Synonyms reported in literature sources	Fermentation of glucose	Traditional Fermented Foods
Torulaspora	*delbrueckii*		+	kombucha, pito
Torulaspora	*delbrueckii*	*Candida colliculosa*	+	amasi, pulque
Torulaspora	*pretoriensis*		+	sugary kefir
Trichosporon	*cutaneum* var. *cutaneum*	*Trichosporon beigelii*	–	idli, tempe
Trichosporon	*cutaneum* var. *cutaneum*		–	pozol
Trichosporon	*pullulans*		–	idli
Yarrowia	*lipolytica*		–	tempe
Zygosaccharomyces	*bailii* var. *bailii*	*Zygosaccharomyces bailii*	+	fufu, kombucha
Zygosaccharomyces	*florentinus*		+	sugary kefir
Zygosaccharomyces	*rouxii*		w	hamanatto, miso
Zygosaccharomyces	*rouxii*	*Saccharomyces rouxii*	w	punjabi waries, ting
Zygosaccharomyces	spp.			agbelima

beer can contribute to human nutrition by the prodution of valuable proteins and amino acids [27]. Phytic acid and polyphenols occur as anti-nutritional factors in cereals such as pearl millet (*Pennisetum typhoideum*). Fermentations of cooked pearl millet [26] with single cultures of *S. cerevisiae* (reported as *S. diastaticus*) result in slight reductions of these compounds. Added impact can be obtained when mixed or sequential yeast-lactic acid bacteria fermentations are carried out, resulting in an improved protein efficiency ratio and higher digestibility values measured in in-vitro rat feeding studies [25].

Another health-related effect is found in the kombucha or "tea-fungus". The mixed yeast-bacterial culture growing on sugary tea extract accumulates lactic (0.1 %), acetic (traces) and gluconic (0.01–0.3 %) acids, and some ethanol (0.3 %). The pH decreases steadily to about 2.5 [49]. The resulting beverage is considered healthy, and it may be expected that in addition to the acids, some vitamins and minerals will be accumulated.

The antimutagenic activity of milk products fermented by various lactic acid bacteria was enhanced by co-culturing it with *Saccharomyces cerevisiae* [62]. Since the mechanisms of these phenomena are not known, it will be of interest to carry out more research on these functional aspects.

Safety improvement by contributing towards the degradation of potentially toxic naturally-occurring substances in food ingredients. For instance, cyanogenic glycosides such as amygdalase can be degraded efficiently by β-glycosidase [EC 3.2.1.21] activities (amygdalase and linamarase) produced by *Saccharomycopsis fibuligera* [6].

17.4 Detrimental aspects of yeasts in (fermented) foods

In particular, oxidative (non-fermenting) yeasts are associated with spoilage of (fermented) foods. Such yeast-associated spoilage can manifest itself as:

Degradation of organic acids, which can be achieved by assimilation or by direct oxidation, causing an increase of pH and concomitant loss of microbial stability [13].

Formation of yeasty off-odours. In fish paste fermentations, it was found that exclusion of oxygen enabled a better control of undesirable levels of yeast growth and resulting off-odours [4]. In tempe, *Cryptococcus humicola*, *Pichia* spp. and *Rhodotorula minuta* were associated with off-odours during the storage of tempe (van Laarhoven, unpubl. observ.). In miso, film-forming yeasts such as *Pichia* sp. and *Zygosaccharomyces rouxii* (reported as *Z. halomembranis*) are detrimental because of their strong unpleasant odour [12].

Formation of discolorations, turbidity and/or gas.

Formation of potentially toxic substances. Ethyl carbamate can accumulate in fermented products undergoing alcoholic fermentation followed by some form of heat treatment. Ethyl carbamate (urethane) is formed by the reaction of carbamic acid with ethanol [11]. Carbamic acid is a yeast metabolite of citrulline. Ethyl carbamate is carcinogenic; the highest levels (100–300 ppb, exceptionally 1 ppm) are found in distilled alcoholic beverages (brandy, bourbon, sake). In view of the relatively limited consumption of these beverages the risks of ethyl carbamate are considered small.

17.5 Physiological key properties

Physiological aspects of importance in the ecology of natural fermentations are the formation of functional enzymes, the assimilation of carbon and nitrogen sources, microbial interactions, tolerance to ethanol and lysis of cells.

Formation of functional enzymes to release assimilable carbon sources. Obviously these enzymes are valuable in brewing and flavour development. For example, glucoamylase (glucan 1,4-alpha-glucosidase) [EC 3.2.1.3] is a key enzyme in rice wine fermentation, converting starch directly into glucose. Glucoamylase from *Pseudozyma* (= Candida) *tsukubaensis* was reported to be constitutive and inducible by glucose, starch, maltose and glycerol [59]. Also, β-glucosidases formed by e. g. *Zygosaccharomyces bailii* [15] occur as extracellular and intracellular enzymes and can degrade a variety of polysaccharides. This ability would enable the yeast to mobilize assimilable carbon sources. On the other hand, glucosidase activity can also contribute to flavour development as a number of flavour precursors in fruits are glycosides.

Assimilation of wide or narrow ranges of carbon and nitrogen sources that correspond with the nutrients available in natural substrates. Using molecular typing techniques for *Saccharomyces cerevisiae*, it was observed that in naturally fermenting African maize dough, *Saccharomyces cerevisiae* strains are involved throughout the fermentation period [20]. The ability to assimilate galactose, saccharose, lactate, raffinose, maltose and glucose was common in these isolates and corresponds very well with the naturally occurring assimilable carbohydrates in uncooked maize. The same group investigated the occurrence of *Issatchenkia orientalis* [21], which is also a fermentative yeast that contributes to the attractive flavour of maize dough.

Microbial interactions that improve chances for survival and growth. Examples are the proto-cooperative interaction of yeasts and lactic acid bacteria in sourdoughs [38] and the formation of killer-toxins by yeasts that reduce competitive yeasts e. g., in wine fermentations [38].

Tolerance to ethanol, especially related to cytoplasmic membrane composition and of practical importance in alcoholic fermentations that must yield high (> 15 % v/v) levels of alcohol content.

Lysis of yeast cells is associated with many of the nutritional benefits but also to enzymatic spoilage phenomena. The lysis is determined by proteases that influence the hydrolysis and solubility of complexes, not involving the cell wall as such [68].

17.6 Future prospects and conclusions

The traditional foods mentioned in this chapter are well accepted, affordable and use local resources. It is important to ensure that their quality and safety meet the requirements of present-day and future consumers. Upgrading traditional home-scale processes is needed so that they can compete successfully with imported products. Whereas small-scale manufacture has advantages of short distribution lines, income generation for families etc., urbanization and the resulting growing demand for ready-to-consume foods requires larger-scale industrial production. Examples of industrialized traditional fermented foods are: alcoholic pastes and rice wines in Asia, such as tapai which is now produced at a small cottage scale in Malaysia using commercially available pure culture starters of the starch degrading mould *Amylomyces rouxii* and the yeast *Saccharomycopsis fibuligera* [31]; palm wine and sorghum beer [16] in S. Africa are prepared at industrial scale in processes involving souring of sorghum mash with *Lactobacillus delbrueckii* at 48–50 °C. After boiling and straining the obtained wort, alcoholic fermentation is performed at 20–35 °C using pure strains of *Saccharomyces cerevisiae*; African doughs such as mawè in Bénin, in which the performance of added starter cultures was tested. Although mawè can be prepared using only lactic acid bacteria such as *Lactobacillus brevis*, the addition of *Issatchenkia orientalis* (commonly found in maize dough) enhances the growth of the lactic acid bacteria and the performance of the fermentation [24]. Although it was observed that traditionally fermented dough had a better flavour, added starter cultures can be very useful in semi-industrial settings to achieve predictability of short fermentation times. Mageu, a non-alcoholic sour maize porridge, is produced at an industrial scale in South Africa. A similar product is known in Kenya as uji. Although these porridges are fermented using lactic acid bacteria only, yeasts mainly *Pichia* spp., as well as *Acetobacter liquefaciens* are involved in the process as spoilage microorganisms [22]. Spoilage yeasts are kept under control using high fermentation temperatures and chemical preservatives such as benzoate, sorbate and propionate. Ogi, a sour fermented starch cake processed from maize, sorghum or millet grains has been industrialized in Nigeria [47]. The fermentation is not inoculated and depends on the natural fermenting flora in which *Lactobacillus plantarum* is considered essential. However, yeasts such as *Saccharomyces cerevisiae* and *Pichia fluxuum* contribute to the acceptability of the flavour. A cost analysis showed that inoculation with pure culture starters would be unacceptably expensive, considering the infrastructure needed to propagate and maintain appropriate quality and safety of such cultures.

Yeast products such as enzymes, vitamins of the B-group, trace elements (Selenium, Chromium), glycans, flavour components and carotenoid pigments [19] occur in traditional foods, but could be exploited more effectively as purified substances and food ingredients. New processing methods including the use of immobilized yeast cells, are promising for obtaining a higher efficiency of starch degradation and ethanol production in rice wines [57]. The development of starters for commercial processes continues. Miso, a fermented salted paste of soybeans, rice and barley, is produced at a large industrial scale. The fermentation takes place in two stages, a mould solid state fermentation initiated by the inoculation with a koji starter, and a brine fermentation during which halophilic lactic acid bacteria (*Tetragenococcus halophila*) and yeasts (*Zygosaccharomyces rouxii* and *Candida versatilis*) are essential for acidity and flavour development. These are grown as defined mixed cultures and are available commercially for processing [12]. Modern molecular biotechnology for starter culture development resulted in the insertion of the α-amylase gene of *Saccharomycopsis fibuligera* into *Saccharomyces cerevisiae* [67] with the advantage of more rapid growth and fermentation. Another example is the insertion of a synthetic gene for lysine into *Saccharomyces* sp. [46]. It was shown that lysine was overproduced and excreted. When used as a fermentation starter culture, the yeast could be used to enrich protein-poor products such as fufu. However, in view of the cost aspects of industrial production, it is doubtful whether such expensive GMO techniques could be applied in practice.

In conclusion, a wide variety of yeasts is involved in traditional fermented foods. Those that contribute to desirable product properties require characterization in view of more efficient exploitation, whereas the undesirable yeasts need further study in order to develop consumer-friendly strategies to avoid their metabolic activity.

17.7 References

[1] ADAMS, M.R.; NOUT, M.J.R. (Eds.): Fermentation and Food Safety. Gaithersburg, Md., U.S.A: Aspen Publishers (2001).

[2] AMOA-AWUA, W.K.; FRISVAD, J.C.; SEFA-DEDEH, S.; JAKOBSEN, M.: The contribution of moulds and yeasts to the fermentation of agbelima cassava dough. J. Appl. Microbiol. **83** (1997) 288–296.

[3] ARMIJO, C.; TABOADA, J.; LAPPE, P.; ULLOA, M.: Products of fermentation by tibicos and associated yeasts. Rev. Latinoam. Microbiol. **33** (1991) 17–23.

[4] AVHURHI, J.B.; OWENS, J.D.: The influence of oxygen accessibility on the growth of yeast in fish/rice fermentation. J. Fd Sci. Technol. India **27** (1990) 104–106.

[5] BATRA, L.R.: Fermented cereals and grain legumes of India and vicinity. In: Advances in biotechnology. Vol. II. Fuels, chemicals, foods and waste treatment. (edited by Moo-Young, M.; Robinson, C.W.). Toronto, Canada: Pergamon Press (1981) 547–553.

[6] BRIMER, L.; NOUT, M.J.R.; TUNCEL, G.: Beta-glycosidase (amygdalase and linamarase) from *Endomyces fibuliger* (LU677): formation and crude enzyme properties. Appl. Microbiol. Biotechnol. **49** (1998) 182–188.

[7] CAMPBELL-PLATT, G.: Fermented foods of the world. A dictionary and a guide. Guildford, Surrey, UK: Butterworth Scientific (1987).

[8] CHAVEZ, N.A.L.; WACHER RODARTE, M.: The fermentation of pozol: the effect of thermal and alkaline treatment of maize. In: Alimentos fermentados indigenas de Mexico. (edited by Wacher-Rodarte, C.; Lappe, P). Ciudad Universitaria, Mexico D.F., Mexico: Universidad Nacional Autonoma de Mexico (1993) 103–108.

[9] CRONK, T.C.; STEINKRAUS, K.H.; HACKLER, L.R.; MATTICK, L.R.: Indonesian tapé ketan fermentation. Appl. Environ. Microbiol. **33** (1977) 1067–1073.

[10] DEMUYAKOR, B.; OHTA, Y.: Characteristics of single and mixed culture fermentation of pito beer. J. Sci. Fd Agric. **62** (1993) 401–408.

[11] DIACHENKO, G.W.; CANAS, B.J.; JOE, F.L.; DINOVI, M.: Ethyl carbamate in alcoholic beverages and fermented foods. In: Food safety assessment (edited by Finley, J.W.; Robinson, S.F.; Armstrong, D.J.). ACS Symposium Series No.484, American Chemical Society (1992) 419–428.

[12] EBINE, H.: Industrialization of Japanese miso fermentation. In: Industrialization of indigenous fermented foods (edited by Steinkraus, K.H.). New York, U.S.A.: Marcel Dekker (1989) 89–126.

[13] FLEET, G.: Spoilage yeasts. CRC Crit. Rev. Biotechnol. **12** (1992) 1–44.

[14] GADAGA, T.H.; MUTUKUMIRA, A.N.; NARVHUS, J.A.: Enumeration and identification of yeasts isolated from Zimbabwean traditional fermented milk. Intern. Dairy J. **10** (2000) 459–466.

[15] GUEGUEN, Y.; CHEMARDIN, P.; ARNAUD, A.; GALZY, P.: Comparative study of extracellular and intracellular beta-glucosidases of a new strain of *Zygosaccharomyces bailii* isolated from fermenting agave juice. J. Appl. Bacteriol. **78** (1995) 270–280.

[16] HAGGBLADE, S.; HOLZAPFEL, W.H.: Industrialization of Africa's indigenous beer brewing. In: Industrialization of indigenous fermented foods (edited by Steinkraus, K.H.). New York, U.S.A.: Marcel Dekker (1989) 191–284.

[17] HALM, M.; LILLIE, A.; SORENSEN, A.K.; JAKOBSEN, M.: Microbiological and aromatic characteristics of fermented maize doughs for Kenkey production in Ghana. Int. J. Fd Microbiol. **19** (1993) 135–143.

[18] HARPER, D.B.; COLLINS, M.A.: Leaf and seed fermentations of western Sudan. In: Applications of biotechnology to traditional fermented foods (edited by Gaden, E.L.; Bokanga, M.; Harlander, S.; Hesseltine, C.W.; Steinkraus, K.H.). Report of an ad-hoc panel of the Board on Science and Technology for International Development. Washington, DC, U.S.A.: National Research Council (1992) 105–113.

[19] HAY, J. D.: Novel yeast products from fermentation processes. J. Chem. Technol. Biotechnol. **58** (1993) 203–205.

[20] HAYFORD, A.E.; JESPERSEN, L.: Characterization of *Saccharomyces cerevisiae* strains from spontaneously fermented maize dough by profiles of assimilation, chromosome polymorphism, PCR and MAL genotyping. J. Appl. Microbiol. **86** (1999) 284–294.

[21] HAYFORD, A.E.; JAKOBSEN, M.: Characterization of *Candida krusei* strains from spontaneously fermented maize dough by profiles of assimilation, chromosome profile, polymerase chain reaction and restriction endonuclease analysis. J. Appl. Microbiol. **87** (1999) 29–40.

[22] HOLZAPFEL, W.H.: Industrialization of mageu fermentation in South Africa.. In: Industrialization of indigenous fermented foods (edited by Steinkraus, K.H). New York, U.S.A.: Marcel Dekker (1989) 285–328.

[23] HOUNHOUIGAN, D.J.; NOUT, M.J.R.; NAGO, C.M.; HOUBEN, J.H.; ROMBOUTS, F.M.: Microbiological changes in mawè during natural fermentation. World J. Microbiol. Biotechnol. **10** (1994) 410–413.

[24] HOUNHOUIGAN, D.J.; NOUT, M.J.R.; NAGO, C.M.; HOUBEN, J.H.; ROMBOUTS, F.M.: Use of starter cultures of lactobacilli and yeast in the fermentation of mawè, an African maize product. Trop. Sci. **39** (1999) 220–226.

[25] KHETARPAUL, N.; CHAUHAN, B.M.: Biological utilisation of pearl millet flour fermented with yeasts and lactobacilli. Pl. Fds Human Nutr. **41** (1991) 309–319.

[26] KHETARPAUL, N.; CHAUHAN, B.M.: Sequential fermentation of pearl millet by yeasts and lactobacilli – effect on the antinutrients and in-vitro digestibility. Pl. Fds Human Nutr. **41** (1991) 321–327.

[27] KONLANI, S.; DELGENES, J.P.; MOLETTA, R.; TRAORE, A.; DOH, A.: Isolation and physiological characterization of yeasts involved in sorghum beer production. Fd Biotechnol. **10** (1996) 29–40.

[28] LAPPE, P.; ULLOA, M.: Microbiology of pulque. In: Alimentos fermentados indigenas de Mexico (edited by Wacher-Rodarte, C.; Lappe, P.). Ciudad Universitaria, Mexico D.F., Mexico: Universidad Nacional Autonoma de Mexico (1993) 75–79.

[29] LEE, C.-H.: Fermentation technology in Korea. Korea University Press (2001).

[30] Lim, G.: Indigenous fermented foods in South East Asia. ASEAN Fd Journ. **6** (1991) 83–101.

[31] MERICAN, Z.; YEOH, Q.-L.: Tapai processing in Malaysia: a technology in transition. In: Industrialization of indigenous fermented foods (edited by Steinkraus, K.H). New York, U.S.A.: Marcel Dekker (1989) 169–189.

[32] MONTANARI, G.; ZAMBONELLI, C.; GRAZIA, L.; KAMESHEVA, G.K.; SHIGAEVA, M.K.: *Saccharomyces unisporus* as the principal alcoholic fermentation microorganism of traditional koumiss. J. Dairy Res. **63** (1996) 327–331.

[33] MORAIS, P.B.; ROSA, C.A.; LINARDI, V.R.; PATARO, C.; MAIA, A.: Characterization and succession of yeast populations associated with spontaneous fermentations during the production of Brazilian sugar cane Aguardente. World J. Microbiol. Biotechnol. **13** (1997) 241–243.

[34] MPUCHANE, S.: Fermentation studies on "ting". In: Alternative uses for *Sorghum* in Botswana, Proc. Workshop. Harare, Zimbabwe; Gaberone, Botswana: Botswana Post Production Food Industry Advisory Unit (1987) 53–73.

[35] NDIR, B.: M'Bannick, acidic fermented milk of Senegal: a microbiological study. In: Development of indigenous fermented foods and food technology in Africa. Proc. IFS/UNU workshop held at Douala, Cameroon Oct 1985 (edited by Prage, L.). IFS provisional report **20** (1991) 398–407.

[36] NOUT, M.J.R.: Microbiological aspects of the traditional manufacture of Busaa, a Kenyan opaque maize beer. Chem. Mikrobiol. Technol. Lebensmittel **6** (1980) 137–142.

[37] NOUT, M.J.R.: Ecological aspects of mixed-culture food fermentations. In: The fungal community: its organization and role in the ecosystem (edited by Carroll, G.C.; Wicklow, D.T.). New York, U.S.A.: Marcel Dekker (1992) 817–851.

[38] NOUT, M.J.R.: Fungal interactions in food fermentations. Can. J. Botany **73** (1995) S1291–S1300.

[39] NOUT, M.J.R.; AIDOO, K.E.: Asian Fungal Fermented Food. In: The mycota, Vol.X: Industrial applications (edited by Osiewacz, H.D.). Berlin, Germany: Springer (2002) 23–47.

[40] NOUT, M.J.R.; BARTELT, R.J.: Attraction of a flying nitidulid (*Carpophilus humeralis*) to volatiles produced by yeasts grown on sweet corn and a corn-based medium. J. Chem. Ecol. **24** (1998) 1217–1239.

[41] NOUT, M. J. R.; ROMBOUTS, F. M.: Recent developments in tempe research. J. Appl. Bacteriol. **69** (1990) 609–633.

[42] NOUT, M.J.R.; ROMBOUTS, F.M.: Fermented and acidified plant foods. In: The microbiological safety and quality of food (edited by Lund, B.M.; Baird-Parker, T.C.; Gould, G.W.) Gaithersburg, U.S.A.: Aspen Publishers (2000) 685–737.

[43] NOUT, M.J.R.; SARKAR, P.K.: Lactic acid food fermentation in tropical climates. Antonie van Leeuwenhoek **76** (1999) 395–401.

[44] NOUT, M.J.R.; DE DREU, M.A.; ZUURBIER, A.M.; BONANTS-VAN LAARHOVEN, T.M.G.: Ecology of controlled soyabean acidification for tempe manufacture. Fd Microbiol. **4** (1987) 165–172.

[45] NURAIDA, L.; WACHER, M.C.; OWENS, J.D.: Microbiology of pozol, a Mexican fermented maize dough. World J. Microbiol. Biotechnol. **11** (1995) 567–571.

[46] OKAFOR, N.; OKEKE, B.C.; UMEH, C.; IBENEGBU, C.: Secretion of lysine in a broth medium by lactic bacteria and yeasts associated with garri production using a synthetic gene. Lett. Appl. Microbiol. **28** (1999) 419–422.

[47] ONYEKWERE, O.O.; AKINRELE, I.A.; KOLEOSO, O.A.: Industrialization of ogi fermentation. In: Industrialization of indigenous fermented foods (edited by Steinkraus, K.H.). New York, U.S.A.: Marcel Dekker. (1989) 329–362.

[48] OYEWOLE, O.B.: Characteristics and significance of yeasts involvement in cassava fermentation for fufu production. Int. J. Fd Microbiol. **65** (2001) 213–218.

[49] REISS, J.: The 'tea fungus' and its metabolic products. Deutsche Lebensmittel-Rundschau **83** (1987) 286–290.

[50] ROMBOUTS, F.M.; NOUT, M.J.R.: Microbial fermentation in the production of plant foods. J. Appl. Bacteriol. Symp. Ser. **79** (1995) 108S–117S.

[51] RUBIO, T.; LAPPE, P.; WACHER, C.: Microbial and biochemical studies of the fermentation of sugary solutions inoculated with tibi grains. In: Alimentos fermentados indigenas de Mexico (edited by Wacher-Rodarte, C.; Lappe, P.). Ciudad Universitaria, Mexico D.F., Mexico: Universidad Nacional Autonoma de Mexico (1993) 93–101.

[52] SAKAI, H.; CALDO, G.A.: Microbiological studies on bubod, a fermentation starter in the Philippines. II. Isolation and identification of yeasts. Philippine Agricult. **68** (1985) 139–144.

[53] SAKAI, H.; CALDO, G.A.: Microbiological studies on bubod, a fermentation starter in the Philippines. III. A new yeast species, *Endomycopsis bubodii* Sakai and Caldo sp. nov., and some enzymological properties of *Endomycopsis* strains. Philippine Agricult. **68** (1985) 181–188.

[54] SANNI, A.I.; ONILUDE, A.A.; FADAHUNSI, I.F.; AFOLABI, R.O.: Microbial deterioration of traditional alcoholic beverages in Nigeria. Fd Res. Int. **32** (1999) 163–167.

[55] SARKAR, P.K.; TAMANG, J.P.; COOK, P.E.; OWENS, J.D.: Kinema – a traditional soybean fermented food: proximate composition and microflora. Fd Microbiol. **11** (1994) 47–55.

[56] SEFA-DEDEH, S.; SANNI, A.I.; TETTEH, G.; SAKYI-DAWSON, E.: Yeasts in the traditional brewing of pito in Ghana. World J. Microbiol. Biotechnol. **15** (1999) 593–597.

[57] SHINDO, S.; KASHIWAGI, Y.; SHIINOKI, S.: Saké brewing from liquefied-rice with immobilised fungal mycelia and immobilised yeast cells. J. Inst. Brew. **104** (1998) 277–281.

[58] SONI, S.K.; SANDHU, D.K.: Role of yeast domination in Indian idli batter fermentation. World J. Microbiol. Biotechnol. **7** (1991) 505–507.

[59] SOON BAI CHUN; HEE YOUNG CHUNG: Induction of glucoamylase in the yeast *Candida tsukubaensis*. J. Biochem. Molec. Biol. **28** (1995) 342–347.

[60] STEINKRAUS, K.H.: Lactic acid fermentations. In: Applications of biotechnology to traditional fermented foods (edited by Gaden, E.L.; Bokanga, M.; Harlander, S.; Hesseltine, C.W.; Steinkraus, K.H.). Report of an ad-hoc panel of the Board on Science and Technology for International Development. Washington, DC, U.S.A..: National Research Council (1992) 43–51.

[61] STEINKRAUS, K.H.: Handbook of indigenous fermented foods. New York, U.S.A.: Marcel Dekker (1995).

[62] TAMAI, Y.; OISHI, H.; NAKAGAWA, I.; WATANABE, Y.; SHINMOTO, H.; KUWABARA, Y.; YAMATO, K.; NAGAI, S.: Antimutagenic activity of the milk fermented by mixed-cultured with various lactic acid bacteria and a yeast. J. Jap. Soc. Fd Sci. Technol. – Nippon Shokuhin Kogyo Gakkaishi **42** (1995) 383–387.

[63] TAMANG, J.P.; SARKAR, P.K.: Microflora of murcha: an amylolytic fermentation starter. Microbios **81** (1995) 115–122.

[64] TEOH, A.; CIX, J.M.; HEARD, G.M.: Yeasts: an essential partner in kombucha fermentation. In: Proc. 19th Int. Spec. Symp.Yeasts: Yeasts in the production and spoilage of food and beverages. Braga, Portugal: Univ. of Minho (1998) 155.

[65] UZOCHUKWU, S.; BALOGH, E.; TUCKNOT, O.G.; LEWIS, M.J.; NGODDY, P.O.: Role of palm wine yeasts and bacteria in palm wine aroma . J. Fd Sci. Technol. Mysore **36** (1999) 301–304.

[66] VENKATASUBBAIAH, P.; DWARAKANATH, C.T.; SREENIVASA MURTHY, V.: Involvement of yeast flora in idli batter fermentation. J. Fd Sci. Technol., India **22** (1985) 88–90.

[67] YIP, C.W.; LIEW, C.W.; NGA, B.H.: Ribosomal RNA genes of *Endomyces fibuliger*: isolation, sequencing and the use of the 26S rRNA gene in integrative transformation of *Saccharomyces cerevisiae* for efficient expression of the alpha-amylase gene of *Endomyces fibuliger*. World J. Microbiol. Biotechnol. **13** (1997) 103–117.

[68] ZAMBONELLI, C.; RAINIERI, S.; CHIAVARI, C.; MONTANARI, G.; BENEVELLI, M.; GRAZIA, L.: Autolysis of yeasts and bacteria in fermented foods. Ital. J. Fd Sci. **12** (2000) 9–21.

18 Index

D

E

F